Convergence Theorems for Lattice Group-Valued Measures

Authored by

Antonio Boccuto

Dipartimento di Matematica e Informatica
via Vanvitelli, 1-06123 Perugia
Italy

Xenofon Dimitriou

Department of Mathematics
University of Athens,
Panepistimiopolis, Athens 15784
Greece

CONTENTS

contd…..

contd....

contd....

FOREWORD

The eBook I am glad to read is a survey of the famous limit theorems for measures (Nikodým convergence theorem, Brooks-Jewett theorem, Vitali-Hahn-Saks theorem, Dieudonné convergence theorem, Schur convergence theorem). The first chapter seems to be the back bone of the eBook's development. Not only it describes the development of the main theorems in the realm of convergence, but also provides a compact review of measures defined on algebras, vector lattice-valued measures and measures defined on abstract structures. The use of these ideas is extensively described in Chapters 3 and 4. The historical development was enthusiastically approached since the second author was preparing his master's thesis. Since then he worked consistently on the subject. Both authors are lovers of the historical development due to their Latin and Greek origin! Therefore the reader has the choice to appreciate an excellent piece of work on this area. The connection of Lattice Theory with Measure is explicitly described in this eBook and therefore the reader can also be addressed to Probability Theory. That is this eBook offers not only a strong background on limit theorems in Measure Theory, but also a solid theoretical insight into the Probability concepts. The norm of a measure, defined in Section 1.2, the definition of a measure on an algebra are essential tools to anybody working not only on Measure Theory but on Probability Theory as well. The next step, the definition of a measure defined on an abstract structure, needs more investigation in future work, while the authors cover completely the subject up to our days.

The definition of a Filter, defined firstly, and its dual notion of Ideal, defined later, are very nicely presented in Chapter 2. The relation between two Ideals is discussed in Section 2.1 as well as the Free Filters and P-Filters. These definitions and results are applied in Chapter 4. Being the authors consistent to their approach to limit theorems, they are extending Filters and Ideals with the corresponding limit theorems to Lattice Groups. Therefore a Lattice-Group-valued Measure is defined and the appropriate results are collected and presented. Nice examples on Filter Convergence in Lattice Groups help the reader to understand common ideas such as limsup or liminf through their development. The relation to Dedekind Complete Space is also discussed and related to Measure Theory. Therefore, I believe, the interested researcher has a compact, solid and rigorous presentation of Filters and Ideals.

The group with structure of lattice, known as (ℓ)-group, is what the authors investigate extensively in Chapter 3. The sense of Integration is very strictly presented under the light of Measure Theory. The convergence theorems for integrals are direct applications to Integration. The theoretical development of this Chapter is applied in Chapter 4, where a number of results is discussed under milder/weaker assumptions. Not only the limit theorems are presented but also interesting decomposition analogues for (ℓ)-group-valued measures are also discussed.

Chapter 4 is devoted to Filter (Ideal) Limit Theorems and their applications. Limit results and convergence theorems are presented in such a way the reader realizes that the authors are the grand masters of this subject. The Regularity of a Measure is discussed on any Dedekind Complete (ℓ)-Group. Topology is hidden everywhere and therefore also in Group-Valued Measures. This part is strongly related to the Preliminaries presented in Section 1.1, where the ideas of Topology, Measure and Banach Lattice are introduced.

The authors have collected more than 750 references on the subject. It is impressive not only for the extensively great number of references covering a wide variety of disciplines, but also for the fact that the authors refer to all of them inside the eBook.

I was glad when the authors asked me to write the preface. Then I realized that it was a hard work to go through this eBook. But I was eventually happy to realize that this excellent eBook covers the subject as well as possible. I did not have the chance to read such a compact review on the subject. I thank the authors for giving me the chance to read it.

Prof. Christos P. Kitsos
Department of Informatics
Technological Educational Institute of Athens
Chair of the ISI Committee on Risk Analysis
Greece

PREFACE

One of the topics of wide interest for several mathematicians, which has been successfully widely studied for more than a century, are the convergence and boundedness theorems for measures, in connection with properties of integrals, double sequences, matrix theorems and interchange of limits. Some related results in this area are the Banach-Steinhaus theorem in the operator setting and integration theory together with its fundamental properties. These topics have several applications in different branches of Mathematics, like for example topology, function spaces and approximation theory.

At the beginning, the case of σ-additive real-valued measures and integrals was treated, together with matrix theorems. These topics have been developed in the literature along several directions. Firstly, by considering not only countably additive, but also finitely additive measures and even set functions which are not necessarily finitely additive. Secondly, dealing with measures with values in abstract structures, like for instance Banach, uniform and locally convex spaces, topological and lattice groups, and so on. Thirdly, investigating measures defined on algebras satisfying suitable properties but which are not necessarily σ-algebras, or more abstract structures like for example *MV*-algebras, orthomodular posets, *D*-posets, minimal clans, which have several applications, for instance to quantum mechanics and multivalued logics.

To prove the main results about these topics, there are two types of techniques: the sliding hump or diagonal argument, which studies properties of the diagonal of an infinite matrix whose rows and columns are convergent, and the Baire category theorem. The sliding hump was known just at the beginning of the last century and was used for the proofs of the first fundamental results about limit theorems. The technique which uses the Baire category theorem is based on certain properties of Fréchet-Nikodým topologies. But this method, in general, is not adaptable in the finitely additive case. So, in most cases, it has been preferable to consider again the sliding hump method, which has been deeply studied in proving limit and boundedness theorems and also in matrix diagonal lemmas, which are very useful for these subjects. Furthermore, two procedures to relate the finitely additive case to the countably additive case have been investigated: the first deals with Stone-type σ-additive extensions of the original measures, and the second uses Drewnowski-type σ-additive restrictions of finitely additive measures on suitable σ-algebras.

The novelty of the research of the authors, which is exposed in Chapter 4, is to study limit theorems in the setting of filter convergence, which is an extension of convergence generated by matrix summability methods and includes as a particular case the statistical convergence, which is related with the filter of all subsets of the natural numbers having asymptotic density one. Note that, in general, it is impossible to expect analogous results corresponding to the classical case, even for σ-additive real-valued measures, because in general filter convergence is not inherited by subsequences. However it has been possible to prove several versions of limit, matrix and boundedness theorems as well as some results about different modes of continuity and convergence for measures, filter exhaustiveness (extending to the filter setting the concept of equicontinuity), continuity properties of the limit measure, weak filter/ideal compactness, and so on. The first chapter contains a historical survey of these topics since the beginning of last century. In Chapter 2 we deal with the basic concepts and tools used, like for instance filters/ideals, lattice group-valued measures, filter/ideal convergence in (ℓ)-groups, and present some fundamental tool, like for example the Maeda-Ogasawara-Vulikh representation theorem and the Stone Isomorphism technique. Chapter 3 contains several versions of limit and boundedness theorems for lattice group-valued measures and some applications to integrals. In the appendix we present an abstract approach on probability theory and random variables in connection with Boolean algebras, metric spaces, σ-additive extensions of finitely additive functions, various kinds of convergence in the lattice setting and tools which can have further developments, and we present some developments of the abstract notion of concept and some applications to Bioassays and related topics investigated by X. Dimitriou and C. P. Kitsos.

The eBook can be used both as a primer on limit theorems and filters/ideals and related topics, for postgraduate and Ph. D. students who want to explore these subjects and their beautifulness, and as a text for advanced researchers, since it exposes some new directions and results, shows some possibilities of further developments and ideas and includes also some open problems in the area.

This eBook includes several topics and developments of the research, started with the Ph. D. thesis of Dr. Xenofon Dimitriou, which was brilliantly discussed on 22th December 2011 under the supervision of Proffs. Nikolaos Papanastassiou and Antonio Boccuto at the National and Kapodistrian University of Athens.

The first author wants to dedicate the eBook to the loving memory of his parents. His father Giuliano died on 8th March 2011, while the authors were cooperating on the topics of the research exposed in this eBook. The second author wants to dedicate the eBook to his parents and to all who support him.

We want to thank Prof. Christos P. Kitsos for writing the foreword and Prof. Władysław Wilczyński for having translated from the Russian the papers by Doubrovsky, which the authors consulted for the preparation of Chapter 1. We thank also the Bentham Science Publishers, in particular Manager Hira Aftab and all her team, for their support and efforts. We also thank the referees for their remarks and suggestons, which improved the exposition of the eBook.

ACKNOWLEDGEMENTS

None declared.

CONFLICT OF INTEREST

The authors confirm that this eBook content has no conflict of interest.

Antonio Boccuto
Dipartimento di Matematica e Informatica
via Vanvitelli, 1
I-06123 Perugia
Italy
Email: antonio.boccuto@unipg.it

Xenofon Dimitriou
Department of Mathematics
University of Athens, Panepistimiopolis
Athens 15784
Greece
E-mails: xenofon11@gmail.com, dxenof@math.uoa.gr

About the Authors

Antonio Boccuto was born in Catanzaro, Italy, on November 1964. He received the degree in Piano from Conservatorio of Perugia in 1985, the degree in Mathematics from the University of Perugia, Italy, in 1987, the "Zertifikat" in German as a foreign language from the Goethe Institut in 1997 and the Ph.D. degree in Mathematical Analysis from the Mathematical Institute of the Slovak Academy of Sciences in Bratislava, Slovakia, in 2000.

He received the habilitation in Mathematics to Associate Professor from the Comenius University in Bratislava in 2008. He has been a researcher in Mathematical Analysis at University of Perugia since the Academic Year 1991/1992. His research interests include Measure Theory and Integration, Real Analysis, Function Theory, Approximation Theory. He has been collaborating with several foreign Universities, among which Bratislava and Athens.

He has published more than 100 papers on Mathematics in journals and conference proceedings with the peer-review process and he is a coauthor of two books on Measure Theory by Bentham Science Publishers. He has participated as a speaker/invited speaker and/or a member of scientific committee in several conferences, meetings and workshops on the fields of his interest.

He is a member of the Editorial Board of scientific journals, a referee for several journals and an evaluator of research proposals/grants.

Xenofon Dimitriou was born in Athens, Greece, on August 1976. He received the First Certificate in English by the University of Cambridge (1990), the Certificate of Proficiency in English by the same university (1992), the Certificate of Proficiency in English by the University of Michigan (1993) and was licensed to teach English as a foreign language by the Greek Ministry of Education and Religious Affairs (1994). He held a Degree in Mathematics by the Department of Mathematics of the University of Athens Greece (2003), a Msc in Pure Mathematics by the same department (2007) and a PhD in Measure Theory and Real Analysis with honors again from the same department (2011).

He has published more than 35 papers on Mathematics and Didactics/Pedagogics in journals and conference proceedings with the peer-review process and he is a coauthor of a book on Measure Theory by Bentham Science Publishers.

He has participated as a speaker/invited speaker and/or a member of scientific committee in more than 15 conferences, meetings and workshops on the fields of his interest.

He was a teaching assistant, for 2009/2010, in ''Calculus 1'' courses at the Department of Mathematics of the University of Athens Greece. He has taught the ''Matlab'' course at the Department of Electrical Engineering of the Technological Educational Institute of Piraeus Greece (winter semester of 2012/2013) and the ''SPSS'' course at the M.A. Program ''Studies in Education'' organized by the Pedagogical Department of the School of Pedagogical and Technological Education of Athens Greece in collaboration with the Roehampton University of London U.K. (spring semester of 2012/2013).

He has a basic computer knowledge (Operating Systems: Linux, Windows. Programming Languages: Turbo Pascal, Mathematica, Matlab. Typography: MS Office, Latex), he is a member of the Greek Mathematical Society, a member of its contests committee and Leader/Deputy Leader of the national Greek teams in Balkan and International Mathematical Olympiads. He is a member of Editorial Board of 4 scientific journals, a referee for various other journals and an evaluator of research proposals/grants.

CHAPTER 1

Historical Survey

Abstract: This chapter contains a historical survey about limit and boundedness theorems for measures since the beginning of the last century. In these kinds of theorems, there are two substantially different methods of proofs: the sliding hump technique and the use of the Baire category theorem. We deal with Vitali-Hahn-Saks, Brooks-Jewett, Nikodým convergence and boundedness theorems, and we consider also some related topics, among which Hahn-Schur-type theorems and some other kind of matrix theorems, the uniform boundedness principle and some (weak) compactness properties of spaces of measures. In this context, the Rosenthal lemma, the biting lemma and the Antosik-Mikusiński-type diagonal lemmas play an important role. We consider the historical evolution of convergence and boundedness theorems for σ-additive, finitely additive and non-additive measures, not only real-valued and defined on σ-algebras, but also defined and/or with values in abstract structures.

Keywords: σ-additive measure, Baire category theorem, biting lemma, Brooks-Jewett theorem, *D*-poset, Drewnowski lemma, finitely additive measure, Hahn-Schur theorem, interpolation property, *k*-triangular set function, matrix theorem, *MV*-algebra, Nikodým boundedness theorem, Nikodým convergence theorem, orthomodular lattice, orthomodular poset, Rosenthal lemma, Sliding hump, Vitali set, Vitali-Hahn-Saks theorem.

1.1. Preliminaries

Throughout this eBook, for the fundamental concepts and their properties and the literature in Mathematical Analysis, in particular for normed, Banach, metric, locally convex and topological spaces, we refer to the books by Banach (1932) and by Dunford and Schwartz (1958, 1963 and 1971). Concerning the recent works about the history of Measure Theory and its fundamental aspects and features, we recall in particular the recent milestone books by Bogachev (2007) and by Fremlin (2000-2003, 2008 and 2011) and their bibliographies. A history of Banach spaces and related topics is found also in the book by Pietsch (2007). We recall also, in particular, the books by Aliprantis and Burkinshaw (1985, 1990 and 2003), Berberian (1999), Cohn (1993), Diestel (1984), Diestel and Uhl (1977), Dinculeanu (1966), Fabian *et al.* (2001), Grothendieck (1973) (see also Cartier *et al.* (1989 and 1990)), Jacobs (1978), König (2009 and 2012), Köthe (1969), Lang (1993), Megginson (1998), Phillips (1984), and the two volumes of the "Handbook of Measure Theory" edited by Pap in 2002, in particular Paunić (2002) and Pap (2002b) concerning the historical evolution and developments.

Antonio Boccuto & Xenofon Dimitriou

Limit and uniform boundedness theorems for measures are among the most important in mathematics, they have been investigated and developed for more than a century and in the context of several abstract structures, and have several applications in several branches of Mathematics, in particular in Topology and in Approximation Theory. In their proofs, there are substantially two different kinds of arguments. The first method follows Dunford and Schwartz (1958, 1963 and 1971) and is due to Fréchet (1921) and Nikodým (1930a and 1930b), see also Weber (2002). It is a topological method, based on the fact that, if Σ is a σ-algebra of subsets of a nonempty set G and m is a non-negative real-valued σ-additive measure defined on Σ, then the set $\{\chi_E : E \in \Sigma\}$ (where χ_E is the *characteristic function* associated with E, namely that function which associates the real number 1 to every element of E and the real number 0 to each element of $G \setminus E$) is a closed subset of the space $L^1(G)$ or $L^1(m)$ of the Lebesgue real-valued functions integrable with respect to m, and hence is a complete metric space. This approach was given by S. Saks (1933a-b and 1937), which was the first to prove the Vitali-Hahn-Saks theorem using the Baire category theorem 1.2 (see also Kalton (1974)). However this method in general does not work, when it is dealt with finitely additive measures. In this case, it is advisable to use the second method, called *sliding hump* or *gliding hump*, that is the study of some property of the diagonal of an infinite matrix, when it is supposed that its rows and its columns converge or are Cauchy. By means of arguments of this kind, it is possible to prove several versions of convergence theorems for finitely additive measures with values in different kinds of structures (for example Banach spaces, locally convex topological vector spaces, topological (semi)groups, vector lattices, lattice groups). The sliding hump technique or diagonal argument, as we will see later, was known in the literature and widely used just in the first third of the last century, and is more advisable and fruitful than the category argument in the context of not necessarily σ-additive measures, defined and/or taking values in abstract structures. In this context the Rosenthal lemma, the biting lemma and Antosik-Mikusiński diagonal-type lemmas will play an important role. A survey on this method and its several applications can be found, for instance, in the book by Antosik and Swartz (1985a), in the book by Swartz (1996b) and in the paper by Swartz (1990), while an overview on the Vitali-Hahn-Saks-Nikodým-type theorems is found, for example, in Choksi (2001). In this chapter we will do a historical survey about the evolution of the limit theorems and the related topics and problems (like for example (weak) compactness of sets of measures, see also the books by Diestel and Uhl (1977) and Diestel (1984)) since the beginning of the last century (see also Dimitriou (2007)). In Chapter 2 and 3 we will deal with the most recent classical like limit theorems in the lattice group setting since the beginning of this century. In Chapter 4 we will consider different kinds of limit theorems in connection with convergence with respect to ideals/filters and modes of (filter) continuity of measures, together with some results on filter weak compactness and filter weak convergence of measures. In Chapter 5 we will present a general discussion about the subjects and results, together with the main techniques and tools used, and the evolution of the research treated in Chapters 2, 3 and 4. In Appendix 1, using lattice theory, we will deal with an abstract approach on probability measures, defined on abstract Boolean algebras, and different types of random variables with respect to various kinds of convergence. In Appendix 2 we will present some topics about concepts, lattices and probabilities and some relations and connections between them.

The concept of finitely additive measure was just known by Jordan (1893 and 1896). Indeed, let B be a bounded subset of the euclidean n-dimensional space and let us define

$$j^*(B) := \inf\{\sum_{j\leq n} | I_j |\} \text{ (resp. } j_*(B) := \sup\{\sum_{j\leq n} | I_j |\}), \tag{1.1}$$

where the infimum (resp. supremum) is intended with respect to all disjoint finite families of

n-dimensional rectangles $\{I_1,\ldots,I_n\}$ such that $B \subset \bigcup_{j=1}^{n} I_j$ (resp. $B \supset \bigcup_{j=1}^{n} I_j$), with the convention $\sup\varnothing = 0$, and $| I_j |$ denotes the n-dimensional (Lebesgue) measure of I_j, $j = 1,\ldots,n$. The set B is said to be *Jordan measurable* iff $j^*(B) = j_*(B)$ (see also Diestel and Spalsbury (2012)).

Finitely additive measures have their importance, first of all because of their dual representation. Indeed, if Σ is a σ-algebra of subsets of an abstract nonempty set G, then the linear space $\mathrm{ba}(\Sigma)$ of all bounded finitely additive real-valued measures defined on Σ is isomorphic to the dual of the Banach space $B(\Sigma)$ of all bounded real valued Σ-measurable functions defined on G (see also Diestel and Spalsbury (2012), Theorem 1.1, Fichtenholtz and Kantorovich (1934) and Hildebrandt (1934)). Moreover, Banach (1923) proved that there exist finitely additive translation invariant extensions on the whole real line of the Lebesgue measure. This is not true if one requires σ-additivity instead of finite additivity (see Vitali (1905)). For a related literature, see also Laczkovich (2002), Paterson (1988), Pier (1984), Wagon (1981 and 1985), Zakrzewski (2002). Furthermore, finitely additive probability measures are used in the subjective approach of Calculus of Probability (see also de Finetti (1974 and 1975)) and in stochastic integration theory (see also Brooks (2002), Brooks and Dinculeanu (1991 and 1995)).

The finitely additive set functions were just treated by von Neumann (1930) and Banach (1932), and among the successive authors we recall, for instance, Bochner (1939 and 1940), Bochner and Phillips (1941), Yosida and Hewitt (1952), Leader (1953), Fefferman (1967a and 1967b), Darst (1962a, 1962b and 1969), Uhl (1967) and Maharam (1976 and 1987). We recall, in particular, the book of Bhaskara Rao and Bhaskara Rao (1983) and the surveys by Candeloro and Volčič (2002), Martellotti (2001) and Diestel and Spalsbury (2012), where some fundamental and many interesting properties of finitely additive measures are found.

In general, as we will see later, the limit theorems do not hold, if it is assumed that the involved set functions are defined only on algebras, without requiring further assumptions. So the problem of finding necessary and/or sufficient conditions for an algebra in order that convergence or boundedness theorems hold has been widely studied in the literature (see for instance Aizpuru (1996) and Schachermayer (1982) for an overview) and it is still an open problem. In connection with this, there have been many studies about limit and uniform boundedness theorems for set functions, defined in abstract structures which are not necessarily Boolean algebras (for example, difference posets, minimal clans, MV-algebras, orthomodular lattices, orthomodular posets). These particular structures have several applications in different fields of Mathematics, like algebra, quantum logics and quantum mechanics (see for instance Barbieri and Weber (2002), Dvurečenskij (2002), Dvurečenskij and Pulmannová (2000), Riečan and Mundici (2002)). Furthermore, there have been also several versions of limit theorems for set functions, which are not necessarily finitely additive. A survey on this and related literature and applications is found, for example, in the book by E. Pap (1995).

1.1.1. Topological Spaces and Groups

We denote by $\mathbb{N}$ and $\mathbb{R}$ the sets of the natural and the real numbers, respectively, and we indicate with $\widetilde{\mathbb{R}} := \mathbb{R} \cup \{+\infty\} \cup \{-\infty\}$ the set of the extended real line

A *quasi-uniformity* on a nonempty set X is a set $\mathcal{D}$ of subsets U of the Cartesian product $X \times X$ satisfying the following axioms:

(i) For each $U \in \mathcal{D}$, $\Delta_X \subset U$, where $\Delta_X = \{(x,x) : x \in X\}$;

(ii) $U \cap V \in \mathcal{D}$ for each U, $V \in \mathcal{D}$;

(iii) If $U \in \mathcal{D}$ and $U \subset V \subset X \times X$, then $V \in \mathcal{D}$;

(iv) For every $U \in \mathcal{D}$ there is $V \in \mathcal{D}$ with $V \circ V \subset U$, where

$V \circ W := \{(x,z) \in X \times X :$ there exists $y \in X$ with $(x,y) \in V, (y,z) \in W\}$.

A *uniform space* $(X, \mathcal{D})$ is a set endowed with a quasi-uniformity $\mathcal{D}$, satisfying the following additional axiom:

(v) for each $U \in \mathcal{D}$ we get $U^{-1} \in \mathcal{D}$, where $U^{-1} := \{(y,x) : (x,y) \in U\}$.

In this case we say that the quasi-uniformity $\mathcal{D}$ is a *uniformity*. If X is a uniform space and $\mathcal{D}$ is the uniformity on X, then the elements of $\mathcal{D}$ are often called *entourages* (see also Engelking (1977), Fletcher and Lindgren (1982), Joshi (1983), Kelley (1955), Kuratowski (1966 and 1968), Pervin (1962) and Willard (1970)). Note that every uniform space is a topological space. Indeed, for every $U \in \mathcal{D}$ and $x \in X$, set $U[x] := \{y \in X : (x,y) \in U\}$. Put $\tau := \{Z \subset X :$ for each $x \in Z$ there is $U \in \mathcal{D}$ such that $U[x] \subset Z\}$. Then τ is a topology on X (see Joshi (1983), Theorem 14.1.9).

Note that a topological space $X = (X, \tau)$ is uniformizable, that is admits a uniformity $\mathcal{D}$ which induces the topology τ, if and only if X is *completely regular*, that is, given any closed set $K \subset X$ and any point $x \notin K$, there is a continuous function $f : X \to \mathbb{R}$ such that $f(x) = 0$ and $f(y) = 1$ for every $y \in K$ (see Joshi (1983), Theorem 14.2.10). Recall that a topological space X is called *Hausdorff* iff for any two distinct elements x_1, $x_2 \in X$ there are two open disjoint subsets U_1, U_2 of X with $x_j \in U_j$, $j = 1,2$; that X is *regular* iff for every point $x \in X$ and for each closed subset $K \subset X$ with $x \in X \setminus K$ there are two open disjoint subsets of X, U_1 and U_2, with $x \in U_1$ and $K \subset U_2$, and that X is *normal* iff for every pair of disjoint closed sets K_1, $K_2 \subset X$ there are two open disjoint subsets of X, U_1 and U_2, with $U_j \supset K_j$, $j = 1,2$. The definition of normal topological space has different equivalent formulations.

Proposition 1.1 (see Dugundji (1966), Proposition VII.3.2, and Munkres (1975), Lemma 2.1 (b)) *The following four properties are equivalent:*

(a) *X is normal;*

(b) *Given any open subset U and any closed subset K of X with $K \subset U$, there is an open subset V of X with $K \subset V \subset \overline{V} \subset U$, where $\overline{V}$ denotes the closure of V in X;*

(c) *For each pair of disjoint closed subsets A, B of X there is an open subset U of X with $A \subset U$ and $\overline{U} \cap B = \emptyset$;*

(d) *For each pair of disjoint closed sets K_1, $K_2 \subset X$ there are two open subsets of X, U_1 and U_2, with $U_j \supset K_j$, $j = 1,2$, and $\overline{U_1} \cap \overline{U_2} = \emptyset$.*

Observe that, if $(X, \mathcal{D})$ is a uniform space and τ is the induced topology, then (X, τ) is Hausdorff if and only if $\Delta_X = \bigcap \{U : U \in \mathcal{D}\}$.

A subset of X is an F_σ *set* in X iff it is a countable union of closed subsets of X, and is a G_δ *set* in X iff it is a countable intersection of open subsets of X.

A *pseudometric space* (X, d) is any nonempty set X, endowed with a *pseudo-distance function* $d : X \times X \to \mathbb{R}$, satisfying the following axioms:

(i) $d(x, y) \geq 0$ for all x, $y \in X$;

(ii) $d(x, y) = d(y, x)$ for all x, $y \in X$;

(iii) $d(x, y) \leq d(x, z) + d(z, y)$ for each x, y, $z \in X$.

A pseudometric space (X, d) is said to be *metric*, iff it is endowed with a function $d : X \times X \to \mathbb{R}$, fulfilling the axioms (i), (ii), (iii) above and such that $d(x, y) = 0$ if and only if $x = y$. In this case we say that d is a *distance*.

Given two metric spaces (X, d), (Y, ρ), a family $f_\lambda : X \to Y$, $\lambda \in \Lambda$, of functions and a fixed point $x \in X$, we say that $(f_\lambda)_\lambda$ is *equicontinuous at* x iff for every $\varepsilon > 0$ there is a $\delta > 0$ (depending on ε and x) with

$\rho(f_\lambda(z), f_\lambda(x)) \leq \varepsilon$ whenever $d(z,x) \leq \delta$ and $\lambda \in \Lambda$. The family $(f_\lambda)_\lambda$ is said to be *equicontinuous on* X iff it is equicontinuous at every point of X. We say that $(f_\lambda)_\lambda$ is *uniformly equicontinuous on* X iff for every $\varepsilon > 0$ there exists a $\delta > 0$ (depending only on ε) with $\rho(f_\lambda(z), f_\lambda(x)) \leq \varepsilon$ whenever x, $z \in X$ with $d(z,x) \leq \delta$ and $\lambda \in \Lambda$.

If $(X, \mathcal{D})$ is a uniform space, $D \in \mathcal{D}$ and $b \in X$, set $D(b) := \{x \in X : (b,x) \in D\}$. If $\emptyset \neq B \subset X$, let $D(B) := \bigcup_{b \in B} D(b)$. We say that $D(B)$ is an *enlargement* of B. When (X,d) is a (pseudo)metric space, an example of a set of the type $D(b)$ is, for every fixed $r > 0$, the ball $S(b,r) = \{x \in X : d(b,x) < r\}$, and the r-*enlargement* of a nonempty set $B \subset X$ is the set $B^r := \bigcup_{b \in B} S(b,r)$ (see also Beer (2010)).

Given a topological vector space X, we denote by X' its *topological dual* and by X'' its *topological bidual.* A *Schauder basis* in a topological vector space X is a sequence $(x_n)_n$ in X such that every element $x \in X$ has a unique representation in the form $x = \sum_{n=1}^{\infty} a_n x_n$, where $(a_n)_n$ is a sequence of real numbers.

A subset S of a topological vector space X is said to be *balanced* iff $aS := \{as : s \in S\} \subset S$ whenever $a \leq 1$.

We say that $S \subset X$ is *absorbing* iff for each $x \in X$, $x \neq 0$, there is a positive real number a with $bx \in S$ whenever $|b| \leq a$.

A *barrelled set* $S \subset X$ is a closed, convex, balanced and absorbing set. A *barrelled space* is a topological vector space for which every barrelled set is a neighborhood of 0 (see also Bourbaki (1950), Schaefer (1971)).

A topological space X is said to be *extremely disconnected* or *Stonian* iff the closure of every open subset of X is open in X; *quasi-Stonian,* iff every subset $U \subset X$ is open whenever it is the closure of an open set of X of type F_σ (that is representable as a countable union of closed sets).

Given a topological space X and a set $N_0 \subset \Omega$, we say that N_0 is *nowhere dense* in X iff its closure has empty interior. We say that a set $N \subset X$ is *meager* or *of first category* in X iff N is a countable union of nowhere dense subsets of X, and that N is *of second category* in X iff it is not of first category in X.

We now recall the Baire category theorem (see for instance Rudin (1973), Theorem 2.2), whose technique will be useful in several lemmas, to prove different types of convergence theorems.

Theorem 1.2 *If X is either a complete metric space or a locally compact Hausdorff topological space, then the intersection of every countable collection of dense open subsets of X is dense in X.*

A *quasi-norm* on an abelian group X endowed with a neutral element 0 is a function $p: X \to \mathbb{R}$ with $p(x) \geq 0$, $p(x) = p(-x)$ and $p(x+y) \leq p(x) + p(y)$ for each x, $y \in X$. A quasi-norm induces a pseudodistance d on X by setting $d(x,y) = p(x-y)$, x, $y \in X$, with respect to which X is a topological group. Note that the topology of an abelian group is generated by a family of quasi-norms (see also Swartz (1996b)). From now on, as no confusion can arise, we denote a quasi-norm by the symbol $\|\cdot\|$. A *norm* on X is a quasi-norm, such that $x = 0$ whenever $x \in X$ and $\|x\| = 0$.

A *Banach space* is a complete normed vector space. A *locally convex topological vector space* is a topological vector space whose topology is generated by translations of balanced, absorbing, convex sets. For the main properties of topological vector spaces, see Dunford and Schwartz (1958).

Given any nonempty set Λ, the space $l^\infty(\Lambda)$ is the space of all bounded functions from Λ to $\mathbb{R}$, endowed with the supremum norm, and l^∞ is the space $l^\infty(\mathbb{N})$. If $0 < p < +\infty$, we denote by $l^p(\Lambda)$ the space of all functions $x: \Lambda \to \mathbb{R}$, such that

$$\sum_{\lambda \in \Lambda} |x(\lambda)| := \sup \{ \sum_{\lambda \in F} |x(\lambda)| : F \text{ is a finite subset of } \Lambda \} < +\infty,$$

by $c_0(\Lambda)$ be the space of all functions $x: \Lambda \to \mathbb{R}$, such that the set $\{\lambda \in \Lambda: |x(\lambda)| \geq \varepsilon\}$ is finite for every $\varepsilon > 0$, by $c_{00}(\Lambda)$ the space of all functions, vanishing in the complement of a finite subset of Λ, by l^p, c_0 and c_{00} the spaces $l^p(\mathbb{N})$, $c_0(\mathbb{N})$ and $c_{00}(\mathbb{N})$ respectively. Note that c_0 is the space of all real sequences, convergent to 0 (see also Fremlin (1974), Luxemburg and Zaanen (1971), Swartz (1989a) and Vulikh (1967)).

If X and Y are two Banach spaces with $X \subset Y$, we say that X is *complemented in* Y iff there exists a projection $\pi : Y \to X$, that is a linear surjective function with $\pi(x) = x$ for every $x \in X$.

A Banach space is *injective* iff it is complemented in every Banach space containing it.

A Banach space X is said to have the *Gould property* (see Gould (1966), Axiom A) iff every sequence $(x_n)_n$ in X, whose norms are lower bounded by a suitable strictly positive real constant C_0, is such that for every $\sigma > 0$ there is a finite subset $E(\sigma) \subset \mathbb{N}$ with

$$\left\| \sum_{k \in E(\sigma)} x_k \right\| \geq \sigma. \qquad \textbf{(1.2)}$$

Note that Gould (1966) (see also Theorem 3.1) showed that every weakly complete Banach space satisfies the Gould property, and so every Hilbert space and the l^p spaces with $1 \leq p < +\infty$ have the Gould property, while l^∞ does not fulfil it.

A Banach space X is said to satisfy the *Saeki fatness condition* iff for each constant $C > 0$ there is a positive integer q such that, given $x_1, x_2, \dots, x_q \in X$ with $\| x_j \| \geq 1$ for each $j = 1, 2, \dots q$, there is a nonempty subset $F \subset \{1, 2, \dots, q\} =: [1, q]$ with $\left\| \sum_{j \in F} x_j \right\| \geq C$ (see also Saeki (1992)).

Given a real number w, we call *signature* of w (shortly, sgn w) the quantity defined by,

$$\text{sgn } w := \begin{cases} 1 & \text{if } w > 0, \\ 0 & \text{if } w = 0, \\ -1 & \text{if } w < 0 \end{cases}$$

(see also Zygmund (1968)).

We say that a Banach space X *has cotype* $p \geq 2$ iff there is a constant $K_p > 0$ such that for every $n \in \mathbb{N}$ and $x_1, x_2, \ldots, x_n \in X$, we have

$$\left(\sum_{j=1}^{n} \| x_j \| \right)^{1/p} \leq K_p \int_0^1 \| \sum_{j=1}^{n} r_j(t) x_j \| dt,$$

where r_j, $j \in \mathbb{N}$, denote the *Rademacher functions*, defined by $r_j(t) = \text{sgn}(\sin(2^j \pi t))$, $t \in [0,1]$.

If $a, b \in \mathbb{R}$, $a < b$, and $L^p([a,b])$, $1 \leq p < +\infty$, denotes the space of all Lebesgue measurable functions $f:[a,b] \to \mathbb{R}$ such that the Lebesgue integral $\int_a^b | f(t) |^p dt < +\infty$, then $L^p([a,b])$ has cotype 2 for $1 \leq p \leq 2$ and cotype p for $p \geq 2$; moreover, every Hilbert space has cotype 2 (see also Diestel (1984)). We say that a Banach space X has *finite cotype* iff there exists a positive real number $p \geq 2$ such that X has cotype p.

For a literature about topological groups and semigroups, see for instance Carruth, Hildebrant and Koch (1983), Comfort (1984), Pontryagin (1986). A survey about the main properties and limit theorems on group- (resp. semigroup-) valued set functions can be found, for instance, in de Lucia (1985) and Sion (1973) respectively.

If R is a semigroup with neutral element 0, $k \in \mathbb{N}$ and A, $A_1, \ldots, A_k \subset R$, set

$A_1 + \cdots + A_k := \{a_1 + \ldots + a_k : a_1 \in A_1, \ldots, a_k \in A_k\}$, and $kA := A + \cdots + A$ (k times).

If R is a topological semigroup, we say that a subset $A \subset R$ is *bounded* (in R) iff for every neighborhood U of 0 there is $k \in \mathbb{N}$ such that $A \subset kU$ (see also de Lucia (1985) and Hejcman (1959)).

A series $\sum_{j=1}^{\infty} x_j$ in an abelian topological group is said to be *subseries convergent* iff for every strictly increasing sequence $(j_k)_k$ in $\mathbb{N}$ the series

$\sum_{k=1}^{\infty} x_{j_k}$ converges. Note that in $\mathbb{R}^n$ a series is absolutely convergent if and only if it is subseries convergent, and in an infinite dimensional normed space there are subseries convergent series, which are not absolutely convergent (see also Swartz (1992b), Theorem 30.1.1). Moreover the series $\sum_{j=1}^{\infty} x_j$ is said to be *unconditionally convergent* to $x \in X$ iff for every neighborhood U of 0 there is a finite set $A_0 \subset \mathbb{N}$ such that for every finite set A with $A_0 \subset A \subset \mathbb{N}$ we get $(\sum_{j\in A} x_j) - x \in U$.

1.1.2. Boolean Algebras, Lattices and Related Structures

In the literature there are many studies about limit theorems for set functions, which are defined on abstract structures, which are not necessarily Boolean algebras, for example orthomodular posets, minimal clans, *D*-posets or effect algebras, which have several applications in quantum logics and quantum mechanics. For a literature on these subjects and related topics see also Barbieri and Weber (2002), Boccuto, Riečan and Vrábelová (2009), de Lucia and Pap (2002), Dvurečenskij and Pulmannová (2000), Pap (1995, 1997a and 2002c), Riečan and Mundici (2002), Schmidt (1998) and the related bibliographies. We now recall some notions about these abstract structures.

A pair $\Lambda = (\Lambda, \geq)$ is called a *directed set* iff Λ is a nonempty set and $\geq$ is a reflexive and transitive binary relation on Λ, such that for any two elements λ_1, $\lambda_2 \in \Lambda$ there is $\lambda_0 \in \Lambda$ with $\lambda_0 \geq \lambda_1$ and $\lambda_0 \geq \lambda_2$.

We say that a triple $L = (L, \oplus, ')$ is a *Boolean algebra* iff L is a set consisting of at least two elements, $\oplus : L \times L \to L$ is a commutative and associative binary operation and $' : L \to L$ is a unary operation with $a \oplus a = a$ and $(a' \oplus b')' \oplus (a' \oplus b)' = a$. In the literature there are several equivalent ways of defining axiomatically Boolean algebras (see also Huntington (1904 and 1933a-c). Choose arbitrarily $a \in L$ and put $1 = a \oplus a'$: in Huntington (1933a) it is shown that this definition does not depend on the chosen element a. Set also $0 = 1'$. Let $\mathcal{L}$ be an algebra of subsets of a nonempty set G: in Huntington (1933a) it is shown that it is possible to define a product binary operation and an order on L, having the same role as the intersection and the inclusion between sets respectively. Thus, $(\mathcal{L}, \cup, ^c)$ (where $A^c = G \setminus A$ for every $A \in \mathcal{L}$) is a Boolean algebra, where the role of $\oplus$, $'$, 1 and 0 is played by $\cup$, ^{c}, G and $\emptyset$ respectively.

Now, for every A, $B \in \mathcal{L}$, let χ_A be the *characteristic function* associated with A, namely the function defined by setting $\chi_A(x)=1$ if $x \in A$ and $\chi_A(x)=0$ if $x \notin A$, set $\chi_A \oplus \chi_B = \chi_{A \cup B}$ and $\chi_A' = \chi_{G \setminus A} = 1 - \chi_A$. It is not difficult to see that $(L, \oplus, ')$, where $L := \{\chi_A : A \in \mathcal{L}\}$, is a Boolean algebra.

We now recall some abstract extensions of the concept of Boolean algebra. A *partially ordered set* L (shortly, *poset*) is a nonempty set endowed with a relation, denoted by $\leq$, satisfying the following properties for every a, b, $c \in L$:

(*i*) $a \leq a$ (*reflexive*);

(*ii*) if $a \leq b$ and $b \leq c$, then $a \leq c$ (*transitive*);

(*iii*) if $a \leq b$ and $b \leq a$, then $a = b$ (*antisymmetric*).

In a poset L, we say that a nonempty subset $A \subset L$ is *bounded from above,* iff there exists $x \in A$ such that $a \leq x$, for all $a \in A$; *bounded from below,* iff there exists $y \in A$ such that $a \geq y$, for every $a \in A$; *bounded,* iff it is bounded both from above and from below.

Given a poset L, a nonempty subset $A \subset L$ and an element $s \in L$, we say that s is the *supremum* of A iff for every element $a \in A$ we have $a \leq s$, and moreover we get $s \leq c$ whenever $c \in L$ is such that $c \geq b$ for all $b \in A$. Analogously, given $j \in L$, we say that j is the *infimum* of A iff for each $a \in A$ we have $a \geq j$, and for every $d \in L$, such that $d \leq b$ for all $b \in A$, we get $j \geq d$. In this case, we write $s = \bigvee A$ and $j = \bigwedge A$ respectively. If Λ is any nonempty set and $(x_\lambda)_{\lambda \in \Lambda}$ is a family of elements in L, we denote also by $\bigvee_{\lambda \in \Lambda} x_\lambda$ and $\bigwedge_{\lambda \in \Lambda} x_\lambda$ the quantities $\bigvee\{x_\lambda : \lambda \in \Lambda\}$ and $\bigwedge\{x_\lambda : \lambda \in \Lambda\}$ respectively, provided that they exist in L. A poset L is said to be a *lattice* iff for every two elements a, $b \in L$ there exist in L the supremum $s := a \vee b$ and the infimum $j := a \wedge b$.

A lattice L is said to be *Dedekind complete* (resp. σ*-Dedekind complete*) iff every (countable) nonempty subset of L, bounded from above, admits supremum in L, and every (countable) nonempty subset of L, bounded from below, has infimum in L. A *Boolean* σ*-algebra* is a σ-Dedekind complete Boolean algebra. A Dedekind complete lattice L is said to be *super Dedekind complete* iff for any nonempty set $A \subset L$, bounded from above, there exists a countable subset $A^* \subset A$, such that $\bigvee A = \bigvee A^*$, and for every nonempty set $B \subset L$, bounded from below, there is a countable subset $B^* \subset B$, with $\bigwedge B = \bigwedge B^*$.

An abelian partially ordered group (resp. semigroup) $L = (L,+,\leq)$ with neutral element 0 is called a *lattice group* (or briefly an (ℓ)*-group*) (resp. a *lattice semigroup*, shortly (ℓ)*-semigroup*) iff it is a lattice and the following implication holds:

$$[a \leq b] \Rightarrow [a + c \leq b + c] \quad \text{for all } a,b,c \in L. \qquad \textbf{(1.3)}$$

Given a (semi)group L, an element $a \in L$ and a positive integer n, we will denote by na the element $a + \ldots + a$ (n times).

An element a of a partially ordered group L is said to be *positive* iff $a \geq 0$. Given any two elements a, $b \in L$, we say that $a < b$ or $b > a$ iff $a \leq b$ and $a \neq b$. If $a > 0$, we say that a is *strictly positive*.

An *order unit* (or *strong order unit*) $u \in L$ is a strictly positive element of a partially ordered group L such that for each $x \in L$ there is $n \in \mathbb{N}$ with $|x| \leq nu$.

For every element x of a lattice group L, set $x^+ := x \vee 0$, $x^- := (-x) \vee 0$; x^+ and x^- are called the *positive* and *negative part* of x respectively. Moreover, setting $|x| := x \vee (-x)$, we say that $|x|$ is the *absolute value* of x. Note that for every element x of an (ℓ)-group L we have $x = x^+ - x^-$ and $x = x^+ - x^-$. Moreover, we get $|x| \geq 0$ for all $x \in L$ and $|x| = 0$ if and only if $x = 0$ (see also Birkhoff (1967) and Boccuto, Riečan and Vrábelová (2009)).

A *Riesz space* or *vector lattice* $L = (L,+,\cdot,\leq)$ is an (ℓ)-group in which an application $\cdot : \mathbb{R} \times L \to L$ is defined, such that L is a real linear space and $\alpha\, x \leq \alpha\, y$ whenever $x, y \in L$, $x \leq y$, and $\alpha \in \mathbb{R}$, $\alpha \geq 0$.

A *normed Riesz space* is a normed space $L = (L, \|\cdot\|)$ endowed with an ordering $\leq$ such that L is a vector lattice and $|x| \leq |y| \Rightarrow \|x\| \leq \|y\|$ for every $x, y \in L$.

If $(L, \| \cdot \|)$ is complete, we say that L is a *Banach lattice*. In a normed Riesz space, we say that $\| \cdot \|$ is *order continuous* iff $\lim_n \| x_n \| = 0$ for every decreasing sequence $(x_n)_n$ in L whose infimum is 0 (see also Meyer-Nieberg (1991) and Schaefer (1974)).

A partially ordered group L is said to be *Archimedean* iff for every choice of $a, b \in L$, with $na \le b$ for all $n \in \mathbb{N}$, we get $a \le 0$.

Proposition 1.3 *Every Dedekind complete* (ℓ)*-group* L *is Archimedean.*

Proof: (see also Birkhoff (1967), Lemma XIII.5, and Boccuto, Riečan and Vrábelová (2009), Proposition 2.10) Let $a, b \in L$, with $na \le b$ for all $n \in \mathbb{N}$. By Dedekind completeness of L, the element $c := \bigvee_{n \in \mathbb{N}} na$ does exist in L. Thus we get

$$c + a = \bigvee_{n \in \mathbb{N}} (n+1)a \le \bigvee_{n \in \mathbb{N}} na = c.$$

From this we obtain $a \le 0$, that is the assertion. □

A *difference poset* (shortly, D*-poset*) is a triple $(L, \le, \ominus)$, where $(L, \le)$ is a poset and $\ominus$ is a partially defined binary operation fulfilling the following conditions for any a, b, $c \in L$.

(*i*) $b \ominus a$ is defined if and only if $a \le b$.

(*ii*) If $a \le b$, then $b \ominus a \le b$ and $b \ominus (b \ominus a) = a$.

(*iii*) If $a \le b \le c$, then $c \ominus b \le c \ominus a$ and $(c \ominus a) \ominus (c \ominus b) = b \ominus a$.

In D-posets it is possible to define a partial operation $\oplus$ as follows:

$c \oplus a$ is defined and $c \oplus b$ iff $b \ominus a$ is defined and $b \ominus a = c$. **(1.4)**

A D-poset which is a lattice is called also a D*-lattice*. The D-posets were introduced by Kôpka and Chovanec (1994). Some fundamental properties and examples can be found, for instance, in Dvurečenskij and Pulmannová (2000).

An alternative structure to a D-poset is that of an *effect algebra*, introduced by Bennet and Foulis (1994) and by Giuntini and Greuling (1989) with the name of "unsharp orthoalgebra".

An *effect algebra* (resp. *orthoalgebra*) is a system $(L,\oplus,0,1)$ consisting of a set L with two special elements 0, $1\in L$ and with a partially defined binary operation $\oplus$, commutative, associative and satisfying the following conditions.

(i) For every $a\in L$ there is a unique element $b\in L$ such that $a\oplus b=1$.

(ii) If $a\in L$ and $a\oplus 1$ (resp. $a\oplus a$) is defined, then $a=0$.

Note that every orthoalgebra is an effect algebra, but in general the converse is not true (see also Dvurečenskij and Pulmannová (2000), Lemma 1.5.2 and Proposition 1.5.3).

Let L be an effect algebra. An element $a\neq 0$, $a\in L$, is said to be an *atom* iff the following condition holds:

if $x\leq a$, $x\in L$, then either $x=a$ or $x=0$.

An effect algebra L is said to be *atomic* iff for every $y\in L$ with $y\neq 0$ there exists an atom $x\in L$ with $x\leq b$ (see also Dvurečenskij and Pulmannová (2000), Kappos (1969)).

The equivalence between the structures of D-poset and effect algebra was proved in Dvurečenskij and Pulmannová (2000), Theorem 1.3.4.

An *orthomodular poset* (shortly, OMP) is a bounded poset $(L,\leq,{}',0,1)$ with a unary operation ${}':L\to L$ (an *orthocomplementation*) satisfying the following conditions for every a, b, $c\in L$.

(i) $a\leq b\Rightarrow b'\leq a'$.

(ii) $(a')'=a$.

(iii) $a\vee a'=1$.

(iv) If $a\leq b'$, then $a\vee b$ exists in L.

(v) If $a \le b$, then there is $c \in L$ with $c \le a'$ and $a \vee c = b$.

An *orthomodular lattice* is a lattice ordered OMP. Observe that every lattice ordered orthoalgebra is an orthomodular lattice, and the converse is also true (see Dvurečenskij and Pulmannová (2000), Proposition 1.5.8).

Note that every orthomodular poset can be regarded as a D-poset (or effect algebra), but in general the converse is not true (see also Dvurečenskij and Pulmannová (2000), Proposition 1.5.7).

If $L = (L, \oplus, 0, 1)$ is an orthoalgebra and a, $b \in L$, then we say that a is *orthogonal* to b (shortly, $a \perp b$), iff $a \oplus b$ is defined in L.

A *Vitali space*, or *commutative minimal clan*, is a lattice $(L, \le)$, endowed with a commutative and associative partial operation $\oplus : S \to L$, defined on a set $S \subset L \times L$, with an element $0 \in L$ such that $(0, a) \in S$ and $a \oplus 0 = a$ for every $a \in L$, and satisfying the following properties:

(i) If $(a,b) \in S$ and $(c,b) \in S$, then $[a \le c] \Leftrightarrow [a \oplus b \le c \oplus b]$ (cancellation law);

(ii) For every a, $b \in L$ there is $d \in L$ such that $(d, a) \in S$ and $(d, a \wedge b) \in S$, and we have,

$d \oplus a = a \vee b$ and $d \oplus (a \wedge b) = b$

(see also Constantinescu (1989), Dvurečenskij and Pulmannová (2000), Schmidt (1985 and 1998) and Wyler (1966)). Note that both Boolean algebras and lattice ordered groups are Vitali spaces, and every Vitali space is a D-lattice (see also Ventriglia (2007)).

Another abstract structure, widely investigated in the literature, is that of MV-algebra (see also Barbieri and Weber (2002), Boccuto, Riečan and Vrábelová (2009), Dvurečenskij and Pulmannová (2000), Riečan and Mundici (2002), Riečan and Neubrunn (1997)). An *MV-algebra* $L = (L, \oplus, ', 0, 1)$ is an abelian semigroup $(L, \oplus)$ with 0, $1 \in L$ and a unary operation $' : L \to L$, such that

$x \oplus 1 = 1$, $(x')' = x$, $0' = 1$ and $(x' \oplus y)' \oplus y = (x \oplus y')' \oplus x$ for every x, $y \in L$.

The MV -algebras were introduced by Chang (1958 and 1959).

Observe that Mundici proved every MV -algebra L is lattice isomorphic to the unit interval of a lattice group R, $[0,u]:=\{x\in R: 0\le x\le u\}$ (see Mundici (1986), Theorems 2.5, 3.8, 3.9; see also Barbieri and Weber (2002), Theorem 1.1.2, Riečan and Mundici (2002), Theorem 1.7). Furthermore, note that every MV -algebra is a D -poset, but in general the converse is not true (see also Dvurečenskij and Pulmannová (2000), Kôpka and Chovanec (1994)).

A **-ring* is an associative ring A with a map $^*: A\to A$ such that

$(x+y)^*=x^*+y^*$, $(xy)^*=y^*x^*$, $(x^*)^*=x$ for every x, $y\in A$.

The map * is called *involution.*

A **-algebra* A is a *-ring with involution *, that is an associative algebra over a commutative *-ring R with involution $'$, such that $(rx)^*=r'x^*$ for every $r\in R$ and $x\in A$.

A Banach algebra that is a *-algebra with respect to the involution * is a C^*-*algebra* iff $\| x^*x \|=\| x\| \, \| x^*\|$ for every $x\in A$ (see also Dixmier (1969), Gel'fand and Naimark (1943) and Segal (1947)).

A *von Neumann algebra* is a *-algebra of bounded operators on a Hilbert space that is closed in the weak operator topology and contains the identity operator (see also von Neumann (1930)).

1.1.3. Set Functions

Let $G\neq\emptyset$ be any abstract set. A *paving* $\mathcal{E}$ is a class of subsets of G containing $\emptyset$. A *set function* is a function defined on a paving $\mathcal{E}$ and with values in $\mathbb{R}$, or a Banach, metric, uniform or topological space, topological or lattice (semi)group, or another abstract structure.

We now recall the main notions about real-valued set functions. In an analogous way it is possible to deal with set functions with values in abstract spaces. Following Pap (1995, 2002c), we say that a family $\mathcal{E}$ of subsets of an abstract nonempty set G is said to be a *lattice* iff $A\cup B$ and $A\cap B$ belong to $\mathcal{E}$ whenever A, $B\in\mathcal{E}$, is a *ring* iff it contains the empty set and for every A, $B\in\mathcal{E}$

we get that $A \cup B$ and $A \setminus B$ belong to $\mathcal{E}$, is an *algebra* iff it is a ring and contains G, is a σ-*ring* (resp. δ-*ring*) iff it is a ring and $\bigcup_{n=1}^{\infty} A_n \in \mathcal{E}$ (resp. $\bigcap_{n=1}^{\infty}) A_n \in \mathcal{E}$) whenever $(A_n)_n$ is a sequence of elements of $\mathcal{E}$, and is a σ-*algebra* iff it is a σ-ring and contains G. We denote by $\mathcal{P}(G)$ the family of all subsets of G.

If $(E_n)_n$ is a sequence in $\mathcal{P}(G)$, then we define

$$\limsup_n E_n := \bigcap_{n=1}^{\infty} \left(\bigcup_{k=n}^{\infty} E_k \right) \text{ and } \liminf_n E_n := \bigcup_{n=1}^{\infty} \left(\bigcap_{k=n}^{\infty} E_k \right).$$

When $\limsup_n E_n = \liminf_n E_n$, we denote this set by $\lim_n E_n$ and we call it the *limit in the set-theoretic sense* of $(E_n)_n$.

Let $\mathcal{E} \subset \mathcal{P}(G)$ be a paving. We say that a set function $m : \mathcal{E} \to \mathbb{R}$ with $m(\emptyset) = 0$ is *monotone* iff $m(A) \le m(B)$ whenever A, $B \in \mathcal{E}$ with $A \subset B$, *finitely additive* (or a *finitely additive measure*) on $\mathcal{E}$ iff $m(A \cup B) = m(A) + m(B)$ whenever A, B, $A \cup B \in \mathcal{E}$ and $A \cap B = \emptyset$, and m is *countably additive* or σ-*additive* (or a *countably* or σ-*additive measure*) on $\mathcal{E}$ iff

$m\left(\bigcup_{n=1}^{\infty} A_n \right) = \sum_{n=1}^{\infty} m(A_n)$ for every disjoint sequence $(A_n)_n$ in $\mathcal{E}$, such that $\bigcup_{n=1}^{\infty} A_n \in \mathcal{E}$.

If m is a non-negative extended real-valued set function, then we say that m is σ-*finite*, iff there exists a sequence $(G_n)_n$ in $\mathcal{L}$ with $G = \bigcup_{n=1}^{\infty} G_n$ and $m(G_n) < +\infty$ for each $n \in \mathbb{N}$.

A *submeasure* η is a monotone increasing non-negative extended real-valued set function, defined on a σ-ring Σ, with $\eta(\emptyset) = 0$ and $\eta(A \cup B) \le \eta(A) + \eta(B)$ whenever A, $B \in \Sigma$. A submeasure η is said to be *order continuous* iff

$\lim_n \eta(H_n) = 0$ whenever $(H_n)_n$ is a decreasing sequence in Σ, with $\bigcap_{n=1}^{\infty} H_n = \emptyset$.

It is well-known that, given a submeasure η on Σ, the function $d(A,B) = \eta(A\Delta B)$, A, $B \in \Sigma$, is a pseudo-distance (see also Weber (2002)).

A topology τ on Σ is said to be a *Fréchet-Nikodým topology* (shortly, *FN-topology*) iff the functions $(A,B) \mapsto A\Delta B$ (symmetric difference) and $(A,B) \mapsto A \cap B$ from $\Sigma \times \Sigma$ (endowed with the product topology) to Σ are continuous, and for each τ-neighborhood V of $\emptyset$ in Σ there is a τ-neighborhood U of $\emptyset$ in Σ with the property that, if $B \in \Sigma$ is contained in some suitable element of U, then $B \in V$ (see Drewnowski (1972a)).

Observe that a topology τ on Σ is a Fréchet-Nikodým topology if and only if there exists a family of submeasures $\mathcal{Z} := \{\eta_i : i \in \Xi\}$, with the property that a base of τ-neighborhoods of $\emptyset$ in Σ is given by the sets of the type $U_{\varepsilon,D} := \{A \in \Sigma : \eta_i(A) < \varepsilon$ for all $i \in D\}$, where $\varepsilon > 0$ and D varies in the class of all finite subsets of Ξ (see also Drewnowski (1972a), Proposition 2.6, Theorem 2.7).

Let $\mathcal{E}$ be a class of subsets of G such that $\emptyset \in \mathcal{E}$, and $k > 0$. A non-negative set function $m : \mathcal{E} \to \mathbb{R}$ is said to be k*-triangular* iff $m(\emptyset) = 0$ and

$$m(A) - km(B) \le m(A \cup B) \le m(A) + km(B)$$

whenever A, $B \in \mathcal{E}$, $A \cap B = \emptyset$ and $A \cup B \in \mathcal{E}$.

The k-triangular functions are examples of non-additive set functions, which are widely studied in the literature in proving limit theorems. For their fundamental properties, we refer to Pap (1995 and 2002c) and their bibliographies.

Given a fixed paving $\mathcal{E}$ and a finitely additive measure $m : \mathcal{E} \to \mathbb{R}$, let us define

$$m^{(+)} = m_{\mathcal{E}}^{(+)}, \quad m^{(-)} = m_{\mathcal{E}}^{(+)}, \; \| \quad m\| = \| m\|_{\mathcal{E}}, \quad v(m) = v_{\mathcal{E}}(m) : \mathcal{E} \to R$$

by setting

$$m^{(+)}(E) = \sup\{m(B) : B \in \mathcal{E}, B \subset E\},$$

$$m^{(-)}(E) = \sup\{-m(B) : B \in \mathcal{E}, B \subset E\},$$

$$\| m\|(E) = m^{(+)}(E) + m^{(-)}(E),$$

$$v(m)(E) = \sup\{| m(B) |: B \in \mathcal{E}, B \subset E\}, \ E \in \mathcal{E}.$$

The quantities $m^{(+)}$, $m^{(-)}$, $\| m\|$, $v(m)$ are called *positive part, negative part, total variation* (or simply *variation*) and *semivariation* of m respectively. Note that, for each $B, E \in \mathcal{E}$ with $B \subset E$ we get,

$$| m(B) |= (m(B))^{+} + (m(B))^{-} = (m(B)) \vee 0 + (-m(B)) \vee 0 \leq$$

$$\leq \sup\{m(B) : B \in \mathcal{E}, B \subset E\} + \sup\{-m(B) : B \in \mathcal{E}, B \subset E\} = \quad \textbf{(1.5)}$$

$$= m^{(+)}(E) + m^{(-)}(E) = \| m\|(E) \leq$$

$$\leq \sup\{| m(B) |: B \in \mathcal{E}, B \subset E\} + \sup\{| m(B) |: B \in \mathcal{E}, B \subset E\} = 2v_{\mathcal{E}}(m)(E).$$

From (1.5) we get,

$$v(m)(E) \leq \| m\|(E) \leq 2\, v(m)(E) \quad \textbf{(1.6)}$$

for all $E \in \mathcal{E}$ (see also Bhaskara Rao and Bhaskara Rao (1983), Diestel and Spalsbury (2012) and Dunford and Schwartz (1958)).

If $\mathcal{L}$ is an algebra of subsets of G, ba$(\mathcal{L})$ will denote the Banach lattice of all finitely additive bounded real-valued measures, endowed with the total variation norm (see also Bhaskara Rao and Bhaskara Rao (1983)).

The concepts of submeasure, finitely and σ-additive set function and the related ones can be formulated analogously as above for (not necessarily additive) set functions defined in an abstract structure which are not necessarily a Boolean algebra, for example taking suitable "orthogonal" elements instead of disjoint sets (see also, for instance, Constantinescu (1989), Convention 1.2, and de Lucia (1985 and 1998)). For example, if L is a lattice with 0 element and $(R,+)$ is an ordered semigroup, we will say that a function $\eta : L \to R$ is a *submeasure* iff $\eta(0) = 0$, $\eta(a) \le \eta(b)$ whenever $a, b \in L$ with $a \le b$ and

$\eta(a \vee b) \le \eta(a) + \eta(b)$ for each $a, b \in L$

(see also Popa (1997)).

1.2. The Evolution of the Limit Theorems

In the literature, some results which are related to convergence and boundedness theorems for set functions are the matrix theorems, the uniform boundedness principle for operators and the compactness properties of sets of measures, which can be used to prove different types of limit theorems, about which there exist also several kinds of direct proofs. The most used tools are the sliding hump, the Baire category theorem, the Fréchet-Nikodým topologies and the construction of σ-additive restrictions or σ-additive Stone-type extensions of finitely additive measures.

1.2.1. The Sliding Hump

The sliding hump technique was known just at the beginning of the last century, for example in proving some theorems about exchange of limits and of series under the sign of integral (see also Hahn (1921 and 1922), Lebesgue (1909), Vitali (1907)) and the uniform boundedness principle (see also Banach (1922 and 1932) and Hahn (1922)), which asserts that, if X is a Banach space, Y is a normed linear space and $T_n : X \to Y$, $n \in \mathbb{N}$, is a sequence of continuous linear operators which is pointwise bounded on X (that is the set $\{T_n(x) : n \in \mathbb{N}\}$ is bounded in Y for every $x \in X$), then the real sequence $\| T_n \|$, $n \in \mathbb{N}$ of operator norms is bounded (see also Banach (1922), Theorem 5, and Hahn (1922)). Banach and Steinhaus (1927) (see Lemma 3), following a technique by S. Saks (1933a and 1937) based on the Baire category theorem, proved that, if $(T_n)_n$ is a sequence of continuous linear operators from a Banach space X into a Banach space Y and if the set $\{x \in X : T_k(x) \text{ is bounded}\}$ is of second Baire category, then the sequence $(T_k)_k$ is bounded in $\mathbb{R}$. A sliding hump-type argument was used by Helly (1912) (see Theorem 3) to prove that, if $(T_n)_n$ is a sequence of continuous linear functionals on the space $C([a,b])$ of the continuous real-valued functions on the compact interval $[a,b]$ such that $\lim_n T_n(f) = T(f)$ exists for each $f \in C([a,b])$ and is a continuous linear functional, then the sequence of norms $(\| T_n \|)_n$ is bounded. A historical survey on the evolution of the uniform boundedness principle is found in Swartz (1990).

Hahn (1922) (see Theorem VI) proved the following "Schur-type" theorem for double sequences, using the sliding hump technique.

Theorem 2.1 *Let $b_n = (v_{n,k})_k$, $n \in \mathbb{N}$, be a sequence of elements of l^1. In order that for every subset $A \subset \mathbb{N}$ the limit*

$$\lim_n \sum_{k \in A} v_{n,k} \tag{1.7}$$

exists in $\mathbb{R}$, the following conditions are necessary and sufficient:

2.1.1) the limit $\lim_n v_{n,k}$ exists in $\mathbb{R}$ for every $k \in \mathbb{N}$;

2.1.2) for each $\varepsilon > 0$ there is $k_0 \in \mathbb{N}$ such that

$$\sum_{k=k_0}^{\infty} | v_{n,k} | \leq \varepsilon \text{ for every } n \in \mathbb{N}.$$

Proof: We first prove that 2.1.1) and 2.1.2) are sufficient. Choose arbitrarily $\varepsilon > 0$ and $A \subset \mathbb{N}$, and let k_0 be according to 2.1.2). Set $A^{'} = A \cap [1, k_0 - 1]$, $A^{''} = A \cap [k_0, +\infty[$ (here, the intervals and halflines are meant in $\mathbb{N}$). Thanks to 2.1.1), the limit

$$\lim_n \sum_{k \in A^{'}} v_{n,k}$$

exists in $\mathbb{R}$, and so, by the Cauchy criterion, there exists $n_0 \in \mathbb{N}$ with

$$\left| \sum_{k \in A^{'}} v_{n,k} - \sum_{k \in A^{'}} v_{n^{'},k} \right| \leq \varepsilon \text{ for all } n,\ n^{'} \geq n_0.$$

By virtue of 2.1.2), we have,

$$\left|\sum_{k\in A''} v_{n,k}\right| \le \varepsilon \text{ for every } n\in\mathbb{N}.$$

So we get,

$$\left|\sum_{k\in A} v_{n,k} - \sum_{k\in A} v_{n',k}\right| \le 3\varepsilon \text{ for all } n,\ n' \ge n_0.$$

that is the limit in (1.7) exists in $\mathbb{R}$.

We now prove that 2.1.1) and 2.1.2) are necessary. The condition 2.1.1) follows immediately from the hypothesis. We now prove 2.1.2) by contradiction, taking into account 2.1.1). If 2.1.2) does not hold, then there is $\varepsilon > 0$ with the following property: for every N, $K\in\mathbb{N}$ there are $n_1 > N$ and $l\in\mathbb{N}$ such that

$$\sum_{k=K}^{K+l} |v_{n_1,k}| > \varepsilon. \qquad \textbf{(1.8)}$$

Indeed, otherwise for every $\varepsilon > 0$ there are $N, K\in\mathbb{N}$ with

$$\sum_{k=K}^{K+l} |v_{n,k}| \le \varepsilon \text{ for all } n > N \text{ and every } l\in\mathbb{N}.$$

From this, letting l tend to $+\infty$, we obtain

$$\sum_{k=K}^{\infty} |v_{n,k}| \le \varepsilon \text{ for every } n > N. \qquad \textbf{(1.9)}$$

Since the sequence $(v_{n,k})_{n,k}$ belongs to l^1 for every $n\in\mathbb{N}$, then all sums of the type

$$\sum_{k=1}^{\infty} |v_{n,k}|$$

are finite, and so there is $K'\in\mathbb{N}$, such that

$$\sum_{k=K'}^{\infty} | v_{n,k} | \leq \varepsilon \text{ for every } n = 1,2,\ldots,N. \qquad \textbf{(1.10)}$$

Let $k_0 := \max\{K, K'\}$. Then from (1.9) and (1.10) it follows that the condition 2.1.2) is fulfilled, against the assumption.

From (1.8) we deduce that for each N, $K \in \mathbb{N}$ there are $N_0 > N$ and a finite set A_1 of numbers greater or equal than K, such that

$$\left| \sum_{k \in A_1} v_{n_1,k} \right| > \frac{\varepsilon}{2}. \qquad \textbf{(1.11)}$$

Indeed, as we will prove in Lemma 1.1 of Chapter 4, if $(x_k)_k \in \ell_1$ and K, $l \in \mathbb{N}$, then

$$\sum_{k=K}^{K+l} | x_k | \leq 2 \max_{A \in \mathcal{P}([K,K+l])} | \sum_{k \in A} x_k |. \qquad \textbf{(1.12)}$$

Since the series

$$\sum_{k=1}^{\infty} | v_{n_1,k} |$$

is convergent, there is an integer k_1, greater than all elements of A_1 and so large, that

$$\sum_{k=k_1}^{\infty} | v_{n_1,k} | < \frac{\varepsilon}{12}.$$

Thanks to condition 2.1.1), there is $N_1 > n_1$, such that

$$\left| \sum_{k \in A_1} v_{n,k} - \sum_{k \in A_1} v_{n',k} \right| < \frac{\varepsilon}{12} \text{ for every } n, \; n' \geq N_1.$$

Moreover there is $K_1 > k_1$, such that

$$\sum_{k=K_1}^{\infty} | v_{N_1,k} | < \frac{\varepsilon}{12}$$

and there are $n_2 > N_1$ and a finite set $A_2 \subset \mathbb{N}$, which are all greater than K_1 (and so greater than all elements of A_1), such that

$$\left| \sum_{k \in A_2} v_{n_2,k} \right| > \frac{\varepsilon}{2}.$$

Put $B_1 := A_1$ and $B_2 := A_1 \cup A_2$. In correspondence with n_2 there is k_2 greater than K_1 and than all the elements of B_2, such that

$$\sum_{k=k_2}^{\infty} | v_{n_2,k} | < \frac{\varepsilon}{12}.$$

Furthermore, there are an integer $N_2 > n_2$, with

$$\left| \sum_{k \in B_2} v_{n,k} - \sum_{k \in B_2} v_{n',k} \right| < \frac{\varepsilon}{12} \text{ for all } n,\ n' \geq N_2.$$

and a natural number $K_2 > k_2$, such that

$$\sum_{k=K_2}^{\infty} | v_{N_2,k} | < \frac{\varepsilon}{12}.$$

By induction, we construct four sequences $(n_i)_i$, $(N_i)_i$, $(k_i)_i$, $(K_i)_i$ with $n_i < N_i < n_{i+1}$ and $k_i < K_i < k_{i+1}$ and two sequences $(A_i)_i$, $(B_i)_i$ of finite subsets of $\mathbb{N}$, with $B_i = \bigcup_{j=1}^{i} A_j$ and $A_{i+1} \cap B_i = \varnothing$ for every $i \in \mathbb{N}$. For $j > i$, A_j contains only numbers greater or equal than K_i. We get,

$$\sum_{k=k_i}^{\infty} | v_{n_i,k} | < \frac{\varepsilon}{12}, \quad \sum_{k=K_i}^{\infty} | v_{N_i,k} | < \frac{\varepsilon}{12}, \tag{1.13}$$

$$\left|\sum_{k\in B_i} v_{n,k} - \sum_{k\in B_i} v_{n',k}\right| < \frac{\varepsilon}{12} \text{ for all } n,\ n' \geq N_2, \tag{1.14}$$

$$\left|\sum_{k\in A_i} v_{n_i,k}\right| > \frac{\varepsilon}{2}. \tag{1.15}$$

The inequality (1.15) means that the sequence $(v_{n_i,k})_k$ has a "hump" in correspondence with the set A_i, which is the only part where the absolute value of the sum of the associated terms of this sequence is "large". Note that, as i tends to $+\infty$, this "hump" "slides" to the right, because the sequence $(n_i)_i$ is strictly increasing.

Now, let us "gather" the "humps" (see also Swartz (1990)). Set $A := \bigcup_{i=1}^{\infty} A_i$, and put $C_i := A \setminus B_i$. We have

$$\left|\sum_{k\in A} v_{N_{i-1},k} - \sum_{k\in A} v_{n_i,k}\right| \geq \left|\sum_{k\in A_i} v_{n_i,k}\right| - \left|\sum_{k\in B_{i-1}} v_{N_{i-1},k} - \sum_{k\in B_{i-1}} v_{n_i,k}\right| - \left|\sum_{k\in C_{i-1}} v_{N_{i-1},k}\right| - \left|\sum_{k\in C_i} v_{n_i,k}\right|. \tag{1.16}$$

By (1.14) we get

$$\left|\sum_{k\in B_{i-1}} v_{N_{i-1},k} - \sum_{k\in B_{i-1}} v_{n_i,k}\right| < \frac{\varepsilon}{12}.$$

Since C_{i-1} contains only numbers greater or equal than K_{i-1}, and C_i contains only numbers greater or equal than $K_i > k_i$, from (1.13) we obtain

$$\left|\sum_{k\in C_{i-1}} v_{N_{i-1},k}\right| < \frac{\varepsilon}{12} \text{ and } \left|\sum_{k\in C_i} v_{n_i,k}\right| < \frac{\varepsilon}{12}.$$

Combining (1.15) and (1.16), we get,

$$\left|\sum_{k\in A} v_{N_{i-1},k} - \sum_{k\in A} v_{n_i,k}\right| > \frac{\varepsilon}{4},$$

and so $\lim_n \sum_{k\in A} v_{n,k}$ does not exist in $\mathbb{R}$, obtaining a contradiction. This ends the proof. □

1.2.2. Vitali-Hahn-Saks-Nikodým, Schur and Dunford-Pettis Theorems

Vitali (1907) (see Theorem 7), using diagonal arguments, proved the following result.

Theorem 2.2 *Let $G \subset \mathbb{R}$ be a set of finite Lebesgue measure, and $\sum_{r=1}^{\infty} u_r(x)$ be a series of integrable real-valued functions defined on G. A necessary and sufficient condition in order that*

$$\sum_{r=1}^{\infty}\int_E u_r(x)\, dx = \int_E \sum_{r=1}^{\infty} u_r(x)\, dx$$

for every measurable set $E \subset G$ is that the series $\sum_{r=1}^{\infty} u_r(x)$ is convergent and the integrals

$$\int \sum_{n=1}^{r} u_n(x)\, dx \quad r \in \mathbb{N}$$

are equiabsolutely continuous, that is: given $\varepsilon > 0$, there exists $\delta > 0$ with

$$\int_E \sum_{n=1}^{r} u_n(x)\, dx \le \varepsilon$$

whenever E is a measurable subset of G, with measure less than δ, and for each $r \in \mathbb{N}$.

Lebesgue (1909) (see Theorem III), using sliding hump techniques, proved the following:

Theorem 2.3 *Let $[a,b]$ be a closed and bounded subinterval of the real line, $h:[a,b]\to\mathbb{R}$ be a bounded integrable function, and $f_n:[a,b]\to\mathbb{R}$, $n\in\mathbb{N}$, be a function sequence, such that the integrals*

$$I_n := \int_a^b h(t) f_n(t)dt, \quad n\in\mathbb{N},$$

exist in $\mathbb{R}$.

Then the following two conditions are necessary and sufficient in order that $\lim_n I_n = 0$:

(a) *The integrals* $\int_E |f_n(t)|\,dt$, $n\in\mathbb{N}$, *are equiabsolutely continuous;*

(b) $\lim_n \int_E f_n(t)dt = 0$, *for each subinterval* $E\subset(a,b)$

(see also Choksi (2001) and Diestel (1991)).

Moreover, again by means of sliding hump methods, Lebesgue proved that a sequence of functions $(f_n)_n$ in $L^2([a,b])$, the space of the Lebesgue square integrable functions, is weakly convergent to 0 if and only if the sequence of the norms $(\|f_n\|_2)_n$ is bounded in $\mathbb{R}$ and $\int_c^d f_n(t)\,dt = 0$ whenever $a\le c<d\le b$ (Lebesgue (1909), Theorem II, see also Swartz (1990)). From this, since

$$\|f_n\|_2 = \sup\{\int_a^b h(t) f_n(t)\,dt : \|h\|_2\le 1\},$$

it follows that a weakly convergent sequence in $L^2([a,b])$ is uniformly bounded on the unit ball of $L^2([a,b])$, that is a version of the uniform boundedness principle for weakly convergent sequences in $L^2([a,b])$ (see also Swartz (1990)).

Some similar results were given for weakly convergent sequences in the space of functions of bounded variation (Lebesgue (1909), Theorem V) and in the space of functions *simply discontinuous* on $[a,b]$, that is such that for every $\varepsilon>0$ there is a finite number of points of discontinuity in correspondence of which the oscillation of f is less or equal to ε (Lebesgue (1909), Theorem IV).

Radon (1919) (see §III) extended to set functions the concept of absolute continuity for real functions.

Hahn (1922) (see Theorem XXI) proved the following theorem on exchange of limits under the sign of integral.

Theorem 2.4 *Let $[a,b] \subset \mathbb{R}$ be a compact interval and $f_n : [a,b] \to \mathbb{R}$, $n \in \mathbb{N}$, be a sequence of Lebesgue integrable functions. In order that the limit*

$$\lim_n \int_E f_n(t)\, dt$$

exists in $\mathbb{R}$ for every measurable subset $E \subset [a,b]$, the following two conditions are necessary and sufficient:

(a) *For each $x \in [a,b)$ the limit $\lim_n \int_x^b f_n(t)\, dt$ exists in $\mathbb{R}$;*

(b) *The integrals $\int_E |f_n(t)|\, dt$, $n \in \mathbb{N}$, are equiabsolutely continuous.*

As a consequence of the Radon-Nikodým theorem and Theorem 2.4 extended to measurable spaces of finite measure, it is possible to prove the following version of the Vitali-Hahn-Saks theorem (see also Choksi (2001), Corollary 2).

Theorem 2.5 *If Σ is a σ-algebra, $m_n : \Sigma \to \mathbb{R}$ is a sequence of σ-additive measures and*

$\lim_n m_n(E)$ exists in $\mathbb{R}$ for every $E \in \Sigma$, then this limit is a σ-additive measure on Σ.

Schur (1920) (see Hilfssatz), using the sliding hump technique, proved the following

Proposition 2.6 *Let $(a_n)_n$ be a sequence of real of complex numbers. Then the series*

$$\sum_{n=1}^{\infty} a_n x_n$$

is convergent for each convergent sequence $(x_n)_n$, *if and only if the series*

$$\sum_{n=1}^{\infty} a_n$$

is absolutely convergent.

An infinite matrix $A=(a_{k,l})_{k,l}$ is said to be *convergence preserving* if and only if it maps convergent sequences into convergent sequences. We say that A is *regular* iff it maps sequences convergent to a real number b into sequences convergent to the same number b. The matrix A is said to be *convergence generating* iff it maps bounded sequences into convergent sequences.

If the matrix A is convergence generating or convergence preserving, then for each $k \in \mathbb{N}$ the series

$$\sum_{l=1}^{\infty} a_{k,l} x_l$$

converges, as soon as $\lim_n x_n$ exists in $\mathbb{R}$. By Proposition 2.6, this can happen only when all the series of rows

$$\sum_{l=1}^{\infty} a_{k,l}$$

are absolutely convergent (this assumption is always done in Schur (1920)). Set now

$$\sigma_k = \sum_{l=1}^{\infty} a_{k,l}, \ \zeta_k = \sum_{l=1}^{\infty} | a_{k,l} |.$$

The numbers σ_k are called the *sums of the rows* and the ζ_k's are called the *norms of the rows* (in the literature sometimes with *norms of the rows* are indicated the quantities

$$\sum_{l=1}^{\infty} | a_{k,l} |^2).$$

The following results hold.

Theorem 2.7 (Schur (1920), Theorem I) *An infinite matrix* $A=(a_{k,l})_{k,l}$ *is convergence preserving if and only if the following conditions are satisfied.*

(a) *For each* $l \in \mathbb{N}$ *the limit* $a_l := \lim_n a_{k,l}$ *exists in* $\mathbb{R}$.

(b) *The sums of the rows* σ_1, σ_2, … *converge to a number* $\sigma \in \mathbb{R}$.

(c) *The norms of the rows* ζ_1, ζ_2, … *are contained in a bounded set.*

If these three conditions hold, then the series $a := \sum_{n=1}^{\infty} a_n$ *is absolutely convergent, and for every convergent sequence* $(x_n)_n$ *with* $x = \lim_n x_n$ *we get:*

$$y = \lim_k \left(\sum_{l=1}^{\infty} a_{k,l} x_l \right) = (\sigma - a)x + \sum_{l=1}^{\infty} a_l x_l .$$

Theorem 2.8 (Schur (1920), Theorem II) *A convergence preserving matrix* A *is regular if and only if the* a_l*'s are all equal to* 0 *and the limit value* σ *of the sums of the rows* σ_k *is* 1.

Theorem 2.9 (Schur (1920), Theorem III) *An infinite matrix* $A=(a_{k,l})_{k,l}$ *is convergence generating if and only if, together with the conditions of Theorem 2.7, also the following condition holds:*

(a) *The series* $\sum_{l=1}^{k} | a_{k,l} |$, $k \in \mathbb{N}$, *are convergent uniformly with respect to* k, *that is for every* $\varepsilon > 0$ *there is a natural number* l *such that for each* $k \in \mathbb{N}$ *we get:*

$$\sum_{p=1}^{\infty} | a_{k,l+p} | \le \varepsilon .$$

In this case, for every bounded sequence $(x_n)_n$ *we get*

$$\lim_{k\to\infty} \left(\sum_{l=1}^{\infty} a_{k,l} x_l \right) = \sum_{n=1}^{\infty} a_n x_n .$$

Mazur and Orlicz (1933) (see Theorem 3) proved that, if X and Y are metric linear spaces with X complete and if $T_n : X \to Y$, $n \in \mathbb{N}$, is a sequence of linear continuous operators such that the sequence $(T_n(x))_n$ is bounded in Y for every $x \in X$, then $(T_n)_n$ is equicontinuous.

Nikodým (1933a) (see Lemme), using the sliding hump method, proved the following:

Lemma 2.10 *Let* $(a_{i,k})_{i,k}$ *be an infinite matrix, satisfying the following conditions:*

(i) The sequence $(a_{i,k})_i$ *is bounded for each* $k \in \mathbb{N}$;

(ii) The series $\sum_{k=1}^{\infty} | a_{i,k} |$ *converges for each* $i \in \mathbb{N}$;

(iii) $\lim_n a_{n,n} = +\infty$.

Then there exists a strictly increasing sequence $(t_j)_j$ *in* $\mathbb{N}$ *such that, if*

$$b_i := \sum_{n=1}^{\infty} a_{t_i,t_n}, \quad i \in \mathbb{N},$$

then

$$\lim_i b_i = +\infty.$$

This result is used for the following version of the Nikodým boundedness theorem (see also Nikodým (1933a) and Cafiero (1968), Theorem 3.2).

Theorem 2.11 *Let* G, Λ *be two nonempty sets,* Σ *be a* σ*-algebra of subsets of* G *and* $m_\lambda : \Sigma \to \mathbb{R}$, $\lambda \in \Lambda$, *be a sequence of* σ*-additive measures. Assume that for every* $E \in \Sigma$ *there exists* $b_E \in \mathbb{R}$ *with* $| m_\lambda(E) | \le b_E$. *Then there is a positive real number* b *with* $| m_\lambda(E) | \le b$ *for each* $E \in \Sigma$ *and* $\lambda \in \Lambda$.

Nikodým (1933b) (see Lemme) proved the following

Lemma 2.12 *Let* $(a_{i,k})_{i,k}$ *be an infinite matrix, satisfying the following conditions:*

(i) $\lim_i a_{i,k} = 0$ *for each* $k \in \mathbb{N}$ *;*

(ii) *The series* $\sum_{k=1}^{\infty} | a_{i,k} |$ *converges for each* $i \in \mathbb{N}$;

(iii) $\lim_i \left(\sum_{k=1}^{\infty} a_{i,k}\right) =: b \in \mathbb{R}$.

Then there exist two strictly increasing sequences $(i_s)_s$, $(k_t)_t$ *in* $\mathbb{N}$ *with*

$$\lim_s \left(\sum_{t=1}^{\infty} a_{i_{2s+1},k_t} \right) = b \text{ and } \lim_s \left(\sum_{t=1}^{\infty} a_{i_{2s},k_t} \right) = 0.$$

This result is used to get the following theorem on σ-additivity of the limit measure (see also Doubrovsky (1945), Theorem 1 and Nikodým (1933b)).

Theorem 2.13 *Let* G *be any nonempty set,* Σ *be a* σ*-algebra of parts of* G *, and* $m_n : \Sigma \to \mathbb{R}$*,* $n \in \mathbb{N}$*, be a sequence of* σ*-additive measures. Moreover, suppose that the limit*

$\lim_n m_n(E) =: m(E)$ *exists in* $\mathbb{R}$ *for each* $E \in \Sigma$ *. Then* m *is a* σ *-additive measure.*

Let now $\mathcal{R}$ be the space of all measurable characteristic functions on $[0,1]$, that is of all functions which almost everywhere assume two values only, 0 and 1. The distance between two functions x, $y \in \mathcal{R}$ is defined by

$$d(x, y) = \| x - y \| = \int_0^1 | x(t) - y(t) | dt.$$

With this definition of distance, $\mathcal{R}$ is a metric complete space (see also Saks (1933a)).

The following result holds (see Saks (1933a), Theorem 2).

Theorem 2.14 *Let $f_n:[0,1]\to\mathbb{R}$, $n\in\mathbb{N}$, be a sequence of Lebesgue integrable functions. If*

$$\limsup_n \left|\int_0^1 f_n(t)x(t)dt\right| < +\infty$$

for all functions $x(t)$ of a set of second category in $\mathcal{R}$, then we get also

$$\limsup_n \int_0^1 |f_n(t)|dt < +\infty.$$

If $m:\Sigma\to\mathbb{R}$ is a σ-additive measure, then a set $A\in\Sigma$ is said to be *singular* (with respect to m) iff for each $B\in\Sigma$ with $B\subset A$ we get either $m(B)=0$ or $m(A\setminus B)=0$. We endow Σ with the *Fréchet-Nikodým topology*, which is generated by the pseudo-distance

$$d_m(E_1,E_2)=|m(E_1)-m(E_2)|,\ E_1,E_2\in\Sigma$$

(see also Diestel (1984)). If $m,\nu:\Sigma\to\mathbb{R}$ are two measures, then m is said to be *absolutely continuous with respect to* ν iff for each $\varepsilon>0$ there is a $\delta>0$ with $|m(E)|\le\varepsilon$ whenever $|\nu(E)|\le\delta$.

Let $\Lambda\neq\emptyset$. A family of measures $m_\lambda:\Sigma\to\mathbb{R}$, $\lambda\in\Lambda$, is *equiabsolutely continuous with respect to* ν iff for each $\varepsilon>0$ there is a $\delta>0$ with $|m_\lambda(E)|\le\varepsilon$ whenever $|\nu(E)|\le\delta$ and $\lambda\in\Lambda$. We say that the family $(m_\lambda)_\lambda$ is *equibounded* or *uniformly bounded* on Σ iff there is a positive real number K_0 with $|m_\lambda(E)|\le K_0$ for any $\lambda\in\Lambda$ and $E\in\Sigma$.

Saks (1933b) (see Theorem 5), using the Baire category theorem, proved the following:

Theorem 2.15 *Let $m_n:\Sigma\to\mathbb{R}$, $n\in\mathbb{N}$, be a sequence of σ-additive measures, absolutely continuous with respect to a non-negative σ-additive measure $\nu:\Sigma\to\mathbb{R}$. If this sequence converges for any set belonging to a class of second category in Σ, then the measures m_n are equiabsolutely continuous with respect to ν, and the sequence $(m_n)_n$ converges for any set $E\in\Sigma$, with*

$E \subset G \setminus (E_1 \cup ... \cup E_q)$, *where* $E_1, ..., E_q$ *are a finite number of singular sets.* *Consequently, if* $(m_n)_n$ *converges pointwise on* Σ, *then the limit function is a* σ-*additive measure, absolutely continuous with respect to* ν.

We recall also the next theorem.

Theorem 2.16 (Saks (1933b), Theorem 6) *Let* $m_n : \Sigma \to \mathbb{R}$, $n \in \mathbb{N}$, *be a sequence of measures,* σ-*additive and absolutely continuous with respect to a non-negative* σ-*additive measure* $\nu : \Sigma \to \mathbb{R}$. *If*

$$\limsup_n | m_n(E) | < +\infty \quad \textbf{(1.17)}$$

for every set E *belonging to a class of the second category in* Σ, *then there is a positive real number* b *with*

$$| m_n(E) | \leq b \quad \textbf{(1.18)}$$

for any set $E \in \Sigma$ *with* $E \subset G \setminus (E_1 \cup ... \cup E_q)$, *where* $E_1, ..., E_q$ *are singular sets.*

Consequently, if (1.17) *holds for every set* $E \in \Sigma$, *then* (1.18) *holds for each set* $E \in \Sigma$.

We now recall the Orlicz-Pettis theorem, which is very useful in the literature in the study of Banach space-valued measures and limit theorems.

Theorem 2.17 (Orlicz (1929), Theorem 3 and Pettis (1938), Theorem 2.32) *If* X *is a Banach space and* $\sum_{j=1}^{\infty} x_j$ *is a series in* X *whose every subseries is weakly convergent, then* $\sum_{j=1}^{\infty} x_j$ *is unconditionally convergent with respect to the norm.*

A result related to the Orlicz-Pettis theorem is the following Bessaga-Pełczyński theorem.

Theorem 2.18 (Bessaga and Pełczyński (1958), Theorem 4, see also Diestel and Uhl (1977), Corollary I.4.6) *If* X *is a Banach space and* X' *contains a subspace* *which is isomorphic to* c_0*, then* X' *contains a subspace which is isomorphic to* l^∞.

Let $\Sigma \subset \mathcal{P}(G)$ be a σ-algebra, X be a Banach space, and X' be its dual space. A finitely additive measure $m:\Sigma \to X$ is said to be *weakly* σ*-additive* iff $x'm$ is σ-additive and has finite total variation for every $x' \in X'$. Using the Nikodým boundedness theorem, it is possible to prove the following

Theorem 2.19 (see Bartle, Dunford and Schwartz (1955), Lemmas 2.2 and 2.3) *If* $m:\Sigma \to X$ *is a weakly* σ*-additive measure, then it is* σ*-additive, namely for each disjoint sequence* $(C_k)_k$ *in* Σ *we get,*

$$m\left(\bigcup_{k=1}^{\infty} C_k\right) = \sum_{k=1}^{\infty} m(C_k), \qquad \textbf{(1.19)}$$

where the series in (1.19) *converges unconditionally in the norm of* X. *Moreover the set*

$\{x'm : x' \in X', \| x' \| \le 1\}$ *is sequentially weakly compact as a subset of* $\mathcal{M}^*$.

Note that the first part of Theorem 2.19 can be viewed as a consequence of the Orlicz-Pettis theorem (see also Diestel and Uhl (1977), Corollary I.4.4). Thanks to Theorem 2.19, in the sequel we will deal with σ-additive Banach space-valued measures without specifying whether in the weak or in the ordinary (that is strong) sense.

Theorem 2.20 (Doubrovsky (1945), Theorem II) *Let* $(m_n)_n$ *be a sequence of* σ*-additive measures, defined on* Σ *and such that* $\lim_n m_n(E) = 0$ *for each* $E \in \Sigma$. *Then the sequence* $(\| m_n \|)_n$ *of the total variations of the* m_n*'s is such that, for each disjoint sequence* $(E_n)_n$ *in* Σ*, we get,*

$$\lim_n \|m_n\|\left(\bigcup_{l=n+1}^{\infty} E_l\right) = 0.$$

Theorem 2.21 (Doubrovsky (1945), Theorem III) *Let $(m_n)_n$ be a pointwise convergent sequence of σ-additive measures. Then the sequence $(\| m_n\|)_n$ is equibounded.*

Theorem 2.22 (Doubrovsky (1945), Theorem IV) *Let $(m_n)_n$ be a sequence of σ-additive measures. Let $A \in \Sigma$ be fixed, and suppose that there exists in $\mathbb{R}$ the limit $\lim_n m_n(E) =: m(E)$ for each $E \subset A$, $E \in \Sigma$. Let $f_n : A \to \mathbb{R}$ be a sequence of equibounded Σ-measurable functions.*

Then m is σ-additive and $\lim_n \int_A f_n(t)dm_n(t) = \int_A f(t)dm(t)$.

If (G, Σ, ν) is a measure space, where $\Sigma \subset \mathcal{P}(G)$ is a σ-algebra and $\nu : \Sigma \to \mathbb{R}$ a finitely or σ-additive measure, then we denote by $L^0(G, \Sigma, \nu)$, $L^0(G)$ or $L^0(\nu)$ the space of all ν-measurable functions, with identification in the complement of a ν-null set. For $0 < p < +\infty$ we denote by $L^p(G, \Sigma, \nu)$, $L^p(G)$ or $L^p(\nu)$ the space of all functions $f \in L^0(\nu)$ such that

$$\int_G | f(t) |^p \, dm(t) < +\infty.$$

The space $L^\infty(G)$ or $L^\infty(\nu)$ is the set of all ν-measurable functions f which are *essentially bounded* on G, that is such that the *essential supremum* $\operatorname{esssup}_{t\in G} | f(t) | < +\infty$, where

$$\operatorname{esssup}_{t\in G} | f(t) | := \inf\{C > 0 : \nu(\{t \in G : | f(t) | > C\}) = 0\}.$$

Using the Baire category theorem, in Dunford and Pettis (1940) (see Theorem 3.2.1) proved the following:

Theorem 2.23 (see also Diestel (1984), Theorem, p. 93, and Diestel (1991), Theorem 3) *Let (G, Σ, ν) be a measure space, where ν is a non-negative extended real-valued σ-additive measure, and $L^1(G)$ be the space of all ν-Lebesgue integrable functions, with identification up to ν-null sets. A subset $\mathcal{K}$ of $L^1(G)$ is weakly compact if and only if the following three conditions hold:*

(a) *$\mathcal{K}$ is bounded;*

(b) *The integrals* $\int_E \phi d\nu$ *are equiabsolutely continuous with respect to* ν *as* ϕ *varies in* $\mathcal{K}$, *namely for every* $\varepsilon > 0$ *there is* $\delta > 0$ *with* $\left|\int_E \phi(t) d\nu(t)\right| < \varepsilon$ *whenever* $\nu(E) < \delta$ *and* $\phi \in \mathcal{K}$;

(c) *For each partition* G_i, $i \in \mathbb{N}$, *of* G *in* Σ, *we get*

$$\lim_n \left(\sum_{i=n}^{\infty} \int_{G_i} |\phi(t) d\nu(t)| \right) = 0$$

uniformly with respect to $\phi \in \mathcal{K}$.

Moreover, if ν *is finite, then* (b) *implies* (c).

Theorem 2.24 (Dunford and Pettis (1940), Theorem 3.2.2) *Under the same notations as in Theorem 2.23, we get that a subset* $\mathcal{K}$ *of* $L^1(G)$ *is weakly compact if and only if the conditions* (a) *and* (b) *in Theorem 2.23 are satisfied together with the following property:*

(c') $\lim_n \left(\sum_{i=n}^{\infty} \int_{G_i} |\phi(t) d\nu(t)| \right) = 0$ *uniformly with respect to* $\phi \in \mathcal{K}$ *for at least one partition* G_i, $i \in \mathbb{N}$, *of* G *in* Σ.

Theorem 2.25 (see also Dunford and Pettis (1940), Theorem 3.1.2 and Diestel (1991), Theorem 14) *Let* G, Σ *be as in Theorem 2.23,* ν *be a non-negative* σ-*additive real-valued measure defined on* Σ, X *be a Banach space, and* $T : L^1(G) \to X$ *be an operator, such that there exists a strongly measurable function* $h : G \to X$ *with* $u(f) = \int_G f h d\nu$ *for each* $f \in L^1(G)$. *Then* T *maps weakly compact sets of* $L^1(G)$ *into norm compact sets of* X.

Phillips (1940b) (see Lemma 3.3) proved, by means of a diagonal technique, the following result, which is very useful to demonstrate different types of limit theorems.

Lemma 2.26 *Let* $(m_n)_n$ *be a sequence of bounded finitely additive real-valued measures, defined on all subsets of* $\mathbb{N}$. *If* $\lim_n m_n(E) = 0$ *for each* $E \subset \mathbb{N}$, *then*

$$\lim_n \left(\sum_{k=1}^{\infty} | m_n(\{k\}) | \right) = 0.$$

Phillips developed also a theory of integration for functions with valued in a convex linear topological space satisfying suitable properties and, using the Baire category theorem, proved a Vitali-Lebesgue-type theorem for such integral (Phillips (1940a), Theorems 6.1 and 6.2). Some theorems of type Vitali-Lebesgue were proved also by Rickart (1942) (see Theorems 9.2 and 9.5) for a suitable kind of integral for multifunctions in the convex linear topological space context.

1.2.3. Finitely additive (s)*-Bounded and (Uniformly) σ-Additive Measures*

Rickart (1943) introduced the concept of (s)-bounded measure, studied some fundamental properties and proved some types of decomposition theorems. Let G be any nonempty set, $\mathcal{E}$ be a lattice of subsets of G and X be a linear normed vector space. A finitely additive measure $m : \mathcal{E} \to X$ is said to be *strongly bounded* (shortly, (s)*-bounded*) iff $\lim_k m(C_k) = 0$ whenever $(C_k)_k$ is a disjoint sequence in $\mathcal{E}$ (with respect to the norm topology). If $\Lambda \neq \emptyset$, then a family $m_\lambda : \mathcal{E} \to X$, $\lambda \in \Lambda$, of finitely additive measures is said to be *uniformly strongly bounded* (shortly, *uniformly* (s)*-bounded*) iff for each disjoint sequence $(C_k)_k$ in $\mathcal{E}$ and $\varepsilon > 0$ there exists $k_0 \in \mathbb{N}$ with $\| m_\lambda(C_k) \| \leq \varepsilon$ whenever $k \geq k_0$ and $\lambda \in \Lambda$. Note that the finitely additive measures satisfy the following property.

Theorem 2.27 (see also Rickart (1943), Theorem 2.3) *If $m : \mathcal{E} \to X$ is a finitely additive and (s)-bounded measure, then the series $\sum_{k=1}^{\infty} m(C_k)$ converges unconditionally for every disjoint sequence $(C_k)_k$ in $\mathcal{E}$.*

A family $m_\lambda : \Sigma \to \mathbb{R}$, $\lambda \in \Lambda$, of σ-additive measures defined on a σ-algebra Σ is said to be *uniformly σ-additive* iff for every disjoint sequence $(E_r)_r$ in Σ we get,

$$\lim_s \left(\sup_\lambda \left| m_\lambda \left(\bigcup_{s=r}^{\infty} E_s \right) \right| \right) = 0$$

(see also Doubrovsky (1947a)).

We now state a characterization of uniform σ-additivity. The following notion was used by Caccioppoli (1929) in studying the problem of the exchange of limits under the sign of integral for Stieltjes-type integrals, which in Cafiero (1953) was developed and extended to the case of functions defined on sets of infinite measure.

A family $m_\lambda : \Sigma \to \mathbb{R}$, $\lambda \in \Lambda$, of σ-additive measures on a σ-algebra Σ is said to be *equi-σ-additive* iff for every sequence $(E_r)_r$ in Σ, convergent to $\varnothing$ in the set-theoretical sense, and for each $\varepsilon > 0$ there is a natural number r_0 with $|m_\lambda(E_r)| \le \varepsilon$ for all $r \ge r_0$ and $\lambda \in \Lambda$.

The following characterization of uniform σ-additivity is given (see also Cafiero (1951 and 1968)).

Proposition 2.28 (Cafiero (1968), Proposition 1.2) *Let $m_\lambda : \Sigma \to \mathbb{R}$, $\lambda \in \Lambda$, be a family of σ-additive measures. Then the following are equivalent.*

(a) *$(m_\lambda)_\lambda$ is uniformly σ-additive.*

(b) *$(m_\lambda)_\lambda$ is equi-σ-additive.*

(c) *For each sequence $(E_r)_r$ in Σ, convergent in the set-theoretical sense, we have*

$$\lim_r (\sup_\lambda | m_\lambda(\lim_r E_r) - m_\lambda(E_r) |) = 0.$$

(d) *For every disjoint sequence $(E_r)_r$ in Σ we get,*

$$\inf_r \left(\sup_\lambda \left| m_\lambda \left(\bigcup_{s=r}^{\infty} E_s \right) \right| \right) = 0.$$

(e) *For every disjoint sequence $(E_r)_r$ in Σ we get,*

$$\inf_r (\sup_\lambda | m_\lambda(E_r) |) = 0.$$

(f) *(in the case $\Lambda = \mathbb{N}$) For every disjoint sequence $(E_r)_r$ in Σ we get,*

$\inf_r (\limsup_n | m_n(E_r) |) = 0.$

The following results on absolute continuity of σ-additive measures hold.

Theorem 2.29 (Doubrovsky (1947b), Theorem 3) *A necessary and sufficient condition in order that there exists a non-negative σ-additive measure ν with respect to which every m_λ is absolutely continuous is that there does not exist an uncountable family $(E_\iota)_{\iota \in I}$ of disjoint sets of Σ, such that for each $\iota \in I$ there is at least a $\lambda \in \Lambda$ such that $m_\lambda(E_\iota)$ is different from 0.*

Theorem 2.30 (Doubrovsky (1947b), Theorem 2 and Cafiero (1968), Theorem 2.1) *If $(m_\lambda)_\lambda$ is a uniformly σ-additive family of σ-additive measures, then there exists a non-negative σ-additive measure ν, with respect to which m_λ is absolutely continuous for each $\lambda \in \Lambda$.*

A measure ν satisfying the thesis of Theorem 2.30 is called also a *basis* (see also Doubrovsky (1947b)).

Theorem 2.31 (see also Caccioppoli (1928 and 1929, Theorem, p. 37) *If $(m_\lambda)_\lambda$ is a family of σ-additive measures, each of which absolutely continuous with respect to a non-negative extended real-valued σ-additive and σ-finite measure ν, then $(m_\lambda)_\lambda$ is uniformly σ-additive if and only if $(m_\lambda)_\lambda$ is equiabsolutely continuous with respect to ν.*

As a consequence, we state a version of the Dunford-Pettis compactness criterion (see also Cafiero (1968), Theorem 4.1), where Λ it is assumed to be a directed set.

Theorem 2.32 *Let $f_\lambda : G \to \mathbb{R}$, $\lambda \in \Lambda$, be a net in $L^1(G,\Sigma,\nu)$, such that the integral functions $\int f_\lambda d\nu$ are equibounded and equiabsolutely continuous.*

Then the convergence of the integrals of the f_λ's on every set $E \in \Sigma$ is a necessary and sufficient condition in order that $(f_\lambda)_\lambda$ converges in the weak topology $\sigma(L^1, L^\infty)$.

We say that $(m_\lambda)_\lambda$ is *equally continuous* with respect to the non-negative measure ν iff for every sequence $(E_n)_n$ in Σ we get that $\lim_n \nu(E_n) = 0$ implies $\lim_n m_\lambda(E_n) = 0$ uniformly with respect to λ (see also Doubrovsky (1947b)).

We now recall the following results.

Theorem 2.33 (Doubrovsky (1947a), Theorems I and III) *Let $m_n : \Sigma \to \mathbb{R}$, $n \in \mathbb{N}$, be a sequence of σ-additive measures. If $(m_n)_n$ is uniformly σ-additive (resp. equiabsolutely continuous with respect to a σ-additive measure ν), then $(\| m_n \|)_n$ is too.*

Theorem 2.34 (Doubrovsky (1947a), Theorem IV) *Suppose that $m_n : \Sigma \to \mathbb{R}$, $n \in \mathbb{N}$, is a sequence of σ-additive measures. If $\lim_n m_n(E)$ exists in $\mathbb{R}$ for each $E \in \Sigma$, then $(m_n)_n$ is uniformly σ-additive.*

Theorem 2.35 (Doubrovsky (1947b), Theorem 4, see also Doubrovsky (1947a), Theorem V) *The properties of uniform σ-additivity and equal continuity are equivalent.*

Theorem 2.36 (Doubrovsky (1947a), Theorem VI) *If $(m_n)_n$ is a sequence of σ-additive measures, pointwise convergent to a set function m, then $(m_n)_n$ is equally continuous with respect to any non-negative σ-additive measure ν, with respect to which all m_n's are absolutely continuous.*

Theorem 2.37 (Doubrovsky (1947a), Theorem VIII) *Let $G = \mathbb{R}^n$ and Σ be the σ-algebra of all measurable rectangles of G, and let $\mathcal{K}$ be a set of σ-additive real-valued measures, defined on the σ-algebra Σ of the Borel subsets of G.*

Then each infinite subfamily of $\mathcal{K}$ contains a sequence which converges at every set $E \in \Sigma$ if and only if:

(a) *The set $\mathcal{K}$ is equibounded;*

(b) *The measures m, $m \in \mathcal{K}$, are uniformly σ-additive.*

Let Σ be a σ-algebra of subsets of a nonempty set G and $m : \Sigma \to \mathbb{R}$ be a σ-additive measure. We say that m is *continuous* on Σ iff

$$\lim_r m(E_r) = m(\lim_r E_r)$$

for each sequence $(E_r)_r$ in Σ, convergent in the set-theoretic sense.

Let $(m_n)_n$ be a sequence of measures, σ-additive on Σ. We say that $(m_n)_n$ is *equicontinuous* iff for every sequence $(E_r)_r$ convergent in the set-theoretic sense and for each $\varepsilon > 0$ there is $r_0 \in \mathbb{N}$ (depending on $(E_r)_r$ and on ε) with $| m(E_r) | < \varepsilon$ for all $r \ge r_0$ and $n \in \mathbb{N}$.

Observe that $(m_n)_n$ is equicontinuous if and only if the m_n's are equiabsolutely continuous (see also Cafiero (1951)).

Theorem 2.38 (Cafiero (1951, Theorem I, and 1952, Theorem I)) *A necessary and sufficient condition in order that a sequence $(m_n)_n$ of σ-additive measures is equicontinuous is that for every disjoint sequence $(E_r)_r$ in Σ and for each $\varepsilon > 0$ there are n_0, $r_0 \in \mathbb{N}$ with $| m_n(E_{r_0}) | \le \varepsilon$ for all $n \ge n_0$.*

Theorem 2.39 (Cafiero (1951, Theorem II, and 1952, Theorem II)) *Let $m_n : \Sigma \to \mathbb{R}$, $n \in \mathbb{N}$, be a sequence of σ-additive measures, and suppose that for every disjoint sequence $(E_r)_r$ in Σ there is a set $\{r_s : s \in \mathbb{N}\} \subset \mathbb{N}$ such that the sequence $\left(m_n\left(\bigcup_{s=1}^{\infty} E_{r_s}\right)\right)_n$ is convergent to a real number. Then the sequence $(m_n)_n$ is equicontinuous.*

Theorem 2.40 (Cafiero (1951), Generalized Vitali-Lebesgue theorem) *Let $m_n : \Sigma \to \mathbb{R}$, $n \in \mathbb{N}$, be a sequence of σ-additive measures, pointwise convergent to 0 at every element of Σ. Then $(m_n)_n$ is equicontinuous.*

1.2.4. Dieudonné, Grothendieck and Related Theorems

We now recall some fundamental results on uniform regularity and convergence of measures, proved by Dieudonné (1951a-b) and Grothendieck (1953), and several related results.

A measure defined on the class of all Borel subsets of a locally compact Hausdorff (resp. normal) topological space G is said to be *regular* iff for every Borel set $E \subset G$ and $\varepsilon > 0$ there are an open set U and a compact (resp. closed) set K with $K \subset E \subset U$ and $|m(U) - m(K)| \leq \varepsilon$.

A family $(m_\lambda)_{\lambda \in \Lambda}$ of finitely additive measures is *uniformly regular* iff for each Borel subset $E \subset G$ and $\varepsilon > 0$ there are an open set U and a compact (resp. closed) set K with $K \subset E \subset U$ and $|m_\lambda(U) - m_\lambda(K)| \leq \varepsilon$ for every $\lambda \in \Lambda$. Analogously as above it is possible to formulate to concept of regularity for measures defined on a paving and that of (uniform) inner regularity of a measure. For example, we say that m is *inner regular* iff for each Borel subset $E \subset G$ and $\varepsilon > 0$ there is a compact (or closed) set K with $K \subset E$ and $|m(E) - m(K)| \leq \varepsilon$. Note that, if G is compact, then a finitely additive measure is regular if and only if it is σ-additive (see also Alexandroff (1941)). Some other properties concerning regularity of locally convex space-valued measures can be found in Dinculeanu and Kluvánek (1967).

We first recall some notions and results about Radon measures. Let E be a locally compact topological space and $K(E)$ be the space of the continuous real-valued functions with compact support, defined on E. A *Radon measure* is a linear form $f \mapsto m(f)$ on $K(E)$ (whose value is often denoted by $\int f\,dm$) such that for every compact set $K \subset E$ there is $a_K \geq 0$ such that, for every function $f \in K(E)$ whose support is contained in K, we have $|m(f)| \leq a_K \cdot \| f \|_\infty$, where $\| f \|_\infty = \sup |f|$. A Radon measure m is said to be *positive* iff $m(f) \geq 0$ for all $f \in K(E)$, $f \geq 0$. The set $\mathcal{M}(E)$ of all Radon measures on E is a vector space, in which $m_1 \leq m_2$ means that $m_2 - m_1$ is a positive measure: it is possible to show that $\mathcal{M}(E)$ is Dedekind complete with respect to this order (see also Luxemburg and Zaanen (1971)).

Let m be a positive Radon measure on E. For each lower semicontinuous function $f \geq 0$ (finite or not), we denote by $m^*(f)$ the supremum of the numbers $m(h)$, where h varies in the set of the functions of $K(E)$ with $0 \leq h \leq f$. For each function $f \geq 0$, we denote by $m^*(f)$ (*lower integral of* f) the infimum of the numbers $m^*(h)$, where h varies in the set of the lower semicontinuous functions $h \geq f$. A function f is said to be *integrable* iff for every $\varepsilon > 0$ there exist a upper semicontinuous function h_1 and a lower semicontinuous function h_2, with $h_1 \le f \le h_2$, $m^*(h_i) < +\infty$, $i = 1, 2$, and $m^*(h_2 - h_1) \le \varepsilon$. With the identification of two equivalent functions, that is $f_1 \sim f_2$ iff $m^*(|f_1 - f_2|) = 0$, the space L^1 of all integrable functions is a Banach space with the norm $N_1(f) = m^*(|f|)$. Moreover $K(E)$ is dense in L^1, and the integral $m(f) = \int f\,dm$ can be extended to the whole of L^1. A set A is said to be *integrable* iff its characteristic function χ_A is integrable, and its *measure* $m(A)$ is equal to $m(\chi_A)$. Observe that every compact set is integrable, and in order that A is integrable it is necessary and sufficient that for every $\varepsilon > 0$ there are a compact set $K \subset A$ and an open set $U \supset A$ with $m(U) - m(K) \le \varepsilon$. Given any set $A \subset E$, its *exterior measure* $m^*(A)$ is, by definition, $m^*(\chi_A)$.

In Dieudonné (1951a-b) different kinds of convergence of a sequence $(m_n)_n$ of Radon measures to a Radon measure m on E are given. The *strong convergence* is characterized by the condition that $m_n(f)$ tends *uniformly* to $m(f)$ in certain parts of $K(E)$, and the problem of finding some other functions h, different from those of $K(E)$, for which $m_n(h)$ converges to $m(h)$, is studied. Moreover, the so-called *weak convergences* are investigated, namely it is supposed only that $m_n(f)$ tends to $m(f)$ for every function f belonging to a set of functions $\Phi \supset K(E)$, but without uniformity, and some relations between different kinds of convergences are investigated.

More precisely, let E be a compact set, $(K(E), \|\cdot\|_\infty)$ be the space of all continuous real-valued functions on E endowed with the supremum norm, and $\mathcal{M}(E)$ be the space of all Radon measures on E, namely the dual space of $K(E)$. On $\mathcal{M}(E)$ it is possible to define the norm $||| m |||$ of a measure m. We get that $||| m ||| = \| m \|(1)$, where $\| m \|$ and 1 denote the total variation of m and the function which assumes the constant value 1 respectively, and $| m(f) | \le ||| m ||| \cdot \| f \|$ for each $f \in K(E)$. The topology generated by this norm on E is called the *strong topology*. Bourbaki (1948) generalized this topology when E is any locally compact space: for each compact set $K \subset E$, let $p_K(m)$ be the supremum of $| m(f) |$ when $f \in K(E)$, $\| f \| \le 1$ and the support of f is contained in K. The p_K's are the *seminorms*, which define on $\mathcal{M}(E)$ the Bourbaki topology, which is still called *strong topology*. If S_K is the set of all functions

$f \in K(E)$ with $\| f \| \leq 1$ and whose support is contained in K, then this topology is the topology of uniform convergence on the sets S_K. It is possible to check that $p_K(m) = p_K(\| m \|)$, and that the function $m \mapsto \| m \|$ is continuous with respect to the strong topology. Let $\mathcal{M}_+(E)$ be the set of all positive measures. We recall the following:

Proposition 2.41 (Dieudonné (1951a), Proposition 1) *Let* $f : E \to \mathbb{R}$, $f \geq 0$. *In order that the function* $m \mapsto m^*(f)$ *from* $\mathcal{M}_+(E)$ *to the extended real line is continuous for the strong topology, it is necessary and sufficient that the support of* f *is compact and that* f *is bounded.*

Corollary 2.42 (Dieudonné (1951a), Corollary) *If* f *is a positive function, which assumes the value* 0 *in the complement of an increasing sequence* $(K_n)_n$ *of compact sets, then the function* $m \mapsto m^*(f)$ *is lower semicontinuous with respect to the strong topology.*

Proposition 2.43 (Dieudonné (1951a), Proposition 2) *Let* $(m_n)_n$ *be an increasing sequence in* $\mathcal{M}_+(E)$, *and let m be its supremum. Then, for each* $f \geq 0$ *with* $f = 0$ *in the complement of an increasing sequence of compact sets,* $m_n^*(f)$ *tends to* $m^*(f)$ *as* n *tends to* $+\infty$.

Proposition 2.44 (Dieudonné (1951a), Proposition 3) *Let* $(m_n)_n$ *be a decreasing sequence in* $\mathcal{M}_+(E)$, *and let m be its infimum. Then, for each* $f \geq 0$ *such that* $m_n^*(f)$ *is finite from a suitable index* n_0, $m_n^*(f)$ *tends to* $m^*(f)$ *as* n *tends to* $+\infty$.

We now suppose that E is a compact set, and consider the Radon measures on E of any sign. Let $(m_n)_n$ be a sequence of Radon measures. The *weak convergence* considered in Dieudonné (1951a) is when the sequence $(m_n(f))_n$ tends to $m(f)$ for each function f belonging to a suitable set Φ. Some examples of such sets Φ are the following:

I. The set Φ_{I} of all continuous functions on E.

II. The set Φ_{II} of all bounded functions, continuous up to a null set depending on the considered function (here, null means with respect to every measure m_n).

III. The set Φ_{III} of all semicontinuous bounded functions (lower or upper semicontinuous).

IV. The set Φ_{IV} of all bounded Borel functions.

First of all note that, if $(m_n(f))_n$ converges to $m(f)$ for each $f \in \Phi$, then this happens for every function, which is a linear combination of a finite number of functions from Φ. Moreover, observe that $\Phi_{\text{II}} \supset \Phi_{\text{I}}$, $\Phi_{\text{III}} \supset \Phi_{\text{I}}$ and $\Phi_{\text{IV}} \supset \Phi_{\text{I}}$. The convergence on Φ_{I} is the *weak convergence* in the dual $\mathcal{M}(E)$ of the Banach space $K(E)$. Following Bourbaki (1948) we say, in this context, that $(m_n)_n$ *vaguely converges* to m: in this case we say that the sequence of norms $(|||\, m_n \,|||)_n$ is *bounded*. We can always suppose that this last condition is satisfied: then, if $(m_n(f))_n$ converges to $m(f)$ for each $f \in \Phi$, it happens the same for every function which is uniform limit of functions from Φ_{j}, j = I, ..., IV. This shows in particular that, in order that $(m_n(f))_n$ tends to $m(f)$ in Φ_{III}, it is necessary and sufficient that $(m_n(A))_n$ tends to $m(A)$ for each closed set A (or each open set A), and that, in order that $(m_n(f))_n$ tends to $m(f)$ in Φ_{IV}, it is necessary and sufficient that $(m_n(A))_n$ tends to $m(A)$ for each Borel subset A. Moreover, note that the convergence of $(m_n(f))_n$ to $m(f)$ in Φ_{IV} is equivalent to weak convergence in $L^1(\nu)$ of the sequence $(h_n)_n$, where ν is a suitable positive measure and $(h_n)_n$ is a suitable sequence in $L^1(\nu)$, whose existence is guaranteed by the Radon-Nikodým theorem. In Dieudonné (1951a) it is shown that, in general, vague convergence does not imply convergence in Φ_{II} and convergence in Φ_{II} does not imply convergence in Φ_{III}, but convergence in Φ_{III} implies convergence in Φ_{IV}, at least when the compact space E involved is metrizable. Furthermore, the following results are proved.

Proposition 2.45 (Dieudonné (1951a), Proposition 4) *In order that a sequence* $(m_n)_n$ *of Radon measures on* E, *vaguely convergent to a measure* m, *converges to* m *on* Φ_{II}, *it is necessary and sufficient that, for each* $\varepsilon > 0$ *and for every* *compact subset* $K \subset E$ *with* $\| m_n \|(K) = 0$ *for each* $n \in \mathbb{N}$, *there is an open set* $U \supset K$ *with* $\| m_n \|(H) \leq \varepsilon$ *for each compact set* $H \subset U \setminus K$ *and* $n \in \mathbb{N}$.

Proposition 2.46 (Dieudonné (1951a), Proposition 5) *In order that a sequence* $(m_n)_n$ *of Radon measures on* E, *vaguely convergent to a measure* m, *converges to* m *on* Φ_{III}, *it is necessary and sufficient that, for each* $\varepsilon > 0$ *and for every compact subset* $K \subset E$, *there is an open set* $U \supset K$ *with* $\| m_n \|(H) \leq \varepsilon$ *for each compact set* $H \subset U \setminus K$ *and* $n \in \mathbb{N}$.

Proposition 2.47 (Dieudonné (1951a), Proposition 6) *If a sequence* $(m_n)_n$ *of Radon measures converges to* m *in* Φ_{III} *and satisfies the condition*

$$\lim_n ||| m_n ||| = ||| m |||, \tag{1.20}$$

then it converges to m *in* Φ_{IV}.

Observe that condition (1.20) is satisfied by any vaguely convergent sequence $(m_n)_n$ of positive measures, because in this case we have

$$||| m ||| = m(1) = \lim_n m_n(1) = \lim_n ||| m_n |||.$$

Proposition 2.48 (Dieudonné (1951a), Proposition 7) *If a sequence* $(m_n)_n$ *of Radon measures converges to* m *in* Φ_{III} *and* E *is a compact metrizable space, then* $(m_n)_n$ *converges to* m *in* Φ_{IV}.

Proposition 2.49 (Dieudonné (1951a), Proposition 8) *Let* $(m_n)_n$ *be a sequence of Radon measures on* E *such that, for every closed set* $A \subset E$, *the sequence* $(m_n(A))_n$ *tends to a finite limit. Then there exists a Radon measure* m *such that* m_n *tends to* m *on* Φ_{III}.

Theorem 2.50 (Dieudonné (1951a), Proposition 9) *Let* $(m_n)_n$ *be a sequence of Radon measures on* E *such that, for each closed set* $A \subset E$, *the sequence* $(m_n(A))_n$ *is bounded by a number depending on* A. *Then the sequence* $(||| m_n |||)_n$ *is bounded.*

Under the same above notations and assumptions, we say that an open set $A \subset E$ is *quadrable* (for all the m_n's) iff its boundary has null measures $\| m_n \|$ for each $n \in \mathbb{N}$.

The following results hold.

Proposition 2.51 (Dieudonné (1951b), Proposition 1) *In order that the sequence* $(||| m_n |||)_n$ *is bounded, it is enough that the sequence* $(m_n(A))_n$ *is bounded for each open set* A *quadrable for every* m_n *and for each finite set* A.

Proposition 2.52 (Dieudonné (1951b), Proposition 2) *In order that* $(m_n)_n$ *converges in* Φ_{II} *to a Radon measure* m *it is sufficient that, for each set* A, *finite or open and quadrable for each* m_n, *the sequence* $(m_n(A))_n$ *has a finite limit.*

Nakamura (1949), using the Baire category theorem, proved that, if a sequence of functionals with values in a Dedekind complete Banach lattice, continuous with respect to the order, is weakly convergent, then the limit functional is also order continuous (Nakamura (1949), Theorem 2), and that, if a sequence $(m_n)_n$ of σ-additive measures with values in a Dedekind complete Banach lattice, each of which absolutely continuous with respect to a positive real-valued σ-additive measure ν, converges pointwise, then the limit measure is absolutely continuous and moreover $(m_n)_n$ is equiabsolutely continuous with respect to ν (Nakamura (1949), Theorem 3).

We now recall the following:

Theorem 2.53 (Alexiewicz (1949), Theorem 1) *Let* X *be a complete metric space,* Y *be a metric space, and* $f_n : X \to Y$, $n \in \mathbb{N}$, *be a sequence of functions, convergent in the complement of a set* $A \subset X$ *of first Baire category. Then there is a set* $B \subset X$ *of first Baire category, such that* $(f_n)_n$ *is equicontinuous on* $X \setminus B$.

Theorem 2.54 (Alexiewicz (1949), Corollary 1.4) *Let* $f_n : X \to Y$, $n \in \mathbb{N}$, *be a sequence of functions, convergent in a set* $A \subset X$ *of second Baire category. Then there is a set* $B \subset X$ *of second category, such that* $(f_n)_n$ *is equicontinuous on* B.

Moreover, in Alexiewicz (1949) some conditions are given, for equicontinuity at 0 of sequences of suitable additive and continuous operators, convergent in the complement of a set of first Baire category (see for instance Alexiewicz (1949), Theorem 12), in connection with the Vitali-Hahn-Saks theorems 2.15 and 2.16.

Pettis (1951) showed that the Vitali-Hahn-Saks theorem can be deduced from some abstract results about suitable properties of sets of first and sets of second Baire category (see for instance Pettis (1951), Theorem 3).

In the literature there are also several studies about (weak) compactness in suitable spaces of functions or measures, related with limit theorems. In Chapter 4 we will deal with some extensions of these concepts and applications will be given in the context of filter/ideal convergence.

Let G be a locally compact topological space, $C(G)$ be the space of all continuous real-valued or complex-valued functions, $C_0(G)$ be the subspace of $C(G)$ of all functions, whose support is a compact subset of G, and $\mathcal{M}^1(G)$ be its topological dual, that is the space of all bounded Radon measures on G. Grothendieck (1953) investigated several properties of weak compactness in the space $\mathcal{M}^1(G)$ and proved the following results.

Theorem 2.55 (Grothendieck (1953), Theorem 2) *Let G be a locally compact space, and $\mathcal{K}$ be a bounded subset of $\mathcal{M}^1(G)$. Then the following are equivalent.*

(a) *$\mathcal{K}$ is weakly compact.*

(b) *For each uniformly bounded sequence $(f_n)_n$ of real-valued (or complex-valued) functions, m-measurable with respect to every $m \in \mathcal{K}$, and pointwise convergent to f on G, we have*

$$\lim_n \int_G f_n \, dm = \int_G f \, dm$$

uniformly as m varies in $\mathcal{K}$. In this condition, it is possible also to suppose, instead of pointwise convergence, that for each compact set $K \subset G$ and $m \in \mathcal{K}$, $(f_n)_n$ tends to f in m-measure on K.

(c) *For each sequence* $(f_n)_n$, *weakly convergent to* 0 *in* $C_0(G)$ *(that is uniformly bounded and pointwise convergent to* 0*), we get,*

$$\lim_n \int_G f_n\, dm = \int_G f\, dm$$

uniformly as m *varies in* $\mathcal{K}$.

(d) *For every disjoint sequence* $(O_n)_n$ *of open subsets of* G, $\lim_n m(O_n) = 0$ *uniformly as* m *varies in* $\mathcal{K}$.

(e) *For each compact set* $K \subset G$ *and every* $\varepsilon > 0$ *there is an open set* U *with* $K \subset U \subset G$ *and* $\| m \| (U \setminus K) \leq \varepsilon$ *whenever* $m \in \mathcal{K}$, *and for every* $\varepsilon > 0$ *there exists a compact set* $K \subset G$ *with* $\| m \| (G \setminus K) \leq \varepsilon$ *for all* $m \in \mathcal{K}$.

Grothendieck (1953) proved also that from Theorem 2.55 it is possible to deduce that, if G is a compact metric space, a sequence $(m_n)_n$ in $\mathcal{M}^1(G)$ is weakly convergent if and only if the sequence $(m_n(O))_n$ is convergent for every open subset $O \subset G$ (see also Dieudonné (1951a), Proposition 8).

Theorem 2.56 (Grothendieck (1953), Theorem 3) *Let* G *be a locally compact topological space and* β_0 *be the space generated by the characteristic functions of the closed subsets of* G. *In order that a bounded subset* $\mathcal{K} \subset \mathcal{M}^1(G)$ *is relatively weakly compact, it is necessary and sufficient that* $\mathcal{K}$ *is relatively compact for the topology* $\sigma(\mathcal{M}^1(G), \beta_0)$.

Given two topological vector spaces X and Y, a continuous linear operator $T : X \to Y$ is said to be (*weakly*) *compact* iff it maps bounded sets of X into relatively (weakly) compact sets (that is subsets of Y whose closure is (weakly) compact). We now recall the following characterization of weak compactness of operators.

Theorem 2.57 (Grothendieck (1953), Theorem 6) *Let* K *be a compact topological space. If* T *is a linear continuous function from* $C(K)$ *in a locally convex complete Hausdorff space* X, *then the following conditions are equivalent:*

(a) T *is weakly compact;*

(b) *For each closed subset* $A \subset K$, *we get* $T''(\chi_A) \in X$, *where* T'' *is the element corresponding to* T *in the bidual space of* $C(K)$ *and* χ_A *is the characteristic function of* A;

(c) T *maps every increasing sequence of positive continuous functions dominated by the constant function* 1 *into a weakly convergent sequence in* X.

Theorem 2.58 (Grothendieck (1953), Theorem 7, see also Diestel and Swart (2002), Theorem 3.3) *Let* K *be a compact topological space. Every linear continuous function from* $C(K)$ *to a space of type* L^1 *is weakly compact, and maps weakly compact subsets of* $C(K)$ *into strongly relatively compact sets, and hence it maps weakly Cauchy sequences into strongly convergent sequences.*

Theorem 2.59 (Grothendieck (1953), Proposition 13) *Let* G *be a locally compact space endowed with a measure* m, X *be a locally convex complete Hausdorff topological space,* ϕ *be a weakly integrable function from* G *to* X. *In order that*

$$\int_G \phi f \, dm \in X$$

for every $f \in L^\infty(m)$, *the space of essentially bounded real-valued functions with respect to* m, *it is necessary and sufficient that*

$$\int_C \phi f \, dm \in X$$

for each $f \in L^\infty(m)$ *and for every closed subset* $C \subset G$.

Theorem 2.60 (Grothendieck (1953), Proposition 14) *Let* K *be a compact space,* X *be a locally convex complete Hausdorff topological space. Then the space of all weakly compact linear functions from* $C(K)$ *to* X *is isomorphic to the space of all (weakly or strongly)* σ*-additive measures defined on the Borel subsets of* K *and with values in* X.

We now recall the following results (see also Velasco and Rodríguez-Palacios (2007), pp. 29-30).

Theorem 2.61 (Grothendieck (1953), Theorem 9) *Let Ω be a Stonian compact topological space. Then every sequence $(m_n)_n$ of Radon measures on Ω, weakly* convergent to 0, converges to 0 weakly in $\mathcal{M}^1(\Omega)$.*

Corollary 2.62 (Grothendieck (1953), Corollary 1) *Let Ω be a Stonian compact topological space and X be a locally convex complete Hausdorff separable space. Then every linear continuous function from $C(\Omega)$ to X is weakly compact.*

Theorem 2.63 (Grothendieck (1953), Lemma 8) *Let X be a Banach space. Then the following conditions are equivalent.*

(a) *Every sequence $(x_n^{'})_n$ in the dual $X^{'}$, weakly convergent to 0, converges to 0 with respect to the topology $\sigma(X^{'},X^{''})$.*

(b) *Every linear continuous function from X in any locally convex Hausdorff complete separable topological space is weakly compact.*

Moreover, Grothendieck (1953) (see Theorem 10) proved that the space c_0 of all real-valued or complex-valued sequence convergent to 0, endowed with the supremum norm, satisfies the following property.

Theorem 2.64 *Let X be a Banach space, which is isomorphic to a subspace of c_0. Then every relatively $\sigma(X^{'},X^{''})$-compact subset of X is relatively strongly compact, and every Cauchy sequence in $\sigma(X^{'},X^{''})$ is strongly convergent. Moreover, every weakly compact linear operator from X to a locally convex space Y is compact.*

Bartle, Dunford and Schwartz (1955) gave a theory of integration for real-valued functions with respect to Banach space-valued measures, and proved the following results concerning weak compactness of subsets of countably additive vector-valued measures. Let $\mathcal{M}^*$ be the space of all σ-additive real-valued measures defined on a σ-algebra Σ of parts of an abstract nonvoid set G, endowed with the total variation norm.

Theorem 2.65 (Bartle, Dunford and Schwartz (1955), Theorem 1.3) *In order that a set $\mathcal{K} \subset \mathcal{M}^*$ is sequentially weakly compact, it is necessary and sufficient that*

$\mathcal{K}$ is bounded, and, if $(H_k)_k$ is any sequence in Σ with $\bigcap_{k=1}^{\infty} H_k = \emptyset$, then $\lim_k \phi(H_k) = 0$ uniformly with respect to $\phi \in \mathcal{K}$.

Theorem 2.66 (Bartle, Dunford and Schwartz (1955), Theorem 1.4) *In order that a set $\mathcal{K} \subset \mathcal{M}^*$, $\mathcal{K} = \{m_\lambda : \lambda \in \Lambda\}$, is sequentially weakly compact, it is necessary and sufficient that $\mathcal{K}$ is bounded and there is a non-negative σ-additive real-valued measure ν such that $(m_\lambda)_\lambda$ is equiabsolutely continuous with respect to ν.*

Corollary 2.67 (Bartle, Dunford and Schwartz (1955), Corollary 2.4, see also Gould (1965), Theorem 3.2) *If $m : \Sigma \to X$ is a σ-additive measure, then there exists a non-negative σ-additive real-valued measure ν, such that m and $\| m \|$ are absolutely continuous with respect to ν and, for every $\varepsilon > 0$, $\nu(E) < \varepsilon$ whenever $\| m \| (E) < \varepsilon$.*

Theorem 2.68 (Bartle, Dunford and Schwartz (1955), Theorem 2.9) *The range of a Banach space-valued σ-additive measure is sequentially weakly compact.*

Nikodým (1955) extended the Helly theorem (see Helly (1912), Theorem 3) to the multidimensional and even to the infinite-dimensional case using the corresponding Helly's monodimensional version and diagonal techniques, and proved the following Helly-type theorem for measures.

Theorem 2.69 (see Nikodým (1955), Theorem 5.7) *Let $\mathcal{E}_\infty$ denote the Cartesian product of countably many copies of $\mathcal{E}_1$, where $\mathcal{E}_1$ is the family of all finite unions of half-open subintervals of $[0,1]$ of the type $]a,b]$, with $0 \le a < b \le 1$. Let $m_n : \mathcal{E}_\infty \to \mathbb{R}$, $n \in \mathbb{N}$, be a sequence of finitely additive measures, and let there exist a positive constant K_0 such that*

$$| m_n (E) | \le K_0 \text{ for each } E \in \mathcal{E}_\infty \text{ and } n \in \mathbb{N}.$$

Then there exists a strictly increasing sequence $(n_k)_k$ in $\mathbb{N}$ such that $m(E) := \lim_k m_{n_k}(E)$ exists in $\mathbb{R}$ for every $E \in \mathcal{E}_\infty$, and m is a finitely additive measure with $| m(E) | \le K_0$ for each $E \in \mathcal{E}_\infty$.

A decomposition theorem and a representation theorem for the dual of a C^*-algebra were proved in Grothendieck (1957), Theorems 1 and 2 respectively.

Dunford and Schwartz (1958) (see Theorem III.7.2 and Corollary III.7.3) gave a Vitali-Hahn-Saks-type theorem for Banach space-valued σ-additive measures, in which the Baire category theorem is used.

Let now $\mathcal{L} \subset \mathcal{P}(G)$ be an algebra and $\mathcal{M}_0$ be the space of finitely additive bounded real-valued measures defined on $\mathcal{L}$, endowed with the total variation norm. Porcelli (1960) gave some characterization of weak convergence and weak compactness in $\mathcal{M}_0$. More precisely, the following results hold.

Theorem 2.70 (Porcelli (1960), Corollary 3.1.1) *A sequence $(m_n)_n$ of $\mathcal{M}_0$ is weakly convergent if and only if for every disjoint sequence $(E_q)_q$ in $\mathcal{L}$ the limit*

$$\lim_n \left(\sum_{q=1}^{\infty} m_n(E_q) \right)$$

exists in $\mathbb{R}$.

Theorem 2.71 (Porcelli (1960), Theorem 3.2) *A subset $\mathcal{K} \subset \mathcal{M}_0$ is sequentially weakly compact if and only if the following conditions are satisfied:*

(a) *$\mathcal{K}$ is equibounded, that is there exists a real number M with $\| m \|(G) \le M$ whenever $m \in \mathcal{K}$;*

(b) *For every sequence $(m_n)_n$ in $\mathcal{K}$ and every disjoint sequence $(E_q)_q$ in $\mathcal{L}$ we get*

$$\lim_n \left(\sum_{q=n}^{\infty} | m_n(E_q) | \right) = 0.$$

Theorem 2.72 (Porcelli (1960), Theorem 3.3) *A subset $\mathcal{K} \subset \mathcal{M}_0$ is sequentially weakly compact if and only if each sequence $(m_n)_n$ in $\mathcal{K}$ contains a subsequence $(m_{n_k})_k$ such that*

$$\lim_k \left(\sum_{q=1}^{\infty} | m_{n_k}(E_q) | \right)$$

exists in $\mathbb{R}$ *for every disjoint sequence* $(E_q)_q$ *in* $\mathcal{L}$.

Corollary 2.73 (Porcelli (1960), Theorem 3.4) *If* $\mathcal{L} = \Sigma$ *is a* σ*-algebra and the measures* m_n, $n \in \mathbb{N}$, *are* σ*-additive on* Σ, *then* $(m_n)_n$ *is weakly convergent if and only if* $\lim_n m_n(E)$ *exists in* $\mathbb{R}$ *for every* $E \in \Sigma$.

When X is a locally compact Hausdorff space, Σ is the σ-algebra of all Borel subsets of X and $(m_n)_n$ is a sequence of Radon measures on X, we have the following:

Corollary 2.74 (Porcelli (1960), Theorem 3.5) *A sequence* $(m_n)_n$ *is weakly convergent if and only if* $\lim_n m_n(O)$ *exists in* $\mathbb{R}$ *for every open set* $O \in \Sigma$.

Theorem 2.75 (Porcelli (1960), Theorem 3.6) *If* Σ *is a* σ*-algebra and* $(m_n)_n$ *is a sequence of* σ*-additive real-valued measures defined on* Σ, *then*

$$\lim_n \int_G f \, dm_n$$

exists in $\mathbb{R}$ *for each bounded measurable function* f *if and only if* $(m_n)_n$ *is a weakly convergent sequence.*

Theorem 2.76 (Porcelli (1960), Theorem 3.7) *If* Σ *is a* σ*-algebra and* $\mathcal{K} \subset \mathcal{M}_0$ *is such that every measure* $m \in \mathcal{K}$ *is* σ*-additive, then a necessary and sufficient condition in order that every sequence* $(m_n)_n$ *in* $\mathcal{K}$ *admits a subsequence* $(m_{n_k})_k$ *such that*

$$\lim_k \int_G f \, dm_{n_k}$$

exists in $\mathbb{R}$ *for every bounded measurable function* f *is that* $\mathcal{K}$ *is sequentially weakly bounded.*

Theorem 2.77 (Porcelli (1960), Corollary 4.1.1) *Let G be a nonempty set and $B(G)$ be the Banach space of all bounded real-valued functions defined on G, endowed with the supremum norm. Every separable subspace of $B(G)$ is isomorphic and isometric to a subset of the space $C([0,1])$ of all continuous real-valued functions defined on $[0,1]$.*

Theorem 2.78 (Porcelli (1960), Theorem 4.2) *A sequence $(f_n)_n$ in $B(G)$ is weakly convergent if and only if the sequence $(\| f_n \|)_n$ is bounded and $\lim_n(\lim_i f_n(x_i))$ exists in $\mathbb{R}$ for all sequences $(x_i)_i$ in G such that $\lim_i f_n(x_i)$ exists in $\mathbb{R}$ for every $n \in \mathbb{N}$.*

A Banach space X is said to be a *Grothendieck space* iff any sequence in X, which converges weakly* to 0, is weakly convergent.

In Theorem 2.61 it was proved that, if K is a compact Hausdorff Stonian topological space, then the space $C(K)$ of all real-valued continuous functions defined on K is a Grothendieck space. From this it follows that, if (G,Σ,ν) is a measure space with ν non-negative, σ-additive and σ-finite, then the space $L^\infty(G,\Sigma,\nu)$ of the essentially bounded real-valued functions defined on G, with the identification of functions with take the same values in the complement of ν-null sets, is a Grothendieck space, because it is isometrically and lattice isomorphic to $C(\Omega)$ for some compact Hausdorff Stonian space Ω, by virtue of the Stone-Krein-Kakutani-Yosida vector lattice representation theorem (see also Kadison (1951), Theorem 4.1; Kakutani (1940 and 1941), Krein and Krein (1940 and 1943), Stone (1940 and 1941), Wright (1971) and Yosida (1941)).

Seever (1968) (see Theorems B and 2.5) proved that, if K is an F-space (that is every pair of disjoint open F_σ set has disjoint closure), then the space $C(K)$ is a Grothendieck space. Observe that in general the converse is not true (see also Seever (1968), Theorem 2.6).

A subset K of a Banach space X is said to be *fundamental* iff X is the closed linear subspace generated by K.

We recall the following results.

Theorem 2.79 (Amir and Lindenstrauss (1968), Main Theorem) *Let X be a Banach space which contains a weakly compact fundamental subset K. Then there exist a set Λ and a bounded one-to-one linear operator T from X into $c_0(\Lambda)$.*

Theorem 2.80 (Amir and Lindenstrauss (1968), Theorem 1) *If K is a weakly compact subset of a Banach space, then K is linearly homeomorphic to a weakly compact subset of $c_0(\Lambda)$ for some set Λ.*

Wells (1969) (see Theorems 3 and 4 and Corollary), using diagonal techniques, extended some Dieudonné and Grothendieck theorems and proved a Nikodým boundedness theorem.

We now recall the following characterization of Grothendieck spaces.

Theorem 2.81 (see also Diestel (1984), Theorem 1 and Diestel and Uhl (1977), Theorem, p. 179) *Let X be a Banach space. Then the following are equivalent.*

(a) *X is a Grothendieck space.*

(b) *For all Banach spaces Y such that Y^* has a weak* sequentially compact unit ball, every bounded linear operator $T : X \to Y$ is weakly compact.*

(c) *For all weakly compactly generated Banach spaces Y, every bounded linear operator $T : X \to Y$ is weakly compact.*

(d) *For any Banach space Y, the weakly compact linear operators from X to Y are closed under the process of taking pointwise weak sequential limits.*

(e) *For any Banach space Y, the weakly compact linear operators from X to Y are closed under the process of taking pointwise norm sequential limits.*

(f) *Every bounded linear operator on X with values in a Banach separable space Y is weakly compact.*

(g) *Every bounded linear operator $T : X \to c_0$ is weakly compact.*

We now recall some results on weak compactness of measures, in connection with limit theorems.

Let G be any nonempty set, Σ be a σ-algebra of parts of G, and ν be non-negative σ-additive real-valued measure. Let us endow Σ with the pseudo-distance d_ν defined by

$$d_\nu(A,B) = |\nu(A) - \nu(B)| = \nu(A \Delta B),\ A, B \in \Sigma.$$

A family $\mathcal{K}$ of finitely additive real-valued measures is *equi-ν-continuous at* $A \in \Sigma$ iff for every $\varepsilon > 0$ there is $\delta > 0$ (depending on ε and A) such that, if $B \in \Sigma$ and $d_\nu(A,B) \leq \delta$, then $|m(A) - m(B)| \leq \varepsilon$; *equi-$\nu$-continuous on* Σ, iff is equi-ν-continuous at every $A \in \Sigma$; *uniformly equi-ν-continuous on* Σ, iff for each $\varepsilon > 0$ there is $\delta > 0$ (depending only on ε) such that $|m(A) - m(B)| \leq \varepsilon$ whenever A, $B \in \Sigma$, $d_\nu(A,B) \leq \delta$.

We say that a finitely additive measure $m : \Sigma \to \mathbb{R}$ is *ν-continuous at* $A \in \Sigma$, *ν-continuous on* Σ, *uniformly ν-continuous on* Σ, iff the single set $\mathcal{K} = \{m\}$ is equi-ν-continuous at $A \in \Sigma$, equi-ν-continuous on Σ, uniformly equi-ν-continuous on Σ respectively.

The following results hold.

Theorem 2.82 (Diestel (1984), Theorem 9) *Let $\mathcal{K} = \{m_\lambda : \lambda \in \Lambda\}$ be a family of finitely additive real-valued measures defined on Σ. Then the following are equivalent:*

(a) *$\mathcal{K}$ is equi-ν-continuous at some $E \in \Sigma$;*

(b) *$\mathcal{K}$ is equi-ν-continuous at $\emptyset$ (namely equiabsolutely continuous with respect to ν);*

(c) *$\mathcal{K}$ is uniformly equi-ν-continuous on Σ.*

Moreover, in this case, the family $(m_\lambda)_\lambda$ is uniformly equi-ν-continuous on Σ.

Theorem 2.83 (Diestel (1984), Theorem 10) *Let $\mathcal{K} = \{m_\lambda : \lambda \in \Lambda\}$ be a family of countably additive real-valued measures, defined on Σ. Then the following are equivalent.*

(a) *If* $(E_n)_n$ *is a disjoint sequence in* Σ, *then for every* $\varepsilon > 0$ *there is* $n_\varepsilon \in \mathbb{N}$ *such that for* $q \geq n \geq n_\varepsilon$,

$$\left|\sum_{i=n}^{q} m(E_i)\right| \leq \varepsilon$$

for each $m \in \mathcal{K}$.

(b) *If* $(E_n)_n$ *is a disjoint sequence in* Σ, *then for any* $\varepsilon > 0$ *there is* $n_\varepsilon \in \mathbb{N}$ *such that for* $n \geq n_\varepsilon$,

$$\left|\sum_{i=n}^{\infty} m(E_i)\right| \leq \varepsilon$$

for all $m \in \mathcal{K}$.

(c) *If* $(E_n)_n$ *is any disjoint sequence in* Σ, *then for each* $\varepsilon > 0$ *there is* $n_\varepsilon \in \mathbb{N}$ *such that for* $n \geq n_\varepsilon$, $|m(E_n)| \leq \varepsilon$ *whenever* $m \in \mathcal{K}$.

(d) *If* $(E_n)_n$ *is an increasing sequence in* Σ, *then for every* $\varepsilon > 0$ *there is* $n_\varepsilon \in \mathbb{N}$ *such that, if* $n, q \geq n_\varepsilon$, *then*

$$|m(E_n) - m(E_q)| \leq \varepsilon$$

for each $m \in \mathcal{K}$.

(e) *If* $(E_n)_n$ *is a decreasing sequence in* Σ, *then for every* $\varepsilon > 0$ *there is* $n_\varepsilon \in \mathbb{N}$ *such that, if* $n, q \geq n_\varepsilon$, *then*

$$|m(E_n) - m(E_q)| \leq \varepsilon$$

for each $m \in \mathcal{K}$.

(f) *The family* $(m_\lambda)_\lambda$ *is uniformly* σ*-additive on* Σ.

Theorem 2.84 (Diestel (1984), Theorem 11) *Let $\mathcal{M}^*$ the space of all σ-additive real-valued measures endowed with the total variation norm. A sequence $(m_n)_n$ in $\mathcal{M}^*$ converges weakly to $m \in \mathcal{M}^*$ if and only if $m(E) = \lim_n m_n(E)$ for each $E \in \Sigma$.*

Theorem 2.85 (Diestel (1984), Theorem 12) *Weakly Cauchy sequences in $\mathcal{M}^*$ are weakly convergent. Moreover, for any non-negative σ-additive real-valued measure ν, weakly Cauchy sequences in $L^1(\nu)$ are weakly convergent.*

Theorem 2.86 (Diestel (1984), Lemma) *Let $\mathcal{L}$ be an algebra of sets generating Σ and suppose that $(m_n)_n$ is a uniformly σ-additive sequence, such that $\lim_n m_n(E)$ exists for each $E \in \mathcal{L}$.*

Then $\lim_n m_n(E)$ exists for each $E \in \Sigma$.

Theorem 2.87 (see also Diestel (1984), Theorem 13 and Diestel (1991), Theorem 22) *Let $\mathcal{K} = \{m_\lambda : \lambda \in \Lambda\}$ be a subset of $\mathcal{M}^*$. Then the following are equivalent:*

(a) *$\mathcal{K}$ is relatively weakly compact;*

(b) *$\mathcal{K}$ is bounded and $(m_\lambda)_\lambda$ is uniformly σ-additive;*

(c) *$\mathcal{K}$ is bounded and there is a non-negative σ-additive real-valued measure ν such that $\mathcal{K}$ is uniformly ν-continuous;*

(d) *Given a sequence $(m_n)_n$ in $\mathcal{K}$, there exists a subsequence $(m_{n_k})_k$ of $(m_n)_n$ such that $\lim_k m_{n_k}(E)$ exists in $\mathbb{R}$ for each $E \in \Sigma$;*

(e) *The set $|\mathcal{K}| := \{\| m \| : m \in \mathcal{K}\}$ is relatively weakly compact.*

We now recall also the following result on weak compactness (see also Diestel (1991), Theorem 8).

Theorem 2.88 *Let $\nu : \Sigma \to \mathbb{R}$ be a non-negative σ-additive measure, X be a closed linear subspace of $L^1(\nu)$, and B_X its associated unit ball.*

Then the following statements about X *are equivalent:*

(a) *the integrals*

$$\int \phi d\nu,\ \phi \in B_{X'}$$

are equiabsolutely continuous;

(b) B_X *is weakly compact;*

(c) X *is reflexive;*

(d) X *contains no subspace isomorphic to* l^1;

(e) X *contains no subspace isomorphic to* l^1 *which is complemented in* $L^1(\nu)$;

(f) *There are* $1 < p \leq 2$ *and a probability measure* P *such that* X *is isomorphic to a closed linear space of* $L^p(P)$.

Successively, some results proved by Diestel were extended and some properties of weakly compact operators and (s)-bounded measures were investigated in the context of Banach lattices (see also Schmidt (1988), Theorem 2.3).

Furthermore, Okada and Ricker (1993-1995) established criteria for weak compactness of the *integration map*, that is the operator which to every real- or complex-valued integrable function f with respect to a locally convex space-valued m associates the value of the integral $\int f\,dm$, which is an element of X, and studied some properties of the range of the integration map.

Zhang (1997) (see Theorem 2.1), using Grothendieck and Seever techniques, gave an extension of the Vitali-Hahn-Saks theorem, proving that, if a set of bounded finitely additive Banach space-valued measures, defined on a σ-algebra, is compact in the strong operator topology and if it has a control real-valued measure ν in the weak sense, then the set of measures is uniformly (s)-bounded and equi-absolutely continuous with respect to ν. Moreover, the following result was established.

Theorem 2.89 (Zhang (1997), Theorem 3.4) *Let* K *be an* F *-space, and* $\mathcal{K} = \{m_\lambda : \lambda \in \Lambda\}$ *be a* $\sigma(C(K)', C(K))$*-compact subset of* $C(K)'$. *If there is a non-negative measure* $\nu \in C(K)'$ *such that* m *is absolutely continuous with resepct to* ν *for every* $m \in \mathcal{K}$, *then* $\mathcal{K}$ *is* $\sigma(C(K)', C(K)'')$*-compact, and* $(m_\lambda)_\lambda$ *is equi-absolutely continuous with respect to* ν.

1.2.5. The Rosenthal Lemma

Rosenthal (1968 and 1970-1971) gave some technical lemmas, by means of which it is possible to deduce Theorem 2.50, Lemma 2.26 and some other property of Banach spaces.

Let Σ be a σ-algebra of a nonempty set G, and $\mathcal{M}^*$ be the space of all σ-additive real-valued measures endowed with the total variation norm.

Let $L^1 := L^1([0,1], \Sigma, \nu)$, where Σ is the σ-algebra of all measurable subsets of $[0,1]$ and ν is the Lebesgue measure. We recall the following result.

Theorem 2.90 (see Kadec and Pełczyński (1962), Theorem 6 and Remark 11) *Let* X *be a non-reflexive space of* L^1. *Then* X *contains a subspace complemented in* L^1 *and isomorphic to* l^1. *Moreover, the embedding operator* $j: X \to L^1$ *is not weakly compact.*

Let $0 < \varepsilon < \delta$. We say that a family $\mathcal{K} = \{m_\lambda : \lambda \in \Lambda\} \subset \mathcal{M}^*$ is (δ, ε)*-disjoint* iff $\mathcal{K}$ is bounded (that is $\sup_{\lambda \in \Lambda} ||| m_\lambda ||| < +\infty$) and there exists a disjoint family $(E_\lambda)_\lambda$ in Σ with

$$\| m_\lambda \| (E_\lambda) > \delta \text{ and } \sum_{\zeta \neq \lambda} \| m_\lambda \| (E_\zeta) < \varepsilon \tag{1.21}$$

for every $\lambda \in \Lambda$. We say that $\mathcal{K}$ is *relatively disjoint* iff $\mathcal{K}$ is (δ, ε)-disjoint for some real numbers ε, δ with $0 < \varepsilon < \delta$.

The following results hold.

Lemma 2.91 (see also Rosenthal (1968), Lemma 1) *Let* $\mathcal{K}$ *be a not weakly sequentially compact subset of* $\mathcal{M}^*$.

Then there are $\delta>0$, *a disjoint sequence* $(E_n)_n$ *in* Σ *and a sequence* $(m_n)_n$ *of measures in* $\mathcal{K}$, *satisfying for every* $n\in\mathbb{N}$ *the following properties:*

(a) $|m_n(E_n)|\geq\delta$,

(b) $\| m_n \| (\overline{\bigcup_{j>n} E_j})\leq\frac{1}{2^n}$,

(c) $\| m_n \| (E_j)\leq\frac{1}{2^j}$, $j=1, 2,\ldots,n-1$.

Lemma 2.92 (see also Rosenthal (1970), Lemma 1.1) *Let* Λ *be an infinite set,* $m_\lambda:\Sigma\to\mathbb{R}$, $\lambda\in\Lambda$, *be an equibounded family of finitely additive measures, and let* $(E_\lambda)_\lambda$ *be a disjoint family of* Σ.

Then for every $\varepsilon>0$ *there is a set* $\Lambda'\subset\Lambda$ *with* $\#(\Lambda')=\#(\Lambda)$ *(where # denotes the cardinality of the set in brackets), and*

$$m_\lambda\left(\bigcup\{E_\zeta:\zeta\in\Lambda',\zeta\neq\lambda\}\right)<\varepsilon$$

for every $\lambda\in\Lambda'$.

The following results can be deduced by Lemmas 2.91 and 2.92 (for other related topics and investigations, see also Behrends (2000)).

Theorem 2.93 (Rosenthal (1968), Corollary 3) *Every infinite-dimensional injective Banach space contains a subspace isomorphic to* l^∞.

Corollary 2.94 (Rosenthal (1968), Corollary 4) *Every infinite-dimensional complemented subspace of* l^∞ *is isomorphic to* l^∞.

Theorem 2.95 (Rosenthal (1970), Proposition 3.2) *Let* $\mathcal{K}:=\{m_\lambda:\lambda\in\Lambda\}$ *be a bounded subset of* $\mathcal{M}^*$ *such that there exist a* $\delta>0$ *and a disjoint family* $(E_\lambda)_\lambda$ *in* Σ *with* $\| m_\lambda \| (E_\lambda)>\delta$ *for every* $\lambda\in\Lambda$.

Then for each $\varepsilon > 0$ *there is a set* $\Lambda' \subset \Lambda$ *with* $\#(\Lambda') = \#(\Lambda)$ *such that* (1.21) *holds for every* $\lambda \in \Lambda'$ *(where the elements* ζ *in* (1.21) *belong to* Λ'*). Thus* $\mathcal{K}$ *contains a relatively disjoint family* $\mathcal{K}'$ *with* $\#(\mathcal{K}') = \#(\mathcal{K})$.

Theorem 2.96 (Rosenthal (1970), Theorem 3.3) *Let* K *be a bounded infinite subset of* $l^1(\Lambda)$, *such that there is a* $\delta > 0$ *with* $\| k_1 - k_2 \| \geq \delta$ *whenever* k_1, $k_2 \in K$, $k_1 \neq k_2$. *Then there is a relatively disjoint set* $K^* \subset K$, *with* $\#(K^*) = \#(K)$.

Lemmas 2.91 and 2.92 were extended to the setting of finitely additive measures (see also Li, Gai and Lu (1988), Corollary 4, Theorems 5 and 6). Moreover some extensions of Lemma 2.92 and a shorter proof were given in Kupka (1974), Theorem 2 and Proposition 4.

1.2.6. Limit Theorems for Finitely and σ-Additive Measures and Matrix Theorems

The first convergence theorem for finitely additive measures was discovered in Andô (1961), who proved the following:

Theorem 2.97 (see also Ando (1961), Main Theorem) *Let* $m_n : \Sigma \to \mathbb{R}$, $n \in \mathbb{N}$, *be a sequence of finitely additive measures, each of which absolutely continuous with respect to a non-negative* σ*-additive measure* $v : \Sigma \to \mathbb{R}$. *If* $(m_n(E))_n$ *converges for every* $E \in \Sigma$, *then the measures* m_n *are equiabsolutely continuous with respect to* v.

Observe that, in general, the Baire category methods seem to be unsuitable in the finitely additive case, and the technique used by Andô is inspired by the Phillips Lemma 2.26 and by the following technical lemma.

Lemma 2.98 (Andô (1961), Lemma 1) *Let* $(m_n)_n$ *be a sequence of finitely additive measures such that* $\lim_{k,j} m^*(E_k \Delta E_j) = 0$ *for every monotone sequence* $(E_k)_k$, *where*

$$m^*(E) = \sup_n \| m_n \| (E)$$

(that is for each $\varepsilon > 0$ there is $n_0 \in \mathbb{N}$ with $|m^*(E_k \Delta E_j)| \leq \varepsilon$ for any k, $j \geq n_0$). Then for each sequence $(B_n)_n$ in Σ and $\varepsilon > 0$ there are two sequences $(G_n)_n$ and $(H_n)_n$ in Σ such that, for every $n \in \mathbb{N}$,

(a) $G_n = B_n \cup B_{n+1} \cup \ldots \cup B_{j(n)}$ *for some integer* $j(n)$,

(b) $G_n \supset H_n \supset H_{n+1}$,

(c) $m^*(G_n \setminus H_n) \leq \varepsilon$.

Gould (1965) studied several properties of integration theory in the Banach space context. Using classical techniques, he proved also the following results about equivalence between the Vitali-Hahn-Saks and the Nikodým convergence theorems for Banach space-valued measures.

Theorem 2.99 (Gould (1965), Theorem 3.9) *Let X be a Banach space, G, Λ be two nonempty sets, Σ be a σ-algebra of parts of G, $v : \Sigma \to \mathbb{R}$ be a σ-additive measure and $m_\lambda : \Sigma \to \mathbb{R}$, $\lambda \in \Lambda$, be a bounded family of measures with values in X, each of which absolutely continuous with respect to v.*

Then the family $(m_\lambda)_{\lambda \in \Lambda}$ is equi-absolutely continuous with respect to v if and only if it is uniformly σ-additive.

Theorem 2.100 (Gould (1965), Theorem 3.10) *Let X, G, Λ, Σ be as in Theorem 2.99 and $m_\lambda : \Sigma \to \mathbb{R}$, $\lambda \in \Lambda$, be a bounded family of uniformly σ-additive real-valued measures.*

Then there is a non-negative σ-additive real-valued measure v such that $(m_\lambda)_\lambda$ is equi-absolutely continuous with respect to v. Moreover, v can be chosen so that

$$v(E) \leq \sup\{v(m_\lambda)(E) : \lambda \in \Lambda\} \text{ for every } E \in \Sigma,$$

where $v(m_\lambda)(E) := \sup\{\| m_\lambda(A)\| : A \in \Sigma, A \subset E\}$, $E \in \Sigma$, $\lambda \in \Lambda$, is the semivariation of m_λ.

Using classical limit theorems, a uniform boundedness theorem and a Vitali-Hahn-Saks theorem for functionals defined in suitable von Neumann algebras were proved in Aarnes (1966), Theorems 1 and 2, respectively.

Darst (1966), using sliding hump techniques, gave some new proofs of some conditions for weak convergence given by Porcelli (1960). Darst (1967), using sliding hump methods, proved a version of the Nikodým boundedness theorem for finitely additive real-valued measures defined on σ-algebras, and deduced some Aarnes-type results in the setting of the von Neumann algebras (Theorems 1 and 2).

Seever (1968) introduced an interpolation property for partially ordered sets (which he called *property (I)* and is often called *Seever property* in the literature): a partially ordered set $(L,\preceq)$ is said to have *property (I)* iff for any pair of sequences $(A_n)_n$, $(B_q)_q$ in L such that $A_n \preceq B_q$ for each n, $q \in \mathbb{N}$, there is an element $C \in L$ with $A_p \preceq C \preceq B_p$ for every $p \in \mathbb{N}$. It is possible to formulate property (I) both for spaces of continuous functions endowed with the usual order and for algebras $\mathcal{L}$ of subsets of an abstract nonempty set G with respect to the usual inclusion $\subset$. The following result holds.

Theorem 2.101 (Seever (1968), Theorem 1.1) *Let K be a compact Hausdorff topological space. Then K is an F-space if and only if $C(K)$ has property (I).*

Moreover, both the uniform boundedness principle and the Vitali-Hahn-Saks theorem were proved, for algebras satisfying property (I).

Theorem 2.102 (see Seever (1968), Theorem 3.2) *Let $\mathcal{L} \subset \mathcal{P}(G)$ be an algebra which satisfies property (I). Then a subset $\mathcal{K} = \{m_\lambda : \lambda \in \Lambda\}$ of the set of all finitely additive bounded measures endowed with the total variation norm is bounded if and only if $\sup_{\lambda \in \Lambda} |m_\lambda(E)| < +\infty$ for every $E \in \mathcal{L}$.*

Corollary 2.103 (see Seever (1968), Corollary of Theorem 3.2) *Let S be a totally disconnected F-space, and $\mathcal{K} = \{m_\lambda : \lambda \in \Lambda\}$ be a subset of the space of all real-valued regular Borel measures on S. Then $\mathcal{K}$ is bounded if and only if $\sup_{\lambda \in \Lambda} |m_\lambda(E)| < +\infty$ for every open-closed set $E \subset S$.*

Theorem 2.104 (Vitali-Hahn-Saks theorem, see Seever (1968), Theorem 4.1) *Let $\mathcal{L} \subset \mathcal{P}(G)$ be an algebra with property (I), and m_n, $n \in \mathbb{N}$, $\nu : \mathcal{L} \to \mathbb{R}$ be finitely* *additive measures, such that each m_n is absolutely continuous with respect to ν, and the limit $\lim_n m_n(E)$ exists in $\mathbb{R}$ for every $E \in \mathcal{L}$. Then $(m_n)_n$ is equi-absolutely continuous with respect to ν.*

Theorem 2.105 (see Rybakov (1970), Theorem) *Let X be a Banach space and Σ be a σ-algebra of subsets of G.*

Then for every σ-additive measure $m : \Sigma \to X$ there exists an $x' \in X'$ such that m is absolutely continuous with respect to $x' \circ m$.

Successively Giannakoulias (1987) (see Lemma 1), using the tool of Schauder bases, proved that, if X is a Hausdorff locally convex space, Σ is a σ-algebra, $\nu : \Sigma \to [0,+\infty)$ is a finitely additive measure and $m : \Sigma \to X$ is a finitely additive (s)-bounded measure such that $x' \circ m$ is absolutely continuous with respect to ν for each $x' \in X'$, then m is absolutely continuous with respect to ν.

Aleksyuk (1970) (see Theorems 1.1-1.6) gave some conditions for the existence and uniqueness of extensions of measures having bounded variation. Moreover the author, by means of the sliding hump technique, proved a Nikodým boundedness theorem (Theorem 2.1) and a Nikodým convergence theorem for σ-additive measures (Theorem 2.2), and deduced a characterization of weak compactness for families of measures (Theorem 3.1). Furthermore, in Section 4, the author gave some applications, dealing with relations between measures and metrics and considering, as examples, the Lebesgue measure, the Lebesgue-Stieltjes measure and lengths of curves.

Using sliding hump-type methods, Brooks and Mikusiński (1970) (see Lemma 2) proved the following:

Lemma 2.106 *Let $(x_{i,j})_{i,j}$ be an infinite matrix of real or complex numbers such that*

$$\sum_{j=1}^{\infty} | x_{i,j} | < +\infty$$

for each $i \in \mathbb{N}$ and the limit

$$\lim_i \left(\sum_{j \in E} x_{i,j} \right)$$

exists for every subset $E \subset \mathbb{N}$. *Then there is a sequence* $(x_j)_j$ *with*

$$\lim_i \left(\sum_{j=1}^{\infty} | x_{i,j} - x_j | \right) = 0.$$

Using Lemma 2.106 and by means of the sliding hump method, without using the Baire category theorem, Brooks (1969b) gave a new proof of the Vitali-Hahn-Saks-Nikodým theorem in the σ-additive setting, and extended it to σ-additive Banach space-valued measures.

Theorem 2.107 (Brooks (1969b), Theorem 3) *Let* $m_n : \Sigma \to \mathbb{R}$, $n \in \mathbb{N}$, *be a sequence of* σ*-additive measures, pointwise convergent on* Σ. *Then* $(m_n)_n$ *is uniformly* σ*-additive and the limit measure is* σ*-additive.*

Moreover, if each m_n *is absolutely continuous with respect to a non-negative (possibly infinite)* σ*-additive measure* ν *defined on* Σ, *then* $(m_n)_n$ *is equi-absolutely continuous with respect to* ν.

This theorem was extended to the Banach space context (where the concepts of (equi)absolute continuity and the related ones are analogous to those concerning real-valued measures), and it was proved by Brooks (1969b) and Mikusiński (1970).

Theorem 2.108 (Brooks (1969b), Corollaries 1 and 2 and Mikusiński (1970), Vitali-Hahn-Saks theorem) *Let* X *be a Banach space and* $m_n : \Sigma \to X$, $n \in \mathbb{N}$, *be a sequence of* σ*-additive measures, pointwise convergent on* Σ. *Then* $(m_n)_n$ *is uniformly* σ*-additive and the limit measure is* σ*-additive.*

If each m_n *is finitely additive and absolutely continuous with respect to a non-negative (extended) real-valued* σ*-additive measure* ν *defined on* Σ *and* $(m_n)_n$ *converges pointwise on* Σ, *then* $(m_n)_n$ *is equi-absolutely continuous with respect to* ν.

We now recall the following diagonal theorem for vector-valued matrices.

Theorem 2.109 (Mikusiński (1970), Diagonal theorem) *Let X be a Banach space. If $x_{i,j}$, i, $j \in \mathbb{N}$, is a double sequence in X such that $\lim_j x_{i,j} = 0$ for every $i \in \mathbb{N}$ and there is a constant $C_0 > 0$ with $\| x_{i,i} \| > C_0$ for each $i \in \mathbb{N}$, then there is an infinite subset $P \subset \mathbb{N}$ with*

$$\sum_{j \in P} \|x_{i,j}\| < +\infty \text{ and } \left\| \sum_{j \in P} x_{i,j} \right\| > \frac{C_0}{2}$$

for every $i \in P$.

Let X, Y, Z be three Banach spaces, and assume that for every $x \in X$ and $y \in Y$ there is a product $xy = z \in Z$. Suppose that the multiplication is bilinear and

$$\|x\| = \sup_{y \in Y, y \neq 0} \frac{\|xy\|}{\|y\|} \text{ for each } x \in X.$$

We now recall the following:

Theorem 2.110 (Mikusiński (1970), Banach-Steinhaus theorem) *If $X_0 \subset X$ and $\sup_{x \in X_0} \|xy\| < +\infty$ for every $y \in Y$, then there is a positive number C' such that $\| x \| \leq C'$ for every $x \in X_0$.*

Using the sliding hump technique, Brooks and Jewett (1970) extended the Vitali-Hahn-Saks and Nikodým convergence theorems in the context of finitely additive Banach space-valued measures. The following result extends Theorem 2.26.

Theorem 2.111 (Brooks and Jewett (1970), Theorem 1, see also Brooks (1975), Theorem 1) *Let X be a Banach space and $m_n : \mathcal{P}(\mathbb{N}) \to X$, $n \in \mathbb{N}$, be a sequence of (s)-bounded finitely additive measures.*

If $\lim_n m_n(A) = 0$ for each $A \subset \mathbb{N}$, then

$$\lim_n \left(\sum_{k \in A} m_n(\{k\}) \right) = 0$$

uniformly with respect to $A \subset \mathbb{N}$.

Corollary 2.112 (Brooks and Jewett (1970), Corollary 1.1) *Let* G *be any nonempty set,* Σ *be a* σ*-ring of parts of* G, $m_n : \Sigma \to X$, $n \in \mathbb{N}$, *be a sequence of finitely additive and* (s)*-bounded measures, such that* $m(E) = \lim_n m_n(E)$ *exists in* X *for every* $E \in \Sigma$. *If* $(E_k)_k$ *is a disjoint sequence in* Σ, *then*

$$\lim_n \left(\sum_{k \in A} (m_{n+1} - m_n)(E_k) \right) = 0$$

uniformly with respect to $A \subset \mathbb{N}$.

Corollary 2.113 (Brooks and Jewett (1970), Corollary 1.2) *Let* $m_n : \Sigma \to X$, $n \in \mathbb{N}$, *be a sequence of finitely additive and* (s)*-bounded measures, such that* $m(E) = \lim_n m_n(E)$ *exists in* X *for every* $E \in \Sigma$. *Then* m *is* (s)*-bounded.*

A sequence $m_n : \Sigma \to X$, $n \in \mathbb{N}$, is said to be *uniformly additive* iff

$$\lim_j \left(\sum_{k \in A, k \geq j} m_n(E_k) \right) = 0$$

uniformly with respect to $A \subset \mathbb{N}$ and $n \in \mathbb{N}$, whenever $(E_k)_k$ is a disjoint sequence in Σ.

The following result holds.

Theorem 2.114 (Brooks and Jewett (1970), Theorem 2) *Let* $m_n : \Sigma \to X$, $n \in \mathbb{N}$, *be a sequence of finitely additive and* (s)*-bounded measures, such that* $m(E) = \lim_n m_n(E)$ *exists in* X *for every* $E \in \Sigma$. *Then* $(m_n)_n$ *is uniformly additive.*

The following Nikodým convergence theorem in the Banach space setting follows from Theorem 2.114.

Corollary 2.115 (Brooks and Jewett (1970), Corollary 2.2) *Let* $m_n : \Sigma \to X$, $n \in \mathbb{N}$, *be a sequence of* σ*-additive measures, such that* $m(E) = \lim_n m_n(E)$ *exists in* X *for every* $E \in \Sigma$. *Then* $(m_n)_n$ *is uniformly* σ*-additive and* m *is* σ*-additive.*

Now we recall the following Vitali-Hahn-Saks-type theorem for finitely additive Banach space-valued measures, whose proof uses a sliding hump-type argument.

Theorem 2.116 (Brooks and Jewett (1970), Theorem 3) *Let* $m_n : \Sigma \to X$, $n \in \mathbb{N}$, *be a sequence of finitely additive* (s)*-bounded measures, each of which absolutely continuous with respect to a non-negative (possibly infinite) extended real-valued finitely additive measure* ν *defined on* Σ. *Suppose that* $m(E) = \lim_n m_n(E)$ *exists in* X *for every* $E \in \Sigma$.

Then $\lim_{\nu(E)\to 0} m_n(E) = 0$ *uniformly with respect to* $n \in \mathbb{N}$, *namely for every* $\varepsilon > 0$ *there is a* $\delta > 0$ *with* $\| m_n(E) \| < \varepsilon$ *whenever* $n \in \mathbb{N}$ *and* $E \in \Sigma$ *with* $\nu(E) < \delta$.

Proof: If the thesis is not true, then there exist an $\varepsilon > 0$ such that for every $\delta > 0$ there exist $E_\delta \in \Sigma$ and $n_\delta \in \mathbb{N}$ such that $\nu(E_\delta) < \delta$ and $\| m_{n_\delta}(E_\delta) \| > \varepsilon$. So, if $\delta_0 > 0$, then there are $E_{\delta_0} = E_0 \in \Sigma$ and $n_{\delta_0} = n_0 \in \mathbb{N}$ with $\nu(E_0) < \delta_0$ and $\| m_{n_0}(E_0) \| > \varepsilon$. Since m_{n_0} is absolutely continuous with respect to ν, then in correspondence with $\frac{\varepsilon}{8}$ there is $\delta(n_0) > 0$ with $\| m_{n_0}(E) \| < \frac{\varepsilon}{8}$ whenever $\nu(E) < \delta(n_0)$. Put $\delta_1 = \delta(n_0)$: since we have denied the thesis, in correspondence with δ_1 there are $E_{\delta_1} = E_1 \in \Sigma$ and $n_{\delta_1} = n_1 \in \mathbb{N}$ with $\nu(E_1) < \delta_1$ and $\| m_{n_1}(E_1) \| > \varepsilon$. Moreover, if $E \subset E_1$, then $\nu(E) \le \nu(E_1) < \delta_1$, and hence $\| m_{n_0}(E) \| < \frac{\varepsilon}{8}$. Proceeding by induction, we construct three sequences $(\delta_k)_k$ in $\mathbb{R}^+$, $(n_k)_k$ in $\mathbb{N}$ and $(E_k)_k$ in Σ, such that $\| m_{n_{k+1}}(E_{k+1}) \| > \varepsilon$, $\nu(E_{k+1}) < \delta_{k+1}$ and

$\| m_{n_i}(E) \| < \varepsilon 2^{-k-3}$ for every $i \le k$, $k \in \mathbb{N}$ and $E \in \Sigma$ with $E \subset E_{k+1}$. Without loss of generality, set $n_i = i$. Thus, we have

$$\| m_{k+1}(E_{k+1}) \| > \varepsilon,\ \nu(E_{k+1}) < \delta_{k+1} \text{ and } \| m_j(E) \| < \varepsilon 2^{-k-3} \qquad \textbf{(1.22)}$$

for every $j \le k$, $k \in \mathbb{N}$ and $E \in \Sigma$ with $E \subset E_{k+1}$.

Put $F_1 = E_2$, and suppose that there is an $i_2 > 2$ with $\| m_{i_2}(F_1 \cap E_{i_2}) \| > \frac{\varepsilon}{4}$. Set $F_2 = F_1 \setminus E_{i_2}$. In general, assume that F_j, i_j, $j = 1, \ldots, k$ have been chosen and that there exists an $i_{k+1} > i_k$ with $\| m_{i_{k+1}}(F_k \cap E_{i_{k+1}}) \| > \frac{\varepsilon}{4}$. Put $F_{k+1} = F_k \setminus E_{i_{k+1}}$. If this process did not end, then there would exist a sequence of disjoint sets $F_k \setminus F_{k+1}$, $k \in \mathbb{N}$, with

$$\| (m_{i_{k+1}} - m_{i_k})(F_k \setminus F_{k+1}) \| \ge$$

$$\| m_{i_{k+1}}(F_k \setminus F_{k+1}) \| - \| m_{i_k}(F_k \setminus F_{k+1}) \| > \varepsilon - \varepsilon 2^{-i_{k+1}-3} > \frac{\varepsilon}{8}$$

for every $k \in \mathbb{N}$.

By virtue of Corollary 2.112, a contradiction is obtained. Hence it is possible to find an $F_k \in \Sigma$ and an $i_k \in \mathbb{N}$ with

$$\| m_j(F_k \cap E_j) \| < \frac{\varepsilon}{4} \text{ for each } j > i_k. \qquad \textbf{(1.23)}$$

Put $p_1 = i_k$, $H_1 = F_k$, $m_i^{(1)} = m_{p_1+i}$, $E_i^{(1)} = E_{p_1+i} \setminus H_1$. Then,

$$\| m_1(H_1) \| < \frac{\varepsilon}{8}. \qquad \textbf{(1.24)}$$

Since $H_1 \subset E_2$, then from (1.22) we obtain $\| m_2(E_2) \| > \varepsilon |$ and $| m_2(E_2 \setminus H_1) \| < \frac{\varepsilon}{4}$. Hence,

$$\| m_2(H_1) \| = \| m_2(E_2) - m_2(E_2 \setminus H_1) \| \geq \| m_2(E_2) \| - \| m_2(E_2 \setminus H_1) \| > \varepsilon - \frac{\varepsilon}{4}. \quad \mathbf{(1.25)}$$

From (1.24) and (1.25) we get:

$$\| (m_2 - m_1)(H_1) \| > \varepsilon - \frac{\varepsilon}{4} - \frac{\varepsilon}{8}.$$

Moreover, from (1.22) and (1.23) we get $\| m_i^{(1)}(E_i^{(1)}) \| > \varepsilon - \frac{\varepsilon}{4}$ and $\| m_j^{(1)}(E) \| < \varepsilon 2^{-i-3}$ for any $E \subset E_i^{(1)}$, $E \in \Sigma$ and $j < i$. Put $F_1^{(1)} = E_1^{(1)}$. By an analogous argument, we find an $F_k^{(1)} \in \Sigma$ and a positive integer i_k with $\| m_j^{(1)}(F_k^{(1)} \cap E_j^{(1)}) \| < \frac{\varepsilon}{8}$ for $j > i_k$. Set $H_2 = F_k^{(1)}$, $p_2 = i_k$, $m_i^{(2)} = m_{p_2+i}^{(1)}$, $E_i^{(2)} = E_{p_2+i}^{(1)} \setminus H_2$. Then $H_1 \cap H_2 = \emptyset$, $\| m_2(H_2) \| < \frac{\varepsilon}{16}$, $\| m_1^{(1)}(H_2) \| > \varepsilon - \frac{\varepsilon}{4} - \frac{\varepsilon}{8}$. Thus,

$$\| (m_1^{(1)} - m_2)(H_2) \| > \varepsilon - \frac{\varepsilon}{4} - \frac{\varepsilon}{8} - \frac{\varepsilon}{16}\text{-}$$

Note that $\| m_i^{(2)}(E_i^{(2)}) \| > \varepsilon - \frac{\varepsilon}{4} - \frac{\varepsilon}{8}$ and $\| m_j^{(2)}(E) \| < \varepsilon 2^{-i-3}$ for each $E \subset E_i^{(2)}$ with $E \in \Sigma$ and $j < i$. Proceeding by induction, suppose that we have defined for $k \geq 2$ disjoint sets H_i, $i \leq k$; $p_1, \ldots, p_k$; $m_i^{(k)}$ (put $m_i^{(0)} = m_{i+1}$), so that $\| m_1^{(k-2)}(H_k) \| < \frac{\varepsilon}{2^{k+2}}$, $\| m_1^{(k-1)}(H_k) \| > \varepsilon - \frac{\varepsilon}{4} - \ldots - \frac{\varepsilon}{2^{k+1}}$, $E_i^{(k)} = E_{p_k+i}^{(k-1)} \setminus H_k$, $\| m_j^{(k)}(E) \| < \varepsilon 2^{-k-i-2}$ for each $E \subset E_i^{(k)}$ and $j < i$. Set $F_1^{(k)} = E_1^{(k)}$. As before, we find a positive integer i_t such that

$$\| m_j^{(k)}(F_t^{(k)} \cap E_j^{(k)}) \| < \varepsilon 2^{-k-2}$$

for every $j > i_t$. Let $p_{k+1} = i_t$, $H_{k+1} = F_t^{(k)}$, $m_i^{(k+1)} = m_{p_{k+1}+i}^{(k)}$, $E_i^{(k+1)} = E_{p_k+i}^{(k)} \setminus H_{k+1}$. Then we have:

$$H_{k+1} \cap \left(\bigcup_{i=1}^{k} H_i \right) = \emptyset,$$

$$\| m_1^{(k-1)}(H_{k+1}) \| < \frac{\varepsilon}{2^{k+3}}, \qquad \| m_1^{(k)}(H_{k+1}) \| > \varepsilon - \frac{\varepsilon}{4} - \ldots - \frac{\varepsilon}{2^{k+2}},$$

$$\| m_i^{(k+1)}(E_i^{(k+1)}) \| > \varepsilon - \frac{\varepsilon}{4} - \ldots - \frac{\varepsilon}{2^{k+2}},$$

and $\| m_j^{(k+1)}(E) \| < \varepsilon 2^{-k-i-3}$ whenever $E \subset E_i^{(k+1)}$ and $j < i$. Proceeding in this way, we get a disjoint sequence $(H_k)_k$ and a sequence of measures $(m_1^{(k)})_k$, satisfying

$$\| (m_1^{(k)} - m_1^{(k-1)})(H_{k+1}) \| > \varepsilon - \frac{\varepsilon}{4} - \ldots - \frac{\varepsilon}{2^{k+1}} - \frac{\varepsilon}{2^{k+2}} - \frac{\varepsilon}{2^{k+3}} > \varepsilon - \frac{\varepsilon}{2} = \frac{\varepsilon}{2}$$

for each $k \geq 2$. From this and Corollary 2.112 we obtain a contradiction. This ends the proof. □

Observe that, if ν is a bounded measure, then every m_n is (s)-bounded, and hence $(m_n)_n$ is uniformly (s)-bounded (see also Brooks and Jewett (1970)). Moreover, Brooks and Dinculeanu (1974) (see Theorem 2.1) proved the thesis of Theorem 2.116 by replacing the hypothesis of pointwise convergence with the condition of uniform (s)-boundedness of the sequence $(m_n)_n$.

In Brooks and Chacon (1980) (see Theorem 1) there is a proof of Theorem 2.116 based on the biting lemma 2.144, which we will see later.

Darst (1970) proved the Nikodým boundedness theorem for finitely additive real-valued measures and the Vitali-Hahn-Saks theorem for finitely additive Banach space-valued measures, using results on weak convergence for finitely additive set functions defined on a σ-algebra.

Topsøe (1970), using diagonal arguments, extended some results of Grothendieck (1953) on weak compactness and relative compactness of measures (see Theorems 7 and 8), and extended the Dieudonné theorem to the case when the measures involved are inner regular and considering two abstract classes of subsets, having a role similar to that of the open and the compact sets (see Corollary 3).

Brooks (1971) proved that a finitely additive Banach space-valued measure m defined on a ring Θ of subsets of G is (s)-bounded if and only if m admits a *control measure*, that is a non-negative finitely additive measure ν defined on Θ, such that m is absolutely continuous with respect to ν and

$$\nu(E) \leq \sup\{\| m(F) \| : F \in \Theta,\ F \subset E\} \text{ for each } E \in \Theta$$

(see Theorem 1), extending Corollary 2.67 (see also Dinculanu and Kluvánek (1967), Theorem 1) for the locally convex space setting, and Gould (1965), Theorem 3.2, together with a corresponding result proved for finitely additive (s)-bounded measures defined on δ-rings (see also Ohba (1972), Theorem 3). Bell, Bilyeu and Lewis (1986) gave some Vitali-Hahn-Saks-type theorems (Theorems 2.3 and 3.1 and Corollary 3.3), Drewnowski-type theorems (Theorem 2.5), Brooks-Jewett-type theorems and some results about the existence of control measures (see also Theorem 2.4) in the Banach lattice context, using suitable properties of the Riesz norm. Some other results about control measures were proved in Drewnowski (1972a) (see Corollary 10.9).

Gänssler (1971a), using diagonal techniques, gave some further results about compactness and subsequential compactness of measure spaces and deduced some Dieudonné-type theorems for uniformly bounded sequences of regular measures on a regular or completely regular Hausdorff space (Theorems 3.7, 3.11, 3.12, see also Gänssler (1971b), Theorem 3.1). In Pfanzagl (1971), Theorem, using a diagonal argument, there is an extension of the Dieudonné theorem for measures defined on an arbitrary Hausdorff space.

Landers and Rogge (1971a), using sliding hump techniques (see Lemmas 9-12), proved a version of the Vitali-Hahn-Saks (Theorem 4) and Nikodým boundedness theorem (Theorem 5) for σ-additive measures defined in σ-rings and with values in an abelian topological group.

Antosik (1971) extended to quasi-normed abelian group-valued double sequences the Mikusiński lemma 2.109, as follows.

Theorem 2.117 (Antosik (1971), Diagonal theorem (II)) *Let* $(X, \|\cdot\|)$ *be an abelian quasi-normed group,* $(x_{i,j})_{i,j}$ *be a double sequence in* X *, such that* $\lim_j x_{i,j} = 0$ *for each* $i \in \mathbb{N}$ *. Then there are an infinite subset* $P \subset \mathbb{N}$ *and a subset (finite or infinite)* $J \subset P$ *with*

$$\sum_{j \in J} \| x_{i,j} \| < +\infty$$

and

$$\left\| \sum_{j \in J} x_{i,j} \right\| \geq \frac{1}{2} \| x_{i,i} \|$$

for every $i \in P$ *.*

Moreover, the constant $\frac{1}{2}$ *is sharp.*

As a consequence of Theorem 2.117, in Antosik (1971) a Banach-type theorem is given in the context of F-normed spaces. An F*-norm* on a vector space X is a function $\|\cdot\| : X \to \mathbb{R}$, such that $\| 0 \| = \| 0 \|$, $\| -x \| = \| x \|$, $\| x + y \| \leq \| x \| + \| y \|$ for every x , $y \in X$, $\| x_n \| \to 0$ implies $\| \alpha x_n \| \to 0$ and $\| \alpha_n \| \to 0$ implies $\| \alpha_n x \| \to 0$. A subset B of an F-normed space X is said to be *bounded* iff $\| \lambda_n x_n \| \to 0$ whenever $\| \lambda_n \| \to 0$ and for every sequence $(x_n)_n$ in B .

The following result holds.

Theorem 2.118 (Antosik (1971), Theorem of Banach, see also Dunford and Schwartz (1958), Theorem II.1.11) *A family* $\mathcal{K}$ *of linear continuous functions from a complete* F *-normed space* X *to an* F *-normed space* Y *is equicontinuous on* X *if and only if the set* $\{\phi(x) : \phi \in \mathcal{K}\}$ *is bounded in* Y *for each* $x \in X$ *.*

Hoffmann-Jørgensen (1971) (see Theorem 7) proved the following:

Theorem 2.119 *Let* X *be a Hausdorff locally convex space,* $\mathcal{L}$ *be an algebra of subsets of* G *and* $m : \mathcal{L} \to X$ *be a bounded finitely additive* X *-valued set* *function. If every bounded closed set in X is complete, then the following are equivalent:*

(a) *m is (s)-bounded,*

(b) *$m(\mathcal{L})$ is conditionally weakly compact, that is every sequence has a weakly Cauchy subsequence.*

Mikusiński (1971) (Theorem N, see also Drewnowski (1973a), Theorem 1) proved the following version of the Nikodým boundedness theorem.

Theorem 2.120 *Let Λ be a nonempty set, $(R, \|\cdot\|)$ be a quasi-normed group, Σ be a σ-algebra and $m_\lambda : \Sigma \to R$, $\lambda \in \Lambda$, be a family of σ-additive measures, with $\sup_{\lambda \in \Lambda} \| m_\lambda(E) \| < +\infty$ for each $E \in \Sigma$.*

Then

$$\sup_{\lambda \in \Lambda, E \in \Sigma} \| m_\lambda(E) \| < +\infty.$$

Landers and Rogge (1972) (see Theorem 1), giving a unified approach, proved a Dieudonné-type theorem for topological group-valued measures defined on a σ-ring $\mathcal{L}$ of subsets of G and concerning inner regularity with respect to two abstract subclasses $\mathcal{G}$, $\mathcal{H}$ of $\mathcal{L}$, such that $\mathcal{G}$ is closed under countable unions, for every $K \in \mathcal{H}$ the system $\{U \in \mathcal{G} : U \supset K\}$ is directed by inclusion, and for every disjoint sets K_1, $K_2 \in \mathcal{H}$ there are disjoint sets U_1, $U_2 \in \mathcal{G}$ with $U_1 \supset K_1$ and $U_2 \supset K_2$.

Ohba (1972) (see Proposition 4), using classical methods, proved that, if a Banach space X has the Gould property (see (1.2)) and m is an X-valued measure, defined on a ring $\mathcal{L}$ of subsets of G, then the following are equivalent:

(a) m is (s)-bounded,

(b) m has bounded semivariation,

(c) For every disjoint sequence $(E_n)_n$ in $\mathcal{L}$ the series

$$\sum_{n=1}^{\infty} m(E_n)$$

converges in X

(see also Bogdan and Oberle (1978), Propositions 2.1.1 and 2.1.2), and a decomposition theorem was deduced (see Ohba (1972), Theorem 4 and Proposition 6).

Stein (1972), using diagonal techniques, proved a uniform boundedness theorem for inner regular real-valued measures (see Theorem 2). In this context, using still diagonal arguments, some Phillips- (Stein (1974), Proposition 2.1), Dieudonné- (Stein (1974), Propositions 2.3 and 3.2) and Vitali-Hahn-Saks-type theorems (see also Stein (1975), Theorem) were given.

We recall here the following:

Proposition 2.121 (Stein (1974), Proposition 2.1) *Let* G *be a regular topological space, such that every subset of* G *of a single point is closed, and* $(m_n)_n$ *be a sequence of bounded regular finitely additive real-valued measures defined on the class of all Borel subsets of* G. *If* $\lim_n m_n(E) = 0$ *for each open set* E, *then there is a maximal open set* U *with* $\lim_n \| m_n \| (U) = 0$. *Moreover, the same conclusion holds if* G *is Hausdorff and* $\lim_n m_n(E) = 0$ *for each open set* E *such that* E *is a countable union of compact subsets of* G.

1.2.7. The Drewnowski Theorem

In the literature, in the study of properties of finitely additive measures, it is often advisable to relate them to corresponding properties of σ-additive measures. In this context, two important tools are the Stone Isomorphism technique, which deals with σ-additive "extensions" of finitely additive measures, and the Drewnowski technique, by means of which it is possible to construct σ-additive restrictions, defined on suitable sub-σ-algebras. Drewnowski (1972b) (see Proposition 2) proved that, if R is an abelian topological group, G, Λ are two arbitrary nonempty sets, Σ is a σ-ring of parts of G and $m_\lambda : \Sigma \to R$, $\lambda \in \Lambda$, is a family of finitely additive (s)-bounded measures, then every disjoint sequence $(C_n)_n$ contains a subsequence $(C_{n_k})_k$ such that every m_λ is σ-additive on the

σ-ring generated by $(C_{n_k})_k$. Drewnowski (1972b) proved also that, in topological group setting, the Brooks-Jewett, Vitali-Hahn-Saks and Nikodým convergence theorems hold and are equivalent, where in the Vitali-Hahn-Saks theorem the concept of absolute continuity is intended with respect to a Fréchet-Nikodým topology. Furthermore in Drewnowski (1973b), Theorems 3.11 and 4.1, some decomposition theorems for additive set functions were proved. Some other decomposition theorems in the topological group context can be found in Traynor (1972a-b) and Traynor (1976). In Chapter 4 we will extend the Drewnowski theorems, giving some results on equivalence between limit theorems in lattice and topological group context with respect to filter convergence.

Some consequences of Drewnowski (1972b), Proposition 2 and the Orlicz-Pettis theorem 2.17 are the following:

Theorem 2.122 (Drewnowski (1972b), Theorem 1) *Let Σ be a σ-ring, X be a locally convex vector space, v be an (s)-bounded finitely additive extended real-valued measure on Σ, $m:\Sigma \to X$ be a finitely additive measure, such that $x'm$ is absolutely continuous with respect to v for every $x' \in X'$. Then m is (s)-bounded and absolutely continuous with respect to v.*

Theorem 2.123 (Drewnowski (1972b), Theorem 2) *Let Σ be a σ-ring, X be a normed vector space and $m_n:\Sigma \to X$, $n \in \mathbb{N}$, be a sequence of finitely additive (s)-bounded measures. If for each $E \in \Sigma$ the sequence $(m_n(E))_n$ is weakly convergent, then the weak limit set function is (s)-bounded and $(m_n)_n$ is equibounded.*

Labuda (1972) gave some extensions of Vitali-Hahn-Saks-Nikodým theorems in the context of Fréchet-Nikodým topologies and submeasures, by means of techniques based on the Baire category theorem.

A σ*-space* is a σ-algebra Σ endowed with a Fréchet-Nikodým topology and the corresponding pseudo-distance generated by a single submeasure, and such that Σ is a topological space of second Baire category.

Let K be a topological space, R be a topological group, $C_s(K,R)$ (resp. $C_u(K,R)$) be the space of all continuous functions from K to R endowed with pointwise (resp. uniform) convergence, and let $\mathbb{N}$ be endowed with the discrete topology. The following results hold.

Theorem 2.124 (see Labuda (1972), Theorem 2.1) *Let* Σ *be a* σ*-space and* $U:\Sigma\to C_s(\mathbb{N},R)$ *be an additive continuous function. Then* U *is an additive continuous function from* Σ *to* $C_u(\mathbb{N},R)$. *Moreover, the limit function* $U_0:\Sigma\to R$ *defined by*

$$U_0(E)=\lim_n U(E)(n),\ E\in\Sigma,$$

is additive and continuous.

Theorem 2.125 (see Labuda (1972), Theorem 2.3) *Let* K *be a sequentially compact topological space,* Σ *be a* σ*-space,* $U:\Sigma\to C_s(K,R)$ *be an additive continuous function. Then* U *is an additive continuous function from* Σ *to* $C_u(K,R)$.

Theorem 2.126 (see Labuda (1972), Theorem 2.4) *Let* K *be a compact metrizable topological space,* $K^{'}\subset K$ *be a dense set,* $C_d(K,R)$ *be the space* $C_s(K,R)$ *endowed with pointwise convergence on* $K^{'}$, Σ *be a* σ*-space,* $U:\Sigma\to C_d(K,R)$ *be an additive continuous function. Then* U *is an additive continuous function from* Σ *to* $C_u(K,R)$.

Theorem 2.127 (see Labuda (1972), Theorem 2.5) *Let* K *be a compact topological space,* Σ *be a separable* σ*-space,* $U:\Sigma\to C_s(K,R)$ *be an additive continuous function. Then* U *is an additive continuous function from* Σ *to* $C_u(K,R)$.

Note that from Theorem 2.124 it is possible to deduce the versions of the Nikodým convergence theorem and the Brooks-Jewett theorem in the topological group setting, given in Drewnowski (1972b).

Theorem 2.128 *Let* Σ *be a* σ*-algebra of a nonempty set* G, R *be a topological group,* $m_n:\Sigma\to R$, $n\in\mathbb{N}$ *be a sequence of* σ*-additive measures, with* $m(E)=\lim_n m_n(E)$ *for every* $E\in\Sigma$. *Then* $(m_n)_n$ *is uniformly* σ*-additive and* m *is* σ*-additive.*

Robertson (1972), using techniques similar to those in Labuda (1972), proved some Nikodým convergence (Theorem 2) and Vitali-Hahn-Saks-type theorem (Theorem 3).

Hoffmann-Jørgensen (1972) investigated several compactness properties of the space of the bounded regular real-valued measures endowed with the weak* topology (viewed as the dual of the space of all bounded continuous real-valued functions).

The following result on weak compactness was given in Brooks (1972).

Theorem 2.129 *Let X be a reflexive Banach space and $\mathcal{L}$ be a ring of sets. A set $\mathcal{K}$ of finitely additive X-valued measures is conditionally weakly compact if and only if $\mathcal{K}$ is bounded and the set $\{\|m\| : m \in \mathcal{K}\}$ is uniformly additive.*

Observe that Theorem 2.129 does not hold, if X is not reflexive (see also Brooks (1972), Remark 2).

An extension of Theorem 2.117 was given in Antosik (1972), Diagonal Theorem. In Antosik (1973), as consequences of such diagonal theorems, there are a theorem on exchange of limits under the sign of series (Corollary 2), a Vitali-Hahn-Saks (Theorem 2) and Nikodým boundedness theorem (Theorem 4) for finitely additive measures defined on a σ-ring and with values in a topological group, some equicontinuity theorems for a sequence of additive and continuous functions defined on suitable groups and taking values in an abelian topological group (Theorem 5) or in a topological vector space (Theorem 6), some result about continuity and equicontinuity of biadditive topological group- or topological vector space-valued functions (Theorems 7, 9, 10) and unconditional convergence (Theorems 11, 12, 13).

Darst (1973), using the sliding hump technique, extended the Nikodým boundedness theorem (Theorem 1) and the Brooks-Jewett theorem (280, Theorem 1) to finitely additive measures with values in an abelian topological group and defined in a σ-algebra.

Labuda (1973) (see Theorems 1-3) gave some extensions of the Orlicz-Pettis theorem in the context of vector space-valued measures defined on σ-rings.

Rogge (1973), using diagonal techniques, extended the Dieudonné theorem to topological group-valued measures defined in topological spaces which are *perfectly normal*, namely normal and such that every open set is a countable union of closed sets (see Theorem 1 and Corollary 3). Similar results were proved in Khurana (1976), by means of the tool of submeasures, for topological group-valued measures defined on regular or normal Hausdorff topological spaces (see Theorem 4 and Corollary 5).

Using diagonal methods, Traynor (1973) gave some versions of Nikodým convergence, Boooks-Jewett-type (Theorems 2.2, 2.3 and 3.4) and Nikodým boundedness-type theorems (Theorem 3.2) for functions defined on a σ-ring and with values in an abelian topological group, and described some property of finitely additive Banach space-valued control measures defined on a ring (Theorem 3.3).

In Brooks (1973) (Theorem 1, see also Brooks and Dinculeanu (1974), Theorems 2.1 and 2.3, the analogous versions of Theorems 2.99 and 2.100 were extended to the case of finitely additive measures defined on rings. Moreover, in Brooks (1973), some extension of the Vitali convergence theorem (Theorem 3) and some relation between Vitali-Hahn-Saks theorems and weak or conditional compactness of sets of measures (Theorems 4 and 5) were given, while in Brooks and Dinculeanu (1974) some results about control measures and some relations between limit theorems and (weak) compactness were proved.

1.2.8. (s)-Bounded Banach Space-Valued Measures

One can ask whether every bounded measure taking values in Banach spaces is (s)-bounded too. The answer is in general negative, as the following example shows.

Example 2.130 (see also Boccuto and Candeloro (2004d)) Let Σ be the σ-algebra of all Lebesgue measurable subsets of $[0,1]$, ν be the Lebesgue measure, $X = L^\infty([0,1],\Sigma,\nu)$ be the space of all essentially bounded functions (with identification up to sets of Lebesgue measure zero), endowed with the essential supremum norm $\|\cdot\|_\infty$. We define $m:\Sigma \to X$ by setting $m(A) = \chi_A$ for all $A \in \Sigma$, where χ_A is the characteristic function associated with A. Then m is a bounded finitely additive measure, since $\| m(A) \|_\infty = \| \chi_A \|_\infty \le 1$ for every $A \in \mathcal{A}$. But m is not (s)-bounded, since for the disjoint sequence $E_n = [1/(n+1), 1/n)$, $n \in \mathbb{N}$, we have $\| m(E_n) \|_\infty = 1$ for all $n \in \mathbb{N}$.

Diestel (1973a), besides some other properties of weak compactness of sets of measures, proved the following result (see also Example 1.7, Theorem 1.8 and Remarks).

Theorem 2.131 *A Banach space X does not contain isomorphic copies of c_0 if and only if every bounded, finitely additive X-valued set function defined on every algebra of subsets of X is (s)-bounded.*

We recall also the following results.

Theorem 2.132 (Diestel and Faires (1974), Theorem 1.1) *A Banach space X does not contain isomorphic copies of l^∞ if and only if every bounded, finitely additive X-valued set function defined on every σ-algebra of subsets of X is (s)-bounded.*

Corollary 2.133 (Diestel and Faires (1974), Corollary 1.2) *A Banach space X does not contain isomorphic copies of l^∞ if and only if every set function defined on every σ-algebra of subsets of X and with values in the topological dual X', which is σ-additive with respect to weak* convergence, is σ-additive also with respect to the norm convergence.*

Moreover in Diestel and Faires (1974) there are also some applications to integration theory, a decomposition result for c_0-valued measures and some versions of the Radon-Nikodým theorem in the Banach space setting.

Let $\mathcal{L} \subset \mathcal{P}(G)$ be an algebra, $\sigma(\mathcal{L})$ be the smallest σ-algebra containing $\mathcal{L}$, X be a Banach space and $m_n : \sigma(\mathcal{L}) \to X$, $n \in \mathbb{N}$, be a sequence of finitely additive measures. We say that $(m_n)_n$ has the *approximation property* relatively to $\sigma(\mathcal{L})$ iff for each subalgebra $\mathcal{L}_0$ of $\mathcal{L}$, $E \in \sigma(\mathcal{L}_0)$ and every finite set $M_0 \subset \{m_n : n \in \mathbb{N}\}$ there is a sequence of sets $(E_k)_k$ in $\mathcal{L}_0$ with $\lim_k v(m_n)(E_k \Delta E) = 0$ for every $n \in \mathbb{N}$, where Δ and $v(\cdot)$ denote the symmetric difference and the semivariation, respectively. Note that, if the m_n's are σ-additive, then $(m_n)_n$ has the approximation property (see also Brooks (1974), Corollary 2).

The following result was proved using the sliding hump technique.

Theorem 2.134 (see Brooks (1974), Theorem 3 and Corollary 4) *Let $\mathcal{L}$ be an algebra of sets, X be a Banach space and $m_n : \sigma(\mathcal{L}) \to X$, $n \in \mathbb{N}$, be a sequence of σ-additive measures (resp. finitely additive (s)-bounded measures having the approximation property), such that*

$$m(E) = \lim_n m_n(E) \text{ exists in } X \text{ for every } E \in \mathcal{L}.$$

Then the limit $m(E) = \lim_n m_n(E)$ exists in X for every $E \in \sigma(\mathcal{L})$ if and only if $(m_n)_n$ is uniformly σ-additive (resp. uniformly (s)-bounded).

Furthermore, in Brooks (1974), Theorem 5, an interchange limit theorem for finitely additive Banach space-valued measures defined on $\mathcal{P}(\mathbb{N})$ is proved. Note that Landers and Rogge (1971b) (see Corollary 4), using the tool of Fréchet-Nikodým topologies, proved a result analogous to Theorem 2.134 for σ-additive measures with values in an abelian topological group.

By means of the sliding hump technique, Faires (1974) extended the Brooks-Jewett theorem (Theorem 1 and Corollary 1) and the Vitali-Hahn-Saks theorem (Theorem 2) to Banach space-valued (s)-bounded measures, defined in algebras, satisfying property (I) of Seever. Moreover, using the Baire category theorem, Kalton (1974) gave a Phillips-type theorem in the lattice group setting (see Theorem 2) and deduced a Nikodým boundedness and a Brooks-Jewett-type theorem for measures defined in σ-algebras and with values in an abelian topological group (see Theorem 6). Kalton (1974-1975) gave also some extension of Theorems 2.131 and 2.132 relatively to linear continuous operators, in the setting of topological vector spaces or F-spaces (see Theorems 4.1 and 4.3). Faires (1974) (see Corollary 4) proved that, if $\mathcal{L}$ is an algebra with property (I) of Seever, X is a weakly compactly generated Banach space and $m : \mathcal{L} \to X$ is bounded and finitely additive, then m is (s)-bounded. Here, "weakly compactly generated" means that there is a weakly compact subset $K \subset X$, such that X coincides with the norm closure of the convex hull of K.

Khurana (1974) extended the Phillips lemma in the context of locally convex Hausdorff topological spaces.

Thomas (1974) extended the Nikodým boundedness theorem and some Grothendieck-type results to the context of vector-valued Radon measures.

We now recall the following extension of the diagonal theorem 2.117.

Theorem 2.135 (Pap (1974), Diagonal Theorem I, see also Pap (1995), Corollary 11.3) *Let* X *be a semigroup,* $f:X\to\mathbb{R}$ *be such that* $f(x+y)\le f(x)+f(y)$ *and* $f(x+y)\ge |f(x)-f(y)|$ *for every* x, $y\in X$, *and* $(x_{i,j})_{i,j}$ *be a double sequence in* X, *with* $\lim_j f(x_{i,j})=0$ *for each* $i\in\mathbb{N}$. *Then there exist an infinite subset* $P\subset\mathbb{N}$ *and a subset (finite or infinite)* $J\subset P$ *with*

$$\sum_{j\in J} f(x_{i,j})<+\infty$$

and

$$f\left(\sum_{j\in J} x_{i,j}\right)\ge\frac{1}{2}f(x_{i,i})$$

for every $i\in P$.

Note that Theorem 2.135 is a proper extension of Theorem 2.117 (see also Pap (1974), Remarks a) and b) on Diagonal Theorem I). Some similar diagonal theorems, which have applications to extensions of the Schur theorem and integration can be found in Hadžić and Pap (1975), Theorems 1-5, and are used to prove some result about equicontinuity of sequences of functionals (see Antosik (1976), Theorem 1) or to give shorter proofs of Theorem 2.131 and of the Bessaga-Pełczyński theorem (see Swartz (1978), Theorem 2 and Lemma 3), or more direct proofs of Rosenthal-type theorems (see Antosik and Swartz (1984b), Corollary 7), of the Orlicz-Pettis theorem (see Antosik and Swartz (1984b), Theorem 9) or of Theorem 2.132 (see Antosik and Swartz (1984b), Corollary 6).

Oberle (1974a) (see Theorem 2) proved that, if $\mathcal{L}$ is a ring of a nonempty set G, X is a Banach space, $m:\mathcal{L}\to X$ is a finitely additive (s)-bounded measure and $\nu:\mathcal{L}\to\mathbb{R}$ is a countably additive measure of bounded variation, then m is absolutely continuous with respect to ν if and only if $x'\circ m$ is absolutely continuous with respect to ν for every $x'\in X'$. Some characterizations of concepts similar to those of uniform (s)-boundedness were given by diagonal methods in Oberle (1974a), Theorem 1 and in Oberle (1974b), Theorem 1, and some result on weak relative compactness was proved in Oberle (1974b), Proposition 3.

Brooks and Dinculeanu (1974) (see Theorem 3.1), together with results on strong and weak compactness for measures, proved the following:

Theorem 2.136 *Let Σ be a δ-ring, and suppose that for each $E \in \Sigma$ there exists a positive real-valued σ-additive measure m_E defined on $\Sigma_E := \{E \cap A: A \in \Sigma\}$ such that m_E and m_F are mutually absolutely continuous on Σ_E whenever $E \subset F$. Then there is a positive σ-additive measure ν on Σ such that m_E is absolutely continuous with respect to ν on Σ_E for each $E \in \Sigma$, and $\nu(A) = 0$ if and only if $m_E(E \cap A) = 0$ for every $E \in \Sigma$.*

Labuda (1975a) (see Theorem 2.8) extended Theorem 2.132 to locally convex complete spaces and to Hausdorff topological vector spaces (see Labuda (1975b), Theorem and Corollaries 1 and 2). Moreover Drewnowski (1975) proved a Rosenthal-type lemma and deduced some extension of Theorem 2.132 in the Hausdorff topological vector space setting, concerning both measures and linear operators (see also Drewnowski (1975), Corollary and Remark 1). Some similar extension of Theorems 2.131 and 2.132 in the Hausdorff topological vector space setting, in which it is dealt with the spaces of the type $c_0(\Lambda)$ and $l^\infty(\Lambda)$, where the cardinality of Λ is not necessarily countable, were proved in Labuda (1976) (see Theorem 1 and Corollary).

Klimkin (1975) (see Theorem 1), using a diagonal argument, established a general and sufficient condition in order that pointwise boundedness of a sequence of measures with values in an abelian topological group and defined on a ring implies equiboundedness of the sequence on the generated σ-ring.

Adamski, Gänssler and Kaiser (1976) (see Theorems 3.1-3.5), using diagonal arguments, gave some extensions of compactness results and Dieudonné-type theorems in the following cases.

(a) G is regular, $(m_n)_n$ is a sequence of measures, pointwise convergent on the class τ of the open subsets of G and τ-smooth (a measure m defined on the Borel subsets of G is said to be *τ-smooth* iff for every $\varepsilon > 0$ and for every subclass τ_0 of τ with the property that for each A, $B \in \tau_0$ there is $C \in \tau_0$ with $C \subset A \cap B$, there exists $E \in \tau_0$ with

$$\sup\{|m(B)|: B \text{ is a Borel set, } B \subset \bigcup_{F \in \tau_0} (F \setminus E)\} < \varepsilon).$$

(b) G is completely regular and $(m_n)_n$ is a sequence of τ-smooth Borel measures and converges on the topology τ_1 generated by the sets of the type $f^{-1}(\{0\})$, as f varies in $C(G)$.

(c) G is regular and $(m_n)_n$ is a sequence of τ-smooth Borel measures, pointwise convergent on the class of regular open subsets of G

(see also Adamski, Gänssler and Kaiser (1976), Corollaries 4.2-4.6, 4.9-4.10).

Faires (1976), by means of the sliding hump argument and a diagonal lemma (Lemma 1.1) extended the Nikodým boundedness theorem for Banach space-valued finitely additive measures, defined in algebras endowed with the Seever property (I) (Theorem 2.1). The same paper contains also a historical survey about the problem of characterizing the algebras for which the classical limit theorems hold. Moreover, in Faires (1976), Theorem 3.1, a Brooks-Jewett-type theorem for topological group-valued measures was proved, even for measures defined on algebras with property (I).

Pap (1976a), using a diagonal theorem, gave an extension of the Nikodým boundedness theorem for 1-triangular (s)-bounded set functions, defined on a σ-ring and taking values in a suitable ordered group. The same author, using an Antosik-type diagonal theorem (see Pap (1976b), Theorems 1-3), proved a version of the Nikodým boundedness and of the Brooks-Jewett theorem for finitely additive measures taking values in an abelian topological group, and defined in an M-lattice (recall that an *M-lattice* is a σ-Dedekind complete lattice $\mathcal{E}$ with the property that

$$x \wedge (\vee_{n\in\mathbb{N}} y_n) = \vee_{n\in\mathbb{N}} (x \wedge y_n)$$

whenever $(y_n)_n$ is a sequence in $\mathcal{E}$).

We now recall the following results.

Lemma 2.137 (Swartz (1977), Lemma 1) *Let $S_n : \mathbb{N} \to X$, $n \in \mathbb{N}$, be such that*

(i) $\sum_{k=1}^{\infty} S_n(k)$ is unconditionally convergent in X for each $n \in \mathbb{N}$ and

(ii) $\sup\{\|\sum_{k\in E} S_n(k)\| : n\in\mathbb{N}\} < +\infty$ *for every* $E\subset\mathbb{N}$.

Then

$$\sup\{\|\sum_{k\in E} S_n(k)\| : E\subset\mathbb{N}, n\in\mathbb{N}\} < +\infty.$$

Observe that in Lemma 2.137 the Baire category theorem is used.

Lemma 2.138 (Swartz (1977), Lemma 3) *Let* G *be a nonempty set,* $\mathcal{L}$ *be an algebra of subsets of* G, $m_n : \mathcal{L}\to X$, $n\in\mathbb{N}$, *be a sequence of* (s)*-bounded measures. Then*

$$\sup\{\|m_n(E)\| : E\subset\mathbb{N}, n\in\mathbb{N}\} < +\infty$$

if and only if

$$\sup\{\|m_n(E_j)\| : j,n\in\mathbb{N}\} < +\infty$$

for every disjoint sequence $(E_j)_j$ *in* $\mathcal{L}$.

The following version of the Nikodým boundedness theorem holds.

Theorem 2.139 (Swartz (1977), Theorem 4) *Let* $m_n : \mathcal{L}\to X$, $n\in\mathbb{N}$, *be a sequence of* (s)*-bounded measures, such that*

$$\sup\{\|\sum_{j=1}^{\infty} m_n(E_j)\| : n\in\mathbb{N}\} < +\infty$$

for every disjoint sequence $(E_j)_j$ *in* $\mathcal{L}$.

Then $\sup\{\|m_n(E)\| : E\in\mathcal{L},\ n\in\mathbb{N}\} < +\infty$.

Moreover, Swartz (1977) (see Remark 6) proved that the Nikodým boundedness theorem for σ-additive real-valued measures defined in σ-algebras is equivalent to the fact that sequences in l^1 are $\sigma(l^1,\mathcal{C})$-bounded if and only if they are norm bounded, where $\mathcal{C}$ is the linear subspace of l^∞ spanned by the set $\{\chi_E : E\subset\mathbb{N}\}$.

Theorem 2.140 (Swartz (1977), Theorem 10) *Let* $S_n : \mathbb{N} \to X$, $n \in \mathbb{N}$, *be such that* $\sum_{k=1}^{\infty} S_n(k)$ *is unconditionally convergent. Suppose that for each* $E \subset \mathbb{N}$ *limit* $\lim_n \sum_{k \in E} S_n(k)$ *exists in* X. *If* $S(k) = \lim_n S_n(k)$, *then*

(a) $\sum_{k=1}^{\infty} S(k)$ *is unconditionally convergent in* X,

(b) *For each* $\varepsilon > 0$ *there is a finite set* $E_0 \subset \mathbb{N}$ *such that* $\|\sum_{k=1}^{\infty} S_n(k)\| \leq \varepsilon$ *for every n and* $E \cap E_0 = \varnothing$,

(c) $\lim_n \sum_{k \in E} S_n(k) = \sum_{k \in E} S(k)$ *uniformly with respect to* $E \subset \mathbb{N}$.

Theorem 2.141 (Swartz (1977), Theorem 11) *Let* G *be a nonempty set,* $\mathcal{L}$ *be an algebra of subsets of* G *and* $m_n : \mathcal{L} \to X$, $n \in \mathbb{N}$, *be a sequence of* (s)*-bounded measures, such that* $\lim_n \sum_{j=1}^{\infty} m_n(E_j)$ *exists in* X *for every disjoint sequence* $(E_j)_j$ *in* $\mathcal{L}$.

Then $m(E) := \lim_n m_n(E)$ *defines an* (s)*-bounded measure on* $\mathcal{L}$, *such that*

$$\lim_n \sum_{j=1}^{\infty} m_n(E_j) = \sum_{j=1}^{\infty} m(E_j)$$

for every disjoint sequence $(E_j)_j$ *in* $\mathcal{L}$.

Moreover in Swartz (1977) it is proved that the Brooks-Jewett theorem for Banach space-valued set functions defined on a σ-algebra is a consequence of Theorem 2.141, using some technical lemmas which are deduced by a diagonal argument. Furthermore Swartz (1977) (see Theorem 15) proved a similar suitable version of the Vitali-Hahn-Saks theorems for absolutely continuous measures defined on algebras.

Khurana (1977) (see Theorem 1) extended Proposition 2.121 to measures with values in a normed abelian group.

Graves (1977) published a survey on the theory of vector measures, containing different kinds of theorems in the locally convex space setting (decomposition and extension theorems, Radon-Nikodým theorem, properties of the range of measures, Bartle-Dunford-Schwartz theorems on weak compactness, Orlicz-Pettis and uniform boundedness theorems).

Simons (1978) proved some Grothendieck-type theorems in the context of locally convex space-valued measures and operators, defined on rings of sets and vector lattices respectively, and gave the following characterization.

Theorem 2.142 (see Simons (1978), Theorem 10) *Let X be a Hausdorff complete locally convex topological vector space, and let $\mathcal{L}$ be a ring. A finitely additive measure $m:\mathcal{L}\to X$ is (s)-bounded if and only if its range $m(\mathcal{L})$ is relatively compact with respect to the topology $\sigma(X,X^{'})$.*

Bogdan and Oberle (1978) published a survey on convergence theorems, weak compactness, decompositions and extensions for finitely and σ-additive vector measures.

Turpin (1978) gave a negative answer to the question whether the Nikodým boundedness theorem holds for every topological group, which is not necessarily a locally convex topological vector space, and gave some counter examples.

Khurana (1979) (see Theorems 3 and 4), using the tool of submeasures, proved the convergence and boundedness Nikodým theorems for finitely additive measure defined on algebras satisfying property (I) of Seever.

Orlicz and Urbański (1980) (see Theorem 2.1), using a diagonal argument, proved a Brooks-Jewett-type theorem for finitely additive measures defined in a σ-ring and with values in an abelian Hausdorff topological semigroup endowed with a suitable uniform structure. This results was improved in Drewnowski (1981) (see Theorems 1 and 2), by studying certain properties of spaces endowed with uniformities.

Kupka (1980) gave an extension of the Rosenthal lemma for regular Banach space-valued measures (Lemma 4.2) and deduced, by sliding hump-type methods, a version of the Nikodým boundedness theorem (Theorem 5.1), of the Nikodým convergence theorem (Theorem 5.2), a characterization of weak convergence (Corollary 5.3) and weak (sequential) compactness (Corollary 5.4).

Graves and Ruess (1980) (see Theorem 7), by means of topological linearizations of measures, universal measure representations and locally convex structure properties of suitable topologies, proved the following:

Theorem 2.143 *For a locally convex space* X *and a subset* $\mathcal{K}$ *of the space of* X*-valued* σ*-additive measures defined on a* σ*-algebra* Σ, $\mathcal{K}$ *is relatively compact in the topology of pointwise convergence on each* $E \in \Sigma$ *if and only if* $\mathcal{K}(E) := \{m(E): m \in \mathcal{K}\}$ *is relatively compact for each* $E \in \Sigma$ *and* $\mathcal{K}$ *is uniformly* σ*-additive.*

Using Drewnowski-type techniques, Bilyeu and Lewis (1980) proved some results on weak compactness of spaces of measures related with (uniform) differentiability (see Theorems 2.5 and 2.8) and deduced Vitali-Hahn-Saks (see Theorems 3.2. 3.3, 3.3L and 3.5), Phillips (see Lemma 3.4) and Brooks-Jewett-type theorems in the Banach space context (see Theorem 3.6).

1.2.9. The Biting Lemma

Brooks and Chacon (1980) proved the *biting lemma*, which is a very useful tool in investigating limit theorems and related topics. Let Σ be a σ-algebra of a nonempty abstract set G. We say that a sequence $m_n : \Sigma \to \mathbb{R}$, $n \in \mathbb{N}$ of absolutely continuous measures with respect to a finitely additive, non-negative and bounded real-valued measure ν, *converges* w^2 to a set function $m : \Sigma \to \mathbb{R}$, iff there exist a decreasing sequence $(B_n)_n$ in Σ and a set $B \in \Sigma$, with $\| m \| (B) = 0$, $\lim_n \nu(B_n) = 0$ and such that, if $A \subset G \setminus B_i$ for some $i \in \mathbb{N}$, then $\lim_n m_n(A) = m(A)$.

Lemma 2.144 (Brooks and Chacon (1980), Biting Lemma) *Let* $m_n : \Sigma \to \mathbb{R}$ *be an equibounded sequence of finitely additive measures, each of which absolutely continuous with respect to a non-negative and bounded finitely additive measure* $\nu : \Sigma \to \mathbb{R}$*. Then the following results hold.*

2.144.1) *If* ε, $\sigma > 0$ *then there exist a set* $C_{\varepsilon,\sigma} \in \Sigma$ *and a subsequence* $(m_{n_i})_i$ *of measures (depending on* ε, σ*) satisfying the following conditions:*

(i) $\nu(C_{\varepsilon,\sigma}) < \varepsilon$;

(ii) There exists a positive real number δ *with* $|m_{n_i}(A)| < \sigma$ *for every* $i \in \mathbb{N}$ *whenever* $\nu(A) < \delta$ *and* $A \subset G \setminus C_{\varepsilon,\sigma}$.

2.144.2) *If* ν *and* m_n, $n \in \mathbb{N}$, *are* σ*-additive, then there is a subsequence* $(m_{n_i})_i$ *such that for every* $\varepsilon > 0$ *there is a set* $B_\varepsilon \in \Sigma$ *with* $\nu(B_\varepsilon) > \varepsilon$ *and* $(m_{n_i})_i$ *is* ν*-equiabsolutely continuous on* $G \setminus B_\varepsilon$.

2.144.3) *If* ν *and* m_n, $n \in \mathbb{N}$, *are* σ*-additive, then there exists a subsequence* $(m_{n_i})_i$, *convergent* w^2 *to a* σ*-additive measure* m, *absolutely continuous with respect to* ν. *Moreover,* m *is the unique* w^2*-limit of this subsequence. If the* m_{n_i}*'s are positive, then* m *is positive.*

In Brooks and Chacon (1980), Theorem 1, it was proved that the Vitali-Hahn-Saks Theorem 2.116 for Banach space-valued measures can be deduced from the biting Lemma 2.144.

Another consequence of Lemma 2.144 is the following Dieudonné-type theorem.

Theorem 2.145 (Brooks (1980), Theorem; Brooks and Chacon (1980), Theorem 2) *Let* G *be a normal Hausdorff topological space,* $(m_n)_n$ *be an equibounded sequence of finitely additive regular Banach space-valued set functions, defined on the Borel subsets of* G. *If* $\lim_n m_n(U)$ *exists for every open set* U, *then* $\lim_n m_n(E)$ *exists for each Borel set* E. *Moreover, if* G *is compact, then the assumption of equiboundedness is superfluous.*

Note that, in Brooks (1980), Theorem 2.145 is proved as a consequence of the following Rosenthal-type lemma.

Lemma 2.146 (Brooks (1980), Lemma) *Let* $m_n : \Sigma \to \mathbb{R}$ *be an equibounded sequence of* σ*-additive measures, each of which absolutely continuous with respect to a countably additive non-negative real-valued measure* ν. *Then there exists a subsequence* $(m_{n_i})_i$ *of* $(m_n)_n$ *such that for every* $\varepsilon > 0$ *there is a set*

$A_\varepsilon \in \Sigma$ such that $\nu(A_\varepsilon) < \varepsilon$ and $(m_{n_i})_i$ is equiabsolutely continuous with respect to ν on $G \setminus A_\varepsilon$.

Luu (1989) (see Lemma 2.2) gave an extension of the biting lemma in the Banach space setting, and deduced also the following theorem on existence of the best approximation in L^1.

Theorem 2.147 (Luu (1989), Theorem 3.1) *Let (G,Σ,P) be a probability space, X be a reflexive Banach space, H be a convex subset of $L^1(G,\Sigma,P)$, closed with respect to convergence in probability. Then for every $f \in L^1(G,\Sigma,P)$ there exist $h \in H$ which minimizes the L^1-distance of f from H, that is $\| f-h \|_1 = \min_{\phi \in H} \int_G \| f-\phi \| \, dP$.*

Moreover, Brooks (2005) (see Theorem 2.1) extended the biting lemma to measures defined in a σ-algebra and with values in a reflexive Banach space (Brooks (2005), Theorem 2.1) and investigated some aspects of equicontinuity, equiabsolute continuity and weak compactness in the context of C^*-algebras (see Brooks (2005), Theorems 3.3-3.5).

Finally, Florescu and Godet-Thobie (2006) (see Theorems 4 and 5) proved a version of the biting lemma for unbounded sequences in L^1 and gave some applications to relative compactness with respect to convergence in measure (see Florescu and Godet-Thobie (2006), Theorem 7) and to a differential problem, in which it is dealt with the temperature of a material of constant conductivity subjected to a heat force (see also Florescu and Godet-Thobie (2006), Theorems 8-10).

Pap (1981) (see Theorems 2 and 3), using diagonal arguments, proved some version of the uniform boundedness theorem for regular (s)-bounded measures defined on a Borel σ-algebra and taking values in commutative semigroups, in the spirit of Dieudonné (1951a).

Guariglia (1982) (see Theorem 2.3 and 3.3), using a Drewnowski-type result and a diagonal argument, proved some versions of the Nikodým uniform boundedness principle for finitely additive measures defined on quasi-σ-rings.

Swartz (1982) (see also Theorems 3 and 4) proved that some Brooks-Jewett theorems, Rosenthal-type theorems, some results given in Labuda (1972) and Theorem 2.132 can be obtained from diagonal-type theorems, by means of sliding hump-type techniques.

Sazhenkov (1982) (see Theorems 1-3), using diagonal methods, proved a version of the Nikodým boundedness theorem according to the approach of Dieudonné (1951a), for measures with values in an abelian topological group, including regularity relative to both compact and closed sets, by means of a unified approach, in which it is dealt with a topological space G, an algebra $\mathcal{L} \subset \mathcal{P}(G)$ and two subclasses of $\mathcal{L}$, $\mathcal{G}$ and $\mathcal{H}$, satisfying the following properties:

(a) If $H_1, H_2 \in \mathcal{H}$ are disjoint, then there are two disjoint sets $U_1, U_2 \in \mathcal{G}$ with $U_1 \supset H_1$ and $U_2 \supset H_2$;

(b) $\emptyset \in \mathcal{H}$, $\{G \setminus H : H \in \mathcal{H}\} \subset \mathcal{G}$;

(c) $\mathcal{G}$ is closed under countable unions and finite intersections.

Sazhenkov (1986) (see Main Theorem), using suitable properties of groups, proved the equivalence between two types of the Nikodým boundedness theorem for (s)-bounded finitely additive topological group-valued measures, defined on rings of subsets of an abstract nonempty set G and with values in an abelian topological group. Some other version of the Nikodým boundedness theorem for real-valued (s)-bounded finitely additive measures defined on rings of sets and of the Brooks-Jewett theorem for real-valued set functions defined on σ-rings can be found in Dobrakov (1984), Theorems 1.R and 2, where diagonal methods are used.

Fox and Morales (1983), using sliding hump techniques, proved some Nikodým convergence theorem (see Theorem 3.1 and Corollary 3.2) and Vitali-Hahn-Saks-type theorem (see Theorem 4.1 and Corollary 4.2) for σ-additive measures, defined on σ-rings and with values in a Hausdorff complete uniform semigroup.

1.2.10. Basic Matrix Theorems

We now present some versions of basic matrix theorems existing in the literature, proved with the sliding hump technique, which extend to the topological group context some Hahn-Schur-type results, and by means of which it is possible to deduce limit theorems for topological group-valued measures.

Lemma 2.148 (see also Swartz (1996b), Lemma 2.2.1, Antosik and Swartz (1985a), Lemma 2.1) *Let R be an abelian topological group and $(x_{i,j})_{i,j}$ be an infinite matrix of elements of R. If $\lim_i x_{i,j} = 0$ for every $j \in \mathbb{N}$, $\lim_j x_{i,j} = 0$ for every $i \in \mathbb{N}$ and if $(U_k)_k$ is a sequence of neighborhoods of 0 in R, then there exists an increasing sequence of positive integers $(p_i)_i$ with $x_{p_i,p_j}, x_{p_j,p_i} \in U_j$ for every $j > i$.*

Theorem 2.149 (Swartz (1996b), Theorem 2.2.2, see also Antosik and Swartz (1983), Theorem 2 for the normed case, Antosik and Swartz (1984a), Theorem 1 and Corollary 2, Antosik and Swartz (1985a), Theorem 2.2) *Let R be an abelian topological group and $(x_{i,j})_{i,j}$ be an infinite matrix of elements of R. Assume that*

(a) $\lim_i x_{i,j} = x_j$ exists for each $j \in \mathbb{N}$ and

(b) For each strictly increasing sequence of positive integers $(q_j)_j$ there is a subsequence $(n_j)_j$ of $(q_j)_j$ such that the sequence $\left(\sum_{j=1}^{\infty} x_{i,n_j}\right)_i$ is Cauchy.

Then $\lim_i x_{i,j} = x_j$ uniformly with respect to $j \in \mathbb{N}$. In particular,

$$\lim_i (\lim_j x_{i,j}) = \lim_j (\lim_i x_{i,j}) = 0 \text{ and } \lim_i x_{i,i} = 0.$$

Proof: If it is not true that $\lim_i x_{i,j} = x_j$ uniformly with respect to $j \in \mathbb{N}$, then there are a closed, symmetric neighborhood U_0 of 0 and two strictly increasing sequences of positive integers $(n_k)_k$ and $(r_k)_k$ with $x_{n_k,r_k} - x_{n_k} \notin U_0$ for each $k \in \mathbb{N}$. Pick a closed, symmetric neighborhood U_1 of 0 with $2U_1 \subset U_0$ and set $i_1 = n_1$, $j_1 = r_1$. Since $x_{i_1,j_1} - x_{j_1} = (x_{i_1,j_1} - x_{i,j_1}) + (x_{i,j_1} - x_{j_1})$, there is a positive integer i_0 with $x_{i_1,j_1} - x_{i,j_1} \notin U_1$ for each $i \geq i_0$. Choose $k_0 \in \mathbb{N}$ with

$n_{k_0} > \max\{i_1, i_0\}$, $r_{k_0} > j_1$ and set $i_2 = n_{k_0}$, $j_2 = r_{k_0}$. Then, $x_{i_1,j_1} - x_{i_2,j_1} \notin U_1$ and $x_{i_2,j_2} - x_{j_2} \notin U_0$. By proceeding in this way, it is possible to find two strictly increasing sequences $(i_k)_k$ and $(j_k)_k$ in $\mathbb{N}$ with $x_{i_k,j_k} - x_{i_{k+1},j_k} \notin U_1$ and $x_{i_k,j_k} - x_{j_k} \notin U_0$ for every $k \in \mathbb{N}$. For each k, $l \in \mathbb{N}$, set $z_{k,l} = x_{i_k,j_l} - x_{i_{k+1},j_l}$. Note that $z_{k,k} \notin U_1$ for every $k \in \mathbb{N}$.

Choose a sequence $(U_n)_n$ of closed, symmetric neighborhoods of 0, with $2U_n \subset U_{n-1}$ for each $n \in \mathbb{N}$. Observe that $\sum_{j=3}^{q} U_j \subset U_2$ for any $q \geq 3$. By hypothesis, we get $\lim_k z_{k,l} = 0$ for each $l \in \mathbb{N}$ and $\lim_l z_{k,l} = 0$ for each $k \in \mathbb{N}$. By Lemma 2.148 there is a strictly increasing sequence $(p_k)_k$ in $\mathbb{N}$ with z_{p_k,p_l}, $z_{p_l,p_k} \in U_{k+2}$ for every $k > l$. Again by hypothesis, the sequence $(p_k)_k$ has a subsequence $(q_k)_k$ such that the sequence $\left(\sum_{k=1}^{\infty} x_{i,q_k}\right)_i$ is Cauchy, and so

$\lim_k \left(\sum_{l=1}^{\infty} z_{q_k,q_l}\right) = 0$. Thus there exists an integer k_0 with $\sum_{l=1}^{\infty} z_{q_{k_0},q_l} \in U_2$. Then for every $r > k_0$ we get:

$$\sum_{l \in [1,r], l \neq k_0} z_{q_{k_0},q_l} = \sum_{l=1}^{k_0-1} z_{q_{k_0},q_l} + \sum_{l=k_0+1}^{r} z_{q_{k_0},q_l} \in \sum_{l=1}^{k_0-1} U_{k_0+2} + \sum_{l=k_0+1}^{r} U_{l+2} \subset \sum_{l=3}^{r+2} U_l \subset U_2,$$

and hence $z_{k_0} := \sum_{l \neq k_0} z_{q_{k_0},q_l} \in U_2$. Thus we obtain

$$z_{q_{k_0},q_{k_0}} = \sum_{l=1}^{\infty} z_{q_{k_0},q_l} - z_{k_0} \in 2U_2 \subset U_1,$$

that is a contradiction. This ends the proof. □

A matrix satisfying the hypotheses of Theorem 2.149 is called also a $\mathcal{K}$*-matrix*. By means of Theorem 2.149, Antosik and Swartz (1983) proved a version of the uniform boundedness principle in the normed space context (Theorem 3) and obtained a Nikodým convergence theorem (Corollary 8) and Nikodým boundedness theorem in the finitely additive case (see Theorem 3) for normed space-valued measures, defined on σ-algebra.

A consequence of Theorem 2.149 is the following:

Corollary 2.150 (Swartz (1996b), Corollary 2.3) *Let* R *be an abelian topological group and* $x_{i,j} \in R$ *for every* i, $j \in \mathbb{N}$. *If every strictly increasing sequence* $(q_j)_j$ *in* $\mathbb{N}$ *has a subsequence* $(n_j)_j$ *such that*

(*a*) $\lim_i x_{n_i,n_j} = 0$ *for every* $j \in \mathbb{N}$ *and*

(*b*) $\lim_i \left(\sum_{j=1}^{\infty} x_{n_i,n_j}\right) = 0,$

then $\lim_i x_{i,i} = 0$.

We now recall the following extension of Theorem 2.149.

Theorem 2.151 (Swartz (1996b), Theorem 2.2.4) *Under the same above notations and assumptions, let* $x_{i,j} \in R$, *and assume that*

(*a*) $\lim_i x_{i,j} = x_j$ *exists for every* $j \in \mathbb{N}$ *and*

(*b*) *For each strictly increasing sequence of positive integers* $(q_j)_j$ *there exist a subsequence* $(n_j)_j$ *and a choice of signs* $s_j \in \{-1,1\}$ *such that the sequence*

$$\left(\sum_{j=1}^{\infty} s_j x_{i,n_j}\right)_i$$

is Cauchy.

Then $\lim_i x_{i,j} = x_j$ *uniformly with respect to* $j \in \mathbb{N}$. *In particular,*

$$\lim_i (\lim_j x_{i,j}) = \lim_j (\lim_i x_{i,j}) = 0 \text{ and } \lim_i x_{i,i} = 0.$$

Observe that Theorem 2.151 is an essential improvement of 2.149, because there are matrices which satisfy the hypotheses of Theorem 2.151 and are not $\mathcal{K}$-matrices, while of course every $\mathcal{K}$-matrix fulfils the hypotheses of Theorem 2.151 (see also Swartz (1996b)).

Swartz (1984) (see Theorem 3) used Theorem 2.149 to prove some extension of the Orlicz-Pettis theorem.

In Swartz (1996b) it is shown that the Nikodým convergence and the Brooks-Jewett theorem for topological group-valued measures can be deduced by Theorem 2.149.

Let R be an abelian Hausdorff topological group, and Σ be a σ-algebra. We first state the following

Lemma 2.152 (Swartz (1996b), Lemma 2.3.1) *Let* $m_i : \Sigma \to R$, $i \in \mathbb{N}$, *be a sequence of* σ*-additive measures. Then the following are equivalent:*

(i) $(m_i)_i$ *is uniformly* σ*-additive;*

(ii) $\lim_j m_i(E_j) = 0$ *uniformly with respect to* $i \in \mathbb{N}$ *for each decreasing sequence* $(E_j)_j$ *in* Σ *with* $\bigcap_{j=1}^{\infty} E_j = \emptyset$;

(iii) $\lim_j m_i(E_j) = 0$ *uniformly with respect to* $i \in \mathbb{N}$ *for any disjoint sequence* $(E_j)_j$ *in* Σ.

We now turn to the Nikodým convergence theorem for topological group-valued measures.

Theorem 2.153 (see Swartz (1996b), Theorem 2.3.2) *If* $m_i : \Sigma \to R$, $i \in \mathbb{N}$, *is a sequence of* σ*-additive measures such that* $m(E) = \lim_i m_i(E)$ *exists in* R *for every* $E \in \Sigma$, *then* $(m_i)_i$ *is uniformly* σ*-additive and* m *is* σ*-additive.*

A consequence of Theorem 2.153 and Drewnowski (1972b), Proposition 2) is the following Brooks-Jewett-type theorem.

Theorem 2.154 (Drewnowski (1972b), Theorem, p. 726, see also Swartz (1996b), Theorem 2.3.5) *Let R be an abelian Hausdorff topological group and $m_i : \Sigma \to R$, $i \in \mathbb{N}$, be a sequence of (s)-bounded measures. If $m(E) = \lim_i m_i(E)$ exists in R for every $E \subset \mathbb{N}$, then $(m_i)_i$ is uniformly (s)-bounded on Σ and m is (s)-bounded on Σ.*

Note that the uniform boundedness principle is a consequence of Theorem 2.149 (see also Antosik and Swartz (1983), Theorem 1 for the normed case). In Antosik and Swartz (1983) there are also some version of the Banach-Steinhaus theorem (see Theorems 3, $3'$, Corollaries 4,5), the Nikodým convergence theorem (Corollary 8) and the Brooks-Jewett theorem (Corollary 9) in the normed space context).

Let us denote the *double limit* of a double sequence $(x_{i,j})_{i,j}$ by $\lim_{i,j} x_{i,j}$ (that is we say that $\lim_{i,j} x_{i,j} = x$ iff for each neighborhood U of 0 there exists an integer n_0 with $x_{i,j} - x \in U$ whenever i, $j \geq n_0$), and the *sum* of a double series $\sum_{i,j=1}^{\infty} x_{i,j}$ by

$$\lim_{m,n} \sum_{i=1}^{m} \left(\sum_{j=1}^{n} x_{i,j} \right).$$

The following Hahn-Schur-type theorem for topological group-valued double sequences holds.

Theorem 2.155 (Swartz (1996b), Theorem 8.1.1, see also Antosik and Swartz (1984a) Theorem 3 and Antosik and Swartz (1985a), Theorem 8.1) *Let R be an abelian Hausdorff topological group and $(x_{i,j})_{i,j}$ be a double sequence in R. Assume that $\sum_{j=1}^{\infty} x_{i,j}$ is subseries convergent for every $i \in \mathbb{N}$. If the limit $\lim_i \sum_{j \in E} x_{i,j}$ exists in R for every $E \subset \mathbb{N}$ and $\lim_i x_{i,j} = x_j$, then the series $\sum_{j=1}^{\infty} x_j$ is subseries convergent and*

$$\lim_i \sum_{j \in E} x_{i,j} = \sum_{j \in E} x_j$$

uniformly with respect to $E \subset \mathbb{N}$.

A consequence of Theorem 2.155 is the following Schur-type theorem.

Corollary 2.156 (see also Antosik and Swartz (1984a), Corollary 5, and Antosik and Swartz (1985a), Corollary 8.2) *Let* $(x_{i,j})_{i,j}$ *be a real-valued double sequence. If the limit* $\lim_i \sum_{j \in E} x_{i,j}$ *exists for every subset* $E \subset \mathbb{N}$ *and if* $\lim_i x_{i,j} = x_j$, *then*

$$(x_j)_j \in l^1,\ \lim_i \sum_{j=1}^{\infty} | x_{i,j} - x_j | = 0$$

and the series $\sum_{j=1}^{\infty} | x_{i,j} |$ *are convergent uniformly with respect to* i.

Another similar Schur-type theorem, which uses analogous techniques, can be found in Swartz (1996a), Theorem 2. We recall also the following interchange theorem for topological group-valued series.

Theorem 2.157 (Swartz (1996b), Theorem 8.5.1, see also Swartz (1992a), Theorem 1)

Let R *be an abelian Hausdorff topological group, and let us assume that the series* $\sum_{i=1}^{\infty} (\sum_{j=1}^{\infty} x_{i,q_j})$

converges in R *for every strictly increasing sequence of positive integers* $(q_j)_j$.

Then the double series $\sum_{i,j=1}^{\infty} x_{i,j}$ *converges in* R, *and*

$$\sum_{i,j=1}^{\infty} x_{i,j} = \sum_{i=1}^{\infty} (\sum_{j=1}^{\infty} x_{i,j}) = \sum_{j=1}^{\infty} (\sum_{i=1}^{\infty} x_{i,j}).$$

Using diagonal techniques, some extension of Theorem 2.157 can be found, for instance, in Stuart (1994), Theorem 2.5 and Swartz (2002), Theorems 1.2 and 1.7, Corollary 1.6. Some other kinds of matrix theorems, uniform boundedness principle, Hahn-Schur, Orlicz-Pettis-type theorems and related topics can be found, for example, in Antosik and Swartz (1985), Boos and Cass (2000), Swartz

(1996b) (see also the related bibliographies), Samaratunga and Sember (1988), Theorems 2.5, 4.7, 4.8; Antosik and Swartz (1991), Theorems 7, 14, 15 and Corollary 10; Li and Swartz (1993), Theorem 15; Qu and Wu (2002), Theorems 1 and 2; Wu, Luo and Lu (2005), Theorem 1, Corollaries 1 and 2; Aizpuru, Armario, García-Pacheco and Pérez-Fernández (2011), Theorems 4.1, 4.2 and Corollary 4.1; Florencio, Paúl and Virués (1992), Máté (1991), Stein (1974), Stuart and Swartz (2005), Swartz (1983, 1985, 1987, 1989b-c, 1991, 1993 and 2013), Swartz and Stuart (1998), Thiago Bernardino (2009) and Wnuk (2009).

We recall also the following version of the Banach-Steinhaus theorem (see also Swartz (1996b), Theorem 5.2.2).

Theorem 2.158 *Let X be a normed group, Y be a Hausdorff topological vector space, $T_i : X \to Y$, $i \in \mathbb{N}$, be a sequence of linear continuous operators such that $\lim_i T_i(x) = T(x)$ exists in Y for every $x \in X$. Then we get:*

(a) *T is continuous,*

(b) $\lim_i T_i(x) = T(x)$ *uniformly with respect to x in a compact subset of X, and*

(c) *If $(x_j)_j$ converges to 0, then $\lim_j T_i(x_j) = 0$ uniformly with respect to $i \in \mathbb{N}$.*

In Antosik and Swartz (1985a) and Swartz (1996b) (see also the related bibliography) there are also some other versions of the Banach-Steinhaus theorem in different contexts. For a related literature, see also Arvanitakis and Stein (2008), Banach and Steinhaus (1927), Carathéorory (1929), Stein (1972 and 1991), Swartz (1990) and the references therein. Some applications of the uniform boundedness principle to Approximation Theory can be found, for instance, in Butzer (1980), Dickmeis and Nessel (1981a-b and 1985).

1.2.11. Measures Defined on Algebras

In general, Brooks-Jewett, Vitali-Hahn-Saks, Nikodým convergence and Nikodým boundedness theorems do not hold, if one supposes that the set of definition of the measures involved is any algebra or any lattice without further assumptions.

In the literature there are several studies on the problem of finding necessary and/or sufficient conditions on the algebra involved in order to fulfil the limit theorems and/or the uniform boundedness theorem. Observe that the problem of finding satisfactory necessary and sufficient conditions is still open.

In the following example we show that, in general, the Nikodým convergence theorem is not true if the involved σ-additive measures are defined on an algebra.

Example 2.159 (see also de Lucia and Pap (2002), Example 2.17) Let $\mathcal{L}$ be the algebra of parts of $[0,1)$ generated by the intervals of type $[a,b)$, with $0 \le a < b \le 1$, put $m_n([a,b)) = b^n - a^n$, $n \in \mathbb{N}$, extend m_n on the whole of $\mathcal{L}$ in the natural way and denote this extension still with m_n. Note that m_n is σ-additive for every n, $\lim_n m_n([a,b)) = 0$ if $0 \le a < b < 1$ and $\lim_n m_n([a,1)) = 0$ if $0 \le a < 1$. Hence, $m(E) = \lim_n m_n(E) = 1$ if and only if 1 belongs to the closure of E and $m(E) = \lim_n m_n(E) = 0$ if and only if 1 does not belong to the closure of E: note that m is finitely additive. We now show that m is not σ-additive. Indeed, set $E_n = [1-1/n, 1-1/(n+1))$, $n \in \mathbb{N}$. Note that the E_n's are pairwise disjoint, $\bigcup_{n=1}^{\infty} E_n = [0,1)$, $m(E_n) = 0$ for every $n \in \mathbb{N}$ but $m([0,1)) = 1$, and hence m is not σ-additive.

Example 2.160 (see also Bhaskara Rao and Bhaskara Rao (1983), Examples 8.2.1 and 8.2.2, and de Lucia and Pap (2002), Examples 2.18 and 2.19) Let $\mathcal{L}$ be the algebra of the finite or cofinite subsets of $\mathbb{R}$ (or of an uncountable set which contains $\mathbb{N}$), and for every $E \in \mathcal{L}$ and $n \in \mathbb{N}$ set $m_n(E) = n$ if $n \in E$ and E is finite, $m_n(E) = -n$ if $n \in \mathbb{R} \setminus E$ and $\mathbb{R} \setminus E$ is finite, $m_n(E) = 0$ otherwise. It is not difficult to see that $\lim_n m_n(E) = 0$ and hence the sequence $(m_n(E))_n$ is order bounded for every $E \in \mathcal{L}$, but the measures m_n are not equibounded. So, in this case, the Nikodým boundedness theorem does not hold. We now claim that m_n is finitely additive for every $n \in \mathbb{N}$. Pick arbitrarily $n \in \mathbb{N}$ and A, $B \in \mathcal{L}$ with $A \cup B \in \mathcal{L}$ and $A \cap B = \emptyset$. If $A \cup B$ is finite, then both A and B are finite. If $n \in A \cup B$, then either $n \in A$ or $n \in B$. Without loss of generality, we can and do suppose that $n \in A$. Then $n \notin B$, $n \in A \cup B$ and so $m_n(A \cup B) = n = m_n(A) + m_n(B)$, since $m_n(A) = n$ and $m_n(B) = 0$. If $n \notin A \cup B$, then $n \notin A$, $n \notin B$, and hence

$m_n(A\cup B)=0=m_n(A)+m_n(B)$. If $A\cup B$ is cofinite, since $A\cap B=\emptyset$, then it is impossible that A and B are both finite or both cofinite. Without loss of generality, assume that A is finite and B is cofinite. If $n\notin A\cup B$, then $m_n(A)=0$, $m_n(B)=m_n(A\cup B)=-n$, and hence $m_n(A\cup B)=-n=m_n(A)+m_n(B)$. If $n\in A\cup B$ and $n\in A$, then $n\notin B$, and so $m_n(A)=n$, $m_n(B)=-n$, $m_n(A\cup B)=0$, and thus $m_n(A\cup B)=0=m_n(A)+m_n(B)$. If $n\in A\cup B$ and $n\notin A$, then $n\in B$, and so $m_n(A)=m_n(B)=m_n(A\cup B)=0$. Hence, $m_n(A\cup B)=0=m_n(A)+m_n(B)$. Thus, m_n is finitely additive for each $n\in\mathbb{N}$. We now claim that, if $(C_k)_k$ is a disjoint sequence in $\mathcal{L}$ with $C:=\bigcup_{k=1}^{\infty}C_k\in\mathcal{L}$, then all the C_k's are empty but a finite number. From this and finite additivity of m_n we will deduce that m_n is σ-additive for every $n\in\mathbb{N}$. If infinitely many C_k are nonempty, then C is necessarily cofinite, and so there exists at least a $k_0\in\mathbb{N}$ such that C_{k_0} is cofinite. Otherwise C should be finite or countable, but C is uncountable, since C is a cofinite subset of $\mathbb{R}$ and $\mathbb{R}$ is uncountable. Since the C_k's are pairwise disjoint, it follows that $\bigcup_{k\neq k_0}C_k$ is finite. But since infinitely many C_k, $k\neq k_0$, are nonempty (and disjoint), then $\bigcup_{k\neq k_0}C_k$ is infinite, a contradiction. Thus we get the claim, and hence σ-additivity of the m_n's. However, $(m_n)_n$ is not uniformly (s)-bounded. Indeed, if $C_k=\{k\}$, $k\in\mathbb{N}$, then $m_n(C_k)=n$ if $n=k$ and $m_n(C_k)=0$ if $n\neq k$, and hence $\sup_{n\in\mathbb{N}} m_n(C_k)=k$. Thus, in general the Brooks-Jewett theorem (and the Nikodým convergence theorem) do not hold. Now, for each $n\in\mathbb{N}$ and $E\in\mathcal{L}$, set

$$\nu(A)=\sum_{n\in A}\frac{1}{2^n}\text{ when } A \text{ is finite, and } \nu(A)=1+\sum_{n\in A}\frac{1}{2^n}\text{ when } A \text{ is cofinite.}\qquad\textbf{(1.26)}$$

We first show that ν is finitely additive: from this, arguing as above, it is possible to deduce that ν is σ-additive. Pick A, $B\in\mathcal{L}$, with $A\cap B=\emptyset$. If $A\cup B$ is finite, then it is easy to see that $\nu(A\cup B)=\nu(A)+\nu(B)$. If $A\cup B$ is cofinite, then without loss of generality suppose that A is finite and B is cofinite. Since A and B are disjoint, we get:

$$\nu(A)+\nu(B)=\sum_{n\in A}\frac{1}{2^n}+1+\sum_{n\in B}\frac{1}{2^n}=1+\sum_{n\in A\cup B}\frac{1}{2^n}=\nu(A\cup B),$$

getting finite additivity of ν. Now, fix arbitrarily $\varepsilon>0$ and $n\in\mathbb{N}$, and pick δ with $0<\delta<\frac{1}{2^n}$. If $E\in\mathcal{L}$ and $\nu(E)<\delta$, then E is finite, $n\notin E$ and hence $m_n(E)=0<\varepsilon$. Thus, every m_n is absolutely continuous with respect to ν. However, note that for every n we get $\nu(\{n\})=\frac{1}{2^n}$, but $\sup_{j\in\mathbb{N}} m_j(\{n\})\geq m_n(\{n\})=n$, and so $(m_n)_n$ is not equiabsolutely continuous with respect to ν. Thus, in general, the Vitali-Hahn-Saks theorem does not hold.

Haydon (1981) introduced the *subsequential completeness property* (shortly, (SCP)) for algebras. We say that an algebra $\mathcal{L}$ of parts of a nonempty set G *satisfies property* (SCP) iff for every disjoint sequence $(E_n)_n$ in $\mathcal{L}$ there exists an infinite subset $P\subset\mathbb{N}$ such that $\bigcup_{n\in P}E_n\in\mathcal{L}$. Note that, analogously as above, property (SCP) can be given in a natural way also for rings of subsets of G, and a ring satisfying property (SCP) is called also a *quasi-σ-ring*. If in a ring $\mathcal{L}$ the property (SCP) is required only for disjoint sequences whose union is contained in some set of $\mathcal{L}$, then we will say that $\mathcal{L}$ is a *quasi-δ-ring* (see also Constantinescu (1981a)). In Haydon (1981), Proposition 1E, an example of an algebra satisfying property (SCP) and which is not a σ-algebra is given by means of the transfinite induction. In Haydon (1981), Theorem 1F, an example of an infinite compact space K, such that $C(K)$ is a Grothendieck space with no subspace isomorphic to l^∞, is given.

Constantinescu (1981a) (see also Theorems 1.7 and 2.7), using sliding hump-type methods, proved some versions of the Nikodým boundedness theorem for regular measures, taking values in an abelian Hausdorff topological group and defined in suitable rings or quasi-δ-rings of subsets of a topological space G, satisfying (SCP). The same author extended to the context of suitable nets and filters the Nikodým convergence theorem (Constantinescu (1981b), Theorem 2.1.3) and the Dieudonné theorem (see also Constantinescu (1981b), Theorem 2.3.1 and Corollaries 2.3.2, 2.3.4, 2.3.17).

Dashiell (1981) (see Theorem 3.5 and Corollary) proved that the space of the bounded Baire one functions on [0,1] is a Grothendieck space. Moreover the same author proved that, if $\mathcal{L}$ is an algebra satisfying property (I) of Seever and such that: for every non-negative σ-additive bounded measure m on $\mathcal{L}$ not identically zero and for each sequence $(E_n)_n$ in $\mathcal{L}$ with $m(E_n)=0$ for every n, there exists $E \in \mathcal{L}$ with $E \cap E_n = \emptyset$ for every $n \in \mathbb{N}$ and $m(E) > 0$, and $(m_n)_n$ is a sequence in $\mathrm{ba}(\mathcal{L})$ such that $(m_n(E))_n$ is bounded (resp. converges) for each $E \in \mathcal{L}$, then $(m_n)_n$ is norm bounded (resp. converges for the weak topology $\sigma(\mathrm{ba}(\mathcal{L}), \mathrm{ba}(\mathcal{L})'))$ (see also Dashiell (1981), Theorem 2.6).

Using the sliding hump technique, some versions of the Brooks-Jewett theorem (Moltó (1981a), Theorem 1.3) and of the Nikodým boundedness theorem (Moltó (1981b), Theorem 1.1 and Remark 1) were given for finitely additive measures with values in an abelian complete metrizable topological group and defined in a ring $\mathcal{L}$, satisfying the following interpolation property: (f) for every pair of disjoint sequences $(A_n)_n$, $(B_n)_n$ in $\mathcal{L}$ with $A_n \cap B_q = \emptyset$ for every n, $q \in \mathbb{N}$, there exist a subsequence $(B_{n_k})_k$ and a set $A \in \mathcal{L}$ with $B_{n_k} \subset A$ and $A_k \cap A = \emptyset$ for every $k \in \mathbb{N}$, and for each subset $P \subset \mathbb{N}$ there exists $V_P \in \mathcal{L}$ with $B_{n_k} \subset V_P$ for all $k \in P$ and $B_{n_k} \cap V_P = \emptyset$ whenever $k \in \mathbb{N} \setminus P$.

As a consequence, it is proved that, if $\mathcal{L}$ is an algebra satisfying property (f) and Q^* is the *Stone space* associated with $\mathcal{L}$, that is a compact totally disconnected space such that $\mathcal{L}$ is algebrically and lattice isomorphic to the algebra $\mathcal{Q}$ of all open-closed subsets of Q^* (see also Chapter 2, Martellotti (2001) and Stone (1936, 1938 and 1949), then the space $C(Q^*)$ of all real-valued functions defined on Q^* is a Grothendieck space (see Haydon (1981), Proposition 1B), and the linear subspace of $C(Q^*)$ generated by the characteristic functions of elements of $\mathcal{Q}$ (see Moltó (1981a), Corollary 1.4). Moltó (1981a)), Example 1.6, proved that the Brooks-Jewett theorem in general does not hold, when one supposes, instead property (f), only that: (f') for any two disjoint sequences $(A_n)_n$, $(B_q)_q$ in the involved algebra $\mathcal{L}$ with $A_n \cap B_q = \emptyset$ for every n, $q \in \mathbb{N}$ there are two subsequences $(A_{n_k})_k$, $(B_{q_k})_k$ and a set $A \in \mathcal{L}$ with $B_{q_k} \subset A$ and $A_{n_k} \cap A = \emptyset$

for every $k \in \mathbb{N}$, and for each subset $P \subset \mathbb{N}$ there exists $V_P \in \mathcal{L}$ with $B_{q_k} \subset V_P$ for every $k \in P$ and $B_{q_k} \cap V_P = \emptyset$ whenever $k \in \mathbb{N} \setminus P$ (note that, assuming $(A_{n_k})_k = (A_n)_n$, we have property (f)).

We now recall some other properties of algebras (or rings) of sets, which are often used in the literature to formulate necessary or sufficient conditions to prove different types of limit theorems. Given an algebra $\mathcal{L}$ of subsets of a nonempty set G, let us denote by $B(\mathcal{L})$ the Banach space of all real-valued bounded $\mathcal{L}$-measurable functions and by $B_s(\mathcal{L})$ the subspace of $B(\mathcal{L})$ of all simple functions, endowed with the supremum norm.

An algebra $\mathcal{L}$ is said to have the *Grothendieck property* (shortly, *property* (*G*)) iff one of the following equivalent conditions is satisfied.

(*i*) Any bounded sequence $(m_n)_n$ in $B(\mathcal{L})'$, convergent pointwise on $\mathcal{L}$, is uniformly (s)-bounded.

(*ii*) The space $B(\mathcal{L})$ is a Grothendieck space (see also Schachermayer (1982)).

An algebra $\mathcal{L}$ has *property* (*BJ*) (*Brooks-Jewett property, property* (*VHS*) in the terminology of Schachermayer (1982)) iff one of the following equivalent conditions is fulfilled.

(*a*) Any sequence $(m_n)_n$ in $B(\mathcal{L})'$, convergent pointwise on $\mathcal{L}$, is uniformly (s)-bounded.

(*b*) Any sequence $(m_n)_n$ in $B(\mathcal{L})'$, which converges with respect to the $\sigma(B(\mathcal{L})', B_s(\mathcal{L}))$-topology, converges also weakly, that is with respect to the $\sigma(B(\mathcal{L})', B_s(\mathcal{L})'')$-topology.

An algebra $\mathcal{L}$ has the *Nikodým boundedness property* (shortly (NB), or *property* (*N*) in the terminology of Schachermayer (1982)) iff one of the following equivalent conditions is fulfilled.

(i) For every set $\Lambda \neq \emptyset$, any family $(m_\lambda)_{\lambda\in\Lambda}$ in $B(\mathcal{L})'$, which is bounded on the members of $\mathcal{L}$ (that is $\sup_\lambda \| m_\lambda(E) \| < +\infty$ for each $E \in \mathcal{L}$) is equibounded (namely $\sup_{\lambda\in\Lambda, E\in\mathcal{L}} \| m_\lambda(E) \| < +\infty$).

(ii) The normed space $B_s(\mathcal{L})$, endowed with the supremum norm, is barrelled.

The following results hold.

Proposition 2.161 (Schachermayer (1982), Proposition 1.2) *A subset* $\mathcal{K} = \{m_\lambda : \lambda \in \Lambda\}$ *of* $B(\mathcal{L})'$ *is relatively weakly compact if and only if* $\mathcal{K}$ *is bounded on the elements of* $\mathcal{L}$ *and* $(m_\lambda)_\lambda$ *is uniformly* (s)*-bounded.*

Theorem 2.162 (see Graves and Wheeler (1983), Theorem 3.5, and Schachermayer (1982), Theorem 2.5) *An algebra* $\mathcal{L}$ *has property* (BJ) *if and only if it has properties* (G) *and* (NB).

An algebra $\mathcal{L}$ is said to have the *Orlicz-Pettis property* (shortly, *property* (OP), see also Schachermayer (1982)) iff one of the following equivalent conditions is fulfilled.

(i) For any Banach space X, every weakly σ-additive measure $m : \mathcal{L} \to X$ is σ-additive too (with respect to the norm topology).

(ii) For each Banach space X, every weakly σ-additive bounded measure $m : \mathcal{L} \to X$ is σ-additive too (see also Schachermayer (1982)).

A Banach space X is said to be a *Rosenthal space* iff for each Banach space Y, every not weakly compact operator $T : X \to Y$ fixes a copy of l^∞, namely there is a continuous linear function $j : l^\infty \to X$ such that $T \circ j$ is an isomorphism into Y.

An algebra $\mathcal{L}$ has the *Rosenthal property* (shortly, *property* (R)) iff $B(\mathcal{L})$ is a Rosenthal space.

The following result holds.

Proposition 2.163 (Schachermayer (1982), Proposition 2.10) *Any algebra $\mathcal{L}$ having property (R) has also property (G).*

Schachermayer (1982) gave an example of an algebra, satisfying (NB) and (OP), but not fulfilling (G), and hence does not having (BJ) and (R) either. Moreover, in Schachermayer (1982), Proposition 3.16, it is proved that, in general, property (OP) does not imply property (NB).

The following result holds.

Theorem 2.164 (Schachermayer (1982), Proposition 4.3) *Every algebra fulfilling property (SCP) satisfies also properties (BJ) and (R).*

As we will see in the sequel, property (SCP) is not a necessary condition to have (BJ) or (R). We now recall the following

Theorem 2.165 (Schachermayer (1982), Proposition 4.6) *Let $\mathcal{L}$ be an algebra, such that there is a strictly increasing sequence $(\mathcal{L}_n)_n$ of subalgebras of $\mathcal{L}$, with $\mathcal{L} = \bigcup_{n=1}^{\infty} \mathcal{L}_n$. Then $\mathcal{L}$ does not satisfy property (G) or property (NB).*

Observe that it may happen that an algebra satisfy the hypothesis of Theorem 2.165 and fulfils property (OP) (see also Schachermayer (1982), Proposition 3.18). Furthermore, observe that (NB) implies neither (BJ) nor (G) nor (OP), (OP) implies neither (BJ) nor (G) nor (NB), while (G) implies (OP) (Schachermayer (1982), Theorem 6.4). Note that Talagrand (1984) (see Theorem (HC)) proved that, under the continuum hypothesis, (G) does not imply (NB).

Let $X = c_0$ be endowed with the supremum norm, and let τ, τ_σ, τ_s denote the norm topology, the weak topology and the topology of coordinate-wise convergence respectively. In this case we have the following result.

Theorem 2.166 (Schachermayer (1982), Proposition 6.4)

(a) *An algebra $\mathcal{L}$ has property (BJ) if and only if every measure $m : \mathcal{L} \to c_0$, (s)-bounded with respect to τ_s, is (s)-bounded with respect to τ too.*

(b) *An algebra* $\mathcal{L}$ *has property* (G) *if and only if every measure* $m : \mathcal{L} \to c_0$, (s)-*bounded with respect to* τ_σ, *is* (s)-*bounded with respect to* τ *too.*

(c) *An algebra* $\mathcal{L}$ *has property* (NB) *if and only if any measure* $m : \mathcal{L} \to c_0$, (s)-*bounded with respect to* τ_s, *is* (s)-*bounded also with respect to* τ_σ.

(d) *An algebra* $\mathcal{L}$ *has property* (OP) *if and only if every measure* $m : \mathcal{L} \to c_0$, σ-*additive with respect to* τ_σ, *is also* σ-*additive with respect to* τ *too.*

We now recall the following characterizations of the Grothendieck property for algebras (see also Diestel (1973b), Theorem 1, Lindenstrauss (1964), Theorem, and Schachermayer (1982), Theorems 5.1, 5.2 and Proposition 5.3).

Theorem 2.167 *For an algebra* $\mathcal{L}$, *the following are equivalent.*

2.167.1) $\mathcal{L}$ *has property* (G).

2.167.2) *Every continuous operator* $T : B(\mathcal{L}) \to c_0$ *is weakly compact.*

2.167.3) *Every continuous operator* $T : B(\mathcal{L}) \to X$ *into a weakly compact generated Banach space* X *is weakly compact.*

2.167.4) *If* $T_n : B(\mathcal{L}) \to X$ *is a sequence of weakly compact operators into a Banach space* X, *converging in the strong operator topology, then the limit* $T = \lim_n T_n$ *is weakly compact.*

2.167.5) *If* $T_n : B(\mathcal{L}) \to X$ *is a sequence of weakly compact operators into a Banach space* X, *converging in the weak operator topology, then the limit* $T = \lim_n T_n$ *is weakly compact.*

2.167.6) *If* $T : B(\mathcal{L}) \to X$ *is a continuous operator into a separable Banach space* X *and* $B(\mathcal{L})$ *is a subspace of a Banach space* Z, *then there is a norm-preserving extension of* T *to* Z.

2.167.7) *If* $T : B(\mathcal{L}) \to X$ *is a continuous operator to a Banach space* X *which is not relatively compact, then* T *fixes a copy of* $C([0,1])$, *that is there is a subspace of* $B(\mathcal{L})$ *isomorphic to* $C([0,1])$ *on which* T *reduces to an isomorphism.*

2.167.8) *For every bounded measure* $m:\mathcal{L}\to X$ *such that there is some bounded positive measure* ν *on* $\mathcal{L}$ *such that* $x'\circ m$ *is absolutely continuous with respect to* ν *for every* $x'\in X'$, *it follows that* m *is absolutely continuous with respect to* ν.

2.167.9) *There is no subspace of* $B(\mathcal{L})$ *isometric to* c_0 *and complemented in* $B(\mathcal{L})$.

Some other results about properties (G) and (NB) for algebras and related topics can be found in Graves and Wheeler (1983).

If $\mathcal{E}$ is a lattice of subsets of an infinite set G, then we say that a disjoint pair $[(A_n)_n,(B_n)_n]$ of sequences of disjoint elements of $\mathcal{E}$ is (SI) iff there exist an infinite set $P\subset\mathbb{N}$ and an element $A\in\mathcal{E}$ with $\bigcup_{n\in P}A_n\subset A$ and $(\bigcup_{n\in\mathbb{N}\setminus P}A_n)\cup(\bigcup_{n\in\mathbb{N}}B_n)\subset G\setminus A$ (see also Aizpuru (1996)).

A lattice $\mathcal{E}$ of subsets of an infinite set G satisfies the *subsequential interpolation property* (shortly, (SIP)) iff every disjoint pair $[(A_n)_n,(B_n)_n]$ of sequences of disjoint elements of $\mathcal{E}$ is (SI), or equivalently iff every subsequence $(A_{j_n})_n$ of each disjoint sequence $(A_j)_j$ in $\mathcal{E}$ has a sub-subsequence $(A_{j_{n_k}})_k$ and there exists a set $B\in\mathcal{E}$, with $A_{j_{n_k}}\subset B$ for every k and $A_j\cap B=\emptyset$ for all $j\in\mathbb{N}\setminus\{j_{n_k}:k\in\mathbb{N}\}$ (see also Aizpuru (1996) and Freniche (1984)).

A lattice $\mathcal{E}\subset\mathcal{P}(G)$ is said to *satisfy property* $(f1)$ iff for each pair $(F_n)_n$, $(H_n)_n$ of sequences in $\mathcal{E}$, such that $F_n\cap F_k=H_n\cap H_k=\emptyset$ for each $n\neq k$ and $H_n\cap F_k=\emptyset$ for all n, $k\in\mathbb{N}$, there are an infinite subset $P\subset\mathbb{N}$ and a set $H\in\mathcal{E}$, with $H_k\subset H$ for every $k\in P$, and $H\cap F_n=H\cap H_k=\emptyset$ for all $n\in\mathbb{N}$ and $k\notin P$.

We now prove the equivalence between property $(f1)$ and (SIP).

Proposition 2.168 *A lattice $\mathcal{E}$ of subsets of G fulfils property $(f1)$ if and only if it has the subsequential interpolation property.*

Proof: We first prove that, if $\mathcal{E}$ satisfies $(f1)$, then it satisfies (SIP). Let $(A_j)_j$ be any disjoint sequence in $\mathcal{E}$ and $(A_{j_n})_n$ be any subsequence of $(A_j)_j$. For each $n\in\mathbb{N}$, put $H_n = A_{j_n}$ and $F_n = A_{k_n}$, where

$$\{k_n : n\in\mathbb{N}\} = \mathbb{N}\setminus\{j_n : n\in\mathbb{N}\}.$$

Then the sequences $(F_n)_n$ and $(H_n)_n$ satisfy the conditions in the definition of property $(f1)$, and so there are an infinite set $P\subset\mathbb{N}$, $P=\{n_k : k\in\mathbb{N}\}$, and a set $H\in\mathcal{E}$, with $A_{j_{n_k}}\subset H$ for every k, and $H\cap F_n = H\cap H_k = \emptyset$ for all $n\in\mathbb{N}$ and $k\notin P$, namely $H\cap A_j = \emptyset$ whenever $j\notin\{j_{n_k} : k\in\mathbb{N}\}$. Putting $B=H$, we obtain that property $(f1)$ implies (SIP).

Conversely, let $(F_n)_n$, $(H_n)_n$ be any two sequences in $\mathcal{E}$, with $F_n\cap F_k = H_n\cap H_k = \emptyset$ for each $n\neq k$ and $H_n\cap F_k = \emptyset$ for all n, $k\in\mathbb{N}$. By (SIP) there are sets H_{j_k}, $k\in\mathbb{N}$, and B in $\mathcal{E}$ with $H_{j_k}\subset B$ for each k and $B\cap F_n = B\cap H_j = \emptyset$ for every $n\in\mathbb{N}$ and $j\notin P$, where $P=\{j_k : k\in\mathbb{N}\}$. Thus, (SIP) implies $(f1)$. $\square$

Let R be a complete Hausdorff topological group R with neutral element 0. The following results hold.

Proposition 2.169 (Candeloro (1985c), Proposition 3.3) *Let $\mathcal{E}$ satisfy $(f1)$, and suppose that $(C_j)_j$ is a disjoint sequence in $\mathcal{E}$. Then there exist a disjoint sequence $(P_k)_k$ of infinite subsets of $\mathbb{N}$, and a disjoint sequence $(F_k)_k$ in $\mathcal{E}$, with $H_n\subset F_k$ for each $n\in P_k$ and $k\in\mathbb{N}$.*

Proposition 2.170 (Candeloro (1985c), Lemma 3.4) *Let $\mathcal{E}$ satisfy $(f1)$, and $(C_p)_p$ be a disjoint sequence in $\mathcal{E}$. Suppose that $(P_k)_k$ is a decreasing sequence* *of subsets of* $\mathbb{N}$, *whose minima are denoted by* p_k, *and with* $p_k \notin P_{k+1}$ *for each* $k \in \mathbb{N}$. *Moreover, let* $(F_k)_k$ *be a decreasing sequence in* $\mathcal{E}$ *such that, for every* $k \in \mathbb{N}$,

$C_p \subset F_k$ *if* $p \in P_k$, *and* $C_p \cap F_{k+1} = \emptyset$ *if* $p \leq p_k$.

Then there exist an increasing sequence $(k_j)_j$ *in* $\mathbb{N}$ *and a set* $F \in \mathcal{E}$ *such that, for every* $j \in \mathbb{N}$,

2.170.1) $C_{p_{k_j}} \subset F$,

2.170.2) $F \setminus (C_{p_{k_1}} \cup \ldots \cup C_{p_{k_j}}) \subset F_{k_j+1}$.

Let $(\Lambda, \geq)$ be a directed set. A net $(B_\lambda)_\lambda$ of subsets of R *converges to* 0 (shortly, $\lim_\lambda B_\lambda = 0$) iff for every neighborhood U of 0 there is $\lambda_0 \in \Lambda$ with $B_\lambda \subset U$ for each $\lambda \geq \lambda_0$. A set $B \subset R$ is said to be *bounded* iff for every $U \in \mathcal{J}(0)$ there is $k \in \mathbb{N}$ with $B \subset kU$.

If $\mathcal{E} \subset \mathcal{P}(G)$ is a lattice, $m : \mathcal{E} \to R$ is any finitely additive measure and $B \in \mathcal{E}$, set

$$m^+(B) := \{m(A) : A \in \mathcal{E}, A \subset B\}. \qquad \textbf{(1.27)}$$

A finitely additive measure $m : \mathcal{E} \to R$ is (s)*-bounded* on $\mathcal{E}$ iff $\lim_k m^+(C_k) = 0$ whenever $(C_k)_k$ is a disjoint sequence in $\mathcal{E}$.

If $m_\lambda : \mathcal{E} \to R$, $\lambda \in \Lambda$, are finitely additive measures, then we say that $(m_\lambda)_\lambda$ is *uniformly* (s)*-bounded* on $\mathcal{E}$ iff $\lim_k \bigcup_{\lambda \in \Lambda} m_\lambda^+(C_k) = 0$ for each disjoint sequence $(C_k)_k$ in $\mathcal{E}$.

Assume now that $\mathcal{L}$ is an algebra, and $\mathcal{G}$, $\mathcal{H} \subset \mathcal{L}$ are two lattices, such that the complement of each element of $\mathcal{H}$ belongs to $\mathcal{G}$ and $\mathcal{G}$ is closed under countable disjoint unions.

A measure $m:\mathcal{L}\to R$ is said to be *regular* on $\mathcal{L}$ iff for each $A\in\mathcal{L}$ and $W\in\mathcal{H}$ there exist four sequences $(F_k)_k$, $(F_k^{'})_k$ in $\mathcal{H}$, $(G_k)_k$, $(G_k^{'})_k$ in $\mathcal{G}$, with

$$F_k \subset F_{k+1} \subset A \subset G_{k+1} \subset G_k, W \subset F_{k+1}^{'} \subset G_k^{'} \subset F_k^{'} \text{ for each } k\in\mathbb{N} \tag{1.28}$$

and

$$\lim_k m^+(G_k \setminus F_k)=0,\ \lim_k m^+(F_k^{'} \setminus W)=0. \tag{1.29}$$

The finitely additive measures $m_\lambda:\mathcal{L}\to R$, $\lambda\in\Lambda$, are *uniformly regular* on $\mathcal{L}$ iff for each $A\in\mathcal{L}$ and $W\in\mathcal{H}$ there exist four sequences $(F_k)_k$, $(F_k^{'})_k$ in $\mathcal{H}$, $(G_k)_k$, $(G_k^{'})_k$ in $\mathcal{G}$, satisfying (1.28) and with

$$\lim_k \bigcup_{\lambda\in\Lambda} m_\lambda^+(G_k \setminus F_k)=\lim_k \bigcup_{\lambda\in\Lambda} m_\lambda^+(F_k^{'} \setminus W)=0. \tag{1.30}$$

Observe that, if G is a topological normal space (resp. locally compact Hausdorff space), and we pick $\mathcal{L}=\{\text{Borel subsets of } G\}$, $\mathcal{H}=\{\text{closed sets}\}$ (resp. $\{\text{compact sets}\}$), $\mathcal{G}=\{\text{open sets}\}$ and $R=\mathbb{R}$, the last equalities in formulas (1.29) and (1.30) are consequences of the respective first ones, and we get the classical definition of regularity and uniform regularity (see also Brooks (1969b), Candeloro (1985b), Candeloro and Letta (1985)). In Chapter 4, we will examine in detail some properties of topological group-valued measures and we will give some related results in the setting of filter convergence.

Freniche (1984) (see Theorem 4), using the sliding hump technique, gave a Brooks-Jewett-type theorem for real- or complex-valued measures defined on Boolean algebras satisfying (*SIP*) and proved that the class of algebras with property (*SIP*) contains strictly the classes of algebras fulfilling properties (I) of Seever, (f) of Moltó and (*SCP*) of Haydon (see Proposition 2 and Theorem 7).

Antosik and Swartz (1985b) (see Theorem 5), using Theorem 2.149, the tool of submeasures and the Drewnowski lemma, gave a version of the Brooks-Jewett theorem for finitely additive measures, taking values in a normed groups and defined on algebras, satisfying the property (*SCP*) of Haydon.

Candeloro (1985c) (Theorem 3.5, see also Candeloro (1985b), Theorem 2.5), using a sliding hump argument, proved the following Nikodým boundedness theorem.

Theorem 2.171 *Let $\mathcal{L}$ be an algebra of subsets of an abstract nonempty set G, satisfying property $(f1)$, and $m_\lambda : \mathcal{L} \to R$, $\lambda \in \Lambda$, be a family of finitely additive measures, (s)-bounded on $\mathcal{L}$. Suppose that for every $A \in \mathcal{L}$ the set $\{m_\lambda(A) : \lambda \in \Lambda\}$ is bounded in R.*

Then the set $\bigcup_{\lambda \in \Lambda} m_\lambda^+(G)$ is bounded in R.

Using the sliding hump technique, in Candeloro (1985b), Theorem 2.6, the following extension of the Brooks-Jewett theorem is given.

Theorem 2.172 *Let $\mathcal{L}$ be an algebra, $\mathcal{E} \subset \mathcal{L}$ be a lattice, satisfying property $(f1)$. Let $m_n : \mathcal{L} \to R$, $n \in \mathbb{N}$, be a sequence of finitely additive measures, whose restrictions on $\mathcal{E}$ are (s)-bounded on $\mathcal{E}$. If the limit $\lim_n m_n(E)$ exists in R for every $E \in \mathcal{E}$, then the measures m_n are uniformly (s)-bounded on $\mathcal{E}$.*

Proof: Let $(H_n)_n$ be a disjoint sequence in $\mathcal{E}$. Proceeding by contradiction, suppose, without loss of generality, that there exists a neighborhood U of 0 with $m_n(H_n) \notin 2U$ for each $n \in \mathbb{N}$. Let now $(U_h)_h$ be a decreasing sequence of neighborhoods of 0, with $U_0 = U$ and $2U_h \subset U_{h-1}$ for every $h \in \mathbb{N}$.

By Proposition 2.169 and (s)-boundedness of m_1, we can find two sets $P_1 \subset \mathbb{N}$ and $F_1 \in \mathcal{E}$, with $H_p \subset F_1$ for each $p \in P_1$ and $m_1^+(F_1) \subset U_1$. Put $p_1 = \min P_1$: in correspondence with H_{p_1} there is an integer $N_1 > p_1$ with $m_i(H_{p_1}) - m_j(H_{p_1}) \in U_1$ for each $i, j \geq N_1$.

Thanks to Proposition 2.169 and (s)-boundedness of m_{p_1}, there are an infinite set $P_2 \subset P_1$, with $p_2 = \min P_2$, and a set $F_2 \in \mathcal{E}$, with $H_p \subset F_2$ for every $p \in P_2$; $p_2 > N_1 + p_1$; $m_{p_1}^+(F_2) \subset U_2$. Moreover, using property $(f1)$, the set F_2 can be chosen in such a way that $F_2 \cap H_p = \varnothing$ for every $p \leq p_1$. In correspondence with H_{p_2} there exists an integer $N_2 > p_2$ with $m_i(H_{p_2}) - m_j(H_{p_2}) \in U_2$ for any $i, j \geq N_2$. Proceeding by induction, we find a decreasing sequence $(P_k)_k$ of infinite subsets of $\mathbb{N}$ and two strictly increasing sequences $(N_k)_k$, $(p_k)_k$ in $\mathbb{N}$, satisfying the following conditions, for every $k \in \mathbb{N}$:

(*i*) $p_k = \min P_k$, $N_k > p_k$, $p_{k+1} > N_k + p_k$;

(*ii*) $H_p \subset F_k$ for every $p \in P_k$ and $H_p \cap F_{k+1} = \emptyset$ for each $p \leq p_k$;

(*iii*) $m^+_{p_k}(F_{k+1}) \subset U_{k+1}$;

(*iv*) $m_i(H_{p_k}) - m_j(H_{p_k}) \in U_k$ for each $i, j \geq N_k$.

Put $m'_k = m_{p_k}$ and $H'_k = H_{p_k}$, $k \in \mathbb{N}$. For each $k \geq 2$ we get $m'_k(H'_k) \notin 2U$, but $m'_{k-1}(H'_k) \in U_k$, and hence

$$m'_k(H'_k) - m'_{k-1}(H'_k) \notin U + U_1. \tag{1.31}$$

Using Proposition 2.170, we can construct a strictly increasing sequence $(k_r)_r$ in $\mathbb{N}$ and a set $F \in \mathcal{E}$, with $H'_{k_r} \subset F$ and $F \setminus (H'_{k_1} \cup \ldots \cup H'_{k_r}) \subset F_{k_r+1}$ for each r. Thus for any i, $r \in \mathbb{N}$ we have

$$m_i(H'_{k_r}) = m_i(F) - \sum_{s=1}^{r-1} m_i(H'_{k_s}) - m_i(F \setminus (H'_{k_1} \cup \ldots \cup H'_{k_r})).$$

As $F_{k_r+1} \subset F_{k_r}$, we get

$$m'_{k_r}(F \setminus (H'_{k_1} \cup \ldots \cup H'_{k_r})) \in U_{r+1} \text{ and } m'_{k_r-1}(F \setminus (H'_{k_1} \cup \ldots \cup H'_{k_r})) \in U_r.$$

Moreover, possibly replacing the sequence $(k_r)_r$ with the sequence $(k_{2r})_r$, without loss of generality we can suppose that $p_{k_r-1} > N_{k_r-2} > N_{k_{r-1}}$, and so

$$m'_{k_r}(H'_{k_s}) - m'_{k_r-1}(H'_{k_s}) \in U_s \text{ for all } s = 1, \ldots, r-1.$$

Finally, for r large enough, we have also $m'_{k_r}(F)-m'_{k_r-1}(F)\in U_2$, since $F\in\mathcal{E}$, and hence

$$m'_{k_r}(H'_{k_r})-m'_{k_r-1}(H'_{k_r})\in U_2+U+U_r+U_{r+1}\subset U+U_1,$$

and this contradicts (1.31). This concludes the proof. □

We have also the following version of the Dieudonné theorem.

Theorem 2.173 (see Candeloro (1985b), Theorem 2.7, and Candeloro and Letta (1985), Theorem 4.1) *Let* $\mathcal{G}$, $\mathcal{H}\subset\mathcal{L}$ *be as above,* $\mathcal{G}$ *satisfy property* $(f1)$, *and* $m_n:\mathcal{L}\to R$, $n\in\mathbb{N}$, *be a sequence of regular measures,* (s)*-bounded on* $\mathcal{G}$. *If the limit* $\lim_n m_n(E)$ *exists in* R *for every* $E\in\mathcal{G}$, *then the measures* m_n *are uniformly regular and uniformly* (s)*-bounded on* $\mathcal{L}$, $m_0(E):=\lim_n m_n(E)$ *exists in* R *for each* $E\in\mathcal{L}$ *and defines a regular measure,* (s)*-bounded on* $\mathcal{L}$.

Weber (1986), using Rosenthal-type lemmas and properties of quasi-norms, gave some extension of Theorem 2.143 (Theorem 3.2.3), some compactness results for topological group-valued measures, some version of the Brooks-Jewett theorem and the Nikodým boundedness theorem for finitely additive measures, defined on rings satisfying property (*SIP*) and with values in abelian Hausdorff topological groups (see also Theorems 4.1 and 5.7), and proved that, if a ring $\mathcal{L}$ satisfies property (*SIP*) and the countable chain condition (shortly (CCC), that is: every set of disjoint elements of $\mathcal{L}$ is at most countable), then it fulfils also property (*SCP*) (see Proposition 7.1.1 (a)). Weber (1985) (see also Theorems 3.4 and 4.1) proved some results about pointwise sequential compactness, weak compactness, weak convergence and uniform (s)-boundedness for measures defined in suitable rings and with values in locally convex topological vector spaces, in connection with properties (BJ) and (G).

Using Antosik-Mikusiński-type diagonal techniques similar to those of Theorems 2.117 and 2.135, the following results was given.

Theorem 2.174 (Pap (1986), Theorem 2, see also Pap (1995), Theorem 11.15) *Let* G *be a locally compact Hausdorff topological space,* Σ *be the* σ*-algebra of all Borel subsets of* G, $\mathcal{K}$ *be a family of* k*-triangular set function defined on* Σ *and* *with regular total variations. If the set* $\{m(U): m \in \mathcal{K}\}$ *is order bounded for every open set* U*, then the set* $\{m(E): m \in \mathcal{K}, E \in \Sigma\}$ *is order bounded.*

By means of diagonal methods, the following Brooks-Jewett and Dieudonné-type theorems for k-triangular set functions were obtained.

Theorem 2.175 (see Pap (1987), Theorems 1 and 2, Pap (1991), Corollary 1, and Pap (1995), Theorems 11.10 and 11.11) *Let* $\mathcal{L}$ *be an algebra of parts of* G*, satisfying property* (SIP)*, and* $m_n : \mathcal{L} \to \mathbb{R}$*,* $n \in \mathbb{N}$*, be a sequence of* k*-triangular* (s)*-bounded set functions. If* $m_0(E) := \lim_n m_n(E)$ *exists in* $\mathbb{R}$ *for all* $E \in \mathcal{L}$ *and* m_0 *is* (s)*-bounded, then* $(m_n)_n$ *is uniformly* (s)*-bounded and* m_0 *is* k*-triangular.*

Theorem 2.176 (Pap (1987), Theorem 3 and Corollary 2, see also Pap (1995), Theorem 11.17 and Corollary 11.18) *Let* G *be a locally compact Hausdorff topological space,* Σ *be the* σ*-algebra of all Borel subsets of* G *and* $m_n : \Sigma \to \mathbb{R}$*,* $n \in \mathbb{N}$*, be a sequence of* k*-triangular set functions with regular total variations. If the limit* $m(U) = \lim_n m_n(U)$ *exists in* $\mathbb{R}$ *for every open set* U*, then* $(m_n)_n$ *is uniformly* (s)*-bounded on* Σ*, the limit* $m(E) = \lim_n m_n(E)$ *exists in* $\mathbb{R}$ *for every* $E \in \Sigma$ *and* m *is* (s)*-bounded on* Σ*.*

Pap (1988), using diagonal techniques, proved another Nikodým uniform boundedness theorem for non-additive set functions, defined on rings satisfying property (SCP) and taking values in a metric space.

De Lucia and Morales (1988) gave some Brooks-Jewett (see Theorem 2.1 and Corollary 2.2), Nikodým convergence (see Theorem 3.1 and Corollary 3.2), Vitali-Hahn-Saks (see Theorem 4.4 and Corollary 4.5), Nikodým boundedness (Theorem 5.3) and Dieudonné boundedness theorem (Theorem 6.6) for set functions taking values in a semigroup with uniformity and defined in a Boolean ring with (SIP). Using a Drewnowski-type theorem, in de Lucia and Morales (1986), Theorem 2.5, it was proved that, in this setting, the Brooks-Jewett, Vitali-Hahn-Saks and Nikodým convergence theorems are equivalent.

Fleischer and Traynor (1989) (see Convergence Lemma, Boundedness Lemma and Corollary) proved some version of the Nikodým convergence and boundedness theorems for finitely additive measures defined in a ring satisfying property (SIP) and with values in non-commutative topological groups.

1.2.12. Vector Lattice-Valued Measures

We now recall some version of matrix theorems, Nikodým convergence and Nikodým boundedness theorems, proved in Swartz (1989a), in the vector lattice context. In Chapter 2 and 3 we recall some other results about limit theorems for lattice group-valued measures given in Boccuto and Candeloro (2002a-c and 2004a-c), and in Chapter 4 we will deal with some of their extensions in the setting of filter convergence.

A lattice group R is said to have the *positive regularity property* (shortly *property* (PR), see also Peressini (1967), 1.5.13) iff there exists an increasing sequence $(t_n)_n$ of positive elements of R such that for each $x \in R$ there is $n \in \mathbb{N}$ with $|x| \leq n t_n$. Observe that property (PR) is fulfilled by every lattice group R having an order unit. However, the space c_{00} of all eventually 0 real-valued sequences does not have any order unit, but it satisfies property (PR) (see also Swartz (1989a)). The following results are proved by means of diagonal arguments.

Theorem 2.177 (Swartz (1989a), Theorem 4) *Let G, $\Lambda \neq \emptyset$, R be a Dedekind complete vector lattice endowed with an order unit, Σ be a σ-algebra of parts of G, $m_\lambda : \Sigma \to R$, $\lambda \in \Lambda$, be a family of finitely additive and order bounded measures. If the set $\{m_\lambda(E) : \lambda \in \Lambda\}$ is order bounded for each $E \in \Sigma$, then the set $\{m_\lambda(E) : \lambda \in \Lambda, E \in \Sigma\}$ is order bounded.*

Note that, in general the hypothesis of existence of an order unit, in Theorem 2.177, cannot be replaced by property (PR) (see also Swartz (1989a), Example 5). However property (PR) is a sufficient condition to get the following

Theorem 2.178 (Swartz (1989a), Theorem 1) *Let G, $\Lambda \neq \emptyset$, $\mathcal{L}$ be an algebra of parts of G, R have property (PR), and $m_\lambda : \Sigma \to R$, $\lambda \in \Lambda$, be a family of finitely additive measures. Then the set $\{m_\lambda(E) : \lambda \in \Lambda, E \in \mathcal{L}\}$ is order bounded iff the set $\{m_\lambda(E_j) : \lambda \in \Lambda, j \in \mathbb{N}\}$ is order bounded for each disjoint sequence $(E_j)_j$ in $\mathcal{L}$.*

Observe that in general, in Theorem 2.178, property (PR) cannot be dropped, as we will see in Example 3.11 of Chapter 2 (see also Swartz (1989a), Example 3). Some other version of the Nikodým boundedness theorem for a family of finitely additive order bounded measures with values in an Archimedean partially ordered vector space with an order unit was given in Giannakoulias (1994).

We now recall a version of the uniform boundedness principle in the vector lattice context, obtained by using diagonal techniques. If X and Y are two vector lattices, then we say that a linear operator $T: X \to Y$ is *order bounded* iff T maps order bounded subsets of X in order bounded subsets of Y, where a subset A of a vector lattice X is *order bounded* iff there is an order interval $[-u,u]=\{x\in X:\ -u\le x\le u\}$ with $A\subset[-u,u]$. The following result holds.

Theorem 2.179 (see Swartz (1989b), Theorem 3) *Let X, Y be two Dedekind complete vector lattices, and suppose that Y has an order unit u. If $T_n : X \to Y$ is a sequence of order bounded linear functions such that the set $\{T_n(x): n\in\mathbb{N}\}$ is order bounded for every $x\in X$, then the set $\{T_n(x): n\in\mathbb{N},\ x\in A\}$ is order bounded for every order bounded set $A\subset X$.*

Note that, without assuming that Y has an order unit, in general Theorem 2.179 does not hold. Indeed, let $X=L^1([0,1])$ and $Y=c_0$ be the space of all real sequences convergent to 0, endowed with the pointwise order. For $f\in X$ and $n\in\mathbb{N}$ set

$$T_n(f)=\sum_{k=1}^{n} f_k e_k, \text{ where } f_k=\int_0^1 f(t)\sin(k\pi t)dt$$

and e_k, $k\in\mathbb{N}$, is the sequence in c_0 with 1 in the k-th coordinate and 0 in the other coordinates. Note that each T_n is order bounded because, if $|f|\le h$ in X, then $|T_n f|\le\left(\sum_{k=1}^{n} e_k\right)\int_0^1 h(t)dt$.

Moreover, the sequence $(T_n)_n$ is pointwise bounded on X: indeed, $|T_n f|\le\sum_{k=1}^{\infty}|f_k|\,e_k\in c_0$ for every $n\in\mathbb{N}$, since $(f_k)_k\in c_0$ by virtue of the Riemann-Lebesgue lemma. However, $(T_n)_n$ is not uniformly order bounded on order bounded subsets of X since, if $P_n(t) = \sin(n\pi t)$, then $|P_n| \le 1$ and hence P_n is order bounded for every $n \in \mathbb{N}$, but $(T_n(P_n))_n = (e_n/2)_n$ is not order bounded in c_0 (see Swartz (1989b), Example 1). Furthermore, in Swartz (1989b), Proposition 4, some condition for equicontinuity of operators in the vector lattice context was given.

We now recall a boundedness result for bilinear vector lattice-valued functions. Given three vector lattices X, Y, Z, we say that an operator $T: X \times Y \to Z$ is *separately order bounded* iff, for each $x \in X$ (resp. $y \in Y$), $T(x,\cdot)$ (resp. $T(\cdot, y)$) is an order bounded linear operator from Y into Z (resp. from X into Z), and that T is *order bounded,* iff it maps the products of order bounded subsets of X and Y into order bounded subsets of Z. The following result holds.

Theorem 2.180 (see also Swartz (1989c), Theorem 4) *Let X, Y, Z be three Dedekind complete vector lattices such that Z is endowed with order unit, and $T: X \times Y \to Z$ be a separately order bounded bilinear operator. Then T is order bounded.*

Similarly as in Theorem 2.179, it is possible to show that in general Theorem 2.180 does not hold, if Z does not have an order unit (see also Swartz (1989c), Example 5).

Antosik and C. Swartz (1992) gave some version of the basic matrix theorem and the Nikodým convergence theorem in the vector lattice context.

A sequence $(x_n)_n$ of elements of a vector lattice R is said to be *relatively uniformly convergent* (shortly, *(r)-convergent*) to $x \in R$ iff there is $u \in R$, $u \ge 0$, such that for every $\varepsilon > 0$ there is n_0 with $|x_n - x| \le \varepsilon u$ for every $n \ge n_0$. The sequence $(x_n)_n$ is said to be *$(*)$-convergent* to $x \in R$ iff every subsequence of $(x_n)_n$ has a sub-subsequence which (r)-converges to x, and in this case we write $(*)\lim_n x_n = x$. In this context, we denote by the symbol $\sum_{n=1}^{\infty} x_n$ the quantity $(*)\lim_n \sum_{i=1}^{n} x_i$. Moreover, following Antosik and Swartz (1992), we say that the *double limit* $(*)\lim_{i,j} x_{i,j} = x \in R$ iff for every pair $(q_i)_i$, $(n_i)_i$ of strictly increasing sequences of positive integers we get $(*)\lim_i x_{q_i,n_i} = x$. Furthermore, if $(*)\lim_j x_{i,j} = x_i$ exists in R for every $i \in \mathbb{N}$, then we say that $(*)\lim_j x_{i,j} = x_i$ *uniformly with respect to* $i \in \mathbb{N}$ iff $(*)\lim_i (x_{q_i,n_i} - x_{q_i}) = 0$ for every pair $(q_i)_i$, $(n_i)_i$ of strictly increasing sequences in $\mathbb{N}$. A sequence of series $\sum_{j=1}^{\infty} x_{i,j}$ is said to be *convergent uniformly with respect to* $i \in \mathbb{N}$ iff

$$(*)\lim_n \sum_{j=n}^{\infty} x_{i,j} = (*)\lim_n \left(\sum_{j=1}^{\infty} x_{i,j} - \sum_{j=1}^{n-1} x_{i,j} \right) = 0$$

uniformly with respect to $i \in \mathbb{N}$.

We say that a lattice group R satisfies *property* (σ) iff for every sequence $(u_n)_n$ of positive elements of R there are a sequence $(\lambda_n)_n$ of positive real numbers and a positive element $u \in R$ with $\lambda_n u_n \leq u$ for every $n \in \mathbb{N}$ (see also Luxemburg and Zaanen (1971)).

Example 2.181 If $\mathcal{M}$ is the σ-algebra of all Lebesgue measurable subsets of $[0,1]$, ν is the Lebesgue measure and $R = L^0([0,1], \mathcal{M}, \nu)$ is the space of all ν-measurable real-valued functions with the identification up to ν-null sets, then R has property (σ), and $(*)$-convergence in R coincides with convergence in ν-measure (see also Kappos (1969), Luxemburg and Zaanen (1971), and Vulikh (1967)).

We now recall the following

Theorem 2.182 (see Antosik and Swartz (1992), Theorem 2.4) *Let* R *be a vector lattice satisfying property* (σ), *and assume that* $(*)\lim_j x_{i,j} = x_i$ *exists in* R *for every* $i \in \mathbb{N}$. *Then* $(*)\lim_{i,j} x_{i,j} = x \in R$ *if and only if* $(*)\lim_i x_i = x$ *exists in* R *and* $(*)\lim_j x_{i,j} = x_i$ *uniformly with respect to* $i \in \mathbb{N}$.

The following Schur-type theorem in the context of vector lattices holds.

Theorem 2.183 (see Antosik and Swartz (1992), Theorem 3.1) *Let R be a vector lattice satisfying property (σ), and suppose that the series $\sum_{j=1}^{\infty} x_{i,j}$ is subseries convergent for each $i \in \mathbb{N}$.*

If $(*)\lim_i \sum_{j=1}^{\infty} x_{i,n_j}$ exists in R for each strictly increasing subsequence $(n_i)_i$ in $\mathbb{N}$, then we get:

(a) $\sum_{j=1}^{\infty} x_j$ *is subseries convergent, where* $x_j = (*)\lim_i x_{i,j}$;

(b) $(*)\lim_i \sum_{j=1}^{\infty} x_{i,j} = \sum_{j=1}^{\infty} x_j$;

(c) *The series* $\sum_{j=1}^{\infty} x_{i,j}$ *are subseries convergent uniformly with respect to* $i \in \mathbb{N}$.

We now recall the lattice version of the Nikodým-type theorem, concerning uniform σ-additivity with respect to $(*)$-convergence.

Theorem 2.184 (see also Antosik and Swartz (1992), Theorem 3.2) *Let R be a vector lattice, endowed with property (σ), G be any nonempty set, and $\Sigma \subset \mathcal{P}(G)$ be a σ-algebra.*

Let $m_n : \Sigma \to \mathbb{R}$, $n \in \mathbb{N}$, be a sequence of σ-additive measures with $(*)\ \lim_n m_n(E) = m(E)$ for every $E \in \Sigma$. Then the measures m_n, $n \in \mathbb{N}$, are uniformly σ-additive and the limit measure m is σ-additive.

Observe that property (σ) is a condition sufficient, but not necessary, for Theorem 2.184, whose assertion is true even if $R = c_{00}$ is the space of all real-valued eventually null sequences endowed with the coordinatewise order, which does not satisfy property (σ) (see also Antosik and Swartz (1992), Remark 3.3, Luxemburg and Zaanen (1971) and Vulikh (1967)).

1.2.13. Measures Defined on Abstract Structures

A subset X of an orthoalgebra L is *jointly orthogonal* iff X is pairwise orthogonal and is contained in a maximal Boolean subalgebra (with respect to the set-theoretic inclusion). We denote by $J(L)$ the class of joint orthogonal subsets of L. We say that an orthoalgebra L has the *weakly subsequential interpolation* (resp. *completeness*) *property*, in short (*WSIP*) (resp. (*WSCP*)) iff for every sequence $(a_n)_n$ in $J(L)$ there exist a subsequence $(a_{n_k})_k$ and a sub-orthomodular lattice of L which contains the set $\{a_{n_k} : k \in \mathbb{N}\}$ and satisfies (*SIP*) (resp. (*SCP*)) (see Habil (1995)). Similar properties and relations between them, the condition (f) of A. Moltó, the Seever and the Grothendieck property were investigated in Aizpuru (1988 and 1996) in the context of Boolean algebras.

Constantinescu (1989) investigated several properties of Vitali spaces and topological group-valued measures, among which the (NB) and (BJ) properties for Vitali spaces with respect to abelian topological groups. In particular, in Constantinescu (1989), Theorem 7.13, it is proved that every Vitali space with the subsequential interpolation property has property (NB) with respect to every topological abelian group. Moreover, the following results hold.

Proposition 2.185 (Constantinescu (1989), Proposition 7.15) *If a Vitali space has the (NB)property with respect to $\mathbb{R}$, then it has the (NB)property with respect to every locally convex space.*

Theorem 2.186 (Constantinescu (1989), Corollary 7.18) *Let Λ be any nonempty set and L be a vector sublattice of $l^\infty(\Lambda)$ containing the constant functions. If L has property (NB) with respect to $\mathbb{R}$, then L endowed with the topology of uniform convergence is barrelled.*

Proposition 2.187 (Constantinescu (1989), Proposition 7.15) *If R is a locally convex space and L has the (BJ)-property with respect to R, then L has the (NB)property with respect to R.*

Theorem 2.188 (Constantinescu (1989), Theorem 8.7) *Every Vitali space with the subsequential interpolation property has property (BJ) with respect to any topological abelian group.*

If L and R are two locally convex spaces and $\mathcal{L}(L,R)$ is the vector space of all continuous linear functions from L to R endowed with the topology of pointwise convergence, then we say that $T \in \mathcal{L}(L,R)$ is *boundedly weakly compact* iff T maps bounded subsets of L into relatively weakly compact subsets of R. We denote by $\mathcal{L}_0(L,R)$ the subspace of $\mathcal{L}(L,R)$ of all boundedly weakly compact functions from L to R, and by $\mathcal{E}(L,R)$ the set of all (s)-bounded functions from L to R endowed with the topology of pointwise convergence. Then we get $\mathcal{L}_0(L,R) \subset \mathcal{E}(L,R)$ (see Constantinescu (1989), Corollary 4.9). Moreover, every subset of $\mathcal{L}_0(L,R)$ having the Grothendieck property has the (NB)-property (see Constantinescu (1989), Proposition 9.3). Conversely, if L has the (BJ)-property and every topologically bounded set of L formed by positive elements of L is order bounded, then L has the Grothendieck property (see also Constantinescu (1989), Proposition 9.5).

Several further results are given, in connection with the Grothendieck property, weak relative compactness and related topics (see Constantinescu (1989)), extending some results of Diestel (1984) and Schachermayer (1982). Some other results on Brooks-Jewett, Nikodým convergence and boundedness theorem for both finitely and σ-additive measures with values in a partially ordered locally convex space with a Schauder basis can be found in Giannakoulias (1996).

Guariglia (1990a), by means of diagonal techniques, proved a Brooks-Jewett-type theorem for k-triangular set functions defined on an orthomodular lattice with (SIP) and with values in an abelian semigroup endowed with a uniformity (see Theorem 4.2) or in an abelian Hausdorff topological group (see Theorem 4.3). Guariglia (1990b) proved a version of the Nikodým boundedness theorem for k-triangular set functions defined in rings satisfying property (SIP) and taking values in a semigroup with uniformity, by using sliding hump methods (see Theorem 3.3). Guariglia (1990c), in Theorem 3.2 (see also Guariglia (1990b), Corollary 3.8), in the same context, by means of techniques similar to those used in Candeloro (1985c), and Candeloro and Letta (1985), proved a version of the Nikodým boundedness theorem, in a form similar to that given in Dieudonné (1951a), for finitely additive inner regular set functions with values in an abelian Hausdorff topological group.

D'Andrea and de Lucia (1991) (see Theorem 5.1) proved a Brooks-Jewett-type theorem for set functions, defined on an orthomodular lattice endowed with (SIP) and with values in a Hausdorff uniform semigroup. D'Andrea, de Lucia and

Morales (1991) gave some Lebesgue decomposition theorem (see Theorems 2.6 and 2.11), a Nikodým convergence (Theorem 3.7), a Vitali-Hahn-Saks (273, Theorem 4.1) and a Nikodým boundedness theorem (Theorem 4.3) for set functions defined on orthomodular posets and taking values in abelian topological or Hausdorff uniform semigroups. Some other kinds of decomposition theorems were proved for set functions defined on orthomodular posets and with values in a normed Riesz space with order continuous norm (see De Lucia and Dvurečenskij (1993a), Theorems 4.1, 6.1) or defined on orthoalgebras and taking values in a Dedekind complete vector lattice (see De Lucia and Dvurečenskij (1993b), Theorems 6.1 - 6.3) respectively.

Pap (1991) (see Theorem 2), using diagonal arguments, proved another Brooks-Jewett-type theorem for non-additive set functions, defined on rings satisfying property (*SCP*) and taking values in a uniform space.

Ferrando and López-Pellicer (1992) (see Theorems 2 and 3) proved some extension of Theorem 2.132 in the locally convex space context, giving some condition under which certain finitely additive measures are σ-additive.

Saeki (1992) (see Theorem 6 and Corollaries 7, 8), using diagonal theorems, proved some Brooks-Jewett and Nikodým convergence-type theorem for 1-triangular set functions.

We now recall that a Banach space X *contains the* l_n^∞ *'s uniformly* iff there is a real number $K_0 > 1$ such that for every finite-dimensional subspace l_* of l^∞ there exist a finite-dimensional subspace X_0 of X and an isomorphism $T : l_* \to X_0$ with $\| T \| \| T^{-1} \| < K_0$. We recall the following:

Theorem 2.189 (see also Abraham (1992), Proposition 3 and Remark 4; Abraham (1996), Theorem 2.3; Maurey and Pisier (1976), Corollary 1.3; Pap (1995), Theorems 11.8 and 11.9; Saeki (1992), Corollary 8) *Let* X *be any Banach space. Then the following are equivalent.*

(a) X *has finite cotype.*

(b) X *satisfies the Saeki fatness condition.*

(c) *If* $(m_n)_n$ *is a sequence of* (s)*-bounded (resp.* σ*-additive)* X*-valued measures on a* σ*-algebra* Σ *such that there exists a real number* $C' \geq 1$ *with*

$\limsup_n \| m_n(E) \| \le C' \liminf_n \| m_n(E) \| < +\infty$ *for each* $E \in \Sigma$ *, then* $(m_n)_n$ *is uniformly* (s)*-bounded (resp. uniformly* σ*-additive).*

(d) *If* $(m_n)_n$ *is a sequence of* X*-valued measures on a* σ*-algebra* Σ*, each of which absolutely continuous with respect to a* σ*-additive non-negative extended real-valued measure* ν*, and if there exists a real number* $C' \ge 1$ *with*

$$\limsup_n \| m_n(E) \| \le C' \liminf_n \| m_n(E) \| < +\infty \text{ for any } E \in \Sigma,$$

then $(m_n)_n$ *is equiabsolutely continuous with respect to* ν.

(e) *If* $(m_n)_n$ *is a sequence of* (s)*-bounded* X*-valued measures on a* σ*-algebra* Σ*, such that* $\lim_n \| m_n(E) \|$ *exists in* X *for all* $E \in \Sigma$*, then* $(m_n)_n$ *is uniformly* (s)*-bounded.*

(f) X *does not contain the* l_n^∞ *'s uniformly.*

In this setting, using Rosenthal-type theorems, some versions of the Brooks-Jewett and Nikodým convergence theorem were proved, for algebras satisfying (SIP) (see also Abraham (1996), Theorem 3.5).

Using some properties of the spaces $C(K)'$, with K compact, some conditions on algebras were given to have property (BJ), (NB) or (G) (see Aizpuru (1992), Corollary 3.4 and Theorem 3.13). Moreover, using diagonal properties, a sufficient condition for property (NB) was found, which is not sufficient for property (G) (see Aizpuru (1992), Theorem 3.8 and Remark 3.10).

Using some property of unconditionally convergent series and Theorem 8.2 in Schachermayer (1982), the following results were established.

Theorem 2.190 (Popa (1992), Corollary 3) *An algebra* $\mathcal{L}$ *has property* (G) *if and only if: for every bounded finitely additive measure* $m : \mathcal{L} \to X$*, where* X *is a Banach space, if there exists a finitely additive non-negative real-valued measure* ν *on* $\mathcal{L}$ *such that* $x' \circ m$ *is absolutely continuous with respect to* ν *for every* $x' \in X'$*, then* m *is* (s)*-bounded.*

Theorem 2.191 (Popa (1992), Theorem 7) *The following assertions are true, and equivalent:*

(a) *The property* (BJ) *for every* σ*-algebra,*

(b) *The Schur* l^1 *and* (NB) *properties for each* σ*-algebra.*

Theorem 2.192 (Popa (1992), Theorem 8) *The following assertions are true, and equivalent:*

(a) *The property* (G) *for real-valued set functions on every* σ*-algebra,*
(b) *The Schur* l^1 *property for each* σ*-algebra.*

Theorem 2.193 (Popa (1992), Theorem 9) *For a given algebra of sets* $\mathcal{L}$ *the following conditions are equivalent.*

(a) $\mathcal{L}$ *has property* (BJ).

(b) $\mathcal{L}$ *has the Phillips property, that is each sequence* $(m_i)_i$ *in* $B(\mathcal{L})'$*, pointwise convergent to* 0*, satisfies the following condition: if* $(E_j)_j$ *is a disjoint sequence in* $\mathcal{L}$*, then*

$$\lim_i \left(\sum_{j=1}^{\infty} \| m_i(F_j) \| \right) = 0.$$

(c) $\mathcal{L}$ *has the property that each sequence* $(m_i)_i$ *in* $B(\mathcal{L})'$*, pointwise convergent, is such that, if* $(E_j)_j$ *is a disjoint sequence in* $\mathcal{L}$*, then* $\lim_i \left(\sum_{j \in E} m_i(F_j) \right)$ *exists in X.*

Drewnowski, Florencio and Paúl (1994) (see Proposition 6), using some properties of barrelled spaces, proved that the family of all subsets of $\mathbb{N}$ of asymptotic density zero fulfils property (NB). Drewnowski, Florencio and Paúl (1996) (see Examples 4.3 and 4.4) obtained the same result by investigating some classes of sets having property (NB) in relation with some properties of submeasures.

Habil (1995), using diagonal arguments, proved some version of the Nikodým boundedness theorem (Theorem 4.1), Brooks-Jewett theorem (Theorem 4.2), Cafiero theorem (Theorem 4.3) and Nikodým convergence theorem (Theorem 4.4) for set functions with values in uniform semigroups and defined in orthoalgebras having property $(WSIP)$.

De Lucia and Pap (1995a) (see Theorem 3.5), using diagonal methods, proved a Nikodým convergence theorem for set functions, defined on D-posets and with values in a uniform topological space.

De Lucia and Pap (1995b) (see Theorem 3), using again diagonal techniques, proved a Nikodým boundedness theorem for set functions, defined in a suitable orthomodular lattice and taking values in a topological space with uniformity.

De Lucia and Salvati (1996) (see Cafiero Theorem), by means of matrix methods and diagonal techniques, proved a Cafiero-type theorem for (s)-bounded set functions defined on an orthomodular poset and with values in an abelian Hausdorff topological group.

Some Vitali-Hahn-Saks-type theorem for sequences of Lebesgue integrals of real functions with compact support was proved in Hernández-Lerma and Lasserre (1996) (see Theorem 2.1), see also Hernández-Lerma and Lasserre (1998), Theorem 2.1. Degla (2000), in Theorems 1, 4, and 6, refined these results, giving some correction and presenting a counterexample.

In Aizpuru (1996) it is dealt with properties of Boolean algebras of *local* character. Let $\mathcal{L}$ be an algebra of subsets of an infinite set G, Q^* be its associated Stone space and $\mathcal{Q}$ be the algebra of all open-closed subsets of Q^*, which is isomorphic to $\mathcal{L}$. We say that $\mathcal{L}$ has the *local subsequential completeness property* (shortly, $(LSCP)$) iff for every $x \in Q^*$ there is a decreasing sequence $(T_n)_n$ of clopen neighborhoods of x such that, if $(A_n)_n$ is any disjoint sequence in $\mathcal{Q}$ with $A_n \subset T_n$ for every $n \in \mathbb{N}$, then there is a strictly increasing sequence $(n_k)_k$ in $\mathbb{N}$ such that $\bigcup_{k=1}^{\infty} A_{n_k} \in \mathcal{Q}$.

We say that $\mathcal{L}$ has the *local subsequential interpolation property* (shortly, $(LSIP)$) iff for every $x \in Q^*$ there is a decreasing sequence $(T_n)_n$ of open-closed neighborhoods of x such that, if $[(A_n)_n, (B_n)_n]$ is any disjoint pair of sequences of disjoint elements of $\mathcal{Q}$ with $A_n \subset T_n$ and $B_n \subset T_n$ for every $n \in \mathbb{N}$, then $[(A_n)_n, (B_n)_n]$ is (SI).

In an analogous way it is possible to define the locally property of Moltó (Lf), the locally property of Seever, and so on (see also Aizpuru (1996)).

In Gassó (1992) property (Lf) is studied, and it is proved that $(Lf) \Rightarrow (NB)$ by means of diagonal arguments (Theorem 3.3), and an example of an algebra fulfilling property (NB) but not property (G) is given (see Proposition 3.19).

In Aizpuru (1996), Proposition 2.1, it was proved that an algebra satisfying (CCC) has property $(LSCP)$ if and only if it satisfies property $(LSIP)$. Again in Aizpuru (1996), in Corollary 2.3 it was proved that property $(LSIP)$ implies property (NB), in Proposition 2.5 it was shown that there exists an algebra with property (NB) and without property (Lf), though to find an algebra with property (NB) and without $(LSIP)$ is still an open problem (see also Remark 2.6).

Let L be a Boolean algebra and $\eta : L \to [0,+\infty)$ be a submeasure. We say that η is a *quasi* σ*-measure* iff for each disjoint sequence $(a_n)_n$ in L there are a subsequence $(b_n)_n$ of $(a_n)_n$ and an element $b \in L$ with $\eta(b_n - b) = 0$ for each $n \in \mathbb{N}$ and $\lim_n \eta(b - \bigvee_{j=1}^{n} b_j) = 0$. We say that L satisfies the *Drewnowski property* (in short, (DR)) iff every (s)-bounded submeasure η on L is a quasi σ-measure. The following implications hold:

- $(SIP) \Rightarrow (DR)$ (Popa (1997), Proposition 5),

- $(DR) \Rightarrow (BJ)$ with respect to each abelian topological group (Popa (1997), Theorem 6).

Avallone and Lepellere (1998) considered modular functions defined in a suitable orthomodular lattice L endowed with property (SCP) and with values in an abelian topological group R (a function $m : L \to R$ is said to be *modular* iff $m(a \vee b) + m(a \wedge b) = m(a) + m(b)$ for every $a, b \in L$) and proved Cafiero (Theorem 2.18 and Corollary 2.20), Brooks-Jewett (Corollary 2.21), Nikodým convergence (Corollary 2.22) and Nikodým boundedness theorems (Corollaries 3.2.7 and 3.2.8), where L is assumed to have only property (SIP)). Moreover, some equivalence results between the Vitali-Hahn-Saks and the Brooks-Jewett theorem were given, together with some relation with properties (NB) and (G) (see Theorem 3.3.10). Furthermore, in Avallone and Lepellere (1998) also some (weak) compactness results in this context were given.

Salvati (2000), in Theorem 7, by means of diagonal techniques, proved a Brooks-Jewett theorem for k-triangular functions defined on an orthomodular poset with the subsequential completeness property and with values in a *triangular semigroup*, that is in an abelian semigroup R endowed with a non-negative function $f : R \to \mathbb{R}$ such that $f(0) = 0$ and $|f(x+y) - f(x)| \leq f(y)$ for every $x, y \in R$.

Some other Vitali-Hahn-Saks-type theorems in the orthomodular poset setting were given in De Simone, Navara and Pták (2002), Theorems 2.1, 2.3-2.5, 2.9.

Matheron (2001) (see Main Lemma) gave new proofs of Schur, Banach-Steinhaus, Nikodým boundedness, Vitali-Hahn-Saks and Orlicz-Pettis-type theorems, deducing them from a lemma on subseries convergence in abelian topological groups.

Bennett, Boos and Leiger (2002) investigated some fundamental properties of spaces of sequences in connection with matrices, and proved again that the class of all subsets of $\mathbb{N}$ having asymptotic density zero satisfies property (NB) (see also Theorem 7.1). Moreover they posed the conjecture whether (NB) is satisfied by the class of all sets $K \subset \mathbb{N}$ having uniform asymptotic density zero, namely the sets K for which

$$\limsup_n \left(\sup_{l \geq 0} \frac{\#(K \cap \{l+1, l+2, \ldots, l+n\})}{n} \right) = 0 .$$

De Lucia and Pap (2003), using matrix methods, diagonal arguments and a Drewnowski-type theorem (see Lemma 2.4), proved a Cafiero-type theorem for finitely additive real-valued measures defined on rings satisfying property (SCP) (Theorem 3.4) and deduced, in this context, some Nikodým convergence, Brooks-Jewett and Nikodým boundedness theorem for the finitely additive case (see Theorems 4.1, 4.2 and 4.3 respectively).

Aizpuru, Gutiérrez-Dávila and Pérez-Fernández (2003) gave some Orlicz-Pettis-type theorems in connection with properties (BJ), (G) and (NB) (see Theorem 3.2), using sliding hump technique in the matrix context and the classical Brooks-Jewett theorem (see also Theorem 2.5).

Brooks, Saitô and Wright (2003) (see Theorem 3.2 and Corollary 3.3) proved a (non-commutative) Dieudonné-type theorem and a weak compactness result for C^*-algebras. In the same setting, there are some Vitali-Hahn-Saks-type theorems (see Brooks, Saitô and Wright (2004), Theorems 3.8, 5.5 and Corollary 3.9, Brooks and Wright (2004), Theorem 7 and Corollary 8), some weak compactness result related with control measures (see also Brooks, Saitô and Wright (2004), Proposition 4.3) and some results on weak compactness together with uniform boundedness theorems versions (see Brooks, Saitô and Wright (2005), Theorems 1.8 and 3.4, Proposition 2.5; Broks and Wright (2004), Lemma 1).

Aizpuru, Gutiérrez-Dávila and Sala (2004) (see Theorems 2 and 3) gave sufficient conditions on convergence and unconditional convergence of series of compact operators.

Aizpuru and Gutiérrez-Dávila (2004), using the sliding hump technique, proved a basic matrix-type theorem for normed space-valued double sequences (Theorem 2.3), and deduced, again with similar methods, some properties of convergence in several spaces of vector sequences (Corollaries 3.8 and 3.10, Theorems 3.12 and 4.1) and some interchange theorem for double series in a normed space (Theorem 3.6).

Stuart and Abraham (2004) (see Theorems 2.7 and 3.4), using diagonal arguments, submeasures and Drewnowski-type techniques, gave some sufficient conditions for a ring of subsets to have properties (NB) and (BJ) in the Banach space context (with respect to a fixed Banach space). Moreover, in the same paper, they deduced that the rings of the subsets of $\mathbb{N}$ having asymptotic density and uniform asymptotic density zero (see Chapter 2 for more details) satisfy property (NB), answering positively to the conjecture posed in Bennett, Boos and Leiger (2002). In Stuart and Abraham (2004) it is shown also that these rings fail (SCP), and consequently do not fulfil property (BJ).

More precisely, let X be a Banach space, $\mathcal{L}$ be a ring of subsets of an abstract set G, $(m_n)_n$ be a sequence of (s)-bounded X-valued measures defined on $\mathcal{L}$, such that the sequence $(\| m_n \|)_n$ of the corresponding total variations is not bounded.

A *bounding sequence* for $(m_n)_n$ is a disjoint sequence $(E_n)_n$ in $\mathcal{L}$ such that $\sup_n \| m_n(E_n) \| = +\infty$. A ring $\mathcal{L}$ is said to have the *bounded subsequential completeness property* (shortly, $(BSCP)$) iff for every sequence $(m_n)_n$ of (s)-bounded X-valued measures such that $(\| m_n \|)_n$ is not bounded and having a bounding sequence, there is a sequence $(E_n)_n$ in $\mathcal{L}$ with the property that every subsequence $(E_{n_k})_k$ of $(E_n)_n$ has a sub-subsequence $(E_{n_{k_l}})_l$ with $\bigcup_{l=1}^{\infty} E_{n_{k_l}} \in \mathcal{L}$.

A *norming sequence* for $(m_n)_n$ is any disjoint sequence $(E_n)_n$ in $\mathcal{L}$ such that there is a positive real number C_0 with $\| m_n(E_n) \| \geq C_0 \, ||| m_n |||$ for every $n \in \mathbb{N}$. A ring $\mathcal{L}$ has the *norm subsequential completeness property* (shortly, $(NSCP)$) iff for each sequence $(m_n)_n$ of (s)-bounded measures from $\mathcal{L}$ to X, such that $\lim_n m_n(E)$ exists in X for every $E \in \mathcal{L}$, the sequence $(m_n)_n$ has a norming sequence $(E_n)_n$ such that every subsequence $(E_{n_k})_k$ of $(E_n)_n$ has a sub-subsequence $(E_{n_{k_l}})_l$ with $\bigcup_{l=1}^{\infty} E_{n_{k_l}} \in \mathcal{L}$. It is proved that $(BSCP) \Rightarrow (NB)$ (see Stuart and Abraham (2004), Theorem 2.7) and that $(NSCP) \Rightarrow (BJ)$ (see Stuart and Abraham (2004), Theorem 3.4). In Boos and Leiger (2008), Theorems 1.5 and 3.1.1, some results of Stuart and Abraham (2004) were extended to ideals generated by non-negative infinite matrices.

Ferrando and Sánchez Ruiz (2004) presented a survey about rings with property (NB) in connection with properties (BJ), (G), (SIP), (I) of Seever and (f), and investigated also some fundamental properties of spaces of vector-valued simple functions and of vector-valued measurable functions.

Aizpuru and Tamayo (2006) (see Theorem 3.1 and Remark 3.2) gave a Brooks-Jewett and a Nikodým boundedness-type theorem for real-valued measures defined on effect algebras and formulated sufficient conditions for (BJ) and (NB) properties. Moreover, in Aizpuru, Gutiérrez-Dávila and Sala (2006) (see Theorems 5.1, 5.5 and Corollary 5.8) the main properties of the Cesàro, weak Cesàro and weak* Cesàro convergence were investigated and some Orlicz-Pettis-type theorems were proved. Some other extensions of the Orlicz-Pettis theorem were given in Swartz and Stuart (1998), Corollary 2.4, using Schauder basis, in Stuart and Swartz (2005), Theorem 9 using the matrix theorem 2.151, in Aizpuru,

Armario and Pérez-Fernández (2008), Theorem 3.2 using almost convergence, in Aizpuru, Pérez-Eslava and Seoane-Sepúlveda (2009), Theorem 4.1 using properties of regular matrices, in Aizpuru, Nicasio-Llach and Sala (2010), Theorem 2.1 using statistical convergence and Cesàro means and in Aizpuru, Nicasio-Llach and Rambla-Barreno (2010), Theorem 2.4 using basic matrix-type theorems for the statistical convergence (see Aizpuru, Nicasio-Llach and Rambla-Barreno (2010), Lemma 2.2).

By means of techniques based to those in Weber (1986), A. Avallone (2006) proved some Nikodým boundedness (Theorem 3.6), Cafiero (Theorem 4.3) and Brooks-Jewett-type theorem (Theorem 4.5) for (s)-bounded set functions, defined on a D-lattice satisfying property (SCP) and taking values in an abelian topological group.

Aizpuru, García-Pacheco and Pérez-Eslava (2007) (see Theorems 3.3 and 3.5) proved some sufficient conditions for uniform convergence of weakly unconditionally Cauchy sequences, by means of regular matrices.

In Aizpuru, Nicasio-Llach and M. Tamayo (2007), Theorem 1, using the sliding hump technique, a Hahn-Schur-type theorem was given for double sequences, taking values in suitable effect algebras.

Avallone, Rinauro and Vitolo (2007), using diagonal techniques, proved some versions of Nikodým boundedness (Theorem 2.18) and Vitali-Hahn-Saks-type theorems (Theorem 3.6 and 3.8) for lattice group-valued measures defined on D-lattices.

Cavaliere and de Lucia (2007), using techniques similar to those in Candeloro and Letta (1985), proved a Cafiero-type (Theorem 4.1) and a Dieudonné-type theorem (Theorem 4.1) for set functions with values in an abelian complete Hausdorff semigroup endowed with a uniformity, in which the role of the class of "closed" (or "compact") sets is played by a joint semilattice (that is a nonempty family closed under finite unions) and the role of the family of "open" sets is played by a meet semilattice (that is a nonempty family closed under finite intersections) satisfying the subsequential interpolation property (*SIP*) (a meet semilattice $\mathcal{G}$ contained in an algebra $\mathcal{L}$ is said to have *property* (*SIP*) iff for any pairwise disjoint sequence $(L_n)_n$ in $\mathcal{G}$ there are a subsequence $(L_{n_k})_k$ and a ring satisfying (SIP), containing every set L_{n_k} and contained in $\mathcal{G}$).

Using diagonal arguments and Constantinescu-type techniques, some versions of Cafiero theorems (see Ventriglia (2007), Theorem 1 and Ventriglia (2008a), Theorem 3), Phillips and Brooks-Jewett-type theorems (see Ventriglia (2007), Theorem 2 and Ventriglia (2008a), Theorem 4) and Nikodým convergence theorems (see Ventriglia (2008a), Theorem 5) were proved for measures, defined in Vitali spaces with the subsequential interpolation property and with values in an abelian Hausdorff topological group. Using diagonal arguments again, a Cafiero-type theorem for non-negative real-valued k-triangular set functions defined on an orthomodular lattice satisfying the subsequential interpolation property was proved in Ventriglia (2008b), Theorem 3.2 and a Brooks-Jewett-type theorem was deduced in Ventriglia (2008b), Theorem 3.3.

Using diagonal techniques similar to those in Candeloro and Letta (2005), P. de Lucia and E. Pap (2008) (see Theorems 2.8, 3.1 and 3.2) proved some other versions of Cafiero and Dieudonné-type theorems for real-valued measures.

Using diagonal methods, some basic matrix theorem for normed space-valued double sequences with respect to the statistical convergence was given in Aizpuru and Nicasio-Llach (2008a), Theorem 3, and in Aizpuru and Nicasio-Llach (2008b), Theorem 4). In Chapter 4, we will see some extensions in the lattice group and filter convergence context.

Using the Baire category theorem, the following extensions of the uniform boundedness principle were given.

Theorem 2.194 (Arvanitakis and Stein (2008), Theorem 13) *Let $n\in\mathbb{N}$, X be a Banach space, Y be a normed vector space, $L(X,Y)$ be the space of all linear continuous operators from X to Y, A_λ, $\lambda\in\Lambda$, be a family of pointwise bounded subsets of $L(X,Y)$ such that $\#(A_\lambda)\le n$ for every $\lambda\in\Lambda$, where $\#(A_\lambda)$ denotes the cardinality of A_λ. Then the family $(\|A_\lambda\|)_\lambda$ is equibounded.*

Theorem 2.195 (Arvanitakis and Stein (2008), Theorem 15) *Let X, Y be two Banach spaces, $L(X,Y)$ be as in Theorem 2.194, A_k, $k\in\mathbb{N}$, be a sequence in $L(X,Y)$, $n\in\mathbb{N}$ be such that $\#(A_k)\le n$ for each $k\in\mathbb{N}$. For every $x\in X$, set*

$$C_x:=\{y\in Y: \text{there is } T_k\in A_k \text{ with } \lim_k T_k(x)=y\}.$$

If $C_x\neq\emptyset$ for each $x\in X$, then there is $T\in L(X,Y)$ with $T(x)\in C_x$ for every $x\in X$.

By means of the sliding hump technique, Cavaliere and de Lucia (2008) proved a Cafiero-type theorem for non-additive topological space-valued set functions, defined on rings having property (*SCP*) (Theorem 3.1) and deduced a Brooks-Jewett-type theorem (Theorem 3.3).

Some Vitali-Hahn-Saks-type theorems and compactness results concerning von Neumann algebras and C^*-algebras were proved in Chetcuti and Hamhalter (2006), Theorems 4.1, 4.6, and in Chetcuti and Hamhalter (2009a), Theorems 2.1 and 2.6. Chetcuti and Hamhalter (2009b), in the same context, gave some other results on weak compactness (see Proposition 5), Vitali-Hahn-Saks-Nikodým-type theorems (see Corollary 7), and the Brooks-Jewett property (see Theorems 9, 11-12 and Corollary 13).

By means of Drewnowski-type techiques, Cavaliere and de Lucia (2009) (see Theorem 2.1) proved some equivalence results between Brooks-Jewett, Vitali-Hahn-Saks and Nikodým convergence theorems for suitable non-additive set functions, defined in a Boolean ring satisfying property (*SCP*) and with values in a Hausdorff topological space.

Barbieri (2009a) (see Theorems 3.6 and 3.8) proved some version of the Nikodým boundedness theorem for regular modular measures, defined in D-lattices with the subsequential interpolation property or the subsequential completeness property and with values in a quasi-normed abelian group. Using techniques similar to those in Candeloro and Letta (1985), some Dieudonné-type theorems were proved in Barbieri (2009b) (see Theorems 4.5 and 5.1) for real-valued measures defined in D-lattices, and in Barbieri (2009c) (see Propositions 10-12 and Theorem 13) for measures defined on MV-algebras and with values in Hausdorff complete topological groups. Note that, in this context, the role of the "open" and "closed" or "compact" sets is played by two sublattices $\mathcal{G}$, $\mathcal{H}$ of L, closed under the operation $\oplus$, where $\oplus$ is as in (1.4), and such that $x \ominus y \in \mathcal{G}$ for each $x \in \mathcal{G}$ and $y \in \mathcal{H}$.

Barbieri (2009d) (see Theorems 4.5 and 4.6) proved some Vitali-Hahn-Saks and Nikodým convergence theorems, using the corresponding classical versions, for measures with values in a Hausdorff locally convex topological vector space and defined in a D-lattice L, where the continuity of the involved measures is intended with respect to a uniform structure on L, which makes the operations $\vee$, $\wedge$ and $\ominus$ uniformly continuous (see also Avallone and Vitolo (2005), Theorems 3.4 and 4.4). In this context, some properties of the range of the measures are studied (see also Barbieri (2009d), Theorems 4.12 and 4.13) and some decomposition theorems are proved (see also Barbieri (2009d), Theorems 4.10 and 4.15). Some other decomposition theorems in similar abstract contexts can be found, for instance, in Avallone, Barbieri and Vitolo (2003, 2008 and 2010), Avallone, Barbieri and Vitolo and Weber (2009), Avallone and Vitolo (2003, 2009 and 2013), Barbieri, Lepellere and Weber (2001), Brooks and Wright (2001), Cavaliere, de Lucia, De Simone and Ventriglia (2013), de Lucia and Morales (1994 and 1998), De Simone (1998), De Simone and Navara (2001-2002), Pap (1994, 1997b, 1998), Weber (2009).

Ventriglia (2010) proved a Dieudonné-type theorem for quasi-triangular set functions with values in a complete Hausdorff uniform space (see Theorem 4.3), using a diagonal argument and a Cafiero-type criterion for uniformly quasi-triangular set functions (see Theorem 4.1).

Cavaliere, de Lucia and Ventriglia (2010) proved, by means of diagonal methods, some Drewnowski- (Lemma 4.1), Brooks-Jewett- (Theorems 5.2 and 6.1) and Cafiero-type theorems (Theorem 6.3) for non-additive set functions, defined in an orthomodular lattice endowed with the subsequential interpolation property and with values in a complete Hausdorff uniform space.

In Aizpuru, Moreno-Pulido and Rambla-Barreno (2013), Theorems 2.1 and 2.5, some versions of the Phillips lemma and the Brooks-Jewett theorem were proved, for Banach-space valued measures, defined on effect algebras.

Finally Barbieri (2014), with techniques similar to those used in Cavaliere and de Lucia (2008), proved a Cafiero-type theorem for topological space-valued functions defined on D-lattices endowed with the subsequential completeness property (see Theorem 3.1) and deduced a Brooks-Jewett-type theorem (see Theorem 3.2).

CHAPTER 2

Basic Concepts and Results

Abstract: In this chapter we recall the fundamental concepts, tools and results which will be used throughout the book, that is filters/ideals, filter/ideal convergence, lattice groups, Riesz spaces and properties of (ℓ)-group-valued measures, and some related fundamental techniques in this setting, like for instance different kinds of convergence, the Fremlin lemma, the Maeda-Ogasawara-Vulikh representation theorem, the Stone Isomorphism technique and the existence of suitable countably additive restrictions of finitely additive strongly bounded measures. We will prove some main properties of filter/ideal convergence and of lattice group-valued measures.

Keywords: (s)-bounded measure, (Uniform) asymptotic density, absolutely continuous measure, additive measure, almost convergence, block-respecting filter, Carathéodory extension, diagonal filter, filter compactness, filter divergence, filter, filter/ideal convergence, Fremlin Lemma, ideal, lattice group, Maeda-Ogasawara-Vulikh theorem, matrix method, P-filter, regular measure, Stone extension.

In this eBook we will present several versions of limit theorems for lattice group-valued measures, namely Schur, Brooks-Jewett, Nikodým, Vitali-Hahn-Saks and Dieudonné-type theorems. We treat both the case of the classical pointwise order or (D)-convergence of the involved measures (often, with respect to a same order sequence or regulator) and the setting of filter pointwise convergence. Since this kind of convergence is in general strictly weaker than the classical one, in general, as we will see in the sequel, one cannot expect to obtain results, analogous to the ones existing in the classical context. However, under certain not restrictive hypotheses, it is possible to get several results also in this direction.

In this chapter we consider the basic notions, tools and properties which will be useful in order to prove the main convergence theorems. First, we deal with some fundamental properties of filters/ideals and some characterization of ultrafilters, and we recall the classical concept of densities, matrix methods, filter/ideal convergence and its fundamental properties. Moreover, we consider almost convergence, giving a characterization and a comparison with filter/ideal convergence, and some filter/ideal compactness properties. Furthermore, we deal with some basic notions on lattice groups and Riesz spaces, and in particular we point out some important mathematical tools in these structures, which will be useful in the sequel. We consider the fundamental properties of order convergence

Antonio Boccuto & Xenofon Dimitriou

and (D)-convergence in lattice groups, which replace the so-called ε-technique. We deal with the Fremlin lemma, which allows to replace countably many order sequences or regulators with a single (O)- or (D)-sequence, and with the Maeda-Ogasawara-Vulikh representation theorem for Archimedean lattice groups, by means of which it is possible to consider the involved lattice group as a suitable subgroup of continuous extended real-valued functions, and to study several properties of lattice group-valued measures relating them with the corresponding ones of the real-valued measures. We investigate also filter/ideal convergence/divergence in the (ℓ)-group setting, and its main properties. We present the main topics on lattice group-valued measures, in particular (uniform) (s)-boundedness, σ-additivity, absolute continuity and continuity with respect to a Fréchet-Nikodým topology, both in the classical sense and with respect to a single order or (D)-sequence. In particular we deal with a characterization of (uniform) global absolute continuity and some relation between regularity and σ-additivity, and we show that the concepts of (s)-boundedness (and the related ones) in the classical like sense and with respect to a single (O)-sequence or regulator are in general different. Finally, we relate finite additivity to countable additivity of lattice group-valued measures, using a Drewnowski-type approach, to find (global) countably additive restrictions of (global) (s)-bounded measures on a suitable σ-algebra, and by means of the Stone Isomorphism technique, which allows to construct some (global) countably additive extensions of (global) (s)-bounded lattice group-valued measures. To this aim, we will use some density properties of suitable σ-algebras. We also present some extension results for lattice group-valued measures, not necessarily finitely additive, but compatible with the operations of supremum and infimum.

2.1. Filters and ideals

2.1.1 Statistical Convergence and Matrix Methods

We begin with the notion of filter/ideal and the related convergences, which are extensions of the statistical convergence. We consider statistical convergence and matrix methods of convergence, which are related with suitable densities. We will associate to them some filters/ideals, by considering the class of all subsets having one/zero density respectively. Statistical convergence was introduced by Fast (1951), Steinhaus (1951) and Schoenberg (1959), and is related with the *asymptotic density* of the natural numbers (see also Buck (1946 and 1953)). Among the historical papers on the evolution of the concept of statistical convergence, we quote Šalát (1980), Freedman and Sember (1981a-b), Fridy (1985 and 1993). In Aizpuru and Nicasio-Llach (2008a) some spaces of vectorial sequences defined by statistical convergence and some of their basic properties are investigated. Further developments are, for instance, in Çakalli (1996), Connor (1988, 1990 and 1992), Connor, Fridy and Kline (1994), Connor and Kline (1996), Fridy and Miller (1991), Fridy and Orhan (1997), Kolk (1991 and 1993), Kostyrko, Mačaj, Šalát and Strauch (2000), Maddox (1988 and 1989), Miller (1995), Mursaleen (2000), Rath and Tripathy (1994), Savaş (1992).

Let $K \subset \mathbb{N}$. If $n \geq 1$ and $l \geq 0$, we denote by $K(l,n)$ the cardinality of the set $K \cap \{l+1, l+2, \ldots, l+n\}$. The *lower* and *upper asymptotic density* of a set $K \subset \mathbb{N}$ are defined by

$$\underline{\delta}(K) = \liminf_n \frac{K(0,n)}{n}, \quad \overline{\delta}(K) = \limsup_n \frac{K(0,n)}{n} \tag{2.1}$$

respectively. If the limit $\lim\limits_n \dfrac{K(0,n)}{n}$ exists in $\mathbb{R}$, then the common value in (2.1) is said to be the *asymptotic density of* K and is denoted by $\delta(K)$.

Let (X,d) be a metric space. If $(x_k)_k$ is a sequence in X, we say that $(x_k)_k$ *converges statistically to* $x_0 \in X$ (shortly, $\lim\limits_k (St)\, x_k = x_0$) iff for each $\varepsilon > 0$ we have $\delta(A(\varepsilon)) = 0$, where $A(\varepsilon) := \{k \in \mathbb{N} : d(x_k, x_0) > \varepsilon\}$.

The concepts of asymptotic density and statistical convergence have been generalized to the context of summability matrices, giving rise to different methods of convergence, which have applications in several branches of Mathematics (see for instance Bell (1973), Bennett and Kalton (1974), Bhardwaj and Bala (2007), Boos and Cass (2000), Boos and Seydel (1999), Connor, Ganichev and Kadets (2000), Connor and Grosse-Erdmann (2003), Dawson (1970), Iwiński (1972), Kolk (1998), Lorentz (1948), Móricz (2004), Petersen (1966), Savaş, Das and Dutta (2012-2013), and in particular in Approximation Theory (see Anastassiou and Duman (2001), Angeloni and Vinti (2005), Bardaro, Boccuto, Dimitriou and Mantellini (2013a-b), Bardaro and Mantellini (2006a-b, 2009 and 2013), Bardaro, Musielak and Vinti (2003), Boccuto, Candeloro and Sambucini (2014), Boccuto and Dimitriou (2013a-b, 2014b), Butzer (1983),

Butzer, Higgins and Stens (2000), Butzer and Nessel (1971), Gadjev and Orhan (2002), Mantellini (2013), Vinti (1998 and 2001) and the literature therein).

Let $A := (a_{j,k})_{j,k}$ be an infinite matrix of elements of $\mathbb{R}$. For a given sequence $\underline{y} = (y_k)_k$ in $\mathbb{R}$, the A*-transform of* $\underline{y}$, denoted by $(A\underline{y})_j$, is given by

$$(A\underline{y})_j := \sum_{k=1}^{\infty} a_{j,k} y_k, \tag{2.2}$$

provided that the series in (2.2) converges for each $j \in \mathbb{N}$.

The infinite matrix A is *regular* iff $\lim_j \lim_j (A\underline{y})_j = L$ whenever $\lim_k y_k = L$, with $L \in \mathbb{R}$.

We now recall the Toeplitz conditions on regularity of a matrix A (Toeplitz (1911), Theorem, see also Anastassiou and Duman (2011), Theorem 1.6).

Theorem 1.1 *An infinite matrix* $A = (a_{j,k})_{j,k}$ *is regular if and only if the following conditions are satisfied:*

(a) $\sup_j (\sum_{k=1}^{\infty} | a_{j,k} |) < +\infty$,

(b) $\lim_j a_{j,k} = 0$ *for each k,*

(c) $\lim_j (\sum_{k=1}^{\infty} a_{j,k}) = 1$.

Using regular matrices, Freedman and Sember (1981a) extended the idea of the statistical convergence to A-statistical convergence as follows.

Let $A = (a_{j,k})_{j,k}$ be a non-negative regular matrix. The A*-density* of a subset $K \subset \mathbb{N}$ is defined by

$$\delta_A(K) := \lim_j \sum_{k \in K} a_{j,k}, \tag{2.3}$$

provided that the limit in (2.3) exists in $\mathbb{R}$. Similarly it is possible to define the concepts of A*-upper density* and A*-lower density* of $K \subset \mathbb{N}$ as follows:

$$\overline{\delta}_A(K) := \limsup_j \sum_{k \in K} a_{j,k}, \quad \underline{\delta}_A(K) := \liminf_j \sum_{k \in K} a_{j,k}$$

respectively.

A sequence $(x_k)_k$ in a metric space (X,d) is A*-convergent* to $x_0 \in \mathbb{R}$ (and in this case we write $(A) \lim_k x_k = x_0$) iff $\delta_A(A(\varepsilon)) = 0$ or equivalently $\lim_j \sum_{k \in A(\varepsilon)} a_{j,k} = 0.$

Observe that, when $A := \mathcal{C}_1 = (c_{j,k})_{j,k}$ is the Cesàro matrix, defined by

$$c_{j,k} := \begin{cases} \dfrac{1}{j} & \text{if } 1 \le k \le j, \\ 0 & \text{otherwise,} \end{cases}$$

the A-density and A-convergence become the usual asymptotic density and statistical convergence respectively.

We now recall the concepts of *uniform* (or *Banach*) (*asymptotic*) *density* for subsets of $\mathbb{N}$. (Note that it is strictly stronger than asymptotic density, see also Banach (1932), Remarks to Chapter II). For a related literature, see also Baayen and Helmberg (1965), Baláž and Šalát (2006), Brown and Freedman (1990), Buck (1946 and 1953), Freedman and Sember (1981a-b), Gáliková, László and Šalát (2002), Giuliano-Antonini and Grekos (2007), Grekos (2005), Grekos, Toma and Tomanová (2010), Jin (2001), Niven, Zuckerman and Montgomery (1991), Raimi (1963).

Given $K \subset \mathbb{N}$ and $n \in \mathbb{N}$, set $a_n(K) = \inf_{l \ge 0} K(l,n), \quad b_n(K) = \sup_{l \ge 0} K(l,n).$

The *upper* and *lower uniform density* of a set $K \subset \mathbb{N}$ are given by

$$\overline{u}(K) = \lim_n \frac{b_n(K)}{n}, \quad \underline{u}(K) = \lim_n \frac{a_n(K)}{n} \qquad \textbf{(2.4)}$$

respectively. If the quantities in (2.4) are equal, then their common value is said to be the *uniform density of* K and is denoted by $u(K)$.

Observe that for every $K \subset \mathbb{N}$ we get: $\underline{u}(K) \le \underline{\delta}(K) \le \overline{\delta}(K) \le \overline{u}(K)$.

Thus, if K has uniform density β, then its asymptotic density is equal to β. However, the set

$$K = \bigcup_{n=1}^{\infty} \{n!+1, n!+2, \ldots, n!+n\}$$

has asymptotic density zero, while its upper uniform density equals to 1, and so its uniform density does not exist (see also Grekos, Toma and Tomanová (2010)).

The classes of the sets of asymptotic density zero and of uniform null density, as well as the family of all subsets $K \subset \mathbb{N}$ with $\delta_A(K) = 0$, where A is a non-negative regular matrix, are particular cases of *ideals* of $\mathbb{N}$.

For other notions of density the reader could be referred to Grekos (2005) and the references contained therein.

2.1.2. Basic Concepts and Properties of Ideals/Filters

We now recall some fundamental notions and properties of ideals and filters.

Let Λ be any nonempty set, and $\mathcal{P}(\Lambda)$ be the class of all subsets of Λ. A family of sets $\mathcal{I} \subset \mathcal{P}(\Lambda)$ is called an *ideal* of Λ iff $A \cup B \in \mathcal{I}$ whenever A, $B \in \mathcal{I}$ and for each $A \in \mathcal{I}$ and $B \subset A$ we get $B \in \mathcal{I}$. A collection of sets $\mathcal{F} \subset \mathcal{P}(\Lambda)$ is a *filter* of Λ iff $A \cap B \in \mathcal{F}$ for all A, $B \in \mathcal{F}$ and for every $A \in \mathcal{F}$ and $B \supset A$ we have $B \in \mathcal{F}$.

An ideal $\mathcal{I}$ (resp. a filter $\mathcal{F}$) of Λ is said to be *non-trivial* iff $\mathcal{I} \neq \emptyset$ and $\Lambda \notin \mathcal{I}$ (resp. $\mathcal{F} \neq \emptyset$ and $\emptyset \notin \mathcal{F}$).

Let $(\Lambda, \ge)$ be a directed set. A non-trivial ideal $\mathcal{I}$ of Λ is said to be (Λ)-*admissible* iff $\Lambda \setminus M_\lambda \in \mathcal{I}$ for all $\lambda \in \Lambda$, where $M_\lambda := \{\zeta \in \Lambda : \zeta \ge \lambda\}$.

A non-trivial filter $\mathcal{F}$ of Λ is (Λ)-*free* iff $M_\lambda \in \mathcal{F}$ for every $\lambda \in \Lambda$.

Given an ideal $\mathcal{I}$ of Λ, we call *dual filter* of $\mathcal{I}$ the family $\mathcal{F} := \{\Lambda \setminus I : I \in \mathcal{I}\}$. In this case we say that $\mathcal{I}$ is the *dual ideal* of $\mathcal{F}$ and we get $\mathcal{I} := \{\Lambda \setminus F : F \in \mathcal{F}\}$.

The following result holds.

Proposition 1.2 (Boccuto, Dimitriou and Papanastassiou (2012a), Proposition 2.3) *Let $(\Lambda, \geq)$ be a directed set, $\mathcal{F}$ be a (Λ)-free filter of Λ and M be an element of $\mathcal{F}$. Then $(M, \geq)$ is a directed set, where the order is that induced by Λ.*

Proof: It is readily seen that $(\Lambda, \geq)$ inherits a reflexive and transitive order on M, said $\geq$ again. Let now λ_1, $\lambda_2 \in M$. There exists $\tau \in \Lambda$ such that $\tau \geq \lambda_1$, $\tau \geq \lambda_2$. Since $\mathcal{F}$ is by hypothesis (Λ)-free, then $M_\tau := \{\lambda \in \Lambda : \lambda \geq \tau\} \in \mathcal{F}$. Since M, $M_\tau \in \mathcal{F}$, we get $M \cap M_\tau \in \mathcal{F}$ too. Pick an element $\lambda_0 \in M$, $\lambda_0 \geq \tau$: we get that $\lambda_0 \geq \lambda_1$, $\lambda_0 \geq \lambda_2$. □

The filter $\mathcal{F}_{\text{cofin}}$ is the filter of all subsets of $\mathbb{N}$ whose complement is finite, and its dual ideal $\mathcal{I}_{\text{fin}}$ is the family of all finite subsets of $\mathbb{N}$. The filter $\mathcal{F}_{\text{st}}$ is the filter of all subsets of $\mathbb{N}$ having asymptotic density one, while its dual ideal $\mathcal{I}_{\text{st}}$ is the family of all parts of $\mathbb{N}$, having null asymptotic density.

A nonempty family $\mathcal{B}' \subset \mathcal{P}(\Lambda)$ is said to be a *filter base of* Λ iff for every A, $B \in \mathcal{B}'$ there is an element $C \in \mathcal{B}'$ with $C \subset A \cap B$. Note that, if $\mathcal{B}'$ is a filter base of Λ, then the family $\mathcal{F} = \{A \subset \Lambda$: there is $B \in \mathcal{B}'$ with $B \subset A\}$ is a filter of Λ. We call it the *filter generated by* $\mathcal{B}'$.

If $(\Lambda, \geq)$ is a directed set and $\mathcal{B}' = \{M_\lambda : \lambda \in \Lambda\}$, then $\mathcal{B}'$ is a filter base, and the filter $\mathcal{F}_\Lambda$ generated by $\mathcal{B}'$ is a (Λ)-free filter (see also Boccuto, Dimitriou and Papanastassiou (2012a)). In general $\mathcal{F}_\Lambda \supsetneq \mathcal{F}_{\text{cofin}}$, but in $\mathbb{N}$ we have $\mathcal{F}_{\mathbb{N}} = \mathcal{F}_{\text{cofin}}$. We denote by $\mathcal{I}_\Lambda$ the dual ideal of $\mathcal{F}_\Lambda$. Of course, $\mathcal{I}_{\mathbb{N}} = \mathcal{I}_{\text{fin}}$.

A (Λ)-admissible ideal $\mathcal{I}$ of Λ is said to be *maximal* iff, for every $A \subset \Lambda$, either A or $\Lambda \setminus A$ belongs to $\mathcal{I}$. A (Λ)-free filter $\mathcal{F}$ of Λ is called an *ultrafilter* iff its dual ideal is maximal. A nonempty family $\mathcal{B}' \subset \mathcal{P}(\Lambda)$ is said to be a *filter base of* Λ iff for every A, $B \in \mathcal{B}'$ there is an element $C \in \mathcal{B}'$ with $C \subset A \cap B$. Note that, if $\mathcal{B}'$ is a filter base of Λ, then the family $\mathcal{F} = \{A \subset \Lambda$: there is $B \in \mathcal{B}'$ with $B \subset A\}$ is a filter of Λ. We call it the *filter generated by* $\mathcal{B}'$.

When $\Lambda = \mathbb{N}$ endowed with the usual order, an ideal of $\mathbb{N}$ is $(\mathbb{N})$-admissible if and only if it is non-trivial and contains all singleton sets of $\mathbb{N}$, if and only if it is non-trivial and contains the ideal $\mathcal{I}_{\text{fin}}$ of all finite subsets of $\mathbb{N}$. Similarly, we get that a filter of $\mathbb{N}$ is $(\mathbb{N})$-free if and only if it is non-trivial and contains the filter $\mathcal{F}_{\text{cofin}}$ of all cofinite subsets of $\mathbb{N}$. From now on, we call shortly *admissible* and *free* the $(\mathbb{N})$-admissible ideals and the $(\mathbb{N})$-free filters of $\mathbb{N}$ respectively.

Let Q be a countable set and $\mathcal{F}$ be a filter of Q. A subset of Q is $\mathcal{F}$*-stationary* iff it has nonempty intersection with every element of $\mathcal{F}$. We denote by $\mathcal{F}^*$ the family of all $\mathcal{F}$-stationary subsets of Q.

A filter $\mathcal{F}$ of Q is said to be *diagonal* iff for every decreasing sequence $(A_n)_n$ in $\mathcal{F}$ and for each $I \in \mathcal{F}^*$ there exists a set $J \subset I$, $J \in \mathcal{F}^*$ such that the set $J \setminus A_n$ is finite for all $n \in \mathbb{N}$ (see also Aviles Lopez, Cascales Salinas, Kadets and Leonov (2007), Boccuto, Dimitriou and Papanastassiou (2012b and 2012e)).

Observe that the definition of diagonal filter can be formulated equivalently even without requiring that the involved sequence $(A_n)_n$ is decreasing. Indeed, if $(A_n)_n$ is any sequence in $\mathcal{F}$ and $A_1^* := A_1$, $A_n^* := \bigcap_{l=1}^{n} A_l$, $n \in \mathbb{N}$, then the sequence $(A_n^*)_n$ is in $\mathcal{F}$ and decreasing, and so for every $I \in \mathcal{F}^*$ there is a set $J \in \mathcal{F}^*$, $J \subset I$ such that $J \setminus A_n^*$ is finite, and then a fortiori $J \setminus A_n \subset J \setminus A_n^*$ is finite too, for each $n \in \mathbb{N}$.

Given an infinite set $I \subset Q$, a *blocking* of I is a countable partition $\{D_k : k \in \mathbb{N}\}$ of I into nonempty finite subsets.

A filter $\mathcal{F}$ of Q is said to be *block-respecting* iff for every $I \in \mathcal{F}^*$ and for each blocking $\{D_k : k \in \mathbb{N}\}$ of I there is a set $J \in \mathcal{F}^*$, $J \subset I$ with $\#(J \cap D_k) = 1$ for all $k \in \mathbb{N}$, where # denotes the number of elements of the set into brackets.

If $I \in \mathcal{F}^*$, then the *trace* $\mathcal{F}(I)$ of $\mathcal{F}$ on I is the family $\{A \cap I : A \in \mathcal{F}\}$.

Observe that $\mathcal{F}(I)$ is a filter of *I*. Indeed, if F_1, $F_2 \in \mathcal{F}(I)$, then $(F_1 \cap F_2) \cap I = (F_1 \cap I) \cap (F_2 \cap I) \in \mathcal{F}$, and hence $F_1 \cap F_2 \in \mathcal{F}(I)$.

Let now $F \in \mathcal{F}$ and $F \cap I \subset F' \subset I$, and set $F^* := F' \cup F$: then $F^* \in \mathcal{F}$ and $F^* \cap I \supset F \cap I$. It is readily seen that $F' \subset F^* \cap I$. To prove the converse inclusion, observe that $F^* \cap I = (F' \cap I) \cup (F \cap I) \subset F'$. Hence, $F' = F^* \cap I$ belongs to $\mathcal{F}(I)$, and thus we get the claim.

Proposition 1.3 *If $\mathcal{F}$ is a block-respecting filter of* $\mathbb{N}$, *then $\mathcal{F}(I)$ is a block-respecting filter of I for every $I \in \mathcal{F}^*$.*

Proof: Let $I \subset \mathbb{N}$ be any $\mathcal{F}$-stationary set, $L \subset I$ be any $\mathcal{F}(I)$-stationary set and $\{D_k : k \in \mathbb{N}\}$ be any blocking of L. If $F \in \mathcal{F}$, then $\emptyset \neq L \cap F \cap I = L \cap F$, and so $L \in \mathcal{F}^*$. By hypothesis, there exists a set $J \in \mathcal{F}^*$, $J \subset L$, with $\#(J \cap D_k) = 1$ for all $k \in \mathbb{N}$. In particular, $\emptyset \neq L \cap F = L \cap F \cap I$. From this it follows that $J \in \mathcal{F}(I)^*$. Thus we get the assertion. □

A free filter $\mathcal{F}$ of $\mathbb{N}$ is a *P-filter* iff for every sequence $(A_n)_n$ in $\mathcal{F}$ there is a sequence $(B_n)_n$ in $\mathcal{F}$, such that the symmetric difference $A_n \Delta B_n$ is finite for all $n \in \mathbb{N}$ and $\bigcap_{n=1}^{\infty} B_n \in \mathcal{F}$.

An admissible ideal $\mathcal{I}$ of $\mathbb{N}$ is a *P-ideal* iff its dual filter $\mathcal{F} = \mathcal{F}(\mathcal{I}) = \{\mathbb{N} \setminus A : A \in \mathcal{I}\}$ is a P-filter, that is for every sequence $(C_n)_n$ in $\mathcal{I}$ there exists a sequence $(D_n)_n$ in $\mathcal{I}$, such that $C_n \Delta D_n$ is finite for every $n \in \mathbb{N}$ and $\bigcup_{n=1}^{\infty} D_n \in \mathcal{I}$.

Observe that every P-filter $\mathcal{F}$ of $\mathbb{N}$ is diagonal. Indeed, let $(A_n)_n$ be a sequence in $\mathcal{F}$ and $I \in \mathcal{F}^*$. As $\mathcal{F}$ is a P-filter, then by Balcerzak, Dems and Komisarski (2007), Proposition 1, there exists $J_0 \in \mathcal{F}$, with the property that $J_0 \setminus A_n$ is finite for every $n \in \mathbb{N}$. We claim that $J := I \cap J_0 \in \mathcal{F}^*$. Indeed, if E is any element of $\mathcal{F}$, then $J_0 \cap E \in \mathcal{F}$. So, as $I \in \mathcal{F}^*$, we have $\emptyset \neq I \cap J_0 \cap E = J \cap E$. By arbitrariness of E, $J \in \mathcal{F}^*$, and thus we get the claim. Therefore, the set J satisfies the condition requested in the definition of diagonal filter.

A particular class of filter, which are block-respecting and diagonal at the same time, is that of the *category respecting* filters. A filter $\mathcal{F}$ of $\mathbb{N}$ is said to be *category respecting* iff for every compact metric space K and for every family of closed subsets $(F_A)_{A\in\mathcal{F}}$ of K, if $F_A \subset F_B$ whenever $B \subset A$ in $\mathcal{F}$ and $K = \bigcup_{A\in\mathcal{F}} F_A$, then there is a set $B \in \mathcal{F}$ such that the interior of F_B is non-empty (see also Aviles Lopez, Cascales Salinas, Kadets and Leonov (2007), Theorem 4.3). Note that the converse is not true. Indeed, let $D = (D_n)_n$ be a disjoint partition of $\mathbb{N}$ into infinite subsets. For each sequence $C = (C_n)_n$ of finite subsets $C_n \subset D_n$, let $B_C := \bigcup_{n=1}^{\infty} (D_n \setminus C_n)$. The sets B_C form a filter base, and let $\mathcal{F}_d$ be the filter generated by it. Observe that $\mathcal{F}_d$ is diagonal, but not category respecting. It is readily seen that every filter with a countable base is category respecting. Furthermore note that, under the assumption of the continuum hypothesis, there exists a diagonal ultrafilter (see Aviles Lopez, Cascales Salinas, Kadets and Leonov (2007), Theorem 4.6).

Let $D = (D_n)_n$ be a disjoint partition of $\mathbb{N}$ into infinite subsets. For each sequence $C = (C_n)_n$ of finite subsets $C_n \subset D_n$ and every $q \in \mathbb{N}$, set $B_{q,C} := \bigcup_{n=q}^{\infty} (D_n \setminus C_n)$. The filter $\mathcal{F}_D$ generated by the sets of type $B_{q,C}$ is a non-diagonal and block-respecting filter.

The filter $\mathcal{F}_{\text{st}}$ is diagonal (since it is a P-filter), but not block-respecting. In Aviles Lopez, Cascales Salinas, Kadets and Leonov (2007), Theorem 5.3, it is proved that the filter $\mathcal{F}_{\text{st}}$ is *weakly l^1-complete*, namely for each bounded sequence $(x_n)_n$ in l^1, $\mathcal{F}$-convergent in the topology $\sigma(l_1^{**}, l_1^*)$, its weak* limit $x \in l_1^{**}$ belongs to l_1. Note that any ultrafilter is not weakly l^1-complete (see Aviles Lopez, Cascales Salinas, Kadets and Leonov (2007), Theorem 5.2) and that, if $\mathcal{F}$ is diagonal and block-respecting, and all traces of $\mathcal{F}$ on $\mathcal{F}$-stationary subsets are not ultrafilters, then $\mathcal{F}$ is weakly l^1-complete (see Aviles Lopez, Cascales Salinas, Kadets and Leonov (2007), Theorem 5.5).

Some examples of P-ideals of $\mathbb{N}$ are the ideal $\mathcal{I}_{\text{fin}}$ of all finite subsets of $\mathbb{N}$ and the ideal $\mathcal{I}_{\text{st}}$ of all subsets of $\mathbb{N}$ having null asymptotic density (see also Farah (2000), Kostyrko, Šalát and Wilczyński (2001)). Observe that $\mathcal{I}_{\text{st}}$ is not maximal. Indeed, if $\mathbb{E}$ is the set of all even integers, then we get $\mathbb{E} \notin \mathcal{I}_{\text{st}}$ and $\mathbb{N} \setminus \mathbb{E} \notin \mathcal{I}_{\text{st}}$. However it is known that, if we assume the continuum hypothesis, then there are several maximal P-ideals of $\mathbb{N}$ (see also Henriksen (1959), §7).

Let $f : \mathbb{N} \to \mathbb{R}^+$ be a function such that $\sum_{n=1}^{\infty} f(n) = +\infty$, and $\mathcal{I}_f$ be the class of all subsets $A \subset \mathbb{N}$ with

$$\sum_{n \in A} f(n) < +\infty .$$

Then $\mathcal{I}_f$ is a P-ideal. Such ideals are called *summable ideals* (see also Farah (2000)).

Some further examples of P-ideals are the *Erdös-Ulam ideals* associated with a function $f : \mathbb{N} \to \mathbb{R}^+$, consisting on the subsets $A \subset \mathbb{N}$ for which

$$\lim_n \frac{\sum_{i \in A \cap [1,n]} f(i)}{\sum_{i=1}^{n} f(i)} = 0,$$

whose $\mathcal{I}_{\text{st}}$ is a particular case, obtained by taking $f(n) = 1$ for each $n \in \mathbb{N}$. Another Erdös-Ulam ideal is the ideal $\mathcal{I}_{\log}$ of logarithmic density zero sets, where $f(n) = 1/n$, that is $\mathcal{I}_{\log}$ is the family of those subsets $A \subset \mathbb{N}$ such that

$$\lim_n \frac{\sum_{i \in A \cap [1,n]} 1/i}{\sum_{i=1}^{n} 1/i} = 0$$

(see also Farah (2000), Example 1.2.3 (d) and Kadets, Leonov and Orhan (2010)).

The ideal $\emptyset \times \mathcal{I}_{\text{fin}}$ of all subsets $A \subset \mathbb{N} \times \mathbb{N}$ such that $\{q \in \mathbb{N} : (n,q) \in A\} \in \mathcal{I}_{\text{fin}}$ *for every* $n \in \mathbb{N}$

is another example of P-ideal.

More generally, an ideal $\mathcal{I}$ is said to be *analytic* iff it is a continuous image of a G_δ subset of the Cantor space.

A set function $\eta : \mathcal{P}(\mathbb{N}) \to [0,+\infty]$ is a *lower semicontinuous submeasure* iff $\eta(\emptyset) = 0$, $\eta(A) \le \eta(A \cup B) \le \eta(A) + \eta(B)$ for every A, $B \subset A$ and $\eta(A) = \lim_n \eta\,(A \cap [1,n])$ for each $A \subset \mathbb{N}$.

If η is a lower semicontinuous submeasure, let $Exh\ (\eta) = \{A \subset \mathbb{N} : \lim_n \eta\,(A \setminus [1,n]) = 0\}$.

In Farah (2000) it is shown that all the above described ideals are of the type $Exh\ (\eta)$, for a suitable lower semicontinuous submeasure. Note that an ideal $\mathcal{I}$ of $\mathbb{N}$ is an analytic P-ideal if and only if $\mathcal{I} = Exh\ (\eta)$ for some lower semicontinuous submeasure η (see also Farah (2000), Theorem 1.2.5, and Solecki (1999), Theorem 3.1).

Let $\mathbb{N} = \bigcup_{k=1}^{\infty} \Delta_k$ be a partition of $\mathbb{N}$ into infinite sets, and $\mathcal{I}_\Delta = \{A \subset \mathbb{N} : A$ intersects at most a finite number of Δ_k's$\}$. The ideal $\mathcal{I}_\Delta$ is not a P-ideal (see also Kostyrko, Šalát and Wilczyński (2001), Example 3.1 (g)).

The ideal $\mathcal{I}_{\text{fin}} \times \emptyset$ of all subsets $A \subset \mathbb{N} \times \mathbb{N}$ such that $\{q \in \mathbb{N} : (n,q) \in A\} = \emptyset$ for every $n \in \mathbb{N}$

is not a P-ideal (see also Farah (2000)).

Let Λ be any nonempty set and $\mathcal{I}_1$, $\mathcal{I}_2$ be any two fixed (Λ)-admissible ideals of Λ.

We say that $\mathcal{I}_2$ *has the* (AP)*-property (additive property) with respect to* $\mathcal{I}_1$ iff for each disjoint sequence $(A_j)_j$ in $\mathcal{I}_2$ there exists a sequence $(B_j)_j$ of subsets of Λ with $A_j \Delta B_j \in \mathcal{I}_1$ for all $j \in \mathbb{N}$ and $\bigcup_{j=1}^{\infty} B_j \in \mathcal{I}_2$.

An ideal $\mathcal{I}$ of Λ has the (AP)*-property* iff it has the (AP)-property with respect to $\mathcal{I}_{\text{fin}}$.

An ideal $\mathcal{K}$ of $\mathbb{N}^2$ satisfies the $(AP2)$*-property* iff it has the (AP)-property with respect to the ideal $\mathcal{K}_1$ of all subsets A of $\mathbb{N}^2$ with the property that there exists a positive integer q_A such that

$$[i, j \geq q_A] \Rightarrow [(i, j) \notin A], \tag{2.5}$$

that is $\mathcal{K}_1$ is the ideal consisting of all subsets of $\mathbb{N}^2$ which are included in a finite numbers of rows and columns of $\mathbb{N}^2$.

A (Λ)-admissible ideal $\mathcal{I}$ of Λ is a (ΛP)*-ideal* iff for any disjoint sequence $(A_j)_j$ in $\mathcal{I}$ there are two sequences $(B_j)_j$ and $(p_j)_j$ in $\mathcal{I}$ and Λ respectively, such that the symmetric difference $A_j \Delta B_j \subset \Lambda \setminus M_{p_j}$ for all $j \in \mathbb{N}$ and $\bigcup_{j=1}^{\infty} B_j \in \mathcal{I}$ (see also Lahiri and Das (2008)).

A (Λ)-free filter $\mathcal{F}$ of Λ is a (ΛP)*-filter* iff its dual ideal is a (ΛP)-ideal.

Note that an ideal of $\mathbb{N}$ has the (AP)-property if and only if it is a P-ideal (see also Balcerzak, Dems and Komisarski (2007)).

We now recall the following characterization of the (AP)-property (see also Mačaj and Sleziak (2011), Lemma 3.9).

Proposition 1.4 *Under the same notations as above, let* $\mathcal{I}_1$ *and* $\mathcal{I}_2$ *be two ideals of* Λ. *Then the following are equivalent:*

1.4.1) $\mathcal{I}_2$ *has the* (AP)*-property with respect to* $\mathcal{I}_1$.

1.4.2) *For every sequence* $(A_j)_j$ *in* $\mathcal{I}_2$ *there exists another sequence* $(B_j)_j$ *in* $\mathcal{I}_2$ *with* $A_j \Delta B_j \in \mathcal{I}_1$ *for each* $j \in \mathbb{N}$ *and* $\bigcup_{j=1}^{\infty} B_j \in \mathcal{I}_2$.

1.4.3) *For every increasing sequence* $(A_j)_j$ *in* $\mathcal{I}_2$, *there exist* $B_j \in \mathcal{I}_2$, $j \in \mathbb{N}$, *such that* $A_j \Delta B_j \in \mathcal{I}_1$ *for each* $j \in \mathbb{N}$ *and* $\bigcup_{j=1}^{\infty} B_j \in \mathcal{I}_2$.

1.4.4) *For each sequence* $(A_j)_j$ *in* $\mathcal{I}_2$, *there is an element* $A \in \mathcal{I}_2$ *with* $A_j \setminus A \in \mathcal{I}_1$ *for all* $j \in \mathbb{N}$.

For other properties of ideals and filters and convergences, see for instance Balcerzak and Dems (2005), Balcerzak, Dems and Komisarski (2007), Baumgartner, Taylor and Wagon (1982), Das and Papanastassiou (2004), Farah (2000), Filipów and Szuca (2010 and 2012), Kadets and Leonov (2007 and 2009), Kadets, Leonov and Orhan (2010), Komisarski (2008), Meza Alcantara (2009), Recław (2012), Solecki (1999), Todorcevic (1997) and the literature therein.

2.1.3. Filter/Ideal Convergence

We now deal with the notion of filter/ideal convergence and its main properties. Filter convergence was introduced in Katětov (1968), while ideal convergence was introduced in Kostyrko, Šalát and Wilczyński (2001), though a primitive version of Kostyrko, Mačaj, Šalát and Sleziak (2005) is also included in the references of Kostyrko, Šalát and Wilczyński (2001), and independently in Nuray and Ruckle (2000) under the name of "cofilter convergence", and was recently developed and investigated in several papers in the context of normed and/or metric spaces. Among the related literature, we quote Balcerzak, Dems and Komisarski (2007), Boccuto and Candeloro (2009b), Boccuto, Das and Dimitriou (2012), Činčura, Šalát, Sleziak and Toma (2005), Demirci (2001), Filipów, Mrożek, Recław and Szuca (2007), Horbaczewska and Skalski (2008), Lahiri and Das (2003 and 2005), Letavaj (2011), Šalát, Tripathy and Ziman (2004-2005), Vrábel and Vrábelová (2013). This notion has been investigated also for double sequences in several papers, among which we quote, for example, Bhunia and Das (2011), Das, Kostyrko, Wilczyński and Malík (2008), Das and Malík (2008a-b), Kumar (2007), Tripathy and Tripathy (2005), extending the concept of *Pringsheim convergence* given in Pringsheim (1897) for double sequences (see also Limaye and Zeltser (2009), Móricz and Stadtmüller (2004)). Among some other related kinds of convergences, widely investigated in the literature, we will recall the *almost convergence,* whose we will give a characterization, and which is not generated by any summability matrix method and by any filter/ideal (see also Antoni (1986), Antoni and Šalát (1980), Borsík and Šalát (1993), Lorentz (1948) and the bibliography therein).

Filter/ideal convergence has been studied also in 2-normed and multiple normed spaces (see *e.g.* Das, Pal and Ghosal (2011), Mursaleen and Alotaibi (2011), Savaş (2012), Tripathy, Sen and Nath (2012)), topological spaces (see for instance Das (2012), Das and Ghosal (2010a), Jasiński and Recław (2008), Lahiri and Das (2003 and 2008)) and lattice groups (see also Boccuto and Candeloro (2009a), Boccuto and Dimitriou (2011a-c, 2013c), Boccuto, Dimitriou and Papanastassiou (2010b, 2011b-c, 2012a and 2012d-e), Boccuto, Dimitriou, Papanastassiou and Wilczyński (2011, 2012 and 2014)). Among a literature about the ideal Cauchy notion, we quote Çakalli and Hazarika (2012), Das and Ghosal (2010b), Dems (2005), Kumar (2007), Nabiev, Pehlivan and Gürdal (2007).

Observe that to consider ideal convergence is equivalent to dealing with filter convergence. We prefer to state these concepts in terms of filters, because the main results on limit theorems in Chapter 4 will be given in terms of filters.

We begin with formulating the concept of filter convergence in the context of metric spaces. A sequence $(x_n)_n$ in a metric space (X,d) $\mathcal{F}$*-converges to* $x_0 \in X$ iff for every $\varepsilon > 0$, $\{n \in \mathbb{N} : d(x_n, x_0) \le \varepsilon\} \in \mathcal{F}$, and in this case we say that $(\mathcal{F})\lim_n x_n = x_0$. A sequence $(x_n)_n$ in (X,d) is said to be $\mathcal{F}$*-Cauchy* iff for each $\varepsilon > 0$ there is $q \in \mathbb{N}$ with $\{n \in \mathbb{N} : d(x_n, x_q) \le \varepsilon\} \in \mathcal{F}$.

If $\mathcal{F}$ is a filter of $\mathbb{N}\times\mathbb{N}$, then a double sequence $(x_{i,j})_{i,j}$ in the space (X,d) is said to be $\mathcal{F}$*-convergent to* $\xi \in X$ iff for every $\varepsilon > 0$, $\{(i,j) \in \mathbb{N}^2 : d(x_{i,j}, \xi) \le \varepsilon\} \in \mathcal{F}$, and we will write $(\mathcal{F})\ \lim_{i,j} x_{i,j} = \xi$. A double sequence $(x_{i,j})_{i,j}$ is $\mathcal{F}$*-Cauchy* iff for each $\varepsilon > 0$ there exists $(p,q) \in \mathbb{N}^2$ with $\{(i,j) \in \mathbb{N}^2 : d(x_{i,j}, x_{p,q}) \le \varepsilon\} \in \mathcal{F}$. If (X,d) is also a vector space, then we define

$(\mathcal{F})\sum_{j=1}^{\infty} x_j := (\mathcal{F})\lim_n \sum_{j=1}^{n} x_j$. (*filter sum* of a series). Note that $\mathcal{F}_{\text{cofin}}$-convergence (or equivalently $\mathcal{I}_{\text{fin}}$-convergence) coincides with ordinary convergence.

Observe that, if $(X,d)=\mathbb{R}$ endowed with the usual distance, then a sequence $(x_n)_n$ in $\mathbb{R}$ converges if and only if there is a real number l with

$$l = \sup_{F\in\mathcal{F}}(\inf_{n\in F} x_n) = \inf_{F\in\mathcal{F}}(\sup_{n\in F} x_n), \qquad \textbf{(2.6)}$$

and in this case we have $l=(\mathcal{F})\lim_n x_n$. Indeed, (2.6) holds if and only if for every $\varepsilon>0$ there is a set $F\in\mathcal{F}$ with

$$\inf_{n\in F} x_n - l \geq -\varepsilon \quad \text{and} \quad \sup_{n\in F} x_n - l \leq \varepsilon,$$

what means $|x_n - l| \leq \varepsilon$ for every $n\in F$.

Similarly as above, it is possible to define filter convergence for real-valued functions, defined on any abstract nonempty set Λ (this will be useful in the sequel, considering applications to measure and integration theory). More precisely, given any fixed (Λ)-free filter $\mathcal{F}$ of Λ, any function $f:\Lambda\to\mathbb{R}$ and $l\in\mathbb{R}$, we say that $l=(\mathcal{F})\lim_{t\in\Lambda} f(t)$ iff for every $\varepsilon>0$,

$$\{t\in\Lambda : |f(t)-l|\leq\varepsilon\}\in\mathcal{F}.$$

In this case, $\mathcal{F}$-convergence of the function f is equivalent to the existence of a real number l with $l=\sup_{F\in\mathcal{F}}(\inf_{t\in F} f(t)) = \inf_{F\in\mathcal{F}}(\sup_{t\in F} f(t))$, and we get:

$$l=(\mathcal{F})\lim_{t\in\Lambda} f(t).$$

Proposition 1.5 *If* $\lim_n x_n = x_0$, *then* $(\mathcal{F})\lim_n x_n = x_0$. *Moreover, if* $(x_n)_n$ *is a monotone sequence in* $\mathbb{R}$ *(endowed with the usual distance) and* $x\in\mathbb{R}$, *then* $(\mathcal{F})\lim_n x_n = x_0$ *if and only if* $\lim_n x_n = x_0$ *in the usual sense.*

Proof: Concerning the first part, see Kostyrko, Šalát and Wilczyński (2001). For the last part, it is enough to prove the "only if" implication. Without loss of generality, assume that $(x_n)_n$ is increasing. By hypothesis, for every $\varepsilon > 0$ there exists $n^* \in \mathbb{N}$ with $0 \le x_0 - x_{n^*} \le \varepsilon$.

By monotonicity we get: $0 \le x_0 - x_n \le x_0 - x_{n^*} \le \varepsilon$ whenever $n \ge n^*$. So the sequence $(x_n)_n$ converges monotonically to x_0 in the ordinary sense. This concludes the proof. □

We now prove the following proposition (see also Boccuto, Das and Dimitriou (2012), Proposition 2.4).

Proposition 1.6 *Let $\mathcal{F}$ be a free filter of $\mathbb{N}$, and $(x_n)_n$ be a sequence in (X,d), such that $(\mathcal{F}) \lim_n x_n = x_0 \in X$. Then there exists a strictly increasing subsequence $(x_{n_q})_q$ of $(x_n)_n$, such that $\lim_q x_{n_q} = x_0$. Moreover, if $\mathcal{F}$ is a P-filter, then the subsequence $(x_{n_q})_q$ can be chosen in such a way that $\{n_1 < n_2 < \ldots < n_q < \ldots\} \in \mathcal{F}$.*

Proof: Choose arbitrarily $\varepsilon > 0$. By definition of filter convergence, there exists a positive integer n_1 with $d(x_{n_1}, x_0) \le \varepsilon$. At the second step, in correspondence with

$$\varepsilon_2 := \frac{\min\{\varepsilon, d(x_1,x_0), d(x_2,x_0), \ldots, d(x_{n_1},x_0)\}}{2}, \tag{2.7}$$

there is $n_2 \in \mathbb{N}$ such that $d(x_{n_2}, x_0) \le \varepsilon_2$. By virtue of (2.7), we get $n_2 > n_1$. Proceeding by induction, supposed that ε_{q-1} and n_{q-1} have been determined, let

$$\varepsilon_q := \frac{\min\{\varepsilon_{q-1}, d(x_1,x_0), d(x_2,x_0), \ldots, d(x_{n_{q-1}},x_0)\}}{2}. \tag{2.8}$$

By $\mathcal{F}$-convergence, there exists an integer $n_q \in \mathbb{N}$ with $d(x_{n_q}, x_0) \le \varepsilon_q$. Thanks to (2.8), we have $n_q > n_{q-1}$. Since $\lim_q \varepsilon_q = 0$, then we get $\lim_q d(x_{n_q}, x_0) = 0$, that is $\lim_q x_{n_q} = x_0$.

For the last part, see Kostyrko, Šalát and Wilczyński (2001), Theorem 3.2. □

We now formulate another kind of convergence in the context of filters/ideals, which is known in the literature as $\mathcal{I}^*$*-convergence* and which we present here as $\mathcal{F}'$*-convergence.*

Definition 1.7 A sequence $(x_n)_n$ in (X,d) $\mathcal{F}'$*-converges to* $x_0 \in X$ iff there exists a set $A_0 \in \mathcal{F}$ with $\lim_{n\in A_0} x_n = x_0$ in the ordinary sense. In this case, we write shortly $(\mathcal{F}')\lim_n x_n = x_0$.

The next result is proved in Kostyrko, Šalát and Wilczyński (2001).

Proposition 1.8 *The* $\mathcal{F}'$*-convergence of sequences implies always the* $\mathcal{F}$*-convergence. Moreover, if* $(x_n)_n$ *is a sequence in* (X,d)*,* $\mathcal{F}$*-convergent to* $x_0 \in X$*, and* $\mathcal{F}$ *is a* P*-filter, then we get* $(\mathcal{F}')\lim_n x_n = x_0$.

We now give the following:

Proposition 1.9 *Let* $(x_{i,j})_{i,j}$ *be a double sequence in* (X,d)*,* $\mathcal{F}$ *be any* P*-filter, and suppose that* $(\mathcal{F})\lim_i x_{i,j} = x_j \in X$ *for every* $j \in \mathbb{N}$.

Then there exists $B_0 \in \mathcal{F}$ *such that* $\lim_{h\in B_0} x_{h,j} = x_j$ *for all* $j \in \mathbb{N}$ *in the usual sense.*

Proof: Since $\mathcal{F}$ is a P-filter, by virtue of Proposition 1.8 we get $(\mathcal{F}')\lim_i x_{i,j} = x_j$ for every $j \in \mathbb{N}$. Hence there is a sequence $(A_j)_j$ in $\mathcal{F}$ such that $\lim_{i\in A_j} x_{i,j} = x_j$ for all $j \in \mathbb{N}$. As $\mathcal{F}$ is a P-filter, there is a sequence of sets $(B_j)_j$ in $\mathcal{F}$ such that $A_j \Delta B_j$ is finite for all $j \in \mathbb{N}$ and $B_0 := \bigcap_{j=1}^{\infty} B_j \in \mathcal{F}$. Since $\lim_{i\in A_j} x_{i,j} = x_j$, then we get also $\lim_{i\in B_j} x_{i,j} = x_j$, for all *j*. Let $B_0 = \{p_1 < \ldots < p_h < \ldots\}$ and choose arbitrarily $j \in \mathbb{N}$: then, since $B_0 \subset B_j$, in correspondence with $\varepsilon > 0$

there is $\overline{h} = \overline{h}(\varepsilon, j) \in \mathbb{N}$ with $d(x_{p_h, j}, x_j) \le \varepsilon$ for any $h > \overline{h}$. This ends the proof. □

We now state the following Cauchy-type criterion (see also Dems (2005), Kumar (2007)), which we will extend in the sequel in the setting of lattice groups.

Proposition 1.10 *A sequence $(x_n)_n$ (resp. double sequence $(x_{i,j})_{i,j}$) in a complete metric space (X,d) is $\mathcal{F}$-convergent if and only if it is $\mathcal{F}$-Cauchy, if and only if for every $\varepsilon > 0$ there is a set $F \in \mathcal{F}$, with $d(x_n, x_q) \le \varepsilon$ whenever n, $q \in F$ (resp. $d(x_{i,j}, x_{p,q}) \le \varepsilon$ whenever (i,j), $(p,q) \in F$).*

We now prove the following characterization of ultrafilters (for a literature on ultrafilters, see also Comfort (1977), Comfort and Negrepontis (1974), García-Ferreira and Kočinac (1996) and the bibliographies therein).

Proposition 1.11 (see also Boccuto and Candeloro (2009b), Proposition 2.2 and Kostyrko, Mačaj, Šalát and Sleziak (2005), Theorem 2.2) *A filter $\mathcal{F}$ of $\mathbb{N}$ is an ultrafilter if and only if every bounded sequence $(x_n)_n$ in $\mathbb{R}$ has $\mathcal{F}$-limit in $\mathbb{R}$.*

Proof: We begin with the "only if" part. Let $(x_n)_n$ be any bounded sequence in $\mathbb{R}$, and let $\mathcal{F}$ be an ultrafilter of $\mathbb{N}$. Since $(x_n)_n$ is bounded, there are two real numbers a, b with $a < x_n < b$ for each $n \in \mathbb{N}$. Set $c_1 := \dfrac{a+b}{2}$, $A_1 := \{n \in \mathbb{N}: a \le x_n < c_1\}$, $B_1 := \{n \in \mathbb{N}: c_1 \le x_n < b\}$. Then $A_1 \cap B_1 = \emptyset$ and $A_1 \cup B_1 = \mathbb{N}$. Since $\mathcal{F}$ is an ultrafilter, then exactly one of the two sets A_1, B_1 does not belong to $\mathcal{F}$. Let us denote it by D_1, and by J_1 that closed interval, such that

$$D_1 = \{n \in \mathbb{N}: x_n \notin J_1\} \notin \mathcal{F}.$$

Proceeding by induction, by means of a bisection technique, we construct a decreasing sequence $J_k = [a_k, b_k]$, $k \in \mathbb{N}$, with $\lim_k (b_k - a_k) = 0$, $\bigcap_{k=1}^{\infty} J_k = \{l\}$ for every k, and a sequence of sets

$$D_k = \{n \in \mathbb{N}: x_n \notin J_k\} \notin \mathcal{F}.$$

Fix now arbitrarily $\varepsilon > 0$. For k large enough we have $J_k \subset (l-\varepsilon, l+\varepsilon)$. Since $D_k \notin \mathcal{F}$, then $\{n \in \mathbb{N} : |x_n - l| > \varepsilon\} \notin \mathcal{F}$. As $\mathcal{F}$ is an ultrafilter, we obtain $\{n \in \mathbb{N} : |x_n - l| \le \varepsilon\} \in \mathcal{F}$. Hence, by arbitrariness of ε, we get $l = (\mathcal{F}) \lim_n x_n$. This ends the proof of the "only if" part.

We now turn to the "if" part. If $\mathcal{F}$ is not an ultrafilter, then there is a set $C \subset \mathbb{N}$ with $C \notin \mathcal{F}$ and $\mathbb{N} \setminus C \notin \mathcal{F}$. Let $x_n = 1$ if $n \in C$ and $x_n = 0$ whenever $n \in \mathbb{N} \setminus C$. Observe that, for every $l \in \mathbb{R}$, there is $\varepsilon > 0$ small enough, such that the set $\{n \in \mathbb{N} : |x_n - l| \le \varepsilon\}$ is equal to C or $\mathbb{N} \setminus C$ or $\emptyset$, and so does not belong to $\mathcal{F}$. So, $(\mathcal{F}) \lim_n x_n$ does not exist in $\mathbb{R}$, though the sequence $(x_n)_n$ is obviously bounded. This completes the proof. □

An *ultrafilter measure* on an abstract nonempty set Λ is a two-valued set function $m : \mathcal{P}(\Lambda) \to \{0,1\}$, such that the family of all sets F with $m(F) = 1$ is an ultrafilter of Λ (see also Boccuto and Candeloro (2009b)).

Proposition 1.12 (see Boccuto and Candeloro (2009b), Proposition 2.2) *Let $m : \mathcal{P}(\Lambda) \to \{0,1\}$ be an ultrafilter measure, and $f : \Lambda \to \mathbb{R}$ any bounded function. Then, if $\mathcal{F}$ denotes the ultrafilter of all sets F such that $m(F) = 1$, the function f has $\mathcal{F}$-limit in $\mathbb{R}$, and*

$$(\mathcal{F}) \lim_{t \in \Lambda} f(t) = \int_\Lambda f(t)\, dm, \tag{2.9}$$

where the integral in (2.9) *is intended in the classical sense.*

Conversely, given any ultrafilter $\mathcal{F}$ of Λ, any bounded function $f : \Lambda \to \mathbb{R}$ has $\mathcal{F}$-limit, which is the integral of f with respect to a suitable ultrafilter measure.

Proof: Proceeding similarly as in Proposition 1.11, we get that $\sup_{F \in \mathcal{F}} \inf_{t \in F} f(t) = \inf_{F \in \mathcal{F}} \sup_{t \in F} f(t)$, and that the common value is the integral $\int_\Lambda f(t)\, dm$. We denote by l this quantity. Fix now arbitrarily $\varepsilon > 0$, and set $A_\varepsilon := \{t \in \Lambda : |f(t) - l| > \varepsilon\}$. If $A_\varepsilon \in \mathcal{F}$, then

$$\inf_{F\in\mathcal{F}}\sup_{t\in F} f(t) \le l-\varepsilon \text{ or } \sup_{F\in\mathcal{F}}\inf_{t\in F} f(t) \ge l+\varepsilon,$$

and both these cases are impossible. Then $\{t\in\Lambda: |f(t)-l|\le\varepsilon\}\in\mathcal{F}$, and so the first part is proved.

Conversely, given any ultrafilter $\mathcal{F}$ of Λ and any bounded function $f:\Lambda\to\mathbb{R}$, the $\mathcal{F}$-limit of f is the integral with respect to that ultrafilter measure whose null sets are exactly the complements of the elements of $\mathcal{F}$ with respect to Λ. This ends the proof. □

We now prove another characterization of ultrafilters, which is a strengthening of Albayrak and Pehlivan (2010), Theorem 2.1.

Proposition 1.13 (see Boccuto, Das and Dimitriou (2012), Proposition 3.4; see also Boccuto, Dimitriou and Papanastassiou (2014)) *A free filter $\mathcal{F}$ of $\mathbb{N}$ is not an ultrafilter if and only if, for any bounded sequence $(a_n)_n$ in $\mathbb{R}$ such that the limit $(\mathcal{F})\lim_h a_{i_h}$ exists in $\mathbb{R}$ for each strictly increasing sequence $(i_h)_h$ in $\mathbb{N}$, we get that the limit $l:=\lim_n a_n$ exists in $\mathbb{R}$ in the classical sense. In this case, the limit $(\mathcal{F})\lim_h a_{i_h}=l$ is the same for every strictly increasing sequence $(i_h)_h$.*

Proof: We begin with the "if" part. Suppose by contradiction that $\mathcal{F}$ is an ultrafilter. Then there exists a bounded real sequence $(a_n)_n$, which does not admit limit in the classical sense. For every strictly increasing sequence $(i_h)_h$ in $\mathbb{N}$, the sequence $(a_{i_h})_h$ is obviously bounded too, and hence, since $\mathcal{F}$ is an ultrafilter, the limit the limit $(\mathcal{F})\lim_h a_{i_h}=l$ exists in $\mathbb{R}$ (see also Boccuto and Candeloro (2009b)). This leads to a contradiction, and concludes the proof of the "if" part.

We now turn to the "only if" part. First of all we claim that, if $\mathcal{F}$ is not an ultrafilter of $\mathbb{N}$, then there exist two disjoint elements $B_1\notin\mathcal{F}$, $B_2\notin\mathcal{F}$ whose union is $\mathbb{N}$. Otherwise, for each partition of $\mathbb{N}$ formed by two elements (B_1,B_2) of $\mathcal{F}$, then either B_1 or B_2 belongs to $\mathcal{F}$. If $B_1\in\mathcal{F}$, then $B_2\notin\mathcal{F}$, otherwise $\emptyset\in\mathcal{F}$, and hence $\mathcal{F}$ should be trivial. Similarly, if $B_2\in\mathcal{F}$, then $B_1\notin\mathcal{F}$. Thus $\mathcal{F}$ should be an ultrafilter. This leads to a contradiction and proves the claim.

Moreover note that, since $\mathcal{F}$ is free, then every cofinite subset of $\mathbb{N}$ belongs to $\mathcal{F}$, and hence B_1 and B_2 are infinite. Thus we can represent them in the form

$$B_1 := \{t_1 < t_2 < \ldots < t_j < \ldots\}, \quad B_2 := \{r_1 < r_2 < \ldots < r_j < \ldots\}. \tag{2.10}$$

Let now $(a_n)_n$ be as in the hypotheses, and suppose by contradiction that $\lim_n a_n$ does not exist in $\mathbb{R}$. As $(a_n)_n$ is bounded, there are two sequences in $\mathbb{N}$, $(p'_h)_h$, $(q'_h)_h$, with $\lim_h a_{p'_h} = l_1$ and $\lim_h a_{q'_h} = l_2$, where $\liminf_n a_n := l_1 < l_2 := \limsup_n a_n$.

Let $P := \{p'_j : j \in \mathbb{N}\}$, $Q := \{q'_l : l \in \mathbb{N}\}$. Let $p_i := p'_1$, and choose $q_1 > p_1$, $q_1 \in Q$: such an element does exist, since Q is infinite. Pick now $p_2 \in P$ such that $p_2 > q_1$: such a choice is possible, because P is infinite. Keeping on by induction, it is possible to construct two sequences $(p_h)_h$, $(q_h)_h$, with $\lim_j a_{p_j} = l_1$, $\lim_s a_{q_s} = l_2$, and

$$p_1 < q_1 < p_2 < \ldots < q_{h-1} < p_h < q_h < p_{h+1} < \ldots$$

For example, if we have just defined $p_1 < q_1 < \ldots < p_{h-1} < q_{h-1}$, let us choose $p_h \in P$ such that $p_h > q_{h-1}$ and $q_h \in Q$ with $q_h > p_h$: this is possible, since P and Q are infinite.

Let now B_1, B_2 be as in (2.10). For every $n \in \mathbb{N}$ there is one natural number j that $n = t_j$, or one positive integer s such that $n = r_s$. In the first case put $b_n := a_{p_j}$, and in the second case set $b_n := a_{q_s}$.

Let now $\mathcal{I}$ be the dual ideal of $\mathcal{F}$. The next step is to prove that for all $l \in \mathbb{R}$ there exists $\delta(l) > 0$, such that

$$\{n \in N : |b_n - l| > \delta\} \notin \mathcal{I}. \tag{2.11}$$

First of all, let us consider the case $l \neq l_1$. Take $\delta := \frac{|l-l_1|}{2}$ and set $\varepsilon := \frac{|l-l_1|}{4} = \frac{\delta}{2}$. By the definition of limit, we get: $|a_{p_j} - l_1| \leq \frac{|l-l_1|}{4}$ in the complement of a finite number of indexes j. So there exists a finite subset $N_1 \subset \mathbb{N}$ such that, if $n \in B_1 \setminus N_1$, then $|b_n - l_1| \leq \frac{|l-l_1|}{4}$. This implies that for all $n \in B_1 \setminus N_1$ we get: $|b_n - l| > \frac{|l-l_1|}{2}$. Otherwise we should have:

$$|l-l_1| \leq |l-b_n| + |b_n - l_1| \leq \frac{|l-l_1|}{2} + \frac{|l-l_1|}{4} = \frac{3}{4}|l-l_1|.$$

This is possible if and only if $l = l_1$, but this is absurd, because it contradicts our assumption. So the set $\{n \in \mathbb{N} : |b_n - l| > \delta\}$ contains $B_1 \setminus N_1$, and so does not belong to $\mathcal{I}$, since $B_1 \notin \mathcal{I}$, N_1 is finite and $\mathcal{I}$ is admissible. Thus (2.11) is proved, at least when $l \neq l_1$.

We now turn to the case $l = l_1$. Take $\delta := \frac{l_2 - l_1}{2}$. Note that $\delta > 0$, since $l_1 < l_2$. Analogously as above, we get $|a_{r_s} - l_2| \leq \frac{l_2 - 1_1}{4}$ in the complement of finitely many indexes s. Thus there is a finite subset $N_2 \subset \mathbb{N}$ such that $|b_n - l_2| \leq \frac{l_2 - l_1}{4}$ whenever $n \in B_2 \setminus N_2$. This implies that for all $n \in B_2 \setminus N_2$ we have: $|b_n - l| > \frac{l_2 - l_1}{2}$. Otherwise, we get:

$$0 < l_2 - l_1 \leq |l_1 - b_n| + |b_n - l_2| \leq \frac{l_2 - l_1}{2} + \frac{l_2 - l_1}{4} = \frac{3}{4}(l_2 - l_1) < l_2 - l_1,$$

a contradiction. Thus the set $\{n \in \mathbb{N} : |b_n - l| > \delta\}$ contains $B_1 \setminus N_1$, and so it does not belong to $\mathcal{I}$, since $B_2 \notin \mathcal{I}$, N_2 is finite and $\mathcal{I}$ contains all the finite subsets of $\mathbb{N}$. From this we get (2.11) in the case $l = l_1$.

From (2.11) it follows that the sequence $(b_n)_n$ does not have $\mathcal{F}$-limit. By construction, it follows easily that the sequence $(a_{p_1}, a_{q_1}, a_{p_2}, \ldots, a_{q_{h-1}}, a_{p_h}, a_{q_h}, a_{p_{h+1}}, \ldots,)$ does not have $\mathcal{F}$-limit. Thus the assertion follows. □

We now prove the next result.

Proposition 1.14 (see Boccuto, Dimitriou and Papanastassiou (2012f), Proposition 2.2)

If $\mathcal{F}$ is any free filter of $\mathbb{N}$, $P \in \mathcal{F}$, $A \subset \mathbb{N}$, $A \notin \mathcal{F}$ and $\mathbb{N} \setminus A \notin \mathcal{F}$, then both $P \cap A$ and $P \setminus A$ are infinite.

Proof: Suppose by contradiction that $P \cap A$ is finite. Note that $\mathbb{N} \setminus (P \cap A) \in \mathcal{F}$, since it is cofinite and $\mathcal{F}$ is free. Then $P \setminus A = P \cap (\mathbb{N} \setminus (P \cap A)) \in \mathcal{F}$. Hence $\mathbb{N} \setminus A \in \mathcal{F}$, which contradicts the hypothesis. Thus, $P \cap A$ is infinite. Analogously it is possible to prove that $P \setminus A$ is infinite. This ends the proof. □

2.1.4. Almost Convergence

We now deal with Single and almost convergence, and the Pringsheim convergence for double sequences (see also Boccuto, Dimitriou and Papanastassiou (2012f), Lorentz (1948), Pringsheim (1897)). Let $(X, \|\cdot\|)$ be a non-trivial normed linear space.

Let $y_{l,n}$, $l \geq 0$, $n \geq 1$ be a double sequence in X. We say that $x_0 \in X$ is the *Pringsheim limit* of $(y_{l,n})_{l,n}$ iff for every $\varepsilon > 0$ there is $\overline{n} \in \mathbb{N}$ with $\| y_{l,n} - x_0 \| \leq \varepsilon$ for all l, $n \geq \overline{n}$.

Given a sequence $(x_n)_n$ in X, we say that $(x_n)_n$ *almost converges* ((A)-*converges*) to $x_0 \in X$, and we write $(A) \lim_n x_n = x_0$, iff $\lim_n y_{l,n} = x_0$ uniformly with respect to l, where

$$y_{l,n} := \frac{1}{n} \sum_{i=1}^{n} x_{l+i}, \quad l \geq 0, n \geq 1. \tag{2.12}$$

A sequence $(x_n)_n$ in X *Pringsheim-Singly converges* ((PS)*-converges*) to $x_0 \in X$, and we write shortly $(PS)\lim_n x_n = x_0$, iff x_0 is the Pringsheim limit of $(y_{l,n})_{l,n}$, where $(y_{l,n})_{l,n}$ is as in (2.12).

A sequence $(x_n)_n$ in X *Singly converges* ((S)*-converges*) to $x_0 \in X$, and we write in brief $(S)\lim_n x_n = x_0$, iff $\lim_n y_{l,n} = x_0$ for all $l \ge 0$, where $(y_{l,n})_{l,n}$ is as in (2.12).

Fix $x_0 \in X$. Given $\varepsilon > 0$, we say that an integer $l \ge 0$ is ε*-stable* with respect to a double sequence $(y_{l,n})_{l,n}$ iff there is an integer $\overline{n} = \overline{n}(\varepsilon, l)$, with the property that $\| y_{l,n} - x_0 \| \le \varepsilon$ for all $n \ge \overline{n}$.

Note that a sequence $(x_n)_n$ Singly converges to x_0 if and only if, for each $\varepsilon > 0$, every integer $l \ge 0$ is ε-stable with respect to the double sequence $(y_{l,n})_{l,n}$ defined in (2.12).

We will prove the equivalence between (A)- and (PS)-convergence. Firstly, we compare almost and Singly convergences. Of course, (A)-convergence implies (S)-convergence. We now give an example of sequence, Singly but not almost convergent (see also Boccuto, Dimitriou and Papanastassiou (2012f), Example 2.4).

Let $X = \mathbb{R}, x_0 = \frac{1}{2}$, and let us construct a sequence $(x_k)_k$ as follows. Set $x_1 = 0$, and for all $n \in \mathbb{N}$ put $x_k = 1$ if $n^2 + 1 \le k \le n^2 + n$ and $x_k = 0$ if $n^2 + n + 1 \le k \le (n+1)^2$. Observe that

$$\frac{1}{k}\sum_{i=1}^{k} x_i \le \frac{1}{2} \quad \text{for all } k \in \mathbb{N}. \tag{2.13}$$

Moreover, for every $n \in \mathbb{N}$ and $n^2 + 1 \le k \le (n+1)^2$, we get:

$$\frac{1}{k}\sum_{i=1}^{k} x_i \geq \frac{1}{(n+1)^2}\sum_{i=1}^{n} i = \frac{n}{2(n+1)}. \quad \textbf{(2.14)}$$

Since $\lim_n \frac{n}{2(n+1)} = \frac{1}{2}$, from (2.13) and (2.14) it follows that

$$\lim_k y_{0,k} = \lim_k \frac{1}{k}\sum_{i=1}^{k} x_i = \frac{1}{2}. \quad \textbf{(2.15)}$$

Set now

$$S_{l,n} := n y_{l,n} = \sum_{i=1}^{n} x_{l+i}, \quad l \geq 0, \ n \geq 1. \quad \textbf{(2.16)}$$

Let us proceed by induction on l. We have proved in (2.15) that $\lim_k y_{0,k} = \frac{1}{2}$. Now suppose that

$\lim_n y_{l-1,n} = \frac{1}{2}$, and we claim that $\lim_n y_{l,n} = \frac{1}{2}$. In this way we will show that $(x_k)_k$ Singly converges to $\frac{1}{2}$.

For all $l \geq 0$, $n \geq 1$ we get:

$$S_{l-1,n+1} - S_{l,n} = \sum_{i=1}^{n+1} x_{l-1+i} - \sum_{i=1}^{n} x_{l+i} = \sum_{i=0}^{n} x_{l+i} - \sum_{i=1}^{n} x_{l+i} = x_l, \quad \textbf{(2.17)}$$

and hence

$$\lim_n y_{l,n} = \lim_n \frac{1}{n} S_{l,n} = \lim_n \frac{1}{n-1} S_{l,n-1} = \lim_n \frac{1}{n-1} S_{l-1,n} - \lim_n \frac{1}{n-1} x_l$$

$$= \lim_n \frac{1}{n} S_{l-1,n} \cdot \lim_n \frac{n}{n-1} = \lim_n y_{l-1,n} = \frac{1}{2}$$

by the inductive hypothesis, which gives the claim.

We now prove that $(x_k)_k$ does not almost converge to $\frac{1}{2}$. Indeed, let $\varepsilon_0 = \frac{1}{2}$, and for every $n \in \mathbb{N}$ and $n^2+1 \le k \le n^2+n$, set $l = l(k) = n^2$. We have:

$$\left| \frac{x_{l+1} + \ldots + x_{l+n}}{n} - \frac{1}{2} \right| = \frac{1}{2} = \varepsilon_0.$$

Thus we get the assertion. □

We now prove that (*PS*)-convergence implies almost convergence. This will give the equivalence between (*PS*)- and almost convergence, since the converse implication is obvious.

Theorem 1.15 (see Boccuto, Dimitriou and Papanastassiou (2012f), Theorem 2.5) *Let $(x_n)_n$ be a sequence in X, (PS)-convergent to $x_0 \in X$. Then $(x_n)_n$ almost converges to x_0 (thus (PS) -convergence is not generated by any matrix summability method, see also* Lorentz (1948), Petersen (1966)).

Proof: Observe that, in order to obtain this result, it will be enough to show that (*PS*)-convergence implies Single convergence, that is

$$\lim_n y_{l,n} = x_0 \quad \text{for every } l \ge 0, \tag{2.18}$$

where $(y_{l,n})_{l,n}$ is as in (2.12). Almost convergence will follow from (*PS*)-convergence and (2.18).

Indeed, for each $\varepsilon > 0$, let $\overline{n}$ be such that

$$\| y_{l,n} - x_0 \| \le \varepsilon \quad \text{for all } l, n \ge \overline{n}.$$

From (2.18) it follows that in correspondence with $l = 0, 1, \ldots, \overline{n}-1$ there is $n_l \in \mathbb{N}$ with

$$\| y_{l,n} - x_0 \| \le \varepsilon \quad \text{for all } n \ge n_l.$$

Let $n^* := \max\{\overline{n}, n_0, n_1, \ldots n_{\overline{n}-1}\}$. For all $l \geq 0$ and $n \geq n^*$, we get $\| y_{l,n} - x_0 \| \leq \varepsilon$, that is almost convergence. To prove (2.18), it will be enough to show that any $l \geq 0$ is 2ε-stable for each $\varepsilon > 0$.

Fix arbitrarily $\varepsilon > 0$, and pick $l = \overline{n} = \overline{n}(\varepsilon)$, where $\overline{n}$ is as in the definition of

(PS)-convergence. So, every integer $q \geq l$ is ε-stable, and hence a fortiori even 2ε-stable.

We now claim that $l-1$ is σ_1-stable for each $\sigma_1 > \varepsilon$. To this aim, let $S_{l,n}$, $l \geq 0$, $n \geq 1$, be as in (2.16). By (2.17) we get:

$$y_{l-1,n+1} = \frac{S_{l-1,n+1}}{n+1} = \frac{x_l}{n+1} + \frac{S_{l,n}}{n+1} = \frac{x_l}{n+1} - \frac{S_{l,n}}{n(n+1)} + \frac{S_{l,n}}{n} = \frac{x_l}{n+1} - \frac{y_{l,n}}{n+1} + y_{l,n}.$$

Therefore, taking into account the ε-stability of l, there is an integer $q=q(l)$ with the property that for each $n \geq q$ we have

$$\| y_{l-1,n+1} - y_{l,n} \| \leq \frac{\| x_l \|}{n+1} + \frac{\| y_{l,n} \|}{n+1} \leq \frac{\| x_l \| + \| x_0 \| + \varepsilon}{n+1}.$$

Observe that for every $\sigma_1 > \varepsilon$ there is an integer r large enough, with

$$\frac{\| x_l \| + \| x_0 \| + \varepsilon}{n+1} \leq \sigma_1 - \varepsilon \text{ whenever } n \geq r.$$

For each $n \geq \max\{q,r\}$, we get:

$$\| y_{l-1,n+1} - x_0 \| \leq \| y_{l-1,n+1} - y_{l,n} \| + \| y_{l,n} - x_0 \| \leq \sigma_1 - \varepsilon + \varepsilon = \sigma_1.$$

Hence, $l-1$ is σ_1-stable.

Proceeding by a similar argument, it is possible to check that $l-2$ is σ_2-stable for each $\sigma_2 > \sigma_1$. By induction, we obtain that $l-j$ is σ_j-stable for all $\sigma_j > \sigma_{j-1}$, $j = 3, \ldots, l$. If we take $\sigma_j = \varepsilon(1 + j/l)$, $j = 1, \ldots, l$, then $\sigma_l = 2\varepsilon$, and so we get that every $l \geq 0$ is 2ε-stable.

Thus we obtain (2.18), and hence almost convergence of the sequence $(x_n)_n$. □

We now see that in general (*A*)- and (*S*)-convergence do not coincide with $\mathcal{F}$-convergence. We begin with giving an example of sequence, (*A*)-convergent to 0 (and a fortiori (*S*)-convergent to 0) but not $\mathcal{F}$-convergent to 0 for any free filter $\mathcal{F}$ of $\mathbb{N}$. Let $X = \mathbb{R}$, and consider the sequence $n \mapsto (-1)^n$. Observe that, if l is even, then we get:

$$\sum_{k=1}^{n}(-1)^{k+l} = \sum_{k=1}^{n}(-1)^{k} = \begin{cases} -1 & \text{if } n \text{ is odd,} \\ 0 & \text{if } n \text{ is even.} \end{cases}$$

If l is odd, then we have

$$\sum_{k=1}^{n}(-1)^{k+l} = -\sum_{k=1}^{n}(-1)^{k} = \begin{cases} 1 & \text{if } n \text{ is odd,} \\ 0 & \text{if } n \text{ is even.} \end{cases}$$

So for every $l \geq 0$, $n \geq 1$ we get $|\sum_{k=1}^{n}(-1)^{k+l}| \leq 1$, and hence the sums $\sum_{k=1}^{n}(-1)^{k+l}$ are bounded uniformly with respect to l. From this it follows that the sequence $n \mapsto (-1)^n$ is almost convergent to 0. On the other hand, this sequence is not $\mathcal{F}$-convergent to 0 for any free filter $\mathcal{F}$ of $\mathbb{N}$. Indeed, let $\varepsilon_0 = \frac{1}{2}$: we get that $\{n \in \mathbb{N} : |(-1)^n| \leq \varepsilon_0\} = \emptyset \notin \mathcal{F}$. □

In a similar way, given a normed linear space $(X, \|\cdot\|)$ and a fixed element $x_0 \in X$, $x_0 \neq 0$, it is possible to show that the sequence $n \mapsto (-1)^n \cdot x_0$ is almost convergent to 0, but it does not $\mathcal{F}$-converge to 0.

In the next proposition we see that for each free filter $\mathcal{F} \neq \mathcal{F}_{\text{cofin}}$ there are some sequences $\mathcal{F}$-convergent to a point x_0, but not (S)-convergent, and a fortiori not (A)-convergent, to x_0.

Proposition 1.16 (see Boccuto, Dimitriou and Papanastassiou (2012f), Proposition 2.6)

Let $\mathcal{F} \neq \mathcal{F}_{\text{cofin}}$ *be a free filter of* $\mathbb{N}$, $(X, \|\cdot\|)$ *be a non-trivial normed linear space, and* $x_0 \in X$.

Then there is a sequence $(x_l)_l$ *in* X *with* $(\mathcal{F})\lim_i x_l = x_0$ *but not* (S)*-convergent to* x_0.

Proof: Let $A = \{k_1 < \ldots < k_n < \ldots\}$ be an infinite set, such that $\mathbb{N} \setminus A \in \mathcal{F}$. Since $\mathcal{F} \neq \mathcal{F}_{\text{cofin}}$, then A does exist. Define $f : A \to \mathbb{N}$ by setting $f(k_n) = \min\{l \in \mathbb{N} : l > k_n,\ l \notin A\}$, $n \in \mathbb{N}$, and let $\varepsilon_0 > \| x_0 \|$. At the first step, pick $z_1 \in X$ with $\| z_1 - x_0 \| > f(k_1)(\varepsilon_0 + \| x_0 \|)$. Note that such an element z_1 does exist, since $\sup\{\| x - y \| : x, y \in X\} = +\infty$. At the second step, choose $z_2 \in X$ with

$$\| z_2 - x_0 \| > \| z_1 - x_0 \| + f(k_1)(\varepsilon_0 + \| x_0 \|).$$

By proceeding analogously, at the n-th step we find an element $z_n \in X$ with

$$\| z_n - x_0 \| > \| z_1 - x_0 \| + \| z_2 - x_0 \| + \ldots + \| z_{n-1} - x_0 \| + f(k_n)(\varepsilon_0 + \| x_0 \|). \quad \textbf{(2.19)}$$

Now, let us define $x_l = z_n$ if $l = k_n$, $n \in \mathbb{N}$, and $x_l = x_0$ if $l \notin A$. By construction, it is easy to see that $(\mathcal{F})\lim_i x_l = x_0$.

We now prove that the sequence $(x_l)_l$ does not (S)-converge to x_0. Choose $l = 0$, fix arbitrarily $n \in \mathbb{N}$ and let $p = k_n$: we have $f(k_n) > k_n = p$. Moreover we get:

$$x_{l+1} + \ldots + x_{l+p} = x_1 + \ldots + x_{k_n} = x_{k_1} + \ldots + x_{k_n} + x_{t_1} + \ldots + x_{t_{n'}} =$$

$$= z_1 + \ldots + z_n + n' x_0,$$

where $t_1, \ldots, t_{n'}$ are suitable elements of $\mathbb{N} \setminus A$.

Taking into account (11), we have:

$$\left\| \frac{x_1 + \ldots + x_p}{p} - x_0 \right\| = \left\| \frac{x_{k_1} + \ldots + x_{k_n} + (n' - p)x_0}{p} \right\|$$

$$= \frac{1}{p} \| z_1 + \ldots + z_n - nx_0 \| = \frac{1}{p} \| (z_1 - x_0) + \ldots + (z_n - x_0) \|$$

$$\geq \frac{1}{p} (\| z_n - x_0 \| - \| z_1 - x_0 \| - \ldots - \| z_{n-1} - x_0 \|)$$

$$\geq \frac{1}{p} (f(k_n)(\varepsilon_0 + \| x_0 \|) + \| z_1 - x_0 \| + \| z_2 - x_0 \| + \ldots + \| z_{n-1} - x_0 \|$$

$$- \| z_1 - x_0 \| - \| z_2 - x_0 \| - \ldots - \| z_{n-1} - x_0 \|)$$

$$> \frac{1}{p} \cdot p \cdot (\varepsilon_0 + \| x_0 \|) = \varepsilon_0 + \| x_0 \| \geq \varepsilon_0 .$$

Thus the sequence $(x_l)_l$ is not (S)-convergent to 0. This ends the proof. □

2.1.5. Filter Compactness

We now deal with the concept of sequential closure and some kinds of compactness (see also Boccuto, Das, Dimitriou and Papanastassiou (2012), Diestel and Uhl (1977) and the bibliography therein). Observe that, in this setting, it is often advisable to deal with *P*-filters.

Let X be a Banach space and $\mathcal{F}$ be a free filter of $\mathbb{N}$. For $F \subset X$ and $u \in X$, we say that u is *in the* $\mathcal{F}$*-closure of* F iff there is a sequence $(x_n)_n$ of points of F such that $(\mathcal{F}) \lim_n x_n = u$. We denote the $\mathcal{F}$-closure of F by $\overline{F}^{\mathcal{F}}$.

A point $u \in X$ is called an $\mathcal{F}$*-limit point* of $F \subset X$ iff there is a sequence $(x_n)_n$ in $F \setminus \{u\}$ with $(\mathcal{F}) \lim_n x_n = u$.

A subset $F \subset X$ is said to be *$\mathcal{F}$-sequentially compact* iff every sequence $(x_n)_n$ in F contains an $\mathcal{F}$-convergent subsequence $(x_{n_k})_k$ with $(\mathcal{F})\lim_k x_{n_k} \in F$.

A set $F \subset X$ is called *$\mathcal{F}$-Fréchet compact* iff every infinite subset of F has an $\mathcal{F}$-limit point.

We now prove the equivalence between filter and classical sequential compactness.

Proposition 1.17 (see Boccuto, Das, Dimitriou and Papanastassiou (2012), Proposition 3.2) *Given a subset F of X, we get that F is $\mathcal{F}$-sequentially compact if and only if it is sequentially compact.*

Proof: If F is $\mathcal{F}$-sequentially compact, then from every sequence $(x_n)_n$ in F it is possible to extract a subsequence $(x_{n_k})_k$ such that the limit $x=(\mathcal{F})\lim_k x_{n_k}$ exists in X. By Proposition 1.6 there exists a sub-subsequence $(x_{n_{k_q}})_q$ of $(x_{n_k})_k$ such that $x = \lim_q x_{n_{k_q}}$. Hence F is sequentially compact. The converse implication is straightforward. □

We now give the following compactness results.

Proposition 1.18 (see Boccuto, Das, Dimitriou and Papanastassiou (2012), Proposition 3.3) *Any $\mathcal{F}$-closed subset of an $\mathcal{F}$-sequentially compact subset of X is $\mathcal{F}$-sequentially compact.*

Proof: Let K be an $\mathcal{F}$-sequentially compact set and $A \subset K$ be $\mathcal{F}$-closed. Let $(x_n)_n$ be any sequence in A. Then it is also in K and it admits an $\mathcal{F}$-convergent subsequence $(y_k)_k := (x_{n_k})_k$ with $(\mathcal{F})\lim_k y_k \in K$. But since A is $\mathcal{F}$-closed, so $(\mathcal{F})\lim_k y_k$ must be in $\overline{A}^{\mathcal{F}} = A$. This shows that A is $\mathcal{F}$-sequentially compact. □

Proposition 1.19 (see Boccuto, Das, Dimitriou and Papanastassiou (2012), Proposition 3.4) *For a P-filter $\mathcal{F}$, any $\mathcal{F}$-sequentially compact subset $K \subset X$ is $\mathcal{F}$-closed.*

Proof: Let K be $\mathcal{F}$-sequentially compact. Let $u \in \overline{K}^{\mathcal{F}}$. Then there is a sequence $(x_n)_n$ in K, $\mathcal{F}$-convergent to u. Since $\mathcal{F}$ is a P-filter, it is also $\mathcal{F}'$-convergent to u, namely there is a subsequence $(y_k)_k := (x_{n_k})_k$ with $\lim_k y_k = u$. Note that consequently $(\mathcal{F})\lim_k y_k = u$.

By virtue of $\mathcal{F}$-sequential compactness of K there exists a subsequence $(z_j)_j := (y_{k_j})_j$ of $(y_k)_k$, $\mathcal{F}$-convergent to $u_1 \in K$. But, as $(z_j)_j$ is a subsequence of $(y_k)_k$, we get $\lim_j z_j = u$, and thus $(\mathcal{F})\lim_j z_j = u$. Hence we must have $u = u_1$, and so $u \in K$. Thus $K = \overline{K}^{\mathcal{F}}$ and K is $\mathcal{F}$-closed. □

Corollary 1.20 (see Boccuto, Das, Dimitriou and Papanastassiou (2012), Corollary 3.5)

If $\mathcal{F}$ is a P-filter, then any $\mathcal{F}$-sequentially compact subset of X is closed in X.

Proposition 1.21 (see Boccuto, Das, Dimitriou and Papanastassiou (2012), Proposition 3.6) *Let $\mathcal{F}$ be a P-filter. A subset K of X is $\mathcal{F}$-sequentially compact iff it is $\mathcal{F}$-Fréchet compact.*

Proof: "$\Rightarrow$" Let K be an $\mathcal{F}$-sequentially compact subset of X and let A be an infinite subset of K. Choose a sequence $(x_n)_n$ of distinct points from A. This sequence has an $\mathcal{F}$-convergent subsequence $(y_k)_k := (x_{n_k})_k$ with $u = (\mathcal{F})\lim_k y_k \in K$. Moreover, since $\mathcal{F}$ is a P-filter, then $(y_k)_k$ is also $\mathcal{F}'$-convergent, and so it has a subsequence $(z_j)_j := (y_{k_j})_j$ with $\lim_j z_j = u$. Note that $(\mathcal{F})\lim_j z_j = u$ too, and hence u is an $\mathcal{F}$-limit point of A. Thus K is $\mathcal{F}$-Fréchet compact.

"$\Leftarrow$" Conversely, let us suppose that K is $\mathcal{F}$-Fréchet compact. Let $(x_n)_n$ be a sequence in K. Note that, if there exists $x \in K$ such that $x_n = x$ for infinitely many integers n, then clearly x forms a constant subsequence, which is obviously $\mathcal{F}$-convergent. So without loss of generality we can assume that $(x_n)_n$ consists of distinct points only (the proof is the same if infinitely many distinct terms are each repeated only finitely many times).

Let now $B := \{x_n : n \in \mathbb{N}\}$. Then, since B is an infinite subset of K, it has an $\mathcal{F}$-limit point $u \in K$. Consequently there exists a sequence $(y_k)_k$ in $B \setminus \{u\}$, such that $(\mathcal{F}) \lim_k y_k = u$. Since $\mathcal{F}$ is a P-filter, so $(\mathcal{F}') \lim_k y_k = u$ and evidently $u \in \overline{B}$, the ordinary closure of B. Since X is Hausdorff, we get that $u \in \overline{B_n}^{\mathcal{F}}$ and so $u \in \overline{B_n}$ for every n, where $B_n := \{x_k : k \geq n\}$. Thus $u \in \bigcap_{n=1}^{\infty} \overline{B_n}$. We can easily construct a subsequence $(z_k)_k$ of $(x_n)_n$, $\mathcal{F}$-convergent to u. □

We now recall the Eberlein-Šmulyan theorem (see also Bruguera, Martín-Peinador and Tarieladze (2004), Megginson (1998)).

Theorem 1.22 *Let Y be a subset of a normed space. Then the following are equivalent:*

1.22.1) *Y is relatively weakly compact;*

1.22.2) *Y is relatively weakly sequentially compact;*

1.22.3) *Y is relatively weakly Fréchet compact;*

1.22.4) *Y is relatively weakly countably compact.*

2.2. Filter Convergence in Lattice Groups

2.2.1. Basic Properties of Lattice Groups

We now recall some preliminary notions and results about (ℓ)-groups and Riesz spaces (for a literature about basic concepts in such structures, see for instance Aliprantis and Burkinshaw (1985 and 2003), Birkhoff (1967), Boccuto, Riečan and Vrábelová (2009), Erné and Gatzke (1985), Fremlin (1974), Ganter and Wille (1999), Kappos (1969), Luxemburg and Zaanen (1971), Peressini (1967), Riečan and Neubrunn (1997), Vulikh (1967), Zaanen (1983), Zabrejko (1997) and the references therein).

A sequence $(p_n)_n$ in R is an *(O)-sequence* iff it is decreasing and $\wedge_n p_n = 0$. A sequence $(x_n)_n$ in R is *order convergent* (or *(O)-convergent*) to $x \in R$ iff there exists an (O)-sequence $(p_n)_n$ in R with $|x_n - x| \leq p_n$ for all $n \in \mathbb{N}$, and in this case we write $(O)\lim_n x_n = x$. If Λ is any nonempty set, $(x_n^{(\lambda)})_n$ are sequences in R and $x^{(\lambda)} \in R$ for all $\lambda \in \Lambda$, we say that $(O)\lim_n x_n^{(\lambda)} = x^{(\lambda)}$ *uniformly with respect to* $\lambda \in \Lambda$ iff there exists an (O)-sequence $(q_n)_n$ in R with $|x_n^{(\lambda)} - x^{(\lambda)}| \le q_n$ for all $n \in \mathbb{N}$ and $\lambda \in \Lambda$. A sequence $(x_n)_n$ in R is (O)*-Cauchy* iff $(O)\lim_n (x_n - x_{n+p}) = 0$ uniformly with respect to $p \in \mathbb{N}$. Given an order bounded sequence $(x_n)_n$ in R (that is, such that the set $\{x_n : n \in \mathbb{N}\}$ is bounded), we set

$$(O)\limsup_n x_n := \bigwedge_n (\bigvee_{q \ge n} x_q); \quad (O)\liminf_n x_n := \bigvee_n (\bigwedge_{q \ge n} x_q).$$

Observe that a bounded sequence $(x_n)_n$ in R (O)-converges to $x \in R$ if and only if

$$x = (O)\limsup_n x_n = (O)\liminf_n x_n.$$

Note that a sequence $(x_n)_n$ (O)-converges to x if and only if there is an (O)-sequence $(\sigma_l)_l$ such that for every $l \in \mathbb{N}$ there is $n_l \in \mathbb{N}$ with $|x_n - x| \le \sigma_l$ whenever $n \ge n_l$. Indeed, concerning the "if" part, without loss of generality we can assume that $n_l < n_{l+1}$ for all $l \in \mathbb{N}$. Let $\sigma_0 := \bigvee_{s=1}^{n_1 - 1} |x_s - x|$. Set $q_n := \sigma_0$ iff $1 \le n < n_1$; $q_n := \sigma_l$ iff $n_l \le n < n_{l+1}$, $l \in \mathbb{N}$. It is easy to check that $(q_n)_n$ is the requested (O)-sequence. The converse implication is straightforward (see also Kitover and Wickstead (2005)). Analogously it is possible to see that $(O)\lim_n x_n^{(\lambda)} = x^{(\lambda)}$ uniformly with respect to $\lambda \in \Lambda$ iff there exists an (O)-sequence $(\tau_l)_l$ in R with the property that for every $l \in \mathbb{N}$ there exists $n_l \in \mathbb{N}$ with $|x_n^{(\lambda)} - x^{(\lambda)}| \le \tau_l$ whenever $n \ge n_l$ and $\lambda \in \Lambda$.

Let $(x_n)_n$ be a sequence in R. We call *series* associated with $(x_n)_n$ the sequence $(S_n)_n$, defined by setting $S_1 = a_1$ and $S_n = S_{n-1} + x_n$, $n \ge 2$, and we denote this series by the symbol $\sum_{n=1}^{\infty} x_n$. We say that the series $\sum_{n=1}^{\infty} x_n$. *converges to* $L \in R$ iff $L = (O)\lim_n S_n$, and in this case we say that L is the *sum* of the series

$\sum_{n=1}^{\infty} x_n$. The space l_R^1 is the set of all sequences of the type $(x_n)_n$, with $x_n \in R$ for all $n \in \mathbb{N}$ and such that $\bigvee_{q\in\mathbb{N}}(\sum_{n=1}^{q} | x_n |)$ exists in R. As R is Dedekind complete, if $(x_n)_n$ belongs to l_R^1, then $S := (O)\lim_q \sum_{n=1}^{q} x_n$ exists in R (see also Boccuto, Dimitriou and Papanastassiou (2012c), Boccuto and Papanastassiou (2007)).

We now recall the concept of (D)-convergence (for its main properties, see also Boccuto, Riečanand Vrábelová (2009), Riečan and Neubrunn (1997)). Observe that (D)-convergence is a very useful tool for our purposes, because in general it quite easy to handle, and its technique allows to prove several results (for instance matrix and convergence theorems) in the context of (ℓ)-groups without assuming super Dedekind completeness. Note that, given any abstract uncountable set D, the space $\mathbb{R}^D$ of the real-valued functions defined on D is a Dedekind complete Riesz space, which is not super Dedekind complete (see also Luxemburg and Zaanen (1971)).

In general, in the (ℓ)-group context, it is impossible to use the so-called ε-technique. Indeed, observe in the real-valued case, if $s = \sup A$ and $\varepsilon > 0$, then there exists $a \in A$ such that $s - \varepsilon < a$. This is in general not true in partially ordered sets. For example let R be the space of all real functions defined on $[0,1]$ with the usual ordering. Put $A = \{f_n : n \in \mathbb{N}\}$, where $f_n : [0,1] \to \mathbb{R}$, $n \in \mathbb{N}$, is defined by $f_n(x) = 1 - x^n$, $n \in \mathbb{N}$, $x \in [0,1]$. Then the lattice supremum of A in R is the function f: $[0,1] \to \mathbb{R}$, defined by $f(x) = 0$ if $x = 1$ and $f(x) = 1$ if $x \in [0,1[$. If ε is considered as a constant function, then $g = f - \varepsilon$ is defined by $g(x) = -\varepsilon$ if $x = 1$ and $g(x) = 1 - \varepsilon$ if $x \in [0,1[$, and there exists no element $h \in A$ such that $f - \varepsilon \le g \le h$. So the double sequence technique, proposed in Fremlin (1975) and by which (D)-convergence is inspired, is used instead of the ε-argument. The use of double sequences is based on the fact that, if $(a_{t,l})_{t,l}$ is a bounded double sequence of real numbers such that $(a_{t,l})_l$ is an (O)-sequence for each $t \in \mathbb{N}$, then for every $t \in \mathbb{N}$ and $\varepsilon > 0$ there exists $\varphi(t) \in \mathbb{N}$ with $a_{t,l} \le \varepsilon$ for any $l \ge \varphi(t)$. Since the inequality $a_{t,\varphi(t)} \le \varepsilon$ holds for every $t \in \mathbb{N}$, we have also

$\bigvee_{t=1}^{\infty} a_{t,\varphi(t)} \leq \varepsilon$. Hence, instead of $\varepsilon > 0$, we can and do consider the quantity $\bigvee_{t=1}^{\infty} a_{t,\varphi(t)}$.

We now follow this idea in defining (D)-convergence in the lattice group context.

A bounded double sequence $(a_{t,l})_{t,l}$ in R is called (D)-*sequence* or *regulator* iff the sequence $(a_{t,l})_l$ is an (O)-sequence for all $t \in \mathbb{N}$. A sequence $(x_n)_n$ in R is said to be (D)-*convergent* to $x \in R$ (and we write $(D)\lim_n x_n = x$) iff there exists a (D)-sequence $(a_{t,l})_{t,l}$ in R, such that to every $\varphi \in \mathbb{N}^{\mathbb{N}}$ there corresponds $n_0 \in \mathbb{N}$ with $|x_n - x| \leq \bigvee_{t=1}^{\infty} a_{t,\varphi(t)}$ whenever $n \geq n_0$. If Λ is as above, $(x_n^{(\lambda)})_n$ are sequences in R and $x^{(\lambda)} \in R$ for all $\lambda \in \Lambda$, we say that $(D)\lim_n x_n^{(\lambda)} = x^{(\lambda)}$ *uniformly* (with respect to $\lambda \in \Lambda$) iff there is a (D)-sequence $(a_{t,l})_{t,l}$ in R, such that for every $\varphi \in \mathbb{N}^{\mathbb{N}}$ there exists $n_0 \in \mathbb{N}$ such that $|x_n^{(\lambda)} - x^{(\lambda)}| \leq \bigvee_{t=1}^{\infty} a_{t,\varphi(t)}$ whenever $n \geq n_0$ and $\lambda \in \Lambda$.

The sequence $(x_n)_n$ is said to be (D)-*Cauchy* iff $(D)\lim_n (x_n - x_{n+p}) = 0$ uniformly with respect to $p \in \mathbb{N}$.

An (ℓ)-group R is (O)-*complete* iff every (O)-Cauchy sequence in R is (O)-convergent, and (D)-*complete* iff every (D)-Cauchy sequence is (D)-convergent. We recall that every Dedekind complete (ℓ)-group is (O)-complete and (D)-complete (see also Boccuto, Riečan and Vrábelová (2009), Chapter 2).

A series $\sum_{n=1}^{\infty} x_n$ in R is said to be *unconditionally convergent* iff there is a regulator $(a_{t,l})_{t,l}$ such that to every $\varphi : \mathbb{N} \to \mathbb{N}$ there is a finite set $A_0 \subset \mathbb{N}$ with $|\sum_{n \in A_1} x_n - \sum_{n \in A_2} x_n| \leq \bigvee_{t=1}^{\infty} a_{t,\varphi(t)}$ whenever A_1, A_2 are two finite subsets of $\mathbb{N}$ with A_1, $A_2 \supset A_0$ (see also Boccuto, Dimitriou and Papanastassiou (2010b)).

An (ℓ)-group R is *weakly σ-distributive* iff, for any regulator $(a_{t,l})_{t,l}$,

$$\bigwedge_{\varphi\in\mathbb{N}^{\mathbb{N}}} (\bigvee_{t=1}^{\infty} a_{t,\varphi(t)}) = 0.$$

In general, the (D)-limit of a sequence is not unique. However, (O)-convergence of sequences implies always (D)-convergence; moreover, if R is weakly σ-distributive, then a sequence is (D)-convergent if and only if it is (O)-convergent, and in this case the limit is unique (see Boccuto (1997); Duchoň and Riečan (1996), Proposition 1).

Observe that (O)- and (D)-convergence do not have in general a topological nature: for example they coincide with almost everywhere convergence in the super Dedekind complete and weakly σ-distributive Riesz space $L^0([0,1],\mathcal{M},\nu)$ as in Example 2.181 of Chapter 1 (see also Boccuto (2003), Vulikh (1967)). Note that Wright (1971) proved that a vector lattice R is weakly σ-distributive if and only if for every nonempty set G, for each algebra $\mathcal{L}\subset\mathcal{P}(G)$ and for every σ-additive measure (with respect to the order convergence) $m:\mathcal{L}\to R$ there exists an R-valued σ-additive extension of m, defined on the σ-algebra $\sigma(\mathcal{L})$ generated by $\mathcal{L}$. An example of a Dedekind complete Riesz space which is not σ-distributive can be found in Wright (1971) (see also Floyd (1955)).

The following proposition will be useful in the sequel, in defining some suitable lattice group-valued measures and proving some of its properties. We formulate it in the setting of order convergence: analogously it is possible to state it with respect to (D)-convergence.

Proposition 2.1 *Let R be a Dedekind complete (ℓ)-group. If $(O)\lim_n \sum_{i=1}^n |x_i|$ exists in R, then the limit $(O)\lim_n \sum_{i\in A, i=1,\dots,n} x_i$ exists in R uniformly with respect to $A\subset\mathbb{N}$.*

Proof: First of all note that $(O)\lim_n \sum_{i=1}^n |x_i| = \bigvee_n \sum_{i=1}^n |x_i|$, since the corresponding sequence of partial sums is increasing. It is not difficult to deduce that the following equations:

$$(O)\lim_n \sum_{i\in A, i=1,\ldots,n} |x_i| = \bigvee_n \sum_{i\in A, i=1,\ldots,n} |x_i|, \quad (O)\lim_n \sum_{i\in A, i=1,\ldots,n} x_i^+ = \bigvee_n \sum_{i\in A, i=1,\ldots,n} x_i^+,$$

$$(O)\lim_n \sum_{i\in A, i=1,\ldots,n} x_i^- = \bigvee_n \sum_{i\in A, i=1,\ldots,n} x_i^-,$$

hold in R uniformly with respect to $A \subset \mathbb{N}$. Since $x_i = x_i^+ - x_i^-$ for all $i \in \mathbb{N}$, then it follows that

$$(O)\lim_n \sum_{i\in A, i=1,\ldots,n} x_i \tag{2.20}$$

exists in R uniformly with respect to $A \subset \mathbb{N}$, that is the assertion. □

We will denote by $\sum_{n\in A} x_n$ the (O)-limit in (2.20), when it exists.

We now recall the following theorem, which links (O)- and (D)-sequences and will be useful to study some properties in lattice group theory and for (ℓ)-group-valued measures.

Theorem 2.2 (Boccuto and Dimitriou (2013c), Theorem 2.3, see also Boccuto (2003), Theorems 3.1 and 3.4) *Given any Dedekind complete (ℓ)-group R and any (O)-sequence $(\sigma_l)_l$ in R, the double sequence defined by $a_{t,l} := \sigma_l$, t, $l \in \mathbb{N}$, is a regulator, such that for every $\varphi \in \mathbb{N}^{\mathbb{N}}$, if $l = \varphi(1)$, then $\sigma_l \le \bigvee_{t=1}^{\infty} a_{t,\varphi(t)}$.*

Conversely, if R is super Dedekind complete and weakly σ-distributive, then for every (D)-sequence $(a_{t,l})_{t,l}$ in R there are an (O)-sequence $(b_k)_k$ in R and a sequence $(\varphi_k)_k$ in $\mathbb{N}^{\mathbb{N}}$, with $\bigvee_{t=1}^{\infty} a_{t,\varphi_k(t)} \le b_k$ for every k.

The following result (Fremlin lemma (Fremlin (1975), Lemma 1C), see also Riečan and Neubrunn (1997), Theorem 3.2.3, Riečan and Volauf (1984), Proposition 2.1) is one of more fundamental tools of (ℓ)-group theory, which allows us to replace a sequence of regulators with a single (D)-sequence.

Lemma 2.3 *Let R be any Dedekind complete (ℓ)-group and $(a_{t,l}^{(n)})_{t,l}$, $n\in\mathbb{N}$, be a sequence of regulators in R. Then for every $u\in R$, $u\geq 0$ there exists a (D)-sequence $(a_{t,l})_{t,l}$ in R with*

$$u\wedge(\sum_{n=1}^{q}(\bigvee_{t=1}^{\infty}a_{t,\varphi(t+n)}^{(n)}))\leq\bigvee_{t=1}^{\infty}a_{t,\varphi(t)}$$

for every $q\in\mathbb{N}$ and $\varphi\in\mathbb{N}^{\mathbb{N}}$.

We now formulate a version of the Fremlin Lemma to the setting of (O)-sequences for super Dedekind complete and weakly σ-distributive (ℓ)-groups, proving that a countable family of (O)-sequences can be replaced by a single (O)-sequence.

Lemma 2.4 (Boccuto and Candeloro (2008), Lemma 2.7) *If R is a super Dedekind complete and weakly σ-distributive lattice group and $(r_n)_n$ is an (O)-sequence, then for each $u\in R$, $u\geq 0$, there is $\omega\in\mathbb{N}^{\mathbb{N}}$ with the property that the sequence c_N, $N\in\mathbb{N}$, is an (O)-sequence, where*

$$c_N:=u\wedge\sum_{n=N}^{\infty}r_{\omega(n)}.$$

Proof: Put $A_{i,k}:=u\wedge(\sum_{n=k}^{k+2^i-1}r_n)$, i, $k\in\mathbb{N}$.

Note that the double sequence $(A_{i,k})_{i,k}$ is a regulator. Next, for every element, set

$$d_{\varphi}:=\bigvee_{t=1}^{\infty}A_{t,\varphi(t)},\quad \varphi\in\mathbb{N}^{\mathbb{N}}.$$

By super Dedekind completeness and weakly σ-distributive of R, there is $(\varphi_h)_h$ in $\mathbb{N}^{\mathbb{N}}$, with $\bigwedge_h d_{\varphi_h}=0$. We can and do suppose that $\varphi_h(n)<\varphi_h(n+1)$ and

$\varphi_h(n) < \varphi_{h+1}(n)$ for every h, $n \in \mathbb{N}$. Hence $g_N := d_{\varphi_N}$, $N \in \mathbb{N}$, is an (O)-sequence. Thus, $g_1 \geq u \wedge (r_{\varphi_1(1)} + r_{\varphi_1(1)+1})$, and also

$$g_1 \geq u \wedge (r_{\varphi_1(2)} + r_{\varphi_1(2)+1} + r_{\varphi_1(2)+2} + r_{\varphi_1(2)+3}), \ldots$$

Note that $\qquad u \wedge 2r_{\varphi_1(1)+1} \leq g_1, u \wedge (r_{\varphi_1(1)+1} + 2r_{\varphi_1(2)+3}) \leq g_1,$

$$u \wedge (r_{\varphi_1(1)+1} + r_{\varphi_1(2)+3} + 2r_{\varphi_1(3)+7}) \leq g_1, \ \ldots$$

Then $u \wedge (\sum_{n=1}^{k} r_{\varphi_1(n)+2^n-1}) \leq g_1$ for each k, and thus $u \wedge (\sum_{n=1}^{\infty} r_{\varphi_1(n)+2^n-1}) \leq g_1$. Analogously, it is possible to show that

$$u \wedge (\sum_{n=1}^{\infty} r_{\varphi_N(n)+2^n-1}) \leq g_N$$

for each N. Let $\omega(N) := \varphi_N(N) + 2^N - 1$, $N \in \mathbb{N}$. For every k we get:

$$u \wedge (\sum_{n=N}^{N+k} r_{\omega(n)}) = u \wedge (\sum_{n=N}^{N+k} r_{\phi_n(n)+2^n-1}) \leq u \wedge (\sum_{n=N}^{N+k} r_{\varphi_N(n)+2^n-1}) \leq u \wedge (\sum_{n=1}^{\infty} r_{\varphi_N(n)+2^n-1}) \leq g_N.$$

Since k is arbitrary, we obtain the requested result. □

Lemma 2.5 (Boccuto and Candeloro (2008), Lemma 2.8) *Let R be a super Dedekind complete and weakly σ-distributive (ℓ)-group, and $\{(r_l^{(t)})_l : t \in \mathbb{N}\}$ be a family of (O)-sequences in R, such that the set $\{(r_l^{(t)})_l : t, l \in \mathbb{N}\}$ is bounded in R.*

Then there exists an (O)-sequence $(b_j)_j$, such that for every $j, t \in \mathbb{N}$ there is $l = l(j,t) \in \mathbb{N}$ with $r_l^{(t)} \leq b_j$.

Proof: For each t, $l \in \mathbb{N}$, set $a_{t,l} = r_l^{(t)}$. Note that the double sequence $(a_{t,l})_{t,l}$ is a (D)-sequence, and hence by Theorem 2.2 there exist a sequence $(\varphi_j)_j$ in $\mathbb{N}^{\mathbb{N}}$ and an (O)-sequence $(b_j)_j$ in R, with $\bigvee_{t=1}^{\infty} a_{t,\varphi_j(t)} \le b_j$ for all j.

We can suppose that $\varphi_j(t) < \varphi_j(t+1)$ and $\varphi_j(t) < \varphi_{j+1}(t)$ for every j, $t \in \mathbb{N}$. Choose arbitrarily j and t and set $l = l(j,t) = \varphi_j(t)$. Then we get:

$$r_l^{(t)} = a_{t,\varphi_j(t)} \le \bigvee_{t=1}^{\infty} a_{t,\varphi_j(t)} \le b_j.$$

This ends the proof. □

The next step is to present the very famous Maeda-Ogasawara-Vulikh representation theorem, which has fundamental importance to prove several results on lattice group-valued measures (see Maeda and Ogasawara (1942), Vulikh (1967), Theorems V.4.2 and V.3.1, Bernau (1965), Theorems 3 and 6; see also Boccuto and Skvortsov (2006 and 2009), and Filter (1994) for a survey related with these topics). This theorem allows us to identify any Archimedean lattice group with a suitable group of continuous extended real-valued functions. To this aim, we will use the following result, which is a consequence of the Baire category theorem.

Proposition 2.6 *Let Ω be a complete metric space or a locally compact Hausdorff topological space. Then the complement of every meager subset Ω is dense in Ω.*

Proof: Let $N := \bigcup_{k=1}^{\infty} N_k$, where N_k is a nowhere dense subsets of Ω for every k. Given $E \subset \Omega$, denote by int (E) and $\overline{E}$ its interior and its closure in Ω respectively. For any $k \in \mathbb{N}$, since int $(\overline{N_k}) = \emptyset$, it follows that $\overline{\text{int}(\Omega \setminus N_k)} = \Omega$, and hence int $(\Omega \setminus N_k)$ is a dense open subset of Ω. By the Baire category theorem, we get $\overline{\bigcap_{k=1}^{\infty} \text{int } (\Omega \setminus N_k)} = \Omega$, and a fortiori

$$\overline{(\Omega \setminus N)} = \overline{\Omega \setminus (\bigcup_{k=1}^{\infty} N_k)} = \overline{\bigcap_{k=1}^{\infty} (\Omega \setminus N_k)} = \Omega,$$

that is $\Omega \setminus N$ is dense in Ω. □

We now are in position to state the Maeda-Ogasawara-Vulikh theorem. From now on, when we regard R as a subset of extended real-valued functions defined on Ω, we will denote by the symbols $\vee$ and $\wedge$ the supremum and infimum in R and by sup and inf the pointwise supremum and infimum, respectively.

Theorem 2.7 *Given an Archimedean (ℓ)-group R, there exists a compact extremely disconnected topological space Ω, unique up to homeomorphisms, such that R can be embedded isomorphically as a subgroup of $C_{\infty}(\Omega) = \{f$ is a continuous extended real-valued function, and $\{\omega : |f(\omega)| = +\infty\}$ is nowhere dense in $\Omega\}$. Moreover, if we denote by a^* an element of $C_{\infty}(\Omega)$ which corresponds to $a \in R$ under the above isomorphism, then for any family $(a_\lambda)_{\lambda \in \Lambda}$ of elements of R with $R \ni a_0 = \bigvee_\lambda a_\lambda$ (where the supremum is taken with respect to R), then $a_0^* = \bigvee_\lambda a_\lambda^*$ with respect to $C_{\infty}(\Omega)$, and we get $a_0^*(\omega) = \sup_\lambda [a_\lambda^*(\omega)]$ in the complement of a meager subset of Ω. The same is true for the infimum of $(a_\lambda)_{\lambda \in \Lambda}$.*

We will prove some properties of measures taking values in our involved (ℓ)-group R, by considering R as embedded in $C_{\infty}(\Omega)$, by "transfering" them in the real-space context, proving the corresponding relations for all $\omega \in \Omega \setminus N$, where N is a suitable meager subset of Ω, and successively "coming back" to R, applying a density argument and taking into account Theorem 1.2.

We now recall some other properties of (ℓ)-groups, which will be useful in the sequel.

A sequence $(t_n)_n$ of positive elements of an (ℓ)-group R is said to be *bounding* iff $2t_n \le t_{n+1}$ for each $n \in \mathbb{N}$. A subset W of an (ℓ)-group R is said to be (PR)-*bounded* by a bounding sequence $(t_n)_n$ iff W is order bounded by t_q for some $q \in \mathbb{N}$, that is iff there is $q \in \mathbb{N}$ with the property that $|w| \le t_q$ for every $w \in W$.

Examples 2.8 (a) The space c_0 of all real sequences convergent to zero, endowed with the coordinatewise order, does not have property (PR) (see Swartz (1989a)).

(b) There are some cases in which the condition of (PR)-boundedness is satisfied in the space $R = L^0([0,1],\mathcal{M},\nu)$ defined as in Example 2.181 of Chapter 1. Note that R does not satisfy property (PR) (see also Boccuto and Candeloro (2005), Example 4.7).

2.2.2. Filter Convergence/Divergence

We now formulate the concepts of order convergence, (D)-convergence and divergence for nets in the (ℓ)-group setting with respect to filters/ideals. These notions were introduced in Boccuto, Dimitriou and Papanastassiou (2012d-e).

Let $(\Lambda,\geq)$ be a directed set, R be a Dedekind complete (ℓ)-group and $\mathcal{F}$ be a (Λ)-free filter of Λ. We say that a net $(x_\lambda)_{\lambda\in\Lambda}$ in R *$(O\mathcal{F})$-converges to* $x \in R$ iff there exists an (O)-sequence $(\sigma_l)_l$ with the property that

$$\{\lambda \in \Lambda : |x_\lambda - x| \leq \sigma_l\} \in \mathcal{F} \qquad \textbf{(2.21)}$$

for all $l \in \mathbb{N}$.

If $\mathcal{I}$ is a (Λ)-admissible ideal of Λ, then we say that a net $(x_\lambda)_{\lambda\in\Lambda}$ in R *$(O\mathcal{I})$-converges to* $x \in R$ iff it $(O\mathcal{F})$-converges to x, where $\mathcal{F}$ denotes the dual filter associated with $\mathcal{I}$.

A net $(x_\lambda)_{\lambda\in\Lambda}$ is said to be *$(O\mathcal{F})$-Cauchy* iff there is an (O)-sequence $(\sigma_l)_l$ such that for each $l \in \mathbb{N}$ there is a $\tau \in \Lambda$ with $\{\lambda \in \Lambda : |x_\lambda - x_\tau| \leq \sigma_l\} \in \mathcal{F}$.

A net $(x_\lambda)_\lambda$ in R *$(D\mathcal{F})$-converges to* $x \in R$ iff there exists a (D)-sequence $(a_{t,l})_{t,l}$ with the property that

$$\{\lambda \in \Lambda : |x_\lambda - x| \leq \bigvee_{t=1}^{\infty} a_{t,\varphi(t)}\} \in \mathcal{F} \qquad \textbf{(2.22)}$$

for all $\varphi \in \mathbb{N}^{\mathbb{N}}$; the net $(x_\lambda)_\lambda$ is said to be $(D\mathcal{F})$-*Cauchy* iff there exists a (D)-sequence $(a_{t,l})_{t,l}$ with the property that for every $\varphi \in \mathbb{N}^{\mathbb{N}}$ there is an element $\tau \in \Lambda$ with

$$\{\lambda \in \Lambda : | x_\lambda - x_\tau | \leq \bigvee_{t=1}^{\infty} a_{t,\varphi(t)}\} \in \mathcal{F}.$$

Observe that, when $R = \mathbb{R}$, the $(O\mathcal{F})$- and $(D\mathcal{F})$-convergence coincide with the filter/ideal convergence existing in the literature, and we denote them simply with $\mathcal{F}$-*convergence*, or equivalently $\mathcal{I}$-*convergence*, where $\mathcal{I}$ is the dual ideal of $\mathcal{F}$.

A net $(x_\lambda)_\lambda$ in R $(O\mathcal{F}')$- [resp. $(D\mathcal{F}')$]-*converges to* $x \in R$ iff there exists a set $M \in \mathcal{F}$ with (O) $\lim_{\lambda \in M} x_\lambda = x$ [(D) $\lim_{\lambda \in M} x_\lambda = x$].

A net $(x_\lambda)_\lambda$ is $\mathcal{F}$-*divergent* iff there are an element $x \in R$ and a bounding sequence $(t_l)_l$ in R, with $\{\lambda \in \Lambda : | x_\lambda - x | \geq t_l\} \in \mathcal{F}$ for each $l \in \mathbb{N}$. In this case we say also that $(x_\lambda)_\lambda$ is $\mathcal{F}$-*divergent at* x.

A net $(x_\lambda)_{\lambda \in \Lambda}$ is said to be $\mathcal{F}'$-*divergent at* $x \in R$ iff there are a set $M \in \mathcal{F}$ and a bounding sequence $(t_l)_l$ in R such that for every $l \in \mathbb{N}$ there is a $\lambda \in M$ with $| x_\zeta - x | \geq t_l$ whenever $\zeta \in M$, $\zeta \geq \lambda$. We say that $(x_\lambda)_\lambda$ is $\mathcal{F}'$-*divergent* iff is $\mathcal{F}'$-divergent at some $x \in R$.

Observe that the concepts given above for nets can be formulated for sequences, by considering $\mathbb{N}$ with the usual order, and also for double sequences, by taking $\Lambda = \mathbb{N}^2$ endowed with the coordinatewise order (see also Boccuto, Dimitriou and Papanastassiou (2012d)). For the sake of clearness, we report them in the case of double sequences, because they will be useful in the sequel.

If $\mathcal{F}$ is an $(\mathbb{N}^2)$-free filter of $\mathbb{N}^2$, then a double sequence $(x_{i,j})_{i,j}$ in R is $(O\mathcal{F})$-*convergent to* $\xi \in R$ iff there is an (O)-sequence $(\sigma_p)_p$ with $\{(i,j) \in \mathbb{N}^2 : | x_{i,j} - \xi | \leq \sigma_p\} \in \mathcal{F}$ for all $p \in \mathbb{N}$; $(x_{i,j})_{i,j}$ is said to be $(O\mathcal{F})$-*Cauchy* if there is an (O)-sequence $(\sigma_p)_p$ such that to every $p \in \mathbb{N}$ there corresponds $(m,n) \in \mathbb{N}$ with $\{(i,j) \in \mathbb{N}^2 : |x_{i,j} - x_{m,n}| \leq \sigma_p\} \in \mathcal{F}$.

A double sequence $(x_{i,j})_{i,j}$ in R is $(D\mathcal{F})$*-convergent to* $\xi \in R$ if there is a regulator $(a_{t,l})_{t,l}$ with $\{(i,j) \in \mathbb{N}^2 : |x_{i,j} - \xi| \leq \bigvee_{t=1}^{\infty} a_{t,\varphi(t)}\} \in \mathcal{F}$ for any $\varphi \in \mathbb{N}^{\mathbb{N}}$; $(x_{i,j})_{i,j}$ is said to be $(D\mathcal{F})$-*Cauchy* if there exists a regulator $(a_{t,l})_{t,l}$ such that to every $\varphi \in \mathbb{N}^{\mathbb{N}}$ there corresponds a pair $(m,n) \in \mathbb{N}^2$ with

$$\{(i,j) \in \mathbb{N}^2 : |x_{i,j} - x_{m,n}| \leq \bigvee_{t=1}^{\infty} a_{t,\varphi(t)}\} \in \mathcal{F}.$$

We now define the *sum of a series* with respect to filter order convergence at follows:

$$(\mathcal{F})\sum_{j=1}^{\infty} x_j := (O\mathcal{F})\lim_n \sum_{j=1}^{n} x_j,$$

provided that this quantity exists in R. It is not difficult to check that, if R has property (PR), then a net $(x_\lambda)_\lambda$ is $\mathcal{F}$-divergent if and only if there is $x \in R$ such that $\{\lambda \in \Lambda : |x_\lambda - x| \geq r\} \in \mathcal{F}$ for every $r \in R$, $r \geq 0$. Indeed, let r be any arbitrary positive element of R. By property (PR) there exist an increasing sequence $(h_l)_l$ in R and an integer $\bar{l}$ such that $r \leq 2^l h_{2^l}$ for every $l \geq \bar{l}$. Setting $t_l := 2^l h_{2^l}$, $l \in \mathbb{N}$, we get that the sequence $(t_l)_l$ is bounding, and thus we get:

$$\{\lambda \in \Lambda : |x_\lambda - x| \geq t_l\} \subset \{\lambda \in \Lambda : |x_\lambda - x| \geq r\}$$

for all $l \geq \bar{l}$. From this one can easily deduce the "only if" part. The "if" part is straightforward.

Analogously it is easy to see that, if R has property (PR), then $(x_\lambda)_\lambda$ is $\mathcal{F}'$-divergent at $x \in R$ iff there is a set $M \in \mathcal{F}$ such that for every $r \in R$, $r \geq 0$, there is a $\lambda \in M$ with $|x_\zeta - x| \geq r$ whenever $\zeta \in M$, $\zeta \geq \lambda$.

If Ξ is any nonempty set, we can formulate the concepts of uniform $(O\mathcal{F})$-convergence, $(O\mathcal{F})$-Cauchy, $(D\mathcal{F})$-convergence and $(D\mathcal{F})$-Cauchy as follows (see also Boccuto, Dimitriou and Papanastassiou (2012d-e)).

We say that $\{(x_{\lambda,\xi})_\lambda : \xi \in \Xi\}$ in R is *uniformly* $(O\mathcal{F})$*-convergent to* $x_\xi \in R$ (with respect to $\xi \in \Xi$) (shortly, $(UO\mathcal{F})$*-converges to* x_ξ), iff there is an (O)-sequence $(\sigma_l)_l$ with

$$\{\lambda \in \Lambda : \bigvee_{\xi\in\Xi} | x_{\lambda,\xi} - x_{\tau,\xi} | \leq \sigma_l\} \in \mathcal{F} \text{ for every } l \in \mathbb{N}.$$

The family $\{(x_{\lambda,\xi})_\lambda : \xi \in \Xi\}$ is said to be *uniformly* $(O\mathcal{F})$*-Cauchy* (with respect to $\xi \in \Xi$) (or $(UO\mathcal{F})$*-Cauchy*) iff there exists an (O)-sequence $(\sigma_l)_l$ such that for all $l \in \mathbb{N}$ there exists an element $\tau \in \Lambda$ with

$$\{\lambda \in \Lambda : \bigvee_{\xi\in\Xi} | x_{\lambda,\xi} - x_{\tau,\xi} | \leq \sigma_l\} \in \mathcal{F}.$$

We say that $\{(x_{\lambda,\xi})_\lambda : \xi \in \Xi\}$ *uniformly* $(D\mathcal{F})$*-convergent to* $x_\xi \in R$ (with respect to $\xi \in \Xi$) (or $(UD\mathcal{F})$*-converges*) iff there is a (D)-sequence $(a_{t,l})_{t,l}$ with

$$\{\lambda \in \Lambda : \bigvee_{\xi\in\Xi} | x_{\lambda,\xi} - x_\xi | \leq \bigvee_{t=1}^{\infty} a_{t,\varphi(t)}\} \in \mathcal{F}$$

for any $\varphi \in \mathbb{N}^{\mathbb{N}}$. The family $\{(x_{\lambda,\xi})_\lambda : \xi \in \Xi\}$ is *uniformly* $(D\mathcal{F})$*-Cauchy* (with respect to $\xi \in \Xi$) (or $(UD\mathcal{F})$*-Cauchy*) iff there is a regulator $(a_{t,l})_{t,l}$ such that for all $\varphi \in \mathbb{N}^{\mathbb{N}}$ there exists an element $\tau \in \Lambda$ with

$$\{\lambda \in \Lambda : \bigvee_{\xi\in\Xi} | x_{\lambda,\xi} - x_{\tau,\xi} | \leq \bigvee_{t=1}^{\infty} a_{t,\varphi(t)}\} \in \mathcal{F}.$$

We now state the concepts of filter convergence with respect to a single (O)-sequence or regulator (see also Boccuto, Dimitriou and Papanastassiou (2012e)).

A family $\{(x_{\lambda,\xi})_\lambda : \xi \in \Xi\}$ $(RO\mathcal{F})$*-converges* to $x_\xi \in R$ iff there exists an (O)-sequence $(\sigma_l)_l$ in R with the property that for each $l \in \mathbb{N}$ and $\xi \in \Xi$ we get:

$\{\lambda \in \Lambda : | x_{\lambda,\xi} - x_\xi | \leq \sigma_l\} \in \mathcal{F}.$

The family $\{(x_{\lambda,\xi})_\lambda : \xi \in \Xi\}$ is said to be $(RD\mathcal{F})$*-convergent* to $x_\xi \in R$ iff there is a regulator $(a_{t,l})_{t,l}$ such that for each $\varphi \in \mathbb{N}^{\mathbb{N}}$ and $\xi \in \Xi$ we get $\{\lambda \in \Lambda : | x_{\lambda,\xi} - x_\xi | \leq \bigvee_{t=1}^{\infty} a_{t,\varphi(t)}\} \in \mathcal{F}.$

We will denote by (RO)-, (RD)-, (UO)-, (UD)- *convergence* the $(RO\mathcal{F}_\Lambda)$-, $(RD\mathcal{F}_\Lambda)$-, $(UO\mathcal{F}_\Lambda)$-, $(UD\mathcal{F}_\Lambda)$-convergence respectively. Observe that the (RO)- and (RD)-convergences are the usual pointwise (O)- and (D)-convergences of a family with respect to a single (O)- and (D)-sequence, while the (UO)-, (UD)-convergences are the corresponding uniform convergences. Note that, if $R = \mathbb{R}$, then $(RO\mathcal{F})$- and $(RD\mathcal{F})$-convergence coincide with usual filter convergence (see also Boccuto, Dimitriou and Papanastassiou (2012d-e)).

We now prove the following relation between filter order and (D)-convergence of nets in (ℓ)-groups, which was proved in Boccuto (2003), Theorem 3.4 in the particular case $\mathcal{F} = \mathcal{F}_{\text{cofin}}$.

Proposition 2.9 (see also Boccuto, Dimitriou and Papanastassiou (2012e), Theorem 2.3)

In any Dedekind complete (ℓ)-group R, every $(RO\mathcal{F})$-convergent net is $(RD\mathcal{F})$-convergent to the same limit. Moreover, if R is a super Dedekind complete and weakly σ-distributive (ℓ)-group, then the converse implication holds too.

Proof: Let $(x_{\lambda,\xi})_{\lambda\in\Lambda}$ be a net, $(RO\mathcal{F})$-convergent to x_ξ, and $(\sigma_p)_p$ be an associated (O)-sequence. For every t, $l \in \mathbb{N}$ set $a_{t,l} := \sigma_l$. It is easy to check that the double sequence $(a_{t,l})_{t,l}$ is a regulator. Pick arbitrarily $\varphi \in \mathbb{N}^{\mathbb{N}}$. By hypothesis, in correspondence with $p = \varphi(1)$ and for every $\xi \in \Xi$ we get $\{\lambda \in \Lambda : | x_{\lambda,\xi} - x_\xi | \leq \sigma_{\varphi(1)} := a_{1,\varphi(1)}\} \in \mathcal{F}$, and a fortiori

$$\{\lambda \in \Lambda : | x_{\lambda,\xi} - x_\xi | \leq \bigvee_{t=1}^{\infty} a_{t,\varphi(t)}\} \in \mathcal{F}.$$

This concludes the first part.

We now turn to the last part. Let $(x_{\lambda,\xi})_{\lambda\in\Lambda,\xi\in\Xi}$ be $(O\mathcal{F})$-convergent to x_ξ, and $(a_{t,l})_{t,l}$ be a regulator, satisfying the condition of $(RD\mathcal{F})$-convergence. Since R is super Dedekind complete and weakly σ-distributive, by Theorem 2.2 there exist an (O)-sequence $(\sigma_p)_p$ in R and a sequence $(\varphi_p)_p$ in $\mathbb{N}^{\mathbb{N}}$ with $\bigvee_{t=1}^{\infty} a_{t,\varphi_p(t)} \le \sigma_p$ for each $p\in\mathbb{N}$. Thus, since by hypothesis

$$\{\lambda\in\Lambda : |x_{\lambda,\xi}-x_\xi| \le \bigvee_{t=1}^{\infty} a_{t,\varphi_p(t)}\}\in\mathcal{F}$$

whenever $p\in\mathbb{N}$, then a fortiori we get $\{\lambda\in\Lambda : |x_{\lambda,\xi}-x_\xi|\le\sigma_p\}\in\mathcal{F}$ for every $p\in\mathbb{N}$. This ends the proof. □

The next result can be proved analogously as Proposition 2.9.

Proposition 2.10 *Every* $(O\mathcal{F})$*-Cauchy net is* $(D\mathcal{F})$*-Cauchy too. Moreover, if* R *is a super Dedekind complete and weakly* σ*-distributive* (ℓ)*-group, then the converse is true.*

Proposition 2.11 *Let* $\mathcal{F}$ *be a* (Λ)*-free filter of* Λ*. If* $(O)\lim_\lambda x_\lambda = x$*, then we get* $(O\mathcal{F})\lim_\lambda x_\lambda = x$*. Moreover, if* $(x_\lambda)_\lambda$ *is an increasing net in* R *and* $x\in R$*, then* $(O\mathcal{F})\lim_\lambda x_\lambda = x$ *if and only if* $(O)\lim_\lambda x_\lambda = x$*.*

Proof: The first part is straightforward.

We now turn to the last part. It is enough to prove the "only if" implication. By hypothesis there is a (O)-sequence $(\sigma_p)_p$ such that for every $p\in\mathbb{N}$ an element $\lambda^*\in\Lambda$ can be found, with

$0\le x-x_{\lambda^*}\le\sigma_p$. By monotonicity of the net $(x_\lambda)_\lambda$ we get: $0\le x-x_\lambda\le x-x_{\lambda^*}\le\sigma_p$ whenever $\lambda\ge\lambda^*$. Thus the net $(x_\lambda)_\lambda$ (O)-converges to x. □

Observe that an easy consequence of Proposition 2.11 is that, if $\sum_{j=1}^{\infty} x_j$ is a series of positive terms in R and S is its sum, then $(\mathcal{F})\sum_{j=1}^{\infty} x_j = S$ (and *vice versa*).

We now prove a Cauchy-type criterion.

Proposition 2.12 *Let $(\Lambda,\geq)$ be as above, R be any Dedekind complete (ℓ)-group, $\mathcal{F}$ be a (Λ)-free filter of Λ and $(x_\lambda)_\lambda$ be a net in R. Then the following are equivalent:*

(a) $(x_\lambda)_\lambda$ is $(O\mathcal{F})$-convergent;

(b) $(x_\lambda)_\lambda$ is $(O\mathcal{F})$-Cauchy;

(c) there exists an (O)-sequence $(\sigma_p)_p$ with the property that for every $p\in\mathbb{N}$ there is $D\in\mathcal{F}$, with $|x_\lambda - x_\tau|\leq\sigma_p$ whenever $\lambda,\ \tau\in D$.

Proof: $(ii)\Rightarrow(i)$. Let $(\varepsilon_p)_p$ be an (O)-sequence, related with the $(O\mathcal{F})$-Cauchy condition: then there is a sequence $(n_p)_p$ in $\mathbb{N}$, with

$$\{\lambda\in\Lambda : |x_\lambda - x_{n_p}|\leq\varepsilon_p\}\in\mathcal{F} \tag{2.23}$$

for all $p\in\mathbb{N}$. For every $p,q\in\mathbb{N}$, we get:

$$\{\lambda\in\Lambda : |x_\lambda - x_{n_p}|\leq\varepsilon_p\}\cap\{\lambda\in\Lambda : |x_\lambda - x_{n_q}|\leq\varepsilon_q\}\in\mathcal{F},$$

and thus there is $l_{p,q}\in\mathbb{N}$ such that $|x_{l_{p,q}} - x_{n_p}|\leq\varepsilon_p$ and $|x_{l_{p,q}} - x_{n_q}|\leq\varepsilon_q$, and hence $|x_{n_p} - x_{n_q}|\leq\varepsilon_p+\varepsilon_q$. As $(\varepsilon_p)_p$ is an (O)-sequence, then $(x_{n_p})_p$ is an (O)-Cauchy sequence in the classical sense. Since every Dedekind complete (ℓ)-group is (O)-complete, there exists $x\in R$ with $x=(O)\lim_p x_{n_p}$. Thanks to (2.23), for every $p\in\mathbb{N}$ we get:

$$\{\lambda\in\Lambda:|x_\lambda-x|\le 2\varepsilon_p\}\supset\{\lambda\in\Lambda:|x_\lambda-x_{n_p}|+|x_{n_p}-x|\le 2\varepsilon_p\}\supset$$

$$\supset\{\lambda\in\Lambda:|x_{n_p}-x|\le\varepsilon_p\}\cap\{\lambda\in\Lambda:|x_\lambda-x_{n_p}|\le\varepsilon_p\}\in\mathcal{F}.$$

$(i)\Rightarrow(iii)$. Since, by hypothesis, $(x_\lambda)_\lambda$ is (O)-convergent to an element $x\in R$, there is an (O)-sequence $(\sigma_p)_p$ with the property that for every $p\in\mathbb{N}$ there is $D\in\mathcal{F}$ with $|x_\lambda-x|\le\sigma_p$ whenever $\lambda\in D$. Let now λ, $\tau\in D$: we get $|x_\lambda-x_\tau|\le|x_\lambda-x|+|x_\tau-x|\le 2\sigma_p$, and hence the implication is proved.

$(iii)\Rightarrow(ii)$. Let $(\sigma_p)_p$ be an (O)-sequence, existing by hypothesis, pick arbitrarily $p\in\mathbb{N}$ and let $D\in\mathcal{F}$ be as in (iii). Let $D\in\mathcal{F}$, and choose arbitrarily $\overline{\lambda}\in D$. By (iii), for all $\lambda\in D$ we get $|x_\lambda-x_{\overline{\lambda}}|\le\sigma_p$. So (ii) follows. □

With similar techniques, it is possible to formulate analogous Cauchy-type criterions for $(D\mathcal{F})$-convergence. We now state the following result, since it will be useful in the sequel.

Proposition 2.13 (see also Boccuto, Dimitriou and Papanastassiou (2012d), Proposition 2.13) *Let R be a super Dedekind complete and weakly σ-distributive (ℓ)-group, $\mathcal{F}$ be any fixed $(\mathbb{N}^2)$-free filter of $\mathbb{N}^2$, and $(x_{i,j})_{i,j}$ be any double sequence in R. Then the following are equivalent:*

(a) $(x_{i,j})_{i,j}$ is $(D\mathcal{F})$-convergent;

(b) $(x_{i,j})_{i,j}$ is $(D\mathcal{F})$-Cauchy;

(c) There exists a (D)-sequence $(a_{t,l})_{t,l}$ with the property that for each $\varphi\in\mathbb{N}^{\mathbb{N}}$ there is $D\in\mathcal{F}$, with $|x_{i_1,j_1}-x_{i_2,j_2}|\le\bigvee_{t=1}^{\infty}a_{t,\varphi(t)}$ whenever (i_1,j_1), $(i_2,j_2)\in D$.

Proposition 2.14 (Boccuto, Dimitriou and Papanastassiou (2012d), Proposition 2.14) *Let R be a super Dedekind complete and weakly σ-distributive (ℓ)-group, and $\mathcal{F}$ be a free filter of $\mathbb{N}$. Then a sequence $(x_n)_n$ in R is $(D\mathcal{F})$-convergent if and only if it is $(D\mathcal{F})$-Cauchy. Moreover, if Ξ is any abstract nonempty set, then* *a family* $\{(x_{n,\xi})_n : \xi \in \Xi\}$ *is* $(D\mathcal{F})$*-convergent uniformly with respect to* $\xi \in \Xi$ *if and only if it is* $(D\mathcal{F})$*-Cauchy uniformly with respect to* $\xi \in \Xi$.

We now give some comparison results between $\mathcal{F}$ - and $\mathcal{F}'$ - convergence in (ℓ)-groups.

Proposition 2.15 (see also Boccuto, Dimitriou and Papanastassiou (2012a), Proposition 2.8) *Let* R *be any Dedekind complete* (ℓ)*-group, and* $(x_\lambda)_{\lambda\in\Lambda}$ *be a net in* R. *If* $(O\mathcal{F}')$ $\lim_\lambda x_\lambda = x$ *with respect to an* (O)*-sequence* $(\sigma_p)_p$, *then* $(O\mathcal{F})$ $\lim_\lambda x_\lambda = x$ *with respect to the same* (O)*-sequence* $(\sigma_p)_p$.

Proof: By hypothesis, there is a set $F \in \mathcal{F}$ with (O) $\lim_{\lambda\in F} x_\lambda = x$ with respect to a suitable (O) -sequence $(\sigma_p)_p$. Then for every $p \in \mathbb{N}$ there is $\lambda \in F$ with $|x_\zeta - x| \leq \sigma_p$ whenever $\zeta \in F$, $\zeta \geq \lambda$. Therefore, we get:

$\{\zeta \in \Lambda : |x_\zeta - x| \leq \sigma_p\} \supset F \cap M_\lambda \in \mathcal{F}$. This ends the proof. □

Analogously as Proposition 2.15 it is possible to prove the following:

Proposition 2.16 (see also Boccuto, Dimitriou and Papanastassiou (2012d), Proposition 2.11) *Let* R *be a Dedekind complete lattice group, and* $(x_\lambda)_{\lambda\in\Lambda}$ *be a net in* R. *Suppose that*

$(D\mathcal{F}')$ $\lim_\lambda x_\lambda = x$ *with respect to a regulator* $(a_{t,l})_{t,l}$. *Then* $(D\mathcal{F})$ $\lim_\lambda x_\lambda = x$ *with respect to the same regulator* $(a_{t,l})_{t,l}$.

Proof: By hypothesis, there exists a set $F \in \mathcal{F}$ such that

$$(D)\lim_{\lambda\in F} x_\lambda = x \tag{2.24}$$

with respect to a suitable regulator $(a_{t,l})_{t,l}$. Fix arbitrarily $\varphi \in \mathbb{N}^{\mathbb{N}}$. Then by (2.24) there exists an element $\lambda \in F$ with $|x_\zeta - x| \leq \bigvee_{t=1}^{\infty} a_{t,\varphi(t)}$ for all $\zeta \in F$, $\zeta \geq \lambda$. Thus the set

$$B_\varphi := \{\zeta \in \Lambda : | x_\zeta - x | \leq \bigvee_{t=1}^{\infty} a_{t,\varphi(t)}\} \supset F \cap M_\lambda \in \mathcal{F},$$

and hence $B_\varphi \in \mathcal{F}$. This concludes the proof. $\square$

Proposition 2.17 (see Boccuto, Dimitriou and Papanastassiou (2012a), Proposition 2.9) *Let R be any Dedekind complete (ℓ)-group, $(x_\lambda)_{\lambda\in\Lambda}$ be a net in R, $(O\mathcal{F})$-convergent to $x \in R$ with respect to an (O)-sequence $(\sigma_j)_j$. If $\mathcal{F}$ is a (ΛP)-filter, then $(x_\lambda)_\lambda$ $(O\mathcal{F}')$-converges to x with respect to the same (O)-sequence $(\sigma_j)_j$.*

Proof: Let $\mathcal{I}$ be the dual ideal of $\mathcal{F}$, and $(\sigma_j)_j$ be an (O)-sequence related to $(O\mathcal{F})$-convergence of the net $(x_\lambda)_{\lambda\in\Lambda}$ to x. Set $O_j := [-\sigma_j, \sigma_j]$, $j \in \mathbb{N}$; $A_1 := \{\lambda \in \Lambda : x_\lambda - x \notin O_1\}$ and $A_j : \{\lambda \in \Lambda : x_\lambda - x \in O_{j-1} \setminus O_j\}$, $j \geq 2$. It is easy to check that $(A_j)_j$ is a disjoint sequence of elements of $\mathcal{I}$. Since $\mathcal{I}$ is a (ΛP)-ideal, in correspondence with $(A_j)_j$ there exist a sequence $(B_j)_j$ in $\mathcal{P}(\Lambda)$ and a sequence $(p_j)_j$ in Λ such that $A_j \Delta B_j \subset \Lambda \setminus M_{p_j}$ for any $j \in \mathbb{N}$ and $B := \bigcup_{j=1}^{\infty} B_j \in \mathcal{I}$. Set $M := \Lambda \setminus B$: then $M \in \mathcal{F}$. Fix arbitrarily $l \in \mathbb{N}$. There is $\lambda_0 \in \Lambda$ with $\lambda_0 \geq p_j$ for any $j = 1, \ldots, l$. By construction we get:

$$(\bigcup_{j=1}^{l} B_j) \cap M_{\lambda_0} = (\bigcup_{j=1}^{l} A_j) \cap M_{\lambda_0}. \tag{2.25}$$

Since $\mathcal{F}$ is (Λ)-free, $M_{\lambda_0} \in \mathcal{F}$. Hence, $M \cap M_{\lambda_0} \in \mathcal{F}$. Pick now $\lambda \in M$ with $\lambda \geq \lambda_0$. Choose arbitrarily $\zeta \in M$, $\zeta \geq \lambda$. Then $\zeta \notin B$, and in particular $\zeta \notin \bigcup_{j=1}^{l} B_j$, $\zeta \in M_{\lambda_0}$. From this and (2.25) it follows that $\zeta \notin \bigcup_{j=1}^{l} A_j$. Thus $x_\zeta - x \in O_l$, namely $-\sigma_l \leq x_\zeta - x \leq \sigma_l$. So we have proved the existence of an element M of $\mathcal{F}$ and of an (O)-sequence $(\sigma_l)_l$, with the property that for every $l \in \mathbb{N}$ there is $\lambda \in M$ with $| x_\zeta - x | \leq \sigma_l$ whenever $\zeta \in M$, $\zeta \geq \lambda$, that is $(O) \lim\limits_{\lambda \in M} x_\lambda = x$. This ends the proof. $\square$

Observe that all results given above hold also for sequences: indeed, it is enough to take $\Lambda = \mathbb{N}$ with the usual order.

Taking into account also Boccuto (2003), Theorem 3.4, a consequence of Proposition 2.17 is the following:

Proposition 2.18 (see Boccuto, Dimitriou and Papanastassiou (2012d), Proposition 2.9) *Let R be a super Dedekind complete and weakly σ-distributive (ℓ)-group, $\mathcal{F}$ be a P-filter and $(x_n)_n$ be a sequence in R, with $(D\mathcal{F})\lim_n x_n = x \in R$. Then there exists a subsequence $(x_{n_q})_q$ of $(x_n)_n$, with $(D)\lim_q x_{n_q} = x$ and $\{n_q : q \in \mathbb{N}\} \in \mathcal{F}$.*

The following technical result will be useful for our versions of limit theorems (see also Boccuto and Dimitriou (2011c), Proposition 3.2).

Proposition 2.19 *Let R be any Dedekind complete (ℓ)-group, $(x_{n,j})_{n,j}$ be a double sequence in R, $\mathcal{F}$ be any P-filter of $\mathbb{N}$, and suppose that $(O\mathcal{F})\lim_n x_{n,j} = x_j \in R$ for every $j \in \mathbb{N}$ with respect to a single (O)-sequence $(\sigma_p)_p$. Then there exists $B_0 \in \mathcal{F}$ with $(O)\lim_{n\in B_0} x_{n,j} = x_j$ for all $j \in \mathbb{N}$, with respect to $(\sigma_p)_p$.*

Proof: By hypothesis and Proposition 2.17 we get the existence of a sequence $(A_j)_j$ in $\mathcal{F}$, with $(O)\lim_{n\in A_j} x_{n,j} = x_j$ for all $j \in \mathbb{N}$ with respect to a single (O)-sequence $(\sigma_p)_p$. Since $\mathcal{F}$ is a P-filter, there is a sequence of sets $(B_j)_j$ in $\mathcal{P}(\mathbb{N})$, such that $A_j \Delta B_j$ is finite for all $j \in \mathbb{N}$ and $B_0 := \bigcap_{j=1}^{\infty} B_j \in \mathcal{F}$. Thus, since $(O)\lim_{n\in A_j} x_{n,j} = x_j$ for all $j \in \mathbb{N}$ with respect to $(\sigma_p)_p$, we get also $(O)\lim_{n\in B_j} x_{n,j} = x_j$, and a fortiori $(O)\lim_{n\in B_0} x_{n,j} = x_j$, for all j, with respect to $(\sigma_p)_p$. □

We now prove that the filter convergence is inherited by the $\mathcal{F}$-stationary sets. We prove the following proposition in the setting of (O)-convergence: an analogous result holds in the context of (D)-convergence.

Proposition 2.20 *If $\mathcal{F}$ is any free filter, x_n, $n\in\mathbb{N}$, is a sequence in R, $(O\mathcal{F})$-convergent to $x\in R$, and $J\in\mathcal{F}^*$, then the sequence x_n, $n\in J$, $(O\mathcal{F}(J))$-converges to x.*

Proof: Let $(\sigma_p)_p$ be an (O)-sequence, related with $(O\mathcal{F})$-convergence, Choose arbitrarily $p\in\mathbb{N}$, and set $F:=\{n\in\mathbb{N}:|x_n-x|\le\sigma_p\}$. We get:

$$\{n\in J:|x_n-x|\le\sigma_p\}=F\cap J\in\mathcal{F}(J),$$

and thus the assertion follows. □

We now prove the following results on filter divergence.

Proposition 2.21 (see Boccuto, Dimitriou and Papanastassiou (2012a), Proposition 2.10) *Let R be a Dedekind complete (ℓ)-group, $(x_\lambda)_\lambda$ be a net in R, $\mathcal{F}$-divergent at $x\in R$ with respect to a bounding sequence $(t_j)_j$. If $\mathcal{F}$ is a (ΛP)-filter, then $(x_\lambda)_\lambda$ $\mathcal{F}'$- diverges at x with respect to the same sequence $(t_j)_j$. Conversely, if $\mathcal{F}$ is any (Λ)-free filter and $(x_\lambda)_\lambda$ $\mathcal{F}'$-diverges at $x\in R$ with respect to a bounding sequence $(t_j)_j$, then $(x_\lambda)_\lambda$ $\mathcal{F}$-diverges at x with respect to $(t_j)_j$.*

Proof: Let $\mathcal{I}$ be the dual ideal of $\mathcal{F}$. For every $j\in\mathbb{N}$, let $H_j:=\{\lambda\in\Lambda:|x_\lambda-x|\not\geq t_j\}$; moreover, set $A_1:=H_1$; $A_j:=H_j\setminus H_{j-1}$, $j\ge 2$. It is easy to check that $(H_j)_j$ is an increasing sequence and $(A_j)_j$ is a disjoint sequence of elements of $\mathcal{I}$. Since $\mathcal{I}$ is a (ΛP)-ideal, in correspondence with $(A_j)_j$ there exist a sequence $(B_j)_j$ in $\mathcal{P}(\Lambda)$ and a sequence $(p_j)_j$ in Λ such that $A_j\Delta B_j\subset\Lambda\setminus M_{p_j}$ for every $j\in\mathbb{N}$ and $B:=\bigcup_{j=1}^{\infty}B_j\in\mathcal{I}$. Set $M:=\Lambda\setminus B$ and fix arbitrarily $l\in\mathbb{N}$. There is $\lambda_0\in\Lambda$ such that $\lambda_0\ge p_j$ for all $j=1,\dots,l$. By construction we get:

$$(\bigcup_{j=1}^{l} B_j) \cap M_{\lambda_0} = (\bigcup_{j=1}^{l} A_j) \cap M_{\lambda_0}. \quad \textbf{(2.26)}$$

As $\mathcal{I}$ is (Λ)-admissible, then we have $M_{\lambda_0} \in \mathcal{F}$, and thus $M \cap M_{\lambda_0} \in \mathcal{F}$. Pick $\lambda \in M$ with $\lambda \geq \lambda_0$. Choose arbitrarily $\zeta \in M$, $\zeta \geq \lambda$. Then $\zeta \notin B$, and in particular $\zeta \notin \bigcup_{j=1}^{l} B_j$, $\zeta \in M_{\lambda_0}$. From this and (2.26) it follows that $\zeta \notin \bigcup_{j=1}^{l} A_j$. So $\zeta \notin H_l$, that is $| x_\zeta - x | \geq t_l$. Thus we have proved that there exist a set $M \in \mathcal{F}$ and a bounding sequence $(t_l)_l$ in R, such that for every $l \in \mathbb{N}$ there is a $\lambda \in M$ with $| x_\zeta - x | \geq t_l$ whenever $\zeta \in M$, $\zeta \geq \lambda$. Hence we get that is $\mathcal{F}'$-divergence of the net $(x_\lambda)_\lambda$. This ends the proof of the first part.

We now turn to the last part. By hypothesis, there are a set $M \in \mathcal{F}$ and a bounding sequence $(t_j)_j$ in R such that for any $j \in \mathbb{N}$ there is $\lambda \in M$ with $| x_\zeta - x | \geq t_j$ for all $\zeta \in M$, $\zeta \geq \lambda$.

Therefore $\{\zeta \in \Lambda : | x_\zeta - x | \geq t_j\} \supset M \cap M_\lambda \in \mathcal{F}$. This ends the proof. □

As an immediate consequence of Proposition 2.21, taking $\Lambda = \mathbb{N}$, we obtain:

Proposition 2.22 *Let R be a Dedekind complete lattice group and $(x_n)_n$ be a net in R, $\mathcal{F}$-divergent at $x \in R$. If $\mathcal{F}$ is a P-filter of $\mathbb{N}$, then $(x_n)_n$ $\mathcal{F}'$-diverges at x. Conversely, if $\mathcal{F}$ is any free filter of $\mathbb{N}$ and $(x_n)_n$ $\mathcal{F}'$-diverges at $x \in R$, then $(x_n)_n$ $\mathcal{F}$-diverges at x.*

Let $\mathcal{U}^*(0)$ be the family of all symmetric subintervals of R of the type $[-u,u] = \{x \in R : -u \leq x \leq u\}$, where $u > 0$, and $\mathcal{F}$ be a filter of $\mathbb{N}$. A sequence $(x_n)_n$ in R is *$\mathcal{F}$-bounded* iff there exists $U \in \mathcal{U}^*(0)$ with $\{n \in \mathbb{N} : x_n \in U\} \in \mathcal{F}$. In this case, we say also that the set $\{x_n : n \in \mathbb{N}\}$ is *$\mathcal{F}$-bounded.*

Note that a sequence $(x_n)_n$ in $\mathbb{R}$ is $\mathcal{F}$-bounded if and only if there exists a positive real number K_0 such that $\{n \in \mathbb{N} : | x_n | \leq K_0\} \in \mathcal{F}$.

Let (X,d) be a metric space and $\mathrm{Bd}(X)$ be the space of all bounded real-valued functions defined on X, endowed with the supremum norm $\|\cdot\|_\infty$. A function sequence $(f_n)_n$ in Bd (X) is $\mathcal{F}$ *-bounded* iff there exists a $K^*>0$ with $\{n\in\mathbb{N}: \| f_n\|_\infty \le K^*\}\in\mathcal{F}$.

If $(U_n)_n$ is any increasing sequence in $\mathcal{U}^*(0)$, then we say that the sequence $(x_n)_n$ is (PR)-$\mathcal{F}$*-bounded* by $(U_n)_n$, or that the set $\{x_n : n\in\mathbb{N}\}$ is (PR)-$\mathcal{F}$-*bounded* by $(U_n)_n$, iff $\{n\in\mathbb{N}: x_n\in U_n\}\in\mathcal{F}$.

We say that a sequence $(x_n)_n$ or the corresponding set $\{x_n\colon n\in\mathbb{N}\}$ is (PR)-$\mathcal{F}$-*bounded* in R iff there is a sequence in $\mathcal{U}^*(0)$ of the type $U_n := [-t_n, t_n]$, $n\in\mathbb{N}$, such that $(x_n)_n$ is (PR)-$\mathcal{F}$-bounded by $(U_n)_n$ and $0\le 2t_n \le t_{n+1}$ for every n.

A sequence $(x_n)_n$ in R is (PR)*-eventually bounded* by $(U_n)_n$ (resp. (PR)-*eventually bounded* in R) iff it is (PR)-$\mathcal{F}_{\text{cofin}}$-bounded by $(U_n)_n$ (resp. (PR)-$\mathcal{F}_{\text{cofin}}$-bounded in R) (see also Boccuto and Candeloro (2004d), Boccuto, Dimitriou and Papanastassiou (2012b)).

The next properties of filter convergence and boundedness will be useful to prove several limit theorems and the uniform boundedness principle in the setting of filter convergence (for similar results existing in the literature, see also Aviles Lopez, Cascales Salinas, Kadets and Leonov (2007), Lemma 3.3; Boccuto, Dimitriou and Papanastassiou (2012e), Lemma 2.2; Boccuto, Dimitriou and Papanastassiou (2012b), Lemma 3.1).

Lemma 2.23 *Let R be a Dedekind complete (ℓ)-group, $(a_{j,n})_{j,n}$ be a double sequence in R, $\mathcal{F}$ be a diagonal filter of $\mathbb{N}$. If $(O\mathcal{F})\lim_{j\in\mathbb{N}} a_{j,n}=0$ for each $n\in\mathbb{N}$ with respect to an (O)-sequence $(\sigma_p^{(n)})_p$ (depending on n), then for every $I\in\mathcal{F}^*$ there is $J\in\mathcal{F}^*$, $J\subset I$, with $(O)\lim_{j\in J} a_{j,n}=0$ for any $n\in\mathbb{N}$ with respect to $(\sigma_p^{(n)})_p$.*

Proof: By hypothesis, for each n, $p\in\mathbb{N}$ we have $A_{n,p} := \{j\in\mathbb{N}: |a_{j,n}| \le \sigma_p^{(n)}\}\in\mathcal{F}$. As $\mathcal{F}$ is diagonal, for any $I\in\mathcal{F}^*$ there is

$J \in \mathcal{F}^*$, $J \subset I$, such that for every n, $p \in \mathbb{N}$ the set $J \setminus A_{n,p}$ is finite. We have $B_{n,p} := \{j \in J : |a_{j,n}| \not\leq \sigma_p^{(n)}\} \subset J \setminus A_{n,p}$, and hence $B_{n,p}$ is finite too. So, for each n, $p \in \mathbb{N}$ there is $\overline{j} \in \mathbb{N}$ (without loss of generality $\overline{j} \in J$) with $|a_{j,n}| \leq \sigma_p^{(n)}$ whenever $j \in J$, $j \geq \overline{j}$. This proves the assertion. □

Observe that a result analogous to Lemma 2.3 holds also when we deal with a single (O)-sequence, independent of n.

Arguing analogously as Lemma 2.23 it is possible to prove the following results.

Lemma 2.24 (Boccuto, Dimitriou and Papanastassiou (2012e), Lemma 2.3) *Let $\mathcal{F}$ be a diagonal filter of $\mathbb{N}$, and $(a_i)_i$ be a sequence in R such that $(O\mathcal{F})\lim_i a_i = 0$ with respect to an (O)-sequence $(\sigma_p)_p$. Then for every $I \in \mathcal{F}^*$ there exists $J \in \mathcal{F}^*$ with $J \subset I$ and $(O)\lim_{i \in J} a_i = 0$ with respect to the same (O)-sequence $(\sigma_p)_p$.*

Lemma 2.25 (see also Boccuto, Dimitriou and Papanastassiou (2012b), Lemma 3.1)

Let $\mathcal{F}$ be a diagonal filter of $\mathbb{N}$. If $(V_i)_i$ is any increasing sequence in $\mathcal{U}^*(0)$ and $(a_{i,n})_i$ is (PR)-$\mathcal{F}$-bounded by $(V_i)_i$ for each $n \in \mathbb{N}$, then for every $I \in \mathcal{F}^*$ there is $J \in \mathcal{F}^*$, $J \subset I$ such that $(a_{i,n})_i$ is (PR)-eventually bounded by $(V_i)_i$.

Note that, in Lemma 2.25, one takes $A_n^* := \{i \in \mathbb{N} : a_{i,n} \in V_i\}$, $n \in \mathbb{N}$, instead of $A_{n,p}$. □

We now extend to the lattice group context the notions of filter (or ideal) convergence/divergence for nets in the lattice group setting with respect to another filter of an abstract nonempty directed set Λ. In the literature, in the topological space setting, they are formulated in terms of ideals ($\mathcal{I}_1^{\mathcal{I}_2}$-convergence, see also Boccuto, Dimitriou, Papanastassiou and Wilczyński (2014), Mačaj and Sleziak (2011)). For a sake of coherence of notations, we prefer to present them in terms of filters, though it is equivalent to treat them dealing with ideals or filters. Of course, these concepts can be given analogously also for sequences, taking $\Lambda = \mathbb{N}$ endowed with the usual order.

Let Λ be any directed set. Given two filters $\mathcal{F}_1$, $\mathcal{F}_2$ of Λ, we say that a net $(x_\lambda)_{\lambda\in\Lambda}$ in R *$(O\mathcal{F}_2^*)$-converges to $x \in X$ with respect to $\mathcal{F}_1$* or *$(O\mathcal{F}_2^{\mathcal{F}_1})$-converges to x* (resp. *$\mathcal{F}_2^{\mathcal{F}_1}$-diverges at x*) iff there exists a set $B \in \mathcal{F}_2$ such that the net $(z_\lambda)_\lambda$, defined by

$$z_\lambda = \begin{cases} x_\lambda, & \text{if } \lambda \in B, \\ x, & \text{if } \lambda \in \Lambda \setminus B, \end{cases} \tag{2.27}$$

$(O\mathcal{F}_1)$-converges to (resp. $\mathcal{F}_1$-diverges at) x. If $(x_\lambda)_\lambda$ $(O\mathcal{F}_2^{\mathcal{F}_1})$-converges to x, then we write

$$(O\mathcal{F}_2^{\mathcal{F}_1})\lim_{\lambda\in\Lambda} x_\lambda = x.$$

Observe that it is not natural to consider $(O\mathcal{F}_1)$-convergence or $\mathcal{F}_1$-divergence of the net $(x_\lambda)_{\lambda\in B}$, where B is a suitable element of $\mathcal{F}_2$. The reason is that, in this case, there may exist a set $B \in \mathcal{F}_2$ such that $\Lambda \setminus B \in \mathcal{F}_1 \setminus \mathcal{F}_\Lambda$. For example, if $R = \mathbb{R}$ with the usual metric, $\Lambda = \mathbb{N}$, $x_n = \dfrac{1}{2^n}$, $t_n = 2^n$ for each $n \in B$, and $x_n = t_n = 1$ for all $n \in \mathbb{N} \setminus B$, then $\lim_{n\in B} x_n = 0$ in the usual sense, and so for each $\varepsilon > 0$ we get $\{n \in B : |x_n| \le \varepsilon\} \in \mathcal{F}_1$. Analogously it is possible to check that the sequence t_n, $n \in B$, $\mathcal{F}_1$-diverges at 0. On the other hand, if

$$z_n = \begin{cases} x_n, & \text{if } n \in B; \\ 1, & \text{if } n \in \mathbb{N} \setminus B; \end{cases} \qquad w_n = \begin{cases} t_n, & \text{if } n \in B; \\ 1, & \text{if } n \in \mathbb{N} \setminus B, \end{cases}$$

then the sequences $(z_n)_n$ and $(w_n)_n$ $\mathcal{F}_1$-converge to 1.

However, when Λ is any directed set and $\mathcal{F}_1 = \mathcal{F}_\Lambda$, we get that $\mathcal{F}_2'$-convergence coincides with $(O\mathcal{F}_2^{\mathcal{F}_1})$-convergence. Indeed it is easy to see that, for every

Dedekind complete (ℓ)-group R, for any net $(x_\lambda)_{\lambda\in\Lambda}$ in R, for each $x\in R$, for any filter $\mathcal{F}_2$ of Λ and for every $B\in\mathcal{F}_2$, we get $(O)\lim_{\lambda\in B} x_\lambda = x$ if and only if the sequence $(z_\lambda)_\lambda$, defined as in (2.27), (O)-converges to x. Analogously it is possible to check that $\mathcal{F}_2'$-divergence of any net $(x_\lambda)_{\lambda\in\Lambda}$ in R is equivalent to its $\mathcal{F}_2^{\mathcal{F}_\Lambda}$-divergence.

We will extend Kostyrko, Šalát and Wilczyński (2001), Theorem 3.2 and Mačaj and Sleziak (2011), Theorems 3.11 and 3.12 to the lattice group context, giving a characterization of the (AP)-property.

We now present some elementary properties of $\mathcal{F}_2^{\mathcal{F}_1}$-convergence in the (ℓ)-group setting.

Note that, in general, the $(O\mathcal{F}_2^{\mathcal{F}_1})$-limit is not unique. The following theorem extends Gordji, Sarabadan and Arani (2012), Corollary 3.9 and gives a necessary and sufficient condition for uniqueness of the $(O\mathcal{F}_2^{\mathcal{F}_1})$-limit.

Theorem 2.26 (see Boccuto, Dimitriou, Papanastassiou and Wilczyński (2014), Theorem 3.4) *Let Λ be a nonempty abstract set, $\mathcal{F}_1$ and $\mathcal{F}_2$ be two fixed (Λ)-free filters of Λ, and suppose that*

2.26.1) *For each $E\in\mathcal{F}_1$ and $A\in\mathcal{F}_2$, $E\cap A\neq\emptyset$.*

If $(x_\lambda)_{\lambda\in\Lambda}$ is a net in R with $(O\mathcal{F}_2^{\mathcal{F}_1})\lim_\lambda x_\lambda = l_1$ and $(O\mathcal{F}_2^{\mathcal{F}_1})\lim_\lambda x_\lambda = l_2$, then $l_1 = l_2$.

Conversely, if the $(O\mathcal{F}_2^{\mathcal{F}_1})$-limit is unique for each $(O\mathcal{F}_2^{\mathcal{F}_1})$-convergent net $(x_\lambda)_\lambda$, then condition 2.26.1) *holds.*

Proof: We begin with the first part. By definition of $(O\mathcal{F}_2^{\mathcal{F}_1})$-limit, there exist two sets B_1, $B_2\in\mathcal{F}_2$ such that the nets $(y_\lambda)_\lambda$, $(z_\lambda)_\lambda$, defined by

$$y_\lambda = \begin{cases} x_\lambda, & \text{if } \lambda\in B_1, \\ l_1, & \text{if } \lambda\in\Lambda\setminus B_1; \end{cases} \qquad z_\lambda = \begin{cases} x_\lambda, & \text{if } \lambda\in B_2, \\ l_2, & \text{if } \lambda\in\Lambda\setminus B_2, \end{cases}$$

$(O\mathcal{F}_1)$-converge to l_1 and l_2 respectively. Let $(\sigma_p)_p$ and $(\xi_p)_p$ be two (O)-sequences, related to $(O\mathcal{F}_1)$-convergence of $(y_\lambda)_\lambda$ and $(z_\lambda)_\lambda$ to l_1 and l_2 respectively.

Set now $B := B_1 \cap B_2$. For each $p \in \mathbb{N}$ we have

$$C_p := \{\lambda \in \Lambda : |y_\lambda - l_1| \le \sigma_p\} \in \mathcal{F}_1, \qquad D_p := \{\lambda \in \Lambda : |z_\lambda - l_2| \le \xi_p\} \in \mathcal{F}_1.$$

Let $F_p := C_p \cap D_p$, $p \in \mathbb{N}$. Since $B \in \mathcal{F}_2$ and $F_p \in \mathcal{F}_1$, thanks to condition 2.26.1), we get $B \cap F_p \neq \emptyset$.

Pick arbitrarily $p \in \mathbb{N}$ and let $\lambda_p \in B \cap F_p$. We have:

$$0 \le |l_1 - l_2| \le |x_{\lambda_p} - l_1| + |x_{\lambda_p} - l_2| = |y_{\lambda_p} - l_1| + |z_{\lambda_p} - l_2| \le \sigma_p + \xi_p. \qquad \textbf{(2.28)}$$

From (2.28) and arbitrariness of p we get $l_1 = l_2$. This concludes the proof of the first part.

We now turn to the last part. Suppose by contradiction that 2.26.1) does not hold. Then there exist two sets $E \in \mathcal{F}_1$, $A \in \mathcal{F}_2$, with $E \cap A = \emptyset$. We get $A \in \mathcal{F}_2$, $\Lambda \setminus A \supset E$, and hence $\Lambda \setminus A \in \mathcal{F}_1$. Pick arbitrarily two different elements l_1, $l_2 \in R$, and set

$$x_\lambda = \begin{cases} l_1, & \text{if } \lambda \in A, \\ l_2, & \text{if } \lambda \in \Lambda \setminus A. \end{cases}$$

If

$$z_\lambda = \begin{cases} x_\lambda, & \text{if } \lambda \in A, \\ l_1, & \text{if } \lambda \in \Lambda \setminus A, \end{cases}$$

then $z_\lambda = l_1$ for each $\lambda \in \Lambda$, and obviously the net $(z_\lambda)_\lambda$ $(O\mathcal{F}_1)$-converges to l_1. On the other hand, if

$$t_\lambda = \begin{cases} x_\lambda, & \text{if } \lambda \in A, \\ l_2, & \text{if } \lambda \in \Lambda \setminus A, \end{cases}$$

then, for every (O) -sequence $(\sigma_p)_p$ in R and $p \in \mathbb{N}$, we get $\{\lambda \in \Lambda : |t_\lambda - l_2| \le \sigma_p\} \supset \Lambda \setminus A \in \mathcal{F}_1$.

Thus the net $(x_\lambda)_\lambda$ $(O\mathcal{F}_2^{\mathcal{F}_1})$-converges both to l_1 and to l_2, and hence the limit of $(x_\lambda)_\lambda$ is not unique, getting a contradiction with the hypothesis. This completes the proof. □

Observe that, if $\mathcal{F}_1 \subset \mathcal{F}_2$ or $\mathcal{F}_2 \subset \mathcal{F}_1$, then condition 2.26.1) is satisfied. Indeed, for each $E \in \mathcal{F}_1$ and $A \in \mathcal{F}_2$ we get that both E and A belong to the *same* filter $\mathcal{F}$, the greatest between $\mathcal{F}_1$ and $\mathcal{F}_2$. Thus $E \cap A \in \mathcal{F}$ and hence $E \cap A \neq \emptyset$.

We now prove the linearity of the $(O\mathcal{F}_2^{\mathcal{F}_1})$-limit.

Proposition 2.27 (see Boccuto, Dimitriou, Papanastassiou and Wilczyński (2014), Proposition 3.6) *With the same notations as in Theorem 2.26, if* $(O\mathcal{F}_2^{\mathcal{F}_1})\lim_\lambda x_\lambda = x$ *and* $(O\mathcal{F}_2^{\mathcal{F}_1})\lim_\lambda y_\lambda = y$, *then* $(O\mathcal{F}_2^{\mathcal{F}_1})\lim_\lambda (x_\lambda + y_\lambda) = x + y$.

Moreover, if $a \in \mathbb{R}$ *and* $(O\mathcal{F}_2^{\mathcal{F}_1})\lim_\lambda x_\lambda = x$, *then* $(O\mathcal{F}_2^{\mathcal{F}_1})\lim_\lambda (a x_\lambda) = ax$.

Proof: We now prove only the first part, since the proof of the last part is analogous. By definition of $(O\mathcal{F}_2^{\mathcal{F}_1})$-limit, there are B_1, $B_2 \in \mathcal{F}_2$ such that the nets

$$z_\lambda = \begin{cases} x_\lambda, & \text{if } \lambda \in B_1, \\ x, & \text{if } \lambda \in \Lambda \setminus B_1, \end{cases} \qquad t_\lambda = \begin{cases} y_\lambda, & \text{if } \lambda \in B_2, \\ y, & \text{if } \lambda \in \Lambda \setminus B_2, \end{cases}$$

$(O\mathcal{F}_1)$-converge to x and y respectively. Hence,

$$(O\mathcal{F}_1)\lim_\lambda (z_\lambda + t_\lambda) = x + y. \qquad \textbf{(2.29)}$$

Put

$$p_\lambda = \begin{cases} x_\lambda + y_\lambda, & \text{if } \lambda \in B_1 \cap B_2, \\ x + y, & \text{if } \lambda \in \Lambda \setminus (B_1 \cap B_2). \end{cases} \quad \textbf{(2.30)}$$

The assertion follows from (2.29) and (2.30). □

The proof of the following results is straightforward.

Proposition 2.28 (Boccuto, Dimitriou, Papanastassiou and Wilczyński (2014), Proposition 3.7) *If* $(O\mathcal{F}_1)\lim_\lambda x_\lambda = x$, *then we get* $(O\mathcal{F}_2{}^{\mathcal{F}_1})\lim_\lambda x_\lambda = x$; *if* $(x_\lambda)_\lambda$ $\mathcal{F}_2$*-diverges at* x, *then* $(x_\lambda)_\lambda$ $\mathcal{F}_2{}^{\mathcal{F}_1}$*-diverges at* x.

Let $\mathcal{F}_1 \subset \mathcal{H}_1$, $\mathcal{I}_2 \subset \mathcal{H}_2$ *be four* (Λ)*-free filters of* Λ. *If* $(O\mathcal{F}_2{}^{\mathcal{F}_1})\lim_\lambda x_\lambda = x$, *then we get also* $(O\mathcal{F}_2{}^{\mathcal{H}_1})\lim_\lambda x_\lambda = x$ *and* $(O\mathcal{H}_2{}^{\mathcal{F}_1})\lim_\lambda x_\lambda = x$.

If $(x_\lambda)_\lambda$ $\mathcal{F}_2{}^{\mathcal{F}_1}$*-diverges at* x, *then* $(x_\lambda)_\lambda$ $\mathcal{F}_2{}^{\mathcal{H}_1}$*- and* $\mathcal{H}_2{}^{\mathcal{F}_1}$*-diverges at* x.

Now we are asking, whether $(O\mathcal{F}_2{}^{\mathcal{F}_1})$-convergence implies $(O\mathcal{F}_2)$-convergence. We give a necessary and sufficient condition, extending Mačaj and Sleziak (2011), Proposition 3.7 to the lattice group setting.

Theorem 2.29 (see Boccuto, Dimitriou, Papanastassiou and Wilczyński (2014), Theorem 3.8) *Let* $\mathcal{F}_1$, $\mathcal{F}_2$ *be two* (Λ)*-free filters of* Λ.

If $\mathcal{F}_1 \subset \mathcal{F}_2$ *and* $(O\mathcal{F}_2{}^{\mathcal{F}_1})\lim_\lambda x_\lambda = x$, *then* $(O\mathcal{F}_2)\lim_\lambda x_\lambda = x$.

Conversely, if $(O\mathcal{F}_2{}^{\mathcal{F}_1})\lim_\lambda x_\lambda = x$ *implies* $(O\mathcal{F}_2)\lim_\lambda x_\lambda = x$, *then* $\mathcal{F}_1 \subset \mathcal{F}_2$.

Proof: We begin with the first part. Let $\mathcal{F}_1 \subset \mathcal{F}_2$, $\mathcal{I}_1$, and $\mathcal{I}_2$ be the dual ideals of $\mathcal{F}_1$ and $\mathcal{F}_2$, respectively, and suppose that $(O\mathcal{F}_2{}^{\mathcal{F}_1})\lim_\lambda x_\lambda = x$. Then there are a set $B \in \mathcal{F}_2$ and an (O)-sequence $(\sigma_p)_p$ with

$$\{\lambda \in B : |z_\lambda - x| \not\leq \sigma_p\} \in \mathcal{F}_1 \subset \mathcal{F}_2 \tag{2.31}$$

for every $p \in \mathbb{N}$, where

$$z_\lambda = \begin{cases} x_\lambda, & \text{if } \lambda \in B, \\ x, & \text{if } \lambda \in \Lambda \setminus B, \end{cases}$$

By virtue of (2.31), for each $p \in \mathbb{N}$ we get:

$$\{\lambda \in \Lambda : |x_\lambda - x| \not\leq \sigma_p\} \subset \{\lambda \in B : |z_\lambda - x| \not\leq \sigma_p\} \cup (\Lambda \setminus B) \in \mathcal{F}_2,$$

namely $(O\mathcal{F}_2) \lim_\lambda x_\lambda = x$. This concludes the proof of the first part.

We now turn to the last part. Let $\mathcal{F}_1 \not\subset \mathcal{F}_2$ and pick an element $A \in \mathcal{F}_1 \setminus \mathcal{F}_2$. Let $y \neq x$ and define

$$x_\lambda = \begin{cases} x, & \text{if } \lambda \in A, \\ y, & \text{if } \lambda \in \Lambda \setminus A. \end{cases}$$

We get $(O\mathcal{F}_1)$ $\lim_\lambda x_\lambda = x$ and hence, thanks to Proposition 2.28, $(O\mathcal{F}_2^{\mathcal{F}_1}) \lim_\lambda x_\lambda = x$.

However, note that for every (O) -sequence $(\sigma_p)_p$ there is a $\overline{p} \in \mathbb{N}$ with $|x - y| \not\leq \sigma_{\overline{p}}$, otherwise we should have $x = y$. Thus, $\{\lambda \in \Lambda : |x_\lambda - x| \leq \sigma_{\overline{p}}\} = A \notin \mathcal{F}_2$, and hence the net $(x_\lambda)_\lambda$ is not $(O\mathcal{F}_2)$-convergent to x. This ends the proof. □

A result analogous to Theorem 2.29 holds in the setting of filter divergence.

Theorem 2.30 (see Boccuto, Dimitriou, Papanastassiou and Wilczyński (2014), Theorem 3.9) *Under the same hypotheses and notations as in Theorem 2.29, if* $\mathcal{F}_1 \subset \mathcal{F}_2$ *and the net* $(x_\lambda)_\lambda$ $\mathcal{F}_2^{\mathcal{F}_1}$*-diverges at* x*, then* $(x_\lambda)_\lambda$ $\mathcal{F}_2$*-diverges at* x.

Conversely, assume that $(x_\lambda)_\lambda$ $\mathcal{F}_2$*-diverges at* x *whenever* $(x_\lambda)_\lambda$ *is a net,* $\mathcal{F}_2^{\mathcal{F}_1}$*-divergent at* x. *Then we get* $\mathcal{F}_1 \subset \mathcal{F}_2$.

We now give a characterization of the (AP)-property in the Riesz space setting in terms of $(O\mathcal{F}_2^{\mathcal{F}_1})$-convergence (or $\mathcal{F}_2^{\mathcal{F}_1}$-divergence).

Theorem 2.31 (see Boccuto, Dimitriou, Papanastassiou and Wilczyński (2014), Theorem 3.10) *Let* $\mathcal{I}_1$ *and* $\mathcal{I}_2$ *be two* (Λ)*-admissible ideals of* Λ, $\mathcal{F}_1$ *and* $\mathcal{F}_2$ *be their dual filters. If* $\mathcal{I}_2$ *has the* (AP)*-property with respect to* $\mathcal{I}_1$ *and* $(x_\lambda)_{\lambda\in\Lambda}$ *is any net in* R, $(O\mathcal{F}_2)$*-convergent to (resp.* $\mathcal{F}_2$*-divergent at)* $x \in R$ *with respect to an order (resp. bounding) sequence* $(\sigma_j)_j$ *in* R, *then* $(x_\lambda)_\lambda$ $(O\mathcal{F}_2^{\mathcal{F}_1})$*-converges to (resp.* $\mathcal{F}_2^{\mathcal{F}_1}$*- diverges at)* x *with respect to* $(\sigma_j)_j$.

Conversely, if every net $(x_\lambda)_\lambda$ *in* R, $(O\mathcal{F}_2)$*-convergent to (resp.* $\mathcal{F}_2$*-divergent at)* $x \in R$ *with respect to an order (resp. bounding) sequence* $(\sigma_j)_j$, *is also* $(O\mathcal{F}_2^{\mathcal{F}_1})$*-convergent to (resp.* $\mathcal{F}_2^{\mathcal{F}_1}$*-divergent at)* x *with respect to* $(\sigma_j)_j$, *then* $\mathcal{I}_2$ *has the* (AP)*-property with respect to* $\mathcal{I}_1$.

Proof: We begin with the first part.

Let $(\sigma_j)_j$ be an (O)-sequence in R, related to $(O\mathcal{F}_2)$-convergence of the net $(x_\lambda)_\lambda$ to x. Set $O_j = [-\sigma_j, \sigma_j]$, $j \in \mathbb{N}$, $A_1 = \{\lambda \in \Lambda : x_\lambda - x \notin O_1\}$, $A_j = \{\lambda \in \Lambda : x_\lambda - x \in O_{j-1} \setminus O_j\}$, $j \geq 2$. It is not difficult to see that $(A_j)_j$ is a disjoint sequence of elements of $\mathcal{I}_2$.

Since $\mathcal{I}_2$ has the (AP)-property with respect to $\mathcal{I}_1$, there is a sequence $(B_j)_j$ in $\mathcal{P}(\Lambda)$, with $A_j \Delta B_j \in \mathcal{I}_1$ for each $j \in \mathbb{N}$ and $B := \bigcup_{j=1}^\infty B_j \in \mathcal{I}_2$. Put now

$$z_\lambda = \begin{cases} x_\lambda, & \text{if } \lambda \in \Lambda \setminus B, \\ x, & \text{if } \lambda \in B. \end{cases} \tag{2.32}$$

For every $j \in \mathbb{N}$, set $D_j := \{\lambda \in \Lambda : z_\lambda - x \notin O_j\}$. We get

$$D_j = \{\lambda \in \Lambda \setminus B : x_\lambda - x \notin O_j\} \subset \bigcup_{i=1}^{j}(A_i \setminus B_i) \subset \bigcup_{i=1}^{j}(A_i \Delta B_i),$$

and so $D_j \in \mathcal{I}_1$. This means that the net $(z_\lambda)_\lambda$ $(O\mathcal{F}_1)$-converges to x, and hence the net $(x_\lambda)_\lambda$ $(O\mathcal{F}_2^{\mathcal{F}_1})$-converges to x, with respect to the (O)-sequence $(\sigma_j)_j$. This concludes the proof of the first part, in the setting of filter convergence.

The case of filter divergence is analogous. Indeed, if $(\sigma_j)_j$ is a bounding sequence, associated with $\mathcal{F}_2$-divergence of the net involved $(x_\lambda)_\lambda$ at x, it is enough to take

$$O_j := \{r \in R : |r| \geq \sigma_j\}, \quad j \in \mathbb{N},$$

and to proceed similarly as above.

We now turn to the last part. Let $u \in R$, $u > 0$, $(\varepsilon_j)_j$ be a strictly decreasing (O)-sequence in $\mathbb{R}$, $\sigma_j := \varepsilon_j u$ for every $j \in \mathbb{N}$, $\sigma_0 := 2\sigma_1$.

Let $(A_j)_j$ be any disjoint sequence in $\mathcal{I}_2$, and for every $\lambda \in \Lambda$ set

$$y_\lambda = \begin{cases} \sigma_{j-1}, & \text{if } \lambda \in A_j, j \in \mathbb{N}; \\ 0, & \text{if } \lambda \in \Lambda \setminus (\bigcup_{j=1}^{\infty} A_j). \end{cases}$$

By construction, for every $j \in \mathbb{N}$ we get:

$$\{\lambda \in \Lambda : y_\lambda \not\leq \sigma_j\} = \bigcup_{i=1}^{j} A_i \in \mathcal{I}_2,$$

and thus $(O\mathcal{F}_2) \lim_\lambda y_\lambda = 0$. By hypothesis, we have

$$(O\mathcal{F}_2^{\mathcal{F}_1}) \lim_\lambda y_\lambda = 0$$

with respect to the (O)-sequence $(\sigma_j)_j$. Hence there is a set $B \in \mathcal{I}_2$ such that the net $(z_\lambda)_{\lambda \in \Lambda}$ defined by

$$z_\lambda = \begin{cases} y_\lambda, & \text{if } \lambda \in \Lambda \setminus B, \\ 0, & \text{if } \lambda \in B, \end{cases}$$

$(O\mathcal{F}_1)$-converges to 0 with respect to $(\sigma_j)_j$. Thus for every $j \in \mathbb{N}$ we get:

$$E_j := \{\lambda \in \Lambda : z_\lambda \not\leq \sigma_j\} \in \mathcal{I}_1 \qquad \textbf{(2.33)}$$

and

$$A_j \setminus B = \{\lambda \in A_j : y_\lambda \not\leq \sigma_j\} \setminus B = \{\lambda \in A_j : z_\lambda \not\leq \sigma_j\} \setminus B \subset E_j. \qquad \textbf{(2.34)}$$

For each $j \in \mathbb{N}$, set $B_j = A_j \cap B$. From (2.33) and (2.34) it follows that $A_j \setminus B \in \mathcal{I}_1$, and a fortiori

$$A_j \Delta B_j = A_j \setminus B_j = A_j \setminus B \in \mathcal{I}_1.$$

Moreover, $B_j \in \mathcal{I}_2$ for all j and

$$\bigcup_{j=1}^{\infty} B_j = (\bigcup_{j=1}^{\infty} A_j) \cap B \subset B,$$

and hence $\bigcup_{j=1}^{\infty} B_j \in \mathcal{I}_2$. Thus, $\mathcal{I}_2$ has the (AP)-property with respect to $\mathcal{I}_1$. This ends the proof, concerning filter convergence.

We now consider filter divergence. Let u be as above, set $t_0 = u$ and $t_j = 2^j u$ for every $j \in \mathbb{N}$. It is easy to see that $(t_j)_j$ is bounding.

If $(A_j)_j$ is any disjoint sequence in $\mathcal{I}_2$, then put

$$v_\lambda = \begin{cases} t_{j-1}, & \text{if } \lambda \in A_j, j \in \mathbb{N}; \\ 0, & \text{if } \lambda \in \Lambda \setminus (\bigcup_{j=1}^{\infty} A_j). \end{cases}$$

For every $j \in \mathbb{N}$, we have $\{\lambda \in \Lambda : v_\lambda \not\geq t_j\} = \bigcup_{i=1}^{j} A_i \in \mathcal{I}_2$. So, the net $(v_\lambda)_\lambda$ $\mathcal{F}_2$-diverges to 0, and by assumption it is also $\mathcal{F}_2^{\mathcal{F}_1}$-divergent at 0. Thus there exists $Q \in \mathcal{I}_2$ such that the net $(w_\lambda)_{\lambda \in \Lambda}$ defined by setting

$$w_\lambda = \begin{cases} v_\lambda, & \text{if } \lambda \in \Lambda \setminus Q, \\ 0, & \text{if } \lambda \in Q, \end{cases}$$

$\mathcal{F}_1$-diverges to 0 with respect to $(t_j)_j$. Therefore for each $j \in \mathbb{N}$ we get:

$$F_j := \{\lambda \in \Lambda : w_\lambda \not\geq t_j\} \in \mathcal{I}_1 \tag{2.35}$$

and

$$A_j \setminus Q = \{\lambda \in A_j : v_\lambda \not\geq t_j\} \setminus Q = \{\lambda \in A_j : w_\lambda \not\geq t_j\} \setminus Q \subset F_j. \tag{2.36}$$

Put $Q_j = A_j \cap Q$, $j \in \mathbb{N}$. From (2.35) and (2.36), proceeding similarly as above, we obtain that $A_j \Delta Q_j \in \mathcal{I}_1$ for all $j \in \mathbb{N}$, and $\bigcup_{j=1}^{\infty} Q_j \in \mathcal{I}_2$. From this it follows that $\mathcal{I}_2$ has the (AP)-property with respect to $\mathcal{I}_1$. This ends the proof. □

We now claim that there exist two ideals $\mathcal{I}_1$, $\mathcal{I}_2$ of $\mathbb{N}$, such that $\mathcal{I}_1 \subset \mathcal{I}_2$ and $\mathcal{I}_2$ has the (AP)-property with respect to $\mathcal{I}_1$, but $\mathcal{I}_1$ and $\mathcal{I}_2$ do not have the (AP)-property with respect to $\mathcal{I}_{\text{fin}}$ (see also Boccuto, Dimitriou, Papanastassiou and Wilczyński (2014), Remarks 3.11). First of all, let $\mathcal{K}_1$ be as in (2.5), and

$$\mathcal{K}_2 := \{A \subset \mathbb{N}^2 : \delta_2(A) = 0\}, \tag{2.37}$$

where

$$\delta_2(A) = \lim_{n,l\to+\infty} \frac{\#(A \cap (\{1,\ldots,n\}\times\{1,\ldots,l\})}{n\cdot l}.$$

In Das, Kostyrko, Wilczyński and Malík (2008) it is shown that $\mathcal{K}_1 \subset \mathcal{K}_2$, $\mathcal{K}_1 \neq \mathcal{K}_2$, $\mathcal{K}_2$ fulfils the (AP)-property with respect to $\mathcal{K}_1$, but neither $\mathcal{K}_1$ nor $\mathcal{K}_2$ have the (AP)-property. Since by Das, Kostyrko, Wilczyński and Malík (2008), Remark 2, any bijection between $\mathbb{N}^2$ and $\mathbb{N}$ maps double sequences into single sequences and ideals of $\mathbb{N}^2$ into ideals of $\mathbb{N}$ and by this transformation the ideal convergence and the (AP)-property are preserved, then it is possible to associate with $\mathcal{K}_1$ and $\mathcal{K}_2$ two ideals $\mathcal{I}_1$ and $\mathcal{I}_2$, satisfying the above claim.

Theorem 2.31 extends Lahiri and Das (2008), Theorem 12 and Mačaj and Sleziak (2011), Theorems 3.11 and 3.12, which deal with filter/ideal convergence in the topological space setting, and Das and Ghosal (2010b), Theorems 3 and 4, which were proved for filter/ideal divergence in the metric space context.

2.3. Lattice Group-Valued Measures

2.3.1. Main Properties of Measures

We now deal with (ℓ)-group-valued measures and some fundamental properties, which will be useful in the sequel in proving different types of limit theorems. These kinds of measures were investigated, for instance, in Antosik and Swartz (1992), Boccuto (1995b-c and 1996a-b), Boccuto and Candeloro (2002b and 2004a), Swartz (1989a-b).

Let G be any nonempty set and $\mathcal{E} \subset \mathcal{P}(G)$ be a lattice. A measure $m : \mathcal{E} \to R$ is said to be *bounded* on $\mathcal{E}$ iff there is a positive element $u \in R$, with $|m(E)| \leq u$ for all $E \in \mathcal{E}$.

The measures $m_j : \mathcal{E} \to R$, $j \in \mathbb{N}$, are *uniformly bounded* or *equibounded* on $\mathcal{E}$ iff there exists a positive $u \in R$, with $|m_j(E)| \leq u$ for every $j \in \mathbb{N}$ and $E \in \mathcal{E}$.

A measure $m : \mathcal{E} \to R$ is said to be *finitely additive* on $\mathcal{E}$ iff $m(E \cup F) = m(E) + m(F)$ whenever E, F are two disjoint elements of $\mathcal{E}$.

Given a finitely additive bounded measure $m:\mathcal{E}\to R$, we define $(m^{(+)})_{\mathcal{E}}$, $(m^{(-)})_{\mathcal{E}}$, $\| m \|_{\mathcal{E}}$, $v_{\mathcal{E}}(m):\mathcal{E}\to R$ by setting, for every $E\in\mathcal{E}$:

$$(m^{(+)})_{\mathcal{E}}(E)=\bigvee\{m(B):B\in\mathcal{E},B\subset E\},$$

$$(m^{(-)})_{\mathcal{E}}(E)=\bigvee\{-m(B):B\in\mathcal{E},B\subset E\},$$

$$\| m \|_{\mathcal{E}}(E)=(m^{(+)})_{\mathcal{E}}(E)+(m^{(-)})_{\mathcal{E}}(E),$$

$$v_{\mathcal{E}}(m)(E)=\bigvee\{| m(B) |: B\in\mathcal{E},B\subset E\}.$$

The quantities $(m^{(+)})_{\mathcal{E}}$, $(m^{(-)})_{\mathcal{E}}$, $\| m \|_{\mathcal{E}}$, $v_{\mathcal{E}}(m)$ are called *positive part, negative part, total variation* and *semivariation* on $\mathcal{E}$ of m respectively. When no confusion can arise, we will often write $m^{(+)}$, $m^{(-)}$, $\| m \|$, $v(m)$ instead of $(m^{(+)})_{\mathcal{E}}$, $(m^{(-)})_{\mathcal{E}}$, $\| m \|_{\mathcal{E}}$, $v_{\mathcal{E}}(m)$ respectively. Observe that

$$v_{\mathcal{E}}(m)(E)\le\| m\|_{\mathcal{E}}(E)\le 2v_{\mathcal{E}}(m)(E) \quad \text{for all } E\in\mathcal{E} \tag{2.38}$$

similarly as in the real case: indeed it is enough to argue analogously as in the proof of formula (1.6).

We now recall the notions of (s)-boundedness, σ-additivity, absolute continuity, continuity with respect to a Fréchet-Nikodým topology and regularity with respect to the semivariation $v_{\mathcal{E}}$.

A finitely additive measure $m:\mathcal{E}\to R$ is said to be (s)*-bounded* on $\mathcal{E}$ iff for every disjoint sequence $(C_k)_k$ in $\mathcal{E}$ we have $(O)\lim_k v_{\mathcal{E}}(m)(C_k)=0$. The finitely additive measures $m_j:\mathcal{E}\to R$, $j\in\mathbb{N}$, are *uniformly* (s)*-bounded* on $\mathcal{E}$ iff $(O)\lim_k (\bigvee_j v_{\mathcal{E}}(m_j)(C_k))=0$ whenever $(C_k)_k$ is a sequence of pairwise disjoint elements of $\mathcal{E}$.

A finitely additive measure $m:\mathcal{E}\to R$ is σ*-additive* on $\mathcal{E}$ iff for every decreasing sequence $(H_k)_k$ in $\mathcal{E}$ with $\bigcap_{k=1}^{\infty} H_k=\emptyset$ we have $(O)\lim_k v_{\mathcal{E}}(m)(H_k)=0$. The σ-additive measures $m_j:\mathcal{E}\to R$, $j\in\mathbb{N}$, are *uniformly* σ*-additive* on $\mathcal{E}$ iff for each decreasing sequence $(H_k)_k$ in $\mathcal{E}$ with $\bigcap_{k=1}^{\infty} H_k = \emptyset$ we get $(O)\lim_k (\bigvee_j v_{\mathcal{E}}(m_j)(H_k)) = 0$.

Let $\nu : \mathcal{E} \to [0,+\infty]$ be a finitely additive measure. A measure $m : \mathcal{E} \to R$ is ν-*absolutely continuous* on $\mathcal{E}$, iff for each decreasing sequence $(H_k)_k$ in $\mathcal{E}$, with $\lim_k \nu(H_k) = 0$ we get $(O)\lim_k v_{\mathcal{E}}(m)(H_k) = 0$. The measures $m_j : \mathcal{E} \to R$, $j \in \mathbb{N}$, are *uniformly* ν*-absolutely continuous* or *equiabsolutely continuous* (with respect to ν) on $\mathcal{E}$ iff $(O)\lim_k (\bigvee_j v_{\mathcal{E}}(m_j)(H_k)) = 0$ for each decreasing sequence $(H_k)_k$ in $\mathcal{E}$ with $\lim_k \nu(H_k) = 0$.

Note that, in general, the above definition of absolute continuity does not coincide with the classical one. Indeed, let $R = L^0([0,1], \mathcal{M}, \nu)$ be as in Example 2.181 of Chapter 1. Define $m : \mathcal{M} \to R$ by setting $m(A) = \chi_A$ for every $A \in \mathcal{M}$, where χ_A is the characteristic function associated with A. Then $\lim_k \nu(A_k) = 0$ if and only if $\chi_{A_k} \to 0$ in L^1, while $(O)\lim_k \chi_{A_k} = 0$ if and only if $\chi_{A_k} \to 0$ almost everywhere. Hence, in general, the implication

$$[\nu(A_k) \to 0] \Rightarrow [(O)\lim_k m(A_k) = 0] \tag{2.39}$$

fails to hold, and so m is not ν-absolutely continuous in the classical sense. However, if $(A_k)_k$ is any decreasing sequence in $\mathcal{M}$, then it is not difficult to check that (2.39) holds, and hence m is ν-absolutely continuous according to the definition above given (see also Boccuto (1996a), Remark 1.13.1).

Let τ be a Fréchet-Nikodým topology on $\mathcal{E}$. A finitely additive measure $m : \mathcal{E} \to R$ is said to be τ*-continuous* on $\mathcal{E}$ iff for each decreasing sequence $(H_k)_k$ in $\mathcal{E}$ with $\tau\text{-}\lim_k H_k = \emptyset$ we get $(O)\lim_k v_{\mathcal{E}}(m)(H_k) = 0$. The finitely additive measures $m_j : \mathcal{E} \to R$, $j \in \mathbb{N}$, are said to be *uniformly* τ*-continuous* on $\mathcal{E}$ iff $(O)\lim_k (\bigvee_j v_{\mathcal{E}}(m_j)(H_k)) = 0$ whenever $(H_k)_k$ is a decreasing sequence in $\mathcal{E}$ with $\tau\text{-}\lim_k H_k = \emptyset$.

Let $\Sigma \subset \mathcal{P}(G)$ is a σ-algebra and $\mathcal{H}$, $\mathcal{G} \subset \Sigma$ are two lattices, such that the complement of every element of $\mathcal{H}$ belongs to $\mathcal{G}$ and $\mathcal{G}$ is closed under countable disjoint unions. A measure $m : \Sigma \to R$ is *regular* on Σ iff for any $A \in \Sigma$ there are two sequences $(F_k)_k$ in $\mathcal{H}$, $(G_k)_k$ in $\mathcal{G}$, with

$$F_k \subset F_{k+1} \subset A \subset G_{k+1} \subset G_k \quad \text{for all } k \in \mathbb{N}, \tag{2.40}$$

and $\bigwedge_k (v_\Sigma(m)(G_k \setminus F_k)) = 0$. The finitely additive measures $m_j : \Sigma \to R$, $j \in \mathbb{N}$, are *uniformly regular* on Σ iff for each $A \in \Sigma$ there exist two sequences $(F_k)_k$, $(G_k)_k$ satisfying (2.40), and with

$$\bigwedge_k \left(\bigvee_j (v_\Sigma(m_j)(G_k \setminus F_k))\right) = 0.$$

Observe that, when G is a topological normal space (resp. locally compact Hausdorff space), and we take $\Sigma = \{$ Borel subsets of $G\}$, $\mathcal{H} = \{$closed sets$\}$ (resp. $\{$compact sets$\}$), $\mathcal{G} = \{$open sets$\}$ and $R = \mathbb{R}$, we obtain the classical definition of regularity for real-valued measures (see also Boccuto (1996b), Brooks (1969b), Candeloro (1985b), Candeloro and Letta (1985), de Lucia and Pap (2002)). Moreover note that the above concepts can be formulated, analogously as before, also with respect to (D)-convergence.

The next result links (ℓ)-group-valued σ-additive and regular measures, and extends Alexandroff (1941), Theorem 5, which was given for $R = \mathbb{R}$ (see also Billingsley (1995), Theorem 1.1). We formulate it in the setting of (D)-convergence, to apply the Fremlin Lemma, to *sums* of (D)-sequences (see also Kawabe (2007), Theorems 1 and 2 and Volauf (1993), Theorem 1, for other related versions existing in the literature in the vector lattice setting).

Theorem 3.1 *Let R be a Dedekind complete weakly σ-distributive (ℓ)-group, (G,d) be a compact metric space, Σ be the σ-algebra of all Borel sets of G, $\mathcal{G}$ and $\mathcal{H}$ be the lattices of all open and all closed subsets of G respectively. Then every bounded regular measure $m : \Sigma \to R$ is σ-additive. Conversely, if R is also super Dedekind complete, then every bounded finitely additive σ-additive measure $m : \Sigma \to R$ is regular.*

Proof: We begin with the first part (see also Boccuto, Minotti and Sambucini (2013), Theorem 2.2). Without loss of generality, we can suppose that m is positive (indeed, in the general case it will be enough to consider $m^{(+)}$ and $m^{(-)}$). So, in order to prove σ-additivity of m, it will be enough to show that, for every disjoint sequence $(A_k)_k$ in Σ,

$$m(\bigcup_{k=1}^{\infty} A_k) \leq \sum_{k=1}^{\infty} m(A_k). \tag{2.41}$$

Let $A := \bigcup_{k=1}^{\infty} A_k$. By regularity of m, for each $k \in \mathbb{N}$ there is a (D)-sequence $(a_{t,l}^{(k)})_{t,l}$ such that, for every $\varphi \in \mathbb{N}^{\mathbb{N}}$, it is possible to find $C_k \in \mathcal{H}$ and $U_k \in \mathcal{G}$ with $C_k \subset A_k \subset U_k$ and $m(U_k \setminus C_k) \leq \bigvee_{t=1}^{\infty} a_{t,\varphi(t+k)}^{(k)}$. Moreover, putting $u = m(G)$, by virtue of the Fremlin Lemma there is a regulator $(a_{t,l})_{t,l}$ with

$$u \wedge (\sum_{k=1}^{q} (\bigvee_{t=1}^{\infty} a_{t,\varphi(t+k)}^{(k)})) \leq \bigvee_{t=1}^{\infty} a_{t,\varphi(t)} \quad \text{for all } q \in \mathbb{N} \text{ and } \varphi \in \mathbb{N}^{\mathbb{N}}. \tag{2.42}$$

Furthermore, again by regularity of m, in correspondence with A there are a (D)-sequence $(b_{t,l})_{t,l}$ and a set $C \in \mathcal{H}$, with $C \subset A$ and

$$m(A) \leq m(C) + \bigvee_{t=1}^{\infty} b_{t,\varphi(t)}. \tag{2.43}$$

We now claim that

$$m(C) \leq \sum_{k=1}^{\infty} m(A_k). \tag{2.44}$$

Choose arbitrarily $\varphi \in \mathbb{N}^{\mathbb{N}}$. In correspondence with φ, there exists a sequence $(U_k)_k$ in $\mathcal{G}$, with $A_k \subset U_k$ and $m(U_k) \leq m(A_k) + \bigvee_{t=1}^{\infty} a_{t,\varphi(t+k)}^{(k)}$ for all k. Since C is closed and G is compact, then C is compact too. As $C \subset \bigcup_{k=1}^{\infty} U_k$, there is $N \in \mathbb{N}$ with $C \subset \bigcup_{k=1}^{N} U_k$. Observe that

$$m(C) \le \sum_{k=1}^{N} m(U_k) \le \sum_{k=1}^{N} m(A_k) + \sum_{k=1}^{N} (\bigvee_{t=1}^{\infty} a_{t,\varphi(t+k)}^{(k)}), \text{ and } m(C) \le u.$$

From the above inequalities and (2.42), we get:

$$m(C) - \sum_{k=1}^{N} m(A_k) \le \bigvee_{t=1}^{\infty} a_{t,\varphi(t)}. \qquad \textbf{(2.45)}$$

Thus (2.44) follows from (2.45) and weak σ-distributivity of R, by arbitrariness of $\varphi \in \mathbb{N}^{\mathbb{N}}$.

From (2.43) and (2.44) we have $m(A) \le \sum_{k=1}^{\infty} m(A_k) + \bigvee_{t=1}^{\infty} b_{t,\varphi(t)}$. Again by weak σ-distributivity of R, we get (2.41), and hence σ-additivity of m. This concludes the proof of the first part.

We now turn to the last part. Set $\mathcal{T} := \{A \in \Sigma :$ there is a regulator $(a_{t,l})_{t,l}$ such that for every $\varphi \in \mathbb{N}^{\mathbb{N}}$

there are $D \in \mathcal{G}, F \in \mathcal{H}$ with $F \subset A \subset D$ and $v_{\Sigma}(m)(D \setminus F) \le \bigvee_{t=1}^{\infty} a_{t,\varphi(t)}\}$.

We claim that $\mathcal{H} \subset \mathcal{T}$. Indeed, choose arbitrarily $W \in \mathcal{H}$, and for each $k \in \mathbb{N}$ set $D_k := \{x \in G : d(x,W) < 1/k\}$, $W_k := D_k \setminus W$. Note that the sequence $(W_k)_k$ is decreasing, and

$\bigcap_{k=1}^{\infty} W_k = \emptyset$. By σ-additivity of m, we find a regulator $(a_{t,l})_{t,l}$, such that for each $\varphi \in \mathbb{N}^{\mathbb{N}}$ there is $k_0 \in \mathbb{N}$, with $v_{\Sigma}(m)(D_{k_0} \setminus W) \le \bigvee_{t=1}^{\infty} a_{t,\varphi(t)}$. Since $D_{k_0} \in \mathcal{G}$, $W \in \mathcal{H}$ and $W \subset D_{k_0}$, then $W \in \mathcal{T}$.

We now prove that $\mathcal{T}$ is a σ-algebra. It is easy to see that, if $A \in \mathcal{T}$, then $G \setminus A \in \mathcal{T}$. Let now $(A_k)_k$ be a disjoint sequence of elements of $\mathcal{T}$, with $A := \bigcup_{k=1}^{\infty} A_k$. We claim that $A \in \mathcal{T}$.

For every $k \in \mathbb{N}$ there are a regulator $(a_{t,l}^{(k)})_{t,l}$ and two sets $D_k \in \mathcal{G}$, $F_k \in \mathcal{H}$, with $F_k \subset A_k \subset D_k$ and $v_\Sigma(m)(D_k \setminus F_k) \le \bigvee_{t=1}^{\infty} a_{t,\varphi(t+k)}^{(k)}$. Let $u := \bigvee_{E \in \Sigma} |m(E)|$. By the Fremlin Lemma there is a (D)-sequence $(a_{t,l})_{t,l}$ in R, with $u \wedge (\sum_{k=1}^{q} (\bigvee_{t=1}^{\infty} a_{t,\varphi(t+k)}^{(k)})) \le \bigvee_{t=1}^{\infty} a_{t,\varphi(t)}$ for each $q \in \mathbb{N}$ and $\varphi \in \mathbb{N}^{\mathbb{N}}$. Since the sequence $(F_k)_k$ is disjoint, by σ-additivity of m there exists a (D)-sequence $(b_{t,l})_{t,l}$ such that to every $\varphi \in \mathbb{N}^{\mathbb{N}}$ there corresponds a $k_0 \in \mathbb{N}$ with

$$v_\Sigma(m)((\bigcup_{k=1}^{\infty} F_k) \setminus (\bigcup_{k=1}^{k_0} F_k)) = v_\Sigma(m)(\bigcup_{k=k_0+1}^{\infty} F_k) \le \bigvee_{t=1}^{\infty} b_{t,\varphi(t)}. \tag{2.46}$$

Set $c_{t,l} := 2(a_{t,l} + b_{t,l})$, t, $l \in \mathbb{N}$, $D := \bigcup_{k=1}^{\infty} D_k$ and $F := \bigcup_{k=1}^{k_0} F_k$. Note that $(c_{t,l})_{t,l}$ is a

(D)-sequence, $F \subset A \subset D$, $D \in \mathcal{G}$, $F \in \mathcal{H}$, and

$$v_\Sigma(m)(D \setminus F) \le v_\Sigma(m)(D \setminus (\bigcup_{k=1}^{\infty} F_k)) + v_\Sigma(m)((\bigcup_{k=1}^{\infty} F_k) \setminus F) \le$$

$$\le v_\Sigma(m)(\bigcup_{k=1}^{\infty} (D_k \setminus F_k)) + \bigvee_{t=1}^{\infty} b_{t,\varphi(t)} \le \sum_{k=1}^{\infty} v_\Sigma(m)(D_k \setminus F_k) + \bigvee_{t=1}^{\infty} b_{t,\varphi(t)} \le$$

$$\bigvee_{t=1}^{\infty} a_{t,\varphi(t)} + \bigvee_{t=1}^{\infty} b_{t,\varphi(t)} \le \bigvee_{t=1}^{\infty} c_{t,\varphi(t)}.$$

Thus $A \in \mathcal{T}$, that is the claim. Therefore, $\mathcal{T}$ is a σ-algebra. Since $\mathcal{T} \supset \mathcal{H}$, then $\mathcal{T} = \Sigma$.

Now choose arbitrarily $A \in \Sigma$. As $\Sigma = \mathcal{T}$, there is a regulator $(a_{t,l})_{t,l}$, satisfying the condition in the definition of $\mathcal{T}$. Since R is super Dedekind complete and weakly σ-distributive, by Theorem 2.2, in correspondence with the (D)-sequence $(a_{t,l})_{t,l}$ we find a sequence $(\varphi_k)_k$ in $\mathbb{N}^{\mathbb{N}}$ and an (O)-sequence $(b_k)_k$ such that, for every $k \in \mathbb{N}$, we get $\bigvee_{t=1}^{\infty} a_{t,\varphi_k(t)} \le b_k$, and hence there are $D_k^* \in \mathcal{G}$,

$F_k^* \in \mathcal{H}$, with $F_k^* \subset A \subset D_k^*$ and $v_\Sigma(m)(D_k^* \setminus F_k^*) \le b_k$. Set $D_k := \bigcap_{i=1}^k D_i^*$, $F_k := \bigcup_{i=1}^k F_i^*$. We get: $F_k \subset F_{k+1} \subset A \subset D_{k+1} \subset D_k$, $D_k \in \mathcal{G}$, $F_k \in \mathcal{H}$, $v_\Sigma(m)(D_k \setminus F_k) \le v_\Sigma(m)(D_k^* \setminus F_k^*) \le b_k$, and so we obtain $\bigwedge_k v_\Sigma(m)(D_k \setminus F_k) = (O)\lim_k v_\Sigma(m)(D_k \setminus F_k) = 0$. Thus, m is regular on Σ. □

Observe that, arguing similarly as above, it is possible to prove that, under the same hypotheses as in Theorem 3.1, given an equibounded sequence $m_j : \Sigma \to R$, $j \in \mathbb{N}$ of measures, the m_j's are uniformly regular if and only if they are uniformly σ-additive.

In general, even when $R = \mathbb{R}$, the concepts of regularity and σ-additivity do not coincide. Indeed, with the above notations, if $\mathcal{G} = \mathcal{H} = \Sigma$, every finitely additive measure is obviously regular, but not necessarily σ-additive. Conversely, if $\mathcal{G} = \mathcal{H} = \{\emptyset, G\}$, and $m : \Sigma \to \mathbb{R}$ is any σ-additive measure such that $m(\emptyset) = 0$, $m(G) = 1$ and there are $E \in \Sigma$ and $a \in (0,1)$ with $m(E) = a$, then it is easy to see that m is not regular.

We now prove the following:

Proposition 3.2 *Let R, G, Σ, $\mathcal{G}$, $\mathcal{H}$ be as above, and $m_j : \Sigma \to R$, $j \in \mathbb{N}$, be a sequence of regular measures. Then for every $A \in \Sigma$ and $k \in \mathbb{N}$ there are $F_k \in \mathcal{H}$, $G_k \in \mathcal{G}$, satisfying* (2.40) *and* $\bigwedge_k v_\Sigma(m_j)(G_k \setminus F_k) = 0$ *for every* $j \in \mathbb{N}$.

Proof: (see also Boccuto (1996b), Proposition 3.2) By hypothesis, for every j and $k \in \mathbb{N}$ there is a pair $(G_k^{(j)}, F_k^{(j)})$ satisfying (2.40). If $G_k := \bigcap_{j=1}^k G_k^{(j)}$, $F_k := \bigcup_{j=1}^k F_k^{(j)}$, then F_k and G_k satisfy (2.40) and are independent of j. □

Using the above notions of (RO)- and (RD)-convergence, it is possible to formulate the concepts of pointwise (D)- and (O)-convergence for measures, with respect to a single regulator and (O)-sequence, respectively. We will often use them in proving several versions of limit theorems in the context of (ℓ)-groups.

Let $\mathcal{E} \subset \mathcal{P}(G)$ be a lattice. Given a sequence of set functions $m_j : \mathcal{E} \to R$, $j \in \mathbb{N}$, we say that the m_j's (RD)*-converge* (or *converge pointwise with respect to a single regulator*) to m_0 on $\mathcal{E}$, or shortly $(RD)\lim_j m_j = m_0$, iff there exists a (D)-sequence $(b_{t,l})_{t,l}$ such that for each $\varphi \in \mathbb{N}^{\mathbb{N}}$ and $E \in \mathcal{E}$ there is an integer j_0 with

$$| m_j(E) - m_0(E) | \leq \bigvee_{t=1}^{\infty} b_{t,\varphi(t)} \quad \text{for every } j \geq j_0. \tag{2.47}$$

We say that $(D)\lim_j m_j = m_0$ *uniformly* on $\mathcal{E}$, or shortly $(UD)\lim_j m_j = m_0$, iff there exists a (D)-sequence $(c_{t,l})_{t,l}$ such that for every $\varphi \in \mathbb{N}^{\mathbb{N}}$ there is $j_0 \in \mathbb{N}$ with

$$| m_j(E) - m_0(E) | \leq \bigvee_{t=1}^{\infty} c_{t,\varphi(t)} \quad \text{for all } E \in \mathcal{E} \text{ and } j \geq j_0, \tag{2.48}$$

iff $(D)\lim_j m_j(E) = m_0(E)$ uniformly with respect to $E \in \mathcal{E}$.

We say that $(m_j)_j$ is (RD)*-Cauchy* on $\mathcal{E}$ iff there is a (D)-sequence $(b_{t,l})_{t,l}$ such that for each $\varphi \in \mathbb{N}^{\mathbb{N}}$ and $E \in \mathcal{E}$ there exists an integer j_0 with

$$| m_j(E) - m_{j+p}(E) | \leq \bigvee_{t=1}^{\infty} b_{t,\varphi(t)} \quad \text{for all } j \in \mathbb{N}, j \geq j_0, \text{ and } p \in \mathbb{N}. \tag{2.49}$$

A sequence of set functions $(m_j)_j$ is said to be (RO)*-convergent* to m_0 on $\mathcal{E}$ iff there exists an (O)-sequence $(b_l)_l$ such that for every $l \in \mathbb{N}$ and $E \in \mathcal{E}$ there is $j_0 \in \mathbb{N}$ with $| m_j(E) - m_0(E) | \leq b_l$ for all $j \geq j_0$. In this case we say that $(RO)\lim_j m_j = m_0$.

We say that $(O)\lim_j m_j = m_0$ *uniformly* on $\mathcal{E}$, or briefly $(UO)\lim_j m_j = m_0$, iff there exists an (O)-sequence $(\tau_l)_l$ such that for every $l \in \mathbb{N}$ there is $j_0 \in \mathbb{N}$ with

$$| m_j(E) - m_0(E) | \le \tau_l \quad \text{for all } E \in \mathcal{E} \text{ and } j \ge j_0, \tag{2.50}$$

iff $(O)\lim_j m_j(E) = m_0(E)$ uniformly with respect to $E \in \mathcal{E}$.

The sequence $(m_j)_j$ is (RO)-*Cauchy* on $\mathcal{E}$ iff there is an (O)-sequence $(q_l)_l$ such that for every $l \in \mathbb{N}$ and $E \in \mathcal{E}$ there is $j_0 \in \mathbb{N}$ with

$$| m_j(E) - m_{j+p}(E) | \le p_l \quad \text{for all } j \in \mathbb{N},\ j \ge j_0, \text{ and } p \in \mathbb{N}. \tag{2.51}$$

Observe that the concepts of (s)-boundedness, σ-additivity and the related ones can be formulated also with respect to a single (O)-sequence or regulator. This will be useful in order to prove several kinds of limit theorems. We begin with considering (D)-convergence.

Let $\mathcal{E}$ be any lattice of subsets of G. Assume that $m : \mathcal{E} \to R$ is any finitely additive measure. We say that m is *globally* σ*-additive* on $\mathcal{E}$ iff there exists a regulator $(a_{t,l})_{t,l}$ such that, for every decreasing sequence $(H_k)_k$ in $\mathcal{E}$ with $\bigcap_{k=1}^{\infty} H_k = \emptyset$, and for every $\varphi \in \mathbb{N}^{\mathbb{N}}$, there exists $k_0 \in \mathbb{N}$ such that $| m(E) | \le \bigvee_{t=1}^{\infty} a_{t,\varphi(t)}$ for all $E \in \mathcal{E}$, $E \subset H_{k_0}$.

Let $m : \mathcal{E} \to R$ be any finitely additive measure. We say that m is *globally* (s)-*bounded* on $\mathcal{E}$ iff there exists a regulator $(a_{t,l})_{t,l}$ such that, for every disjoint sequence $(C_k)_k$ in $\mathcal{E}$ and any $\varphi \in \mathbb{N}^{\mathbb{N}}$, there exists an integer $\overline{k}$ with

$$| m(C_k) | \le \bigvee_{t=1}^{\infty} a_{t,\varphi(t)} \tag{2.52}$$

for all $k \ge \overline{k}$. Note that formula (2.52) can be equivalently replaced by

$$\bigvee \{ | m(E) | : E \in \mathcal{E}, E \subset C_k \} \le \bigvee_{t=1}^{\infty} a_{t,\varphi(t)} \tag{2.53}$$

for every $k \geq \overline{k}$. Indeed, if (2.53) were not true for some disjoint sequence $(C_k)_k$ and some $\varphi \in \mathbb{N}^{\mathbb{N}}$, then there would exist a strictly increasing sequence $(k_h)_h$ in $\mathbb{N}$ and a corresponding sequence $(E_h)_h$ in $\mathcal{E}$ of subsets of H_{k_h}, with $| m(E_h) | \not\leq \bigvee_{t=1}^{\infty} a_{t,\varphi(t)}$ for every $h \in \mathbb{N}$, and this contradicts global (s)-boundedness of m (see also Boccuto and Candeloro (2002b), Remark 2.11).

Observe that the concept of global (s)-boundedness above formulated is different than the one of (s)-boundedness. Indeed, we will see later that every bounded (ℓ)-group-valued measure is (s)-bounded too, but not necessarily globally (s)-bounded.

We now formulate global σ-additivity, global (s)-boundedness and the related notions in the setting of (O)-convergence. Observe that, when R is a super Dedekind complete and weakly σ-distributive (ℓ)-group, thanks to Theorem 2.2, to deal with (O)-convergence and to consider (D)-convergence are equivalent.

A finitely additive measure $m:\mathcal{E} \to R$ is said to be *globally σ-additive with respect to (O)-convergence* on $\mathcal{E}$ iff there exists an (O)-sequence $(b_l)_l$ in R such that, for each decreasing sequence $(H_k)_k$ in $\mathcal{E}$ with $\bigcap_{k=1}^{\infty} H_k = \emptyset$, and for every $l \in \mathbb{N}$, one has $v_{\mathcal{E}}(m)(H_k) \leq b_l$ for k large enough. We say that m is *globally (s)-bounded with respect to (O)-convergence* on $\mathcal{E}$ iff there exists an (O)-sequence $(b_l)_l$ in R such that, for every disjoint sequence $(C_k)_k$ in $\mathcal{E}$ and for every integer l, one has $v_{\mathcal{E}}(m)(C_k) \leq b_l$ for l sufficiently large.

The following technical result links global σ-additivity and global (s)-boundedness. We point out that in this context these concepts are formulated with respect to a same regulator, and hence the proof is not similar to the classical one, but presents some more technical difficulties.

Proposition 3.3 (see Boccuto and Candeloro (2004a), Proposition 2.12) *Let $\Sigma \subset \mathcal{P}(G)$ be a σ-algebra, $m:\Sigma \to R$ be a measure, globally (s)-bounded on*

Σ, and $(a_{t,l})_{t,l}$ be a regulator, working for global (s)-boundedness. Assume that $m(\bigcup_{k=1}^{\infty} C_k) = \sum_{n=1}^{\infty} m(C_k)$ for every disjoint sequence $(C_k)_k$ in Σ.

Then m is globally σ-additive on Σ, and $(a_{t,l})_{t,l}$ works for that property.

Proof: Without loss of generality, we can assume that m is positive: indeed, in the general case, it will be enough to consider $m^{(+)}$ and $m^{(-)}$.

Let $(C_k)_k$ be a disjoint sequence in Σ. We get:

$$m(\bigcup_{k=j}^{\infty} C_k) = \bigvee_{p\in\mathbb{N}} (m(\bigcup_{k=j}^{j+p} C_k)) \quad \text{for every } j \in \mathbb{N}.$$

We now show that for every $\varphi \in \mathbb{N}^{\mathbb{N}}$ there exists an integer k_0 with $m(\bigcup_{k=k_0}^{\infty} C_k) \le \bigvee_{t=1}^{\infty} a_{t,\varphi(t)}$. Indeed, fix $\varphi \in \mathbb{N}^{\mathbb{N}}$ and assume by contradiction that there is a strictly increasing sequence $(k_s)_s$ in $\mathbb{N}$, with $m(\bigcup_{k=k_s}^{\infty} C_k) \not\le \bigvee_{t=1}^{\infty} a_{t,\varphi(t)}$ for each $s \in \mathbb{N}$. Note that for each s there exists an integer $p(s)$ with

$$m(\bigcup_{k=k_s}^{k_s+p(s)} C_k) \not\le \bigvee_{t=1}^{\infty} a_{t,\varphi(t)}.$$

Now, by choosing a subsequence if necessary, we can assume that $k_{s+1} > k_s + p(s)$. So, we obtain that the sets $E_s := \bigcup_{k=k_s}^{k_s+p(s)} C_k$ are pairwise disjoint and satisfy $m(E_s) \not\le \bigvee_{t=1}^{\infty} a_{t,\varphi(t)}$ for each $s \in \mathbb{N}$, which contradicts global (s)-boundedness of m. □

We now state the following technical lemma. Here and in the sequel, the intervals of the type $[N, M]$ are meant in $\mathbb{N}$.

Lemma 3.4 (see also Boccuto and Papanastassiou (2007), Lemma 3.2) *Let* $m:\mathcal{P}(\mathbb{N})\to R$ *be any finitely additive measure. Fix* N, $M\in\mathbb{N}$, *with* $N<M$, *and let* $\mathcal{P}([N,M])$ *be the class of all subsets of* $[N,M]$. *Then we get:*

$$\bigvee_{S\in\mathcal{P}([N,M])} |m(S)| \leq \sum_{l=N}^{M} |m(\{l\})| = \| m\|([N,M]) \leq 2 \bigvee_{S\in\mathcal{P}([N,M])} |m(S)|.$$

Proof: Let M,N be as in the hypothesis. For each $l\in\mathbb{N}$ we have

$$m^{(+)}(\{l\})=\vee\{m(\{l\}),0\}=[m(\{l\})]^{+},\qquad m^{(-)}(\{l\})=-\wedge\{m(\{l\}),0\}=[m(\{l\})]^{-},$$

and hence

$$\sum_{l=N}^{M}|m(\{l\})|=\sum_{l=N}^{M}[m(\{l\})]^{+}+\sum_{l=N}^{M}[m(\{l\})]^{-}=\sum_{l=N}^{M}m^{(+)}(\{l\})+\sum_{l=N}^{M}m^{(-)}(\{l\})$$

$$=m^{(+)}([N,M])+m^{(-)}([N,M])=\| m\|([N,M]). \qquad \textbf{(2.54)}$$

The assertion follows from (2.38) and (2.54). □

We now have the following properties of globally (s)-bounded lattice group-valued measures. We formulate them in the setting of (D)-convergence: analogously they can be proved in the context of (O)-convergence.

Proposition 3.5 (see Boccuto and Candeloro (2002b), Proposition 2.12) *Let* $m:\mathcal{E}\to R$ *be a finitely additive bounded measure. Then* m *is globally* (s)-*bounded if and only if* $m^{(+)}+m^{(-)}$ *is.*

Proof: The "if" part is straightforward.

We turn to the "only if" part. By (2.53), global (s)-boundedness (and boundedness) of m imply that $m^{(+)}$ and $m^{(-)}$ are globally (s)-bounded, and so $m^{(+)}+m^{(-)}$ is too. □

Proposition 3.6 (Boccuto and Candeloro (2002b), Proposition 2.13) *Let* $\Sigma\subset\mathcal{P}(G)$ *be a* σ-*algebra, and* $m:\Sigma\to R$ *be a globally* σ-*additive measure. Then* m *is globally* (s)-*bounded.*

Proof: Given any disjoint sequence $(C_k)_k$ in Σ, set $H_k = \bigcup_{l=k}^{\infty} C_l$. Then $(H_k)_k$ is a decreasing sequence in Σ, with $\bigcap_{n=1}^{\infty} H_k = \varnothing$. The assertion follows from global σ-additivity and the inclusion $C_k \subset H_k$, $k \in \mathbb{N}$. □

Proposition 3.7 (see also Boccuto and Candeloro (2002b), Proposition 2.14) *Let $m : \Sigma \to R$ be any globally (s)-bounded measure. Then, for every monotone sequence $(H_k)_k$ in Σ and for each $\varphi \in \mathbb{N}^{\mathbb{N}}$ there is $N \in \mathbb{N}$ with $v(m)(H_k \Delta H_p) \leq \bigvee_{t=1}^{\infty} a_{t,\varphi(t)}$ for all $k, p \geq N$, and so there exists in R the limit*

$$(D)\lim_k m(H_k). \qquad \textbf{(2.55)}$$

Proof: Without loss of generality, we can suppose that $(H_k)_k$ is decreasing.

Let $(a_{t,l})_{t,l}$ be a regulator related to global (s)-boundedness of m, and let us show that it works for (2.55). If this is not true, there exist: an element $\varphi \in \mathbb{N}^{\mathbb{N}}$, two sequences $(k_h)_h$, $(p_h)_h$ in $\mathbb{N}$, with $\lim_h k_h = +\infty$ and $k_{h+1} > k_h + p_h$ for every h, and a sequence $(B_h)_h$ in Σ, with $B_h \subset H_{k_h} \setminus H_{k_h + p_h}$ and $| m(B_h) | \not\leq \bigvee_{t=1}^{\infty} a_{t,\varphi(t)}$ for every $h \in \mathbb{N}$. Since the B_h's are pairwise disjoint, this contradicts global (s)-boundedness of m with respect to the regulator $(a_{t,l})_{t,l}$.
□

We now show that every globally (s)-bounded measure is bounded too.

Lemma 3.8 (Boccuto and Candeloro (2002b), Lemma 2.15) *Let $m : \Sigma \to R$ be any globally (s)-bounded measure, and let $(a_{t,l})_{t,l}$ be the regulator corresponding to global (s)-boundedness. Let u be such that $a_{t,l} \leq u$ for each t, $l \in \mathbb{N}$, and let there exist $B \in \Sigma$, with $| m(B) | \not\leq u$. Then there is $A \in \Sigma$, $A \subset B$, with $| m(A) | \not\leq u$, but such that the set $\{m(E) : E \in \Sigma, E \subset A\}$ is bounded in R.*

Proof: If the set $\{m(E): E \in \Sigma, E \subset B\}$ is bounded in R, we take $A = B$, and we have finished. Otherwise, there exists a set $B_1 \in \Sigma$, $B_1 \subset B$, such that $| m(B_1) | \not\leq u + | m(B) |$.

Therefore, we get $| m(B \setminus B_1) | \not\leq u$. If the set $\{m(E): E \in \Sigma, E \subset B_1\}$ is bounded in R, we can take $A = B_1$, and we have finished. Otherwise, there exists a set $B_2 \in \Sigma$, $B_2 \subset B_1$, with $| m(B_2) | \not\leq u + | m(B_1) |$. Thus, we get $| m(B_1 \setminus B_2) | \not\leq u$. Proceeding by induction, either we find a subset $B_n \in \Sigma$, $B_n \subset B$, satisfying the requirement of the lemma, or we construct a decreasing sequence $(B_n)_n$ in Σ, with $| m(B_n \setminus B_{n+1}) | \not\leq u$ for each $n \in \mathbb{N}$. Since the sets $B_n \setminus B_{n+1}$ are pairwise disjoint and $u \geq a_{t,l}$ for every t, $l \in \mathbb{N}$, we get a contradiction. Thus, only the first case can occur. This ends the proof. □

Theorem 3.9 (Boccuto and Candeloro (2002b), Theorem 2.16) *Let $m: \Sigma \to R$ be any globally (s)-bounded finitely additive measure. Then m is bounded.*

Proof: By contradiction, assume that the set $\{m(A): A \in \Sigma\}$ is not bounded in R. Define u as in Lemma 3.8, and choose any set $B_1 \in \Sigma$, with $| m(B_1) | \not\leq u$. Thanks to Lemma 3.8, there exists in Σ a set $A_1 \subset B_1$, such that $| m(A_1) | \not\leq u$, but the set $\{m(E): E \in \Sigma, E \subset A_1\}$ is bounded in R.

Now, the set $\{m(E): E \in \Sigma,\ E \subset G \setminus A_1\}$ is not bounded in R, and hence there is an element $B_2 \in \Sigma$, $B_2 \subset G \setminus A_1$, such that $| m(B_2) | \not\leq u$. Then, by Lemma 3.8, there exists in Σ an element $A_2 \subset B_2$, with $| m(A_2) | \not\leq u$, but such that $\{m(E): E \in \Sigma, E \subset A_2\}$ is bounded in R.

Now, the set $\{m(E): E \in \Sigma,\ E \subset G \setminus (A_1 \cup A_2)\}$ is not bounded in R, and so there exists an element $B_3 \in \Sigma$, $B_3 \subset G \setminus (A_1 \cup A_2)$, with $| m(B_3) | \not\leq u$. So we find an element $A_3 \in \Sigma$, disjoint from $A_1 \cup A_2$, such that $| m(A_3) | \not\leq u$. Proceeding in this way, we construct a disjoint sequence $(A_n)_n$ in Σ, with $| m(A_n) | \not\leq u$ for all n, which is a contradiction with the meaning of u in the condition of global (s)-boundedness. □

Example 3.10 (Boccuto and Candeloro (2002b), Example 2.17) We now show that, in general, boundedness does not imply global (s)-boundedness.

Let $G=[0,1]$, Σ be the Borel σ-algebra on $[0,1]$, and $R=L^0([0,1],\Sigma,\nu)$, where ν denotes the Lebesgue measure. Set $m(A)=\chi_A$ for every $A\in\Sigma$. Then, m is a bounded finitely additive measure. If m were (s)-bounded, then there should exist a regulator $(a_{t,l})_{t,l}$ in R such that, for each $\varphi\in\mathbb{N}^{\mathbb{N}}$ and for every disjoint sequence $(C_k)_k$ in Σ, one should have $\bigvee_{t=1}^{\infty} a_{t,\varphi(t)}(x)\geq 1$ for almost all $x\in\bigcup_{k\geq k_0} C_k$, for some integer k_0. Now, for every $x\in[0,1]$ it is not difficult to construct a disjoint sequence $(C_k)_k$ in Σ, such that for every k_0 the set $\{x\}\cup(\bigcup_{k\geq k_0} C_k)$ contains a neighborhood of x. This means that for each $\varphi\in\mathbb{N}^{\mathbb{N}}$ and $x\in[0,1]$ there exists $\delta>0$ such that $\bigvee_{t=1}^{\infty} a_{t,\varphi(t)}(y)\geq 1$ for almost all $y\in]x-\delta,x+\delta[$. By compactness, we deduce that $\bigvee_{t=1}^{\infty} a_{t,\varphi(t)}\geq 1$. Thus, we get

$$\bigwedge_{\varphi\in\mathbb{N}^{\mathbb{N}}}\left(\bigvee_{t=1}^{\infty} a_{t,\varphi(t)}\right)\geq 1. \tag{2.56}$$

But it is known that the space $L^0([0,1],\Sigma,\nu)$ is weakly σ-distributive (see also Boccuto (2003)). This contradicts (2.56). □

The following example shows that (s)-boundedness (intended similarly as in the classical context, and not with respect to a single (O)- or (D)-sequence) does not imply boundedness, for Riesz space-valued measures (see Swartz (1989a), Example 3).

Example 3.11 Let (G,Σ,ν) be the measure space consisting of the interval $[0,1]\subset\mathbb{R}$, endowed with the Borel σ-algebra and the Lebesgue measure. Let R be the Riesz space c_0 of all real sequences convergent to 0 with the usual ordering, and define $m:\Sigma\to R$ as follows:

$$m(A)=(m_1(A),\ldots,m_n(A),\ldots)$$

for every set $A \in \Sigma$, where $m_n(A) = \int_0^1 \sin(n\pi t)dt$. It is easy to see that m is not order bounded in c_0, because $\bigvee\{|m_n(A)|: A \in \Sigma\} = 1$ for each n. However, in Swartz (1989a), Example 3 it is shown that m is σ-additive, and a fortiori (s)-bounded (in the classical sense). □

We now see that, in the (ℓ)-group context, boundedness implies (s)-boundedness (in the classical sense). Note that this assertion is in general not true in the Banach space context (see Example 2.130 of Chapter 1). In what follows, suppose that (uniform) (s)-boundedness is formulated with respect to (O)-convergence.

Theorem 3.12 (Boccuto, Dimitriou and Papanastassiou (2010c), Theorem 3.1, see also Boccuto and Candeloro (2004d), Theorem 3.2) *Let R be a Dedekind complete (ℓ)-group, G be any nonempty set, $\mathcal{L} \subset \mathcal{P}(G)$ be an algebra, $\mathcal{G} \subset \mathcal{L}$ be a lattice, closed under countable disjoint unions. Then every finitely additive measure $m: \mathcal{L} \to R$, bounded on $\mathcal{G}$, is (s)-bounded on $\mathcal{G}$ too (in the classical meaning).*

Proof: Let R be a subgroup of $\mathcal{C}_\infty(\Omega)$, where Ω is as in the Maeda-Ogasawara-Vulikh representation theorem 2.7. Then there is a nowhere dense set $N_0 \subset \Omega$ such that the real-valued set functions $m(\cdot)(\omega), \omega \in \Omega \setminus N_0$, are finitely additive and bounded on $\mathcal{G}$. By the classical results (see also Bhaskara Rao and Bhaskara Rao (1983) Corollary 2.1.7), they are (s)-bounded on $\mathcal{G}$.

Fix now an arbitrary disjoint sequence $(C_k)_k$ in $\mathcal{G}$. Then, again by the Maeda-Ogasawara-Vulikh theorem, there is a meager set N, depending on $(C_k)_k$, without loss of generality with $N \supset N_0$, and such that

$$[(O)\lim_k v_{\mathcal{G}}(m)(C_k)](\omega) = (\bigwedge_k \bigvee_{s \geq k} v_{\mathcal{G}}(m)(C_s))(\omega)$$

$$= (\bigwedge_n \bigvee_{s \geq k} \bigvee \{|m(A)|: A \in \mathcal{G}, A \subset C_s\})(\omega)$$

$$= \inf_k \sup_{s \geq k} (\sup\{|m(A)(\omega)|: A \in \mathcal{G}, A \subset C_s\}) = \inf_k \sup_{s \geq k} v_{\mathcal{G}}(m(\cdot)(\omega))(C_s)$$

$$= \lim_k v_{\mathcal{G}}(m(\cdot)(\omega))(C_k) = 0 \quad \text{for all } \omega \in \Omega \setminus N.$$

Since the complement of a meager set of Ω is dense in Ω, we obtain $(O)\lim_k v_{\mathcal{G}}(m)(C_k) = 0$. Thus we get the assertion. □

Let $m_j : \mathcal{E} \to R$, $j \in \mathbb{N}$, be any sequence of finitely additive globally (s)-bounded measures on $\mathcal{E}$. We say that they are *globally uniformly* (s)*-bounded* on $\mathcal{E}$ iff there exists a regulator $(a_{t,l})_{t,l}$ such that, for each disjoint sequence $(C_k)_k$ in $\mathcal{E}$ and $\varphi \in \mathbb{N}^{\mathbb{N}}$, there is a $k_0 \in \mathbb{N}$, with

$$\bigvee_{j \in \mathbb{N}} v(m_j)(C_k) \leq \bigvee_{t=1}^{\infty} a_{t,\varphi(t)} \tag{2.57}$$

for all $k \geq k_0$.

If we consider (O)-convergence, then we say that the measures m_j are *globally uniformly* (s)*-bounded with respect to* (O)*-convergence* iff there is an (O)-sequence $(b_l)_l$ such that, for every disjoint sequence $(C_k)_k$ in $\mathcal{E}$ and any $l \in \mathbb{N}$, there is $k_l \in \mathbb{N}$, with

$$\bigvee_{j \in \mathbb{N}} v(m_j)(C_k) \leq b_l \quad \text{whenever } k \geq k_l. \tag{2.58}$$

Contrarily to intuition, global uniform (s)-boundedness for a sequence of measures does not imply equiboundedness, even in the case $R = \mathbb{R}$. For instance, take $G = \{1\}$, $\Sigma = \mathcal{P}(G)$, and define $m_j(\{1\}) = j$, $j \in \mathbb{N}$. Clearly, the measures m_j are globally uniformly (s)-bounded (since G is finite), and also globally uniformly σ-additive, but they are not equibounded (see also Boccuto and Candeloro (2002b), Remark 3.2).

We now state the following result.

Theorem 3.13 (see Boccuto and Candeloro (2002b), Theorem 3.3) *Let* $m_j : \Sigma \to R$, $j \in \mathbb{N}$, *be a sequence of globally uniformly* (s)*-bounded measures.*

Then they are equibounded on Σ if and only if the set $\{m_j(E) : j \in \mathbb{N}\}$ is bounded in R for every $E \in \Sigma$.

Proof: Clearly, only one implication have to be be proved. So, set $u(E) = \bigvee\{| m_j(E) | : \; j \in \mathbb{N}\}$, and assume by contradiction that the set $\{u(E) : E \in \Sigma\}$ is not bounded in R. Let $(a_{t,l})_{t,l}$ be a regulator related to uniform (s)-boundedness, and let u be such that $a_{t,l} \leq u$ for each t, $l \in \mathbb{N}$. Then, there exists a set $H_1 \in \Sigma$, with $u(H_1) \not\leq u$.

Let us show that H_1 can be chosen so that the set $\{v(m_j)(H_1) : j \in \mathbb{N}\}$ is bounded. Otherwise there exists a set $B_1 \subset H_1$, $B_1 \in \Sigma$, with $u(B_1) \not\leq u + u(H_1)$, and subsequently a set $B_2 \subset B_1$, $B_2 \in \Sigma$, with $u(B_2) \not\leq u + u(B_1)$, and so on: this yields a decreasing sequence $(B_k)_k$ in Σ and a subsequence $(m_{j_k})_k$ with $| m_{j_k}(B_{k+1}) | \not\leq u + | m_{j_k}(B_k) |$, and so $| m_{j_k}(B_k \setminus B_{k+1}) | \not\leq u$ for every $k \in \mathbb{N}$, which contradicts global uniform (s)-boundedness.

So, let us take $H_1 \in \Sigma$, with $u(H_1) \not\leq u$, but such that the set $\{v(m_j)(H_1) : j \in \mathbb{N}\}$ is bounded: then we deduce that in $G \setminus H_1$ there exists a set $H_2 \in \Sigma$, with $u(H_2) \not\leq u$. Again, we obtain that the set $\{v(m_j)(H_2) : j \in \mathbb{N}\}$ is bounded. By induction, we construct a disjoint sequence $(H_k)_k$, with $u(H_k) \not\leq u$ for each $k \in \mathbb{N}$, which contradicts global uniform (s)-boundedness with respect to the regulator $(a_{t,l})_{t,l}$. □

We now prove the following:

Proposition 3.14 (see Boccuto and Candeloro (2002b), Proposition 3.4) *Let $(m_j)_j$ be a globally uniformly (s)-bounded sequence of finitely additive R-valued measures on Σ, and $(a_{t,l})_{t,l}$ be a regulator related to this property. Then, for every monotone sequence $(S_h)_h$ in Σ, the limit $r_j := (RD)\lim_h m_j(S_h)$ exists uniformly with respect to j, and the regulator $(a_{t,l})_{t,l}$ works for this property.*

Proof: The existence of the limit is due to Proposition 3.7.

If the last assertion were not true, then by virtue of the Cauchy condition there would exist some (say) decreasing sequence $(S_h)_h$, a $\varphi \in \mathbb{N}^{\mathbb{N}}$ and strictly increasing sequences $(j_k)_k$, $(p_k)_k$, $(h_k)_k$ in $\mathbb{N}$, such that $h_k + p_k < h_{k+1}$ and

$$|m_{j_k}(S_{h_k} \setminus S_{h_k+p_k})| \not\leq \bigvee_{t=1}^{\infty} a_{t,\varphi(t)} \quad \text{for each } k.$$

This contradicts global uniform (s)-boundedness with respect to the regulator $(a_{t,l})_{t,l}$. □

Corollary 3.15 (see Boccuto and Candeloro (2002b), Corollary 3.5) *Given a sequence $(m_j)_j$ of globally uniformly (s)-bounded and globally σ-additive measures on Σ, then they are globally uniformly σ-additive, namely there is a regulator $(a_{t,l})_{t,l}$ with the property that, for every decreasing sequence $(H_k)_k$ in Σ with $\bigcap_{k=1}^{\infty} H_k = \emptyset$ and every $\varphi \in \mathbb{N}^{\mathbb{N}}$, there is a $k_0 \in \mathbb{N}$ with $v(m_j)(H_{k_0}) \leq \bigvee_{t=1}^{\infty} a_{t,\varphi(t)}$ for every $j \in \mathbb{N}$.*

We now deal with global absolute continuity of measures with respect to a finitely additive measure λ. We will formulate this notion by considering decreasing sequences $(H_k)_k$ in Σ with $\lim_k \lambda(H_k) = 0$ and, differently than in the classical like case, we will see that this concept can be equivalently formulated by taking arbitrary sequences $(H_k)_k$ in Σ with $\lim_k \lambda(H_k) = 0$, because in this case we deal just with a single regulator or (O)-sequence. Here we consider (D)-convergence, since for technical reasons, in this framework, is easier to handle, though it is possible also to deal with order convergence.

Let $m : \Sigma \to R$, $\lambda : \Sigma \to [0,+\infty]$ be two finitely additive measures. We say that m is *globally λ-absolutely continuous* (and write $m \ll \lambda$) iff there exists a (D)-sequence $(a_{t,l})_{t,l}$ such that for every decreasing sequence $(H_k)_k$ in Σ, satisfying $\lim_k \lambda(H_k) = 0$, we get $(D) \lim_k v(m)(H_k) = 0$ with respect to $(a_{t,l})_{t,l}$.

If $m_j : \Sigma \to R$, $j \in \mathbb{N}$, is a sequence of finitely additive measures, then the m_j's are said to be *globally uniformly λ-absolutely continuous* iff there exists a (D)-

sequence $(a_{t,l})_{t,l}$ such that for every decreasing sequence $(H_k)_k$ in Σ with $\lim_k \lambda(H_k)=0$, we have $(D)\lim_k \bigvee_j v(m_j)(H_k)=0$ with respect to $(a_{t,l})_{t,l}$.

We now show that, if we consider global absolute continuity, then to deal with decreasing sequences and to proceed in terms of arbitrary sequences is equivalent. This result, as seen before, is in general not true when absolute continuity is intended in the classical like sense. We first extend Drewnowski (1972a), Lemma 4.6 to the lattice group setting.

Lemma 3.16 *Let $m:\Sigma\to R$ be a globally (s)-bounded measure. Then there is a regulator $(a_{t,l})_{t,l}$ such that for any sequence $(E_k)_k$ in Σ and for every $\varphi\in\mathbb{N}^{\mathbb{N}}$ there is a positive integer q with $v(m)(E_k\setminus\bigcup_{l=1}^{q}E_l)\le\bigvee_{t=1}^{\infty}a_{t,\varphi(t)}$ for each $k\ge q$.*

Proof: Let $(a_{t,l})_{t,l}$ be a regulator, related with global (s)-boundedness of m, and let us prove that it satisfies the thesis. Otherwise we can find a $\varphi\in\mathbb{N}^{\mathbb{N}}$ and construct a strictly increasing sequence $(r_h)_h$ in $\mathbb{N}$, with $v(m)(B_h)\not\le\bigvee_{t=1}^{\infty}a_{t,\varphi(t)}$ for every $h\in\mathbb{N}$, where $B_h:=E_{r_{h+1}}\setminus\bigcup_{l=1}^{r_h}E_l$. As the B_h's are pairwise disjoint, we get a contradiction with global (s)-boundedness of m with respect to the (D)-sequence $(a_{t,l})_{t,l}$. □

We now prove the following characterization of global absolute continuity, using a technique similar to that of Drewnowski (1972a), Lemma 6.1 (a).

Theorem 3.17 *A globally (s)-bounded measure $m:\Sigma\to R$ is globally λ-absolutely continuous if and only if there is a regulator $(a_{t,l})_{t,l}$ in R, such that $(D)\lim_n v(m)(C_n)=0$ with respect to $(a_{t,l})_{t,l}$, for any sequence $(C_n)_n$ in Σ with $\lim_n \lambda(C_n)=0$.*

Proof: The "if" part is straightforward.

We now turn to the "only if" part. Let $(a_{t,l})_{t,l}$ be a regulator, related with global absolute continuity of m. Since m is globally (s)-bounded, then, by virtue of Theorem 3.9, m is bounded too. Set $u := \bigvee_{E \in \Sigma} |m(E)|$. By the Fremlin Lemma, a (D)-sequence $(b_{t,l})_{t,l}$ can be found, with

$$u \wedge \left(\sum_{h=1}^{q} \left(\bigvee_{t=1}^{\infty} a_{t,\varphi(t+h)}\right)\right) \leq \bigvee_{t=1}^{\infty} b_{t,\varphi(t)}$$

for every $\varphi \in \mathbb{N}^{\mathbb{N}}$ and $q \in \mathbb{N}$. Put $c_{t,l} := 2(a_{t,l} + b_{t,l})$, t, $l \in \mathbb{N}$. Then, the double sequence $(c_{t,l})_{t,l}$ is a regulator. Let us show that it satisfies the thesis. Otherwise there exist an element $\varphi \in \mathbb{N}^{\mathbb{N}}$ and a sequence $(C_n)_n$ in Σ, with $\lim_n \lambda(C_n) = 0$ and $v(m)(C_n) \not\leq \bigvee_{t=1}^{\infty} c_{t,\varphi(t)}$ for each $n \in \mathbb{N}$. Let $(C_{n_k})_k$ be a subsequence of $(C_n)_n$, with $\lambda(C_{n_k}) \leq 2^{-k}$ for every $k \in \mathbb{N}$.

Let $E_k := C_{n_k}$. At the first step, by Lemma 3.16 applied to the sequence E_k, $k \in \mathbb{N}$, there exists $k_1 \in \mathbb{N}$, with

$$v(m)(E_k) \leq v(m)(E_k \setminus \bigcup_{l=1}^{k_1} E_l) + v(m)(E_k \cap (\bigcup_{l=1}^{k_1} E_l)) \leq$$

$$\leq \bigvee_{t=1}^{\infty} a_{t,\varphi(t+1)} + v(m)(E_k \cap (\bigcup_{l=1}^{k_1} E_l)) \leq \tag{2.59}$$

$$\leq \bigvee_{t=1}^{\infty} b_{t,\varphi(t)} + v(m)(E_k \cap (\bigcup_{l=1}^{k_1} E_l)) \quad \text{for every } k \geq k_1.$$

Put $B_1 := \bigcup_{l=1}^{k_1} E_l$. From (2.59) we deduce

$$v(m)(E_k \cap B_1) \not\leq \bigvee_{t=1}^{\infty} a_{t,\varphi(t)}, \tag{2.60}$$

otherwise we get $v(m)(E_k) \leq \bigvee_{t=1}^{\infty} b_{t,\varphi(t)} + \bigvee_{t=1}^{\infty} a_{t,\varphi(t)} \leq \bigvee_{t=1}^{\infty} c_{t,\varphi(t)}$, a contradiction. Thus from (2.60) it follows that $v(m)(B_1) \not\leq \bigvee_{t=1}^{\infty} a_{t,\varphi(t)}$.

Proceeding by induction, at the $h+1$-th step, let us suppose that we have found $k_1 < k_2 < \ldots < k_h \in \mathbb{N}$, and $B_1, \ldots B_h \in \Sigma$, with

$$B_0 := G,\ B_h = B_{h-1} \cap \Big(\bigcup_{l=k_{h-1}+1}^{k_h} E_l \Big), \quad v(m)(E_k \cap B_h) \not\leq \bigvee_{t=1}^{\infty} a_{t,\varphi(t)} \tag{2.61}$$

for all $h \in \mathbb{N}$ and $k \geq k_h$. By Lemma 3.16 applied to the sequence $E_k \cap B_h$, $k = k_h + 1,\ k_h + 2, \ldots$, there is an integer $k_{h+1} > k_h$, with

$$v(m)(E_k \cap B_h) \leq v(m)\Big((E_k \cap B_h) \setminus \bigcup_{l=k_h+1}^{k_{h+1}} E_l\Big) +$$

$$+ v(m)\Big((E_k \cap B_h) \cap \Big(\bigcup_{l=k_h+1}^{k_{h+1}} E_l\Big)\Big) \leq \bigvee_{t=1}^{\infty} a_{t,\varphi(t+h)} + v(m)(E_k \cap B_{h+1}) \leq \tag{2.62}$$

$$\leq \bigvee_{t=1}^{\infty} b_{t,\varphi(t)} + v(m)(E_k \cap B_{h+1})$$

whenever $k \geq k_{h+1}$, where $B_{h+1} = B_h \cap \big(\bigcup_{l=k_h+1}^{k_{h+1}} E_l\big)$. From (2.62) we obtain that

$$v(m)(E_k \cap B_{h+1}) \not\leq \bigvee_{t=1}^{\infty} a_{t,\varphi(t)},$$

otherwise we get $v(m)(E_k \cap B_h) \leq \bigvee_{t=1}^{\infty} b_{t,\varphi(t)} + \bigvee_{t=1}^{\infty} a_{t,\varphi(t)} \leq \bigvee_{t=1}^{\infty} c_{t,\varphi(t)}$, which contradicts (2.61). Hence, $v(m)(B_{h+1}) \not\leq \bigvee_{t=1}^{\infty} a_{t,\varphi(t)}$. By construction, $(B_h)_h$ is a decreasing sequence in Σ, $\lim_h \lambda(B_h)=0$ and $v(m)(B_h)\not\leq \bigvee_{t=1}^{\infty} a_{t,\varphi(t)}$ for every $h\in\mathbb{N}$, which contradicts λ-absolute continuity. This ends the proof. □

Arguing as in Theorem 3.17 it possible to prove the following:

Theorem 3.18 *A sequence $m_j:\Sigma\to R$, $j\in\mathbb{N}$, of equibounded globally uniformly (s)-bounded measures is globally uniformly λ-absolutely continuous on Σ if and only if there is a regulator $(a_{t,l})_{t,l}$ in R with $(D)\lim_k (\bigvee_j v(m_j)(H_k))=0$ with respect to $(a_{t,l})_{t,l}$, for each sequence $(H_k)_k$ in Σ with $\lim_k \lambda(H_k)=0$.*

A consequence of Theorem 3.17 is the following

Corollary 3.19 *A finitely additive measure $m:\Sigma\to R$ is globally λ-absolutely continuous if and only if there is a regulator $(a_{t,l})_{t,l}$ such that for any sequence $(C_n)_n$ in Σ with $\lim_n \lambda(C_n)=0$ and for each $\varphi\in\mathbb{N}^{\mathbb{N}}$ there is $\overline{n}\in\mathbb{N}$, with*

$$|m(C_n)|\leq \bigvee_{t=1}^{\infty} a_{t,\varphi(t)} \quad \textit{for any } n\geq \overline{n}. \tag{2.63}$$

Proof: The "only if" part is straightforward.

Concerning the "if" part, observe that, if $(a_{t,l})_{t,l}$ is a regulator satisfying the condition in (2.63), then from Theorem 3.17 we get that for any sequence $(C_n)_n$ in Σ with $\lim_n \lambda(C_n)=0$ and for every $\varphi\in\mathbb{N}^{\mathbb{N}}$ there exists $\overline{n}\in\mathbb{N}$, with $v(m)(C_n)\leq \bigvee_{t=1}^{\infty} a_{t,\varphi(t)}$ for any $n\geq\overline{n}$. This will imply global λ-absolute continuity. Otherwise there are a sequence $(D_k)_k$ in Σ and a $\varphi\in\mathbb{N}^{\mathbb{N}}$, with

$$v(m)(D_k)\not\leq \bigvee_{t=1}^{\infty} a_{t,\varphi(t)} \quad \text{for any } k\in\mathbb{N}, \tag{2.64}$$

and, thanks to (2.64) and the definition of semivariation, it is possible to find a sequence $(E_k)_k$ with $E_k\subset D_k$ for every k and

$$| m(E_k) | \not\leq \bigvee_{t=1}^{\infty} a_{t,\varphi(t)} \quad \text{for any } k \in \mathbb{N}, \tag{2.65}$$

which contradicts (2.63). □

Analogously it is possible to prove that the measures $m_j : \Sigma \to R$, $j \in \mathbb{N}$, are globally uniformly λ-absolutely continuous on Σ if and only if there is a (D)-sequence $(a_{t,l})_{t,l}$ with the property that for each sequence $(C_n)_n$ in Σ, with $\lim_n \lambda(C_n) = 0$ and for every $\varphi \in \mathbb{N}^{\mathbb{N}}$ there is $n_0 \in \mathbb{N}$, with $\bigvee_j v(m_j)(C_n) \leq \bigvee_{t=1}^{\infty} a_{t,\varphi(t)}$ whenever $n \geq n_0$.

In the next proposition we give a property of global absolute continuity.

Proposition 3.20 (see also Boccuto and Candeloro (2002b), Proposition 4.2)

Let $m : \Sigma \to R$ be any finitely additive measure, globally λ-absolutely continuous.

Then there exists a regulator $(a_{t,l})_{t,l}$ such that, for every $\varphi \in \mathbb{N}^{\mathbb{N}}$ a positive real number δ can be found, satisfying the following implication:

$$A \in \Sigma,\ \lambda(A) \leq \delta \Rightarrow | m(A) | \leq \bigvee_{t=1}^{\infty} a_{t,\varphi(t)}. \tag{2.66}$$

Proof: Let $(a_{t,l})_{t,l}$ be a regulator as in Corollary 3.19, and let us see that it works for the assertion. If this is not the case, then there exists a $\varphi \in \mathbb{N}^{\mathbb{N}}$ such that for every integer k there is $C_k \in \Sigma$, with $\lambda(C_k) \leq 1/k$, and $| m(C_k) | \not\leq \bigvee_{t=1}^{\infty} a_{t,\varphi(t)}$. This contradicts global absolute continuity, and concludes the proof. □

Observe that, arguing analogously as in Corollary 3.19, it is possible to prove that the expression in (2.66) can be equivalently replaced with the following:

$$A \in \Sigma,\ \lambda(A) \leq \delta \Rightarrow v(m)(A) \leq \bigvee_{t=1}^{\infty} a_{t,\varphi(t)}. \tag{2.67}$$

Indeed, if $(C_k)_k$ is a sequence in Σ, with $\lambda(C_k) \le 1/k$ and

$$v(m)(C_k) \not\le \bigvee_{t=1}^{\infty} a_{t,\varphi(t)} \quad \text{for every } k \in \mathbb{N},$$

then there is a sequence $(D_k)_k$ with $D_k \subset C_k$ and hence $\lambda(D_k) \le 1/k$ and

$$| m(D_k) | \not\le \bigvee_{t=1}^{\infty} a_{t,\varphi(t)} \quad \text{for every } k \in \mathbb{N}.$$

In the global σ-additive case, we get the following characterization.

Proposition 3.21 (see Boccuto and Candeloro (2002b), Proposition 4.3) *Assume that $\lambda : \Sigma \to [0,+\infty]$ and $m : \Sigma \to R$ are globally σ-additive. Then m is globally λ-absolutely continuous if and only if*

$$\lambda(A) = 0 \Rightarrow m(A) = 0 \textit{ for every } A \in \Sigma. \tag{2.68}$$

Proof: Using weak σ-distributivity of R, we see easily that global λ-absolute continuity of m implies (2.68). Conversely, assume that (2.68) holds, and let $(a_{t,l})_{t,l}$ be a regulator, according to global σ-additivity of m. We prove that $(a_{t,l})_{t,l}$ works for global absolute continuity. To this aim, let $(C_k)_k$ be any sequence in Σ, with $\lim_k \lambda(C_k) = 0$, and assume by contradiction that there exists $\varphi \in \mathbb{N}^{\mathbb{N}}$ with $v(m)(C_k) \not\le \bigvee_{t=1}^{\infty} a_{t,\varphi(t)}$ for infinitely many integers k. By taking a subsequence if necessary, we may assume that $v(m)(C_k) \not\le \bigvee_{t=1}^{\infty} a_{t,\varphi(t)}$ for all k, and $\sum_{k=1}^{\infty} \lambda(C_k) < +\infty$. Set $B_k = \bigcup_{j=k}^{\infty} C_j$ and $B = \bigcap_{k=1}^{\infty} B_k$. We get $\lim_k \lambda(B_k) = 0$, and hence $\lambda(B) = 0$. Therefore $m(B) = 0$, by (2.68). As the sequence $(B_k \setminus B)_k$ decreases to $\varnothing$, there exists $\bar{k} \in \mathbb{N}$ with

$$v(m)(B_k \setminus B) = v(m)(B_k) \le \bigvee_{t=1}^{\infty} a_{t,\varphi(t)}$$

for all $k \ge \bar{k}$. As $C_k \subset B_k$, this is absurd. This ends the proof. □

In case $(m_j)_j$ is a sequence of finitely additive R-valued measures on Σ, the *global uniform λ-absolute continuity* of $(m_j)_j$ can be defined in a similar way, but clearly by requiring that the integer $\overline{k}$ involved is independent of j.

If we deal with (O)-convergence, we say that m is *globally λ-absolutely continuous* iff there is an (O)-sequence $(b_l)_l$ such that for each sequence $(C_k)_k$ in Σ, satisfying $\lim_k \lambda(C_k) = 0$ and for every $l \in \mathbb{N}$ there is $k_l \in \mathbb{N}$, with $|m(C_k)| \leq b_l$ whenever $k \geq k_l$.

Similarly as above it is possible to formulate the concept of global uniform λ-absolute continuity of a sequence of measures $(m_j)_j$, where the integer involved k_l is requested to be independent of j.

If a finitely additive measure m is globally λ-absolutely continuous, it is not difficult to see that m is globally (s)-bounded in any case, and globally σ-additive if λ is.

An extension of the concept of global (uniform) absolute continuity is the notion of global (uniform) continuity related to a Fréchet-Nikodým topology τ on a σ-algebra $\Sigma \subset \mathcal{P}(G)$ and with respect to a single (O)-sequence or regulator.

We say that a finitely additive measure $m:\Sigma \to R$ is *globally τ-continuous* iff there exists a regulator $(a_{t,l})_{t,l}$ such that, for every decreasing sequence $(H_k)_k$ in Σ with τ - $\lim_k H_k = \varnothing$ and for each $\varphi \in \mathbb{N}^{\mathbb{N}}$ there is $\overline{k} \in \mathbb{N}$, with $|m(H_k)| \leq \bigvee_{t=1}^{\infty} a_{t,\varphi(t)}$ for all $k \geq \overline{k}$.

If $(m_j)_j$ is a sequence of finitely additive measures on Σ, the *global uniform τ-continuity* of $(m_j)_j$ can be defined analogously, by requiring that $\overline{k}$ is independent of j.

If we deal with (O)-convergence, we say that m is *globally τ-continuous* iff there is an (O)-sequence $(b_l)_l$ such that for each sequence $(H_k)_k$ in Σ, satisfying τ - $\lim_k H_k = \varnothing$ and for every $l \in \mathbb{N}$ there is $k_0 \in \mathbb{N}$, with $|m(H_k)| \leq b_l$

whenever $k \geq k_0$. Similarly as above it is possible to give the notion of global uniform τ-continuity of a sequence of measures $(m_j)_j$.

Finally we formulate the concept of global (uniform) regularity for lattice group-valued measures.

Let $\Sigma \subset \mathcal{P}(G)$ be a σ-algebra, and $\mathcal{H}$, $\mathcal{G} \subset \Sigma$ be two lattices, such that the complement of every element of $\mathcal{H}$ belongs to $\mathcal{G}$ and $\mathcal{G}$ is closed under to countable disjoint unions.

We say that a finitely additive measure $m: \Sigma \to R$ is *globally regular* on Σ with respect to (D)-convergence (resp. (O)-convergence) iff there exists a regulator $(a_{t,l})_{t,l}$ (resp. an (O)-sequence $(b_l)_l$) such that for each $A \in \Sigma$ there are two sequences $(F_k)_k$ in $\mathcal{H}$, $(G_k)_k$ in $\mathcal{G}$, satisfying (2.40) and for every $\varphi \in \mathbb{N}^{\mathbb{N}}$ (resp. for each $l \in \mathbb{N}$) there is $k_0 \in \mathbb{N}$ with

$$v_\Sigma(m)(G_k \setminus F_k) \leq \bigvee_{t=1}^{\infty} a_{t,\varphi(t)} \quad (\text{resp.} v_\Sigma(m)(G_k \setminus F_k) \leq b_l) \text{ whenever } k \geq k_0.$$

In an analogous way it is possible to state the notion of global uniform regularity with respect to a single (O)-sequence or regulator of a sequence of measures $(m_j)_j$, by requiring that the index n_0 involved does not depend on j.

Analogously as in Theorem 3.1 it is possible to prove the following:

Theorem 3.22 *Let R be a Dedekind complete and weakly σ-distributive (ℓ)-group, G, Σ, $\mathcal{G}$ and $\mathcal{H}$ be as in Theorem* 3.1. *Then every globally regular measure $m: \Sigma \to R$ is globally σ-additive. Conversely, if R is also super Dedekind complete, then every globally σ-additive measure $m: \Sigma \to R$ is globally regular.*

2.3.2. Countably Additive Restrictions

We now consider some relations between finite and countable additivity for lattice group-valued measures. Note that many problems involving finitely additive measures can be solved by finding suitable σ-additive measures in connection with them, and then applying some corresponding results concerning the σ-additive case.

One of the main tools in this framework is to find suitable σ-additive restrictions of the given measure, while another tool, the so-called *Stone extension*, provides to construct a countably additive measure, defined on a larger σ-algebra.

We begin with σ-additive restrictions. We now present a Drewnowski-type theorem, about existence of globally σ-additive restrictions of globally (s)-bounded finitely additive measures with respect to (O)-convergence, inspiring us to Drewnowski (1972a) and Faires (1976). We first treat the case of (s)-boundedness and σ-additivity in a classical like sense, and after we deal with global (s)-boundedness and global σ-additivity.

Theorem 3.23 (see also Boccuto, Dimitriou and Papanastassiou (2010c), Theorem 3.2)

Let R be a Dedekind complete (ℓ)-group, Ω be as in the Maeda-Ogasawara-Vulikh theorem 2.7 and $m:\Sigma \to R$ be a finitely additive bounded measure.

Then there is a meager set $N \subset \Omega$ such that for every disjoint sequence $(H_n)_n$ there is a subsequence $(H_{n_k})_k$ with

$$\lim_k [v(m)(\bigcup_{j\geq k} H_{n_j})](y) = \inf_k [v(m)(\bigcup_{j\geq k} H_{n_j})](y) = 0 \quad \textit{for each } y \in \Omega \setminus N.$$

Moreover, m and $m(\cdot)(y)$, $y \in \Omega \setminus N$, are σ-additive along the σ-algebra generated by $(H_{n_k})_k$.

Proof: Let $r(\omega) := (\bigvee_{A \in \Sigma} |m(A)|)(\omega)$, $\omega \in \Omega$, and put $N := \{\omega \in \Omega : r(\omega) < +\infty\}$. By the Maeda-Ogasawara-Vulikh representation theorem, N is meager in Ω. By Theorem 3.12, m is (s)-bounded and the measures $m(\cdot)(y)$, $y \in \Omega \setminus N$, are (s)-bounded too. For every $n \in \mathbb{N}$ set $T_n := \{\omega \in \Omega : r(\omega) \leq n\}$. Of course, $T_n \subset N$ for every $n \in \mathbb{N}$, $\bigcup_{n=1}^{\infty} T_n$ is meager in Ω and, by Proposition 2.6, the set $T := \Omega \setminus \bigcup_{n=1}^{\infty} T_n$ is dense in Ω.

Let $\{P_k^1 : k \in \mathbb{N}\}$ be an infinite partition of $\mathbb{N}$, whose elements are infinite sets; put $H_k^1 := \bigcup_{j \in P_k^1} H_j$ for every $k \in \mathbb{N}$ and $\phi_j^1 := \bigvee_{k \geq j} v(m)(H_k^1)$. By (s)-boundedness of m, we get that the sequence $(\phi_j^1)_j$ is decreasing, and $\bigwedge_j \phi_j^1 = 0$. Since T_1 is compact, in correspondence with every $\omega \in T_1$ there exist $j(\omega) \in \mathbb{N}$ with $\phi_{j(\omega)}^1(\omega) < 1$ and an open neighborhood U_ω with $\phi_{j(\omega)}^1(y) < 1$ whenever $y \in U_\omega$. Since the family $\{U_\omega : \omega \in T_1\}$ covers T_1, we can find a finite subfamily $\{U_{\omega_1}, \ldots, U_{\omega_q}\}$, with associated indexes $j(\omega_1), \ldots, j(\omega_q)$. Let $j_1^* := \max_{i \in [1,q]} j(\omega_i)$: then $\phi_{j_1^*}^1(y) < 1$ whenever $y \in T_1$. Indeed for each fixed $y \in T_1$ there is (at least) an element ω_l, with $l \in \{1, \ldots, q\}$, such that $y \in U_{\omega_l}$, and hence $0 \leq \phi_{j_1^*}^1(y) \leq \phi_{j(\omega_l)}^1(y) < 1$.

Set now $\overline{H}_1 := H_{j_1^*}^1$: then $0 \leq v(m(\cdot)(y))(\overline{H}_1) \leq v(m)(\overline{H}_1)(y) < 1$ for each $y \in T_1$, since the pointwise suprema are less or equal than the corresponding lattice suprema.

At the second step, define $P_1^* := P_{j_1^*}^1$, and denote by $\{P_k^2 : k \in \mathbb{N}\}$ an infinite partition of P_1^*, whose elements are infinite sets. For each $k \in \mathbb{N}$, set $H_k^2 := \bigcup_{j \in P_k^2} H_j$. Observe that the sets H_k^2, $k \in \mathbb{N}$, form a disjoint partition of $\overline{H}_1$.

Put now $\phi_j^2 := \bigvee_{k \geq j} v(m)(H_k^2)$ for every $j \in \mathbb{N}$. Again by (s)-boundedness of m, the sequence $(\phi_j^2)_j$ is decreasing, and $\bigwedge_j \phi_j^2 = 0$. Proceeding analogously as in the previous step, let us consider the compact set T_2 and the elements ϕ_j^2: so, we find an integer j_2^* with $\phi_{j_2^*}^2(y) < 1/2$ for all $y \in T_2$. Set $P_2^* := P_{j_2^*}^2$ and $\overline{H}_2 := \bigcup_{j \in P_2^*} H_j$: then $\overline{H}_2 \subset \overline{H}_1$, and

$$0 \leq v(m(\cdot)(y))(\overline{H}_2) \leq v(m)(\overline{H}_2)(y) < 1/2 \quad \text{for any } y \in T_2.$$

By induction, we construct a decreasing sequence $(P_n^*)_n$ of infinite subsets of $\mathbb{N}$ such that, if $\overline{H}_n := \bigcup_{j\in P_n^*} H_j$, $n\in\mathbb{N}$, then $(\overline{H}_n)_n$ is decreasing, and $0 \le v(m(\cdot)(y))(\overline{H}_n) \le v(m)(\overline{H}_n)(y) < 1/n$ whenever $y\in T_n$, $n\in\mathbb{N}$. Set now $p_n := \min\{P_n^*,\ n\in\mathbb{N}\}$: without loss of generality, we can assume $p_n < p_{n+1}$ for all $n\in\mathbb{N}$. We claim that the sequence $(H_{p_n})_n$ is the requested one. Indeed, for any $n\in\mathbb{N}$, $\bigcup_{j\ge n} H_{p_j} \subset \overline{H}_n$ and $[v(m)(\bigcup_{j\ge n} H_{p_j})](y) < 1/n$ for each $y\in T_n$. If we put $\phi_n := v(m)(\bigcup_{j\ge n} H_{p_j})$, $n\in\mathbb{N}$, and $\phi := \bigwedge_n \phi_n$, then $0\le\phi(y)\le\phi_j(y) < 1/j$ for all $y\in T_j$. Since $T_j\subset T_{j+1}$ for all $j\in\mathbb{N}$, then $0\le\phi(y)<1/k$ for all $k\ge j$ and $y\in T_j$, and hence $\phi(y)=0$ for each $y\in T_j$, $j\in\mathbb{N}$. By arbitrariness of j, we obtain $\phi(y)=0$ for all $y\in\Omega\setminus N$ and so, by a density argument, we get $\phi=0$. This proves that

$$(O)\lim_n v(m)(\bigcup_{j\ge n} H_{p_j}) = 0, \tag{2.69}$$

and a fortiori

$$\lim_n v(m(\cdot)(y))(\bigcup_{j\ge n} H_{p_j}) = 0, \tag{2.70}$$

since the pointwise suprema are always less or equal than the corresponding lattice ones.

The σ-additivity of the measures m and $m(\cdot)(y)$, $y\in\Omega\setminus N$, on the σ-algebra generated by the sets H_{p_n}, $n\in\mathbb{N}$, follows from (2.69) and (2.70), by virtue of monotonicity of $v(m)$ and $v(m(\cdot)(y))$, $y\in\Omega\setminus N$.

We now show only σ-additivity of m, since the proof of the other property is analogous. Let $(b_n)_n$ be an (O)-sequence, related with the (O)-limit in (2.69), and set $H^* := \bigcup_{n=1}^\infty H_{p_n}$. Pick any decreasing sequence $(F_h)_h$ in the σ-algebra generated in H^* by the sets H_{p_n}, with $\bigcap_{h=1}^\infty F_h = \emptyset$. For every $h\in\mathbb{N}$, let $j(h)$ be the largest integer with $F_h \subset \bigcup_{l \geq j(h)} H_{p_l}$. We get that $\lim_h j(h) = +\infty$, and hence $(b_{j(h)})_h$ is an (O)-sequence. Now, for all $h \in \mathbb{N}$ we have

$$v(m)(F_h) \leq v(m)(\bigcup_{l \geq j(h)} H_{p_l}) \leq b_{j(h)}.$$

From this we get σ-additivity of m. □

A consequence of Theorem 3.23 is the following

Theorem 3.24 *Let* R, G, Σ, Ω *be as in Theorem* 3.23, *and* $m_n : \Sigma \to R$, $n \in \mathbb{N}$, *be a sequence of bounded measures. Then there exists a meager set* $N \subset \Omega$ *such that for every disjoint sequence* $(H_s)_s$ *in* Σ *there is a subsequence* $(H_{s_k})_k$ *of* $(H_s)_s$ *with*

$$\lim_k v(m_n)(\bigcup_{j \geq k} H_{s_j})(y) = 0$$

for each $y \in \Omega \setminus N$ *and* $n \in \mathbb{N}$. *In particular, the measures* m_n *and* $m_n(\cdot)(y)$, $n \in \mathbb{N}$, $y \in \Omega \setminus N$, *are* σ*-additive on the* σ*-algebra generated by* $(H_{s_k})_k$.

Proof: By virtue of Theorem 3.23, for every $n \in \mathbb{N}$ there is a meager set $N_n \subset \Omega$, satisfying the thesis of Theorem 3.23. Set $N := \bigcup_{n=1}^{\infty} N_n$: we claim that N is the requested set. By Theorem 3.23, there exists an infinite subset $P_1 \subset \mathbb{N}$ with

$$\lim_h v(m_1)(\bigcup_{j \in P_1, j \geq h} H_j)(y) = 0 \quad \text{for all } y \in \Omega \setminus N.$$

Moreover there is an infinite subset $P_2 \subset P_1$ such that

$$\lim_h v(m_2)(\bigcup_{j \in P_2, j \geq h} H_j)(y) = 0 \quad \text{for all } y \in \Omega \setminus N.$$

Let $p_1 := \min P_1$: without loss of generality, we can suppose that P_2 does not contain p_1.

By induction, it is possible to construct a decreasing sequence of infinite subsets $P_n \subset \mathbb{N}$, $n \in \mathbb{N}$, with

$$\lim_h v(m_n)(\bigcup_{j \in P_n, j \geq h} H_j)(y) = 0 \tag{2.71}$$

for each $y \in \Omega \setminus N$ and $n \in \mathbb{N}$, and $p_n < p_{n+1}$ for all $n \in \mathbb{N}$, where $p_n := \min P_n$. Let now $P := \{p_n : n \in \mathbb{N}\}$. For every $n \in \mathbb{N}$, $h \geq p_n$ and $y \in \Omega \setminus N$, we get:

$$v(m_n)(\bigcup_{j \in P, j \geq h} H_j)(y) \leq v(m_n)(\bigcup_{j \in P_n, j \geq h} H_j)(y). \tag{2.72}$$

From (2.71), (2.72) and by a density argument it follows that

$$(O)\lim_h v(m_n)(\bigcup_{j \in P, j \geq h} H_j) = 0, \tag{2.73}$$

and a fortiori

$$\lim_h v(m_n(\cdot)(y))(\bigcup_{j \in P, j \geq h} H_j) = 0 \tag{2.74}$$

for every $n \in \mathbb{N}$ and $y \in \Omega \setminus N$, since the pointwise suprema are less or equal than the corresponding lattice ones. The σ-additivity of the set functions m_n and $m_n(\cdot)(y)$, $n \in \mathbb{N}$, $y \in \Omega \setminus N$, on the σ-algebra generated by the sets H_{p_n}, $n \in \mathbb{N}$, follows from (2.73) and (2.74), arguing analogously as at the end of the proof of Theorem 3.23. □

We now deal with a Drewnowski-type theorem with respect to global σ-additivity, in the setting of (O)-convergence.

Theorem 3.25 (see also Boccuto and Candeloro (2004a), Theorem 5.3, Boccuto and Candeloro (2004d), Theorem 4.1) *Let R be any Dedekind complete Riesz space, and $m : \Sigma \to R$ be any globally (s)-bounded measure. Then for any disjoint sequence $(H_n)_n$ in Σ there is a subsequence $(H_{n_k})_k$ on which m is globally σ-additive.*

Proof: Without loss of generality, by replacing m by $v_\Sigma(m)$, we can assume that m is positive. Let $(b_k)_k$ be any (O)-sequence, with respect to which m is globally (s)-bounded. Choose a disjoint sequence $(P_r^1)_r$ of infinite subsets of $\mathbb{N}$, and set $H_r^1 := \bigcup\{H_n : n \in P_r^1\}$ for each r.

The sequence $(H_r^1)_r$ is disjoint, and so, by global (s)-boundedness, we find an $r_1 \in \mathbb{N}$ with $m(H_r^1) \le b_1$ for all $r \ge r_1$.

We now consider $P_{r_1}^1$, and choose any infinite partition of it into a sequence of disjoint infinite subsets $(P_r^2)_r$. For each r let us define $H_r^2 := \bigcup\{H_n : n \in P_r^2\}$. Note that the sets H_r^2, $r \in \mathbb{N}$, are pairwise disjoint, and are all contained in $H_{r_1}^1$ by construction.

Again by global (s)-boundedness, there exists an integer $r_2 > r_1$ with $m(H_r^2) \le b_2$ for all $r \ge r_2$. We now proceed in the same fashion, finding an infinite partition $(P_r^3)_r$ of $P_{r_2}^2$ into disjoint infinite subsets, and setting $H_r^3 := \bigcup\{H_n : n \in P_r^3\}$.

In this way, we obtain a decreasing sequence $(P_{r_l}^l)_l$ of infinite subsets of $\mathbb{N}$, and a corresponding sequence $(H_{r_l}^l)_l$, $H_{r_l}^l = \bigcup\{H_n : n \in P_{r_l}^l\}$, satisfying $(H_{r_l}^l) \le b_l$ for all l. Let us denote by n_1 the first element of $P_{r_1}^1$, by n_2 the first element of $P_{r_2}^2$ larger than n_1, and so on. We claim that the sequence $(H_{n_k})_k$ is the required one.

Indeed, observe that $H_{n_{k+p}} \subset H_{n_k}^k$ for all k, $p \in \mathbb{N}$. Let H^* be the union of all sets H_{n_k}, choose any decreasing sequence $(F_h)_h$ in the σ-algebra generated in H^* by the sets H_{n_k}, and assume $\bigcap_{h=1}^{\infty} F_h = \emptyset$. For every $h \in \mathbb{N}$, let $j(h)$ be the largest integer with $F_h \subset \bigcup_{l \ge j(h)} H_{n_l}$. Note that

$\lim_h j(h) = +\infty$, and hence $(b_{j(h)})_h$ is an (O)-sequence. Now, for all $h \in \mathbb{N}$ we have

$$m(F_h) \le m(\bigcup_{l \ge j(h)} H_{n_l}) \le m(H^{j(h)}_{r_{j(h)}}) \le b_{j(h)}.$$

This ends the proof. □

As a consequence of Theorem 3.25, we prove the following

Corollary 3.26 (see also Boccuto and Candeloro (2004d), Corollary 4.3) *Let $m_n : \Sigma \to R$, $n \in \mathbb{N}$, be a sequence of globally (s)-bounded measures with respect to an (O)-sequence $(b_l)_l$. Then, for every disjoint sequence $(H_k)_k$ in Σ there is a subsequence $(H_{k_n})_n$ on which every m_n is globally σ-additive with respect to the same (O)-sequence $(b_l)_l$.*

Proof: By Theorem 3.25, there exists an infinite subset $P_1 \subset \mathbb{N}$ with

$$\lim_h v(m_1)(\bigcup_{j \in P_1, j \ge h} H_j) = 0$$

with respect to $(b_l)_l$. There is also an infinite subset $P_2 \subset P_1$ with $\lim_h v(m_2)(\bigcup_{j \in P_2, j \ge h} H_j) = 0,$

again with respect to $(b_l)_l$. Without loss of generality, we suppose $\min P_2 > \min P_1$.

Proceeding by induction, we find a decreasing sequence $(P_n)_n$ of infinite subsets of $\mathbb{N}$, a strictly increasing sequence $(p_n)_n$ in $\mathbb{N}$, with $p_n = \min P_n$ for all n, and a double sequence $(h_n^{(l)})_{l,n}$ with $h_n^{(l)} < h_{n+1}^{(l)}$ for each l, $n \in \mathbb{N}$, and $v(m_n)(\bigcup_{j \in P_n, j \ge h} H_j) \le b_l$ for every $h \ge h_n^{(l)}$. Let $P := \{p_n : n \in \mathbb{N}\}$, and for every l, $n \in \mathbb{N}$ set $k_n^{(l)} := \max\{h_n^{(l)}, p_n\}$. For each l, n and $h \ge k_n^{(l)}$, we get $v(m_n)(\bigcup_{j \in P, j \ge h} H_j) \le v(m_n)(\bigcup_{j \in P_n, j \ge h} H_j) \le b_l$. From this, arguing analogously as at the end of the proof of Theorem 3.25, it follows that each m_n is globally σ-additive with respect to $(b_l)_l$ on the σ-algebra generated by the sets H_{p_n}, $n \in \mathbb{N}$. □

2.3.3. Carathéodory and Stone Extensions

We now deal with Stone-type extensions for lattice group-valued measures. These constructions are investigated in Boccuto and Candeloro (2002b, 2004a) when it is dealt with respect to a single regulator or order sequence, and in Boccuto (1995b) when σ-additivity is considered not necessarily with respect to a common regulator.

Given a Dedekind complete weakly σ-distributive lattice group R, a σ-additive positive R-valued measure defined on an algebra $\mathcal{L}$ of subsets of an abstract nonempty set G, we will define the *Carathéodory extension* of m, defined on the σ-algebra $\sigma(\mathcal{L})$ generated by $\mathcal{L}$ (see also König (2009), Chapter II).

Let us define $m^* : \mathcal{P}(G) \to R$ by

$$m^*(A) = \bigwedge \{\sum_{k=1}^{\infty} m(A_k) : (A_k)_k \text{ in } \mathcal{L},\ A \subset \bigcup_{k=1}^{\infty} A_k\}, A \subset G.$$

Let Σ be the family of all subsets $U \subset G$ with

$$m^*(A) = m^*(A \cap U) + m^*(A \setminus U) \quad \text{for any } A \subset G.$$

The previous definitions are well-known in the classical theory of Carathéodory extensions, for real-valued non-negative measures. With analogous techniques as in the classical case and taking into account weak σ-distributivity of R, it is possible to see that Σ is a σ-algebra containing $\mathcal{L}$, that $m^*(A) \le \sum_{k=1}^{\infty} m^*(A_k)$ whenever $(A_k)_k$ is any sequence of subsets of G, with $A \subset \bigcup_{k=1}^{\infty} A_k$, that $m^*(E) = m(E)$ for every $E \in \mathcal{L}$, and finally that

$$m^*(A) = \sum_{k=1}^{\infty} m^*(A_k) \tag{2.75}$$

for every disjoint sequence $(A_k)_k$ in Σ, with $A = \bigcup_{k=1}^{\infty} A_k$ (see also Bartle (1966), Boccuto and Candeloro (2004a), Definitions 4.2, Diestel and Spalsbury (2012), Theorem 4.3).

The measure defined as the restriction of the measure m^* to Σ is called the *Carathéodory extension* of m to Σ.

The next step is to prove global σ-additivity of the Carathéodory extension. We now formulate the following extension and density theorem (for the case $R=\mathbb{R}$, see also Maharam (1987), Theorem 2).

Theorem 3.27 (see Boccuto and Candeloro (2004a), Theorem 4.4) *Let R be a super Dedekind complete and weakly σ-distributive lattice group, $m:\mathcal{L}\to R$ be any σ-additive measure, defined on an algebra $\mathcal{L}$. Then there exists a (unique) σ-additive measure $\mu:\sigma(\mathcal{L})\to R$, with $m(F)=\mu(F)$ for all $F\in\mathcal{L}$. Moreover there is a regulator $(c_{t,l})_{t,l}$, such that, for any $A\in\Sigma(\mathcal{L})$ and $\varphi\in\mathbb{N}^{\mathbb{N}}$, an element $F\in\mathcal{L}$ can be found, with $v(\mu)(A\Delta F)\le\bigvee_{t=1}^{\infty}c_{t,\varphi(t)}$.*

Proof: Replacing if necessary m by its positive and negative parts, it is possible to assume, without loss of generality, that m is positive.

We now define μ on the σ-algebra $\sigma(\mathcal{L})$ as the restriction of the Carathéodory extension m^* of m. So, we will prove the final assertion, which can be described as a density property of $\mathcal{L}$ in $\sigma(\mathcal{L})$, and from it we will deduce σ-additivity of μ. First of all, let us denote by $(a_{t,l})_{t,l}$ a regulator working for global (s)-boundedness of m. According to the Fremlin lemma 2.3, a regulator $(A_{t,l})_{t,l}$ can be found, with $m(G)\wedge\sum_{s=1}^{q}(\bigvee_{t=1}^{\infty}a_{t,\varphi(t+s)})\le\bigvee_{t=1}^{\infty}A_{t,\varphi(t)}$ for every $\varphi\in\mathbb{N}^{\mathbb{N}}$ and $q\in\mathbb{N}$. Set $b_{t,l}=2(a_{t,l}+A_{t,l})$ and $c_{t,l}=2b_{t,l}$, t, $l\in\mathbb{N}$. We show that $(c_{t,l})_{t,l}$ is the regulator needed for the density property, which we are proving. To this end, we first observe that, whenever $A\in F_\sigma$, say $A=\bigcup_{n=1}^{\infty}H_n$, where the sets H_n are disjoint elements from $\mathcal{L}$, by Proposition 3.7 we deduce that for every $\varphi\in\mathbb{N}^{\mathbb{N}}$ there is $n\in\mathbb{N}$ with

$$\mu(A\Delta(\bigcup_{j=1}^{n}H_j))\le\bigvee_{t=1}^{\infty}a_{t,\varphi(t)}. \qquad \textbf{(2.76)}$$

Moreover, since R is super Dedekind complete, we see that for every set $A \subset G$ there is a set $B \in F_{\sigma,\delta}$ (that is $B = \bigcap_{n=1}^{\infty} B_n$, where the sets B_n are F_σ), with $A \subset B$ and $m^*(A) = m^*(B)$.

Thus it will be enough to prove that for every $B \in F_{\sigma,\delta}$ and $\varphi \in \mathbb{N}^{\mathbb{N}}$ there exists an element $F \in \mathcal{L}$ such that $m^*(B\Delta F) \le \bigvee_{t=1}^{\infty} c_{t,\varphi(t)}$.

Thanks to (2.76), it is sufficient to show that for each $B \in F_{\sigma,\delta}$ and $\varphi \in \mathbb{N}^{\mathbb{N}}$ there is an element $V \in F_\sigma$ with

$$m^*(V\Delta B) \le \bigvee_{t=1}^{\infty} b_{t,\varphi(t)}. \tag{2.77}$$

So, choose any element $B \in F_{\sigma,\delta}$ and let $(V_n)_n$ be any decreasing sequence in F_σ such that $B = \bigcap_{n=1}^{\infty} V_n$. Now, fix any element $\varphi \in \mathbb{N}^{\mathbb{N}}$: then we see that for each integer n there exists an element $E_n \in \mathcal{L}$ such that $E_n \subset V_n$ and

$$m^*(V_n \setminus E_n) \le \bigvee_{t=1}^{\infty} a_{t,\varphi(t+n)}.$$

Set now $F_1 = E_1$ and $F_n = E_n \cap F_{n-1}$ for every $n \ge 2$. The sequence $(F_n)_n$ is decreasing in $\mathcal{L}$, and hence from Theorem 3.7 we find a positive integer N such that

$$m(F_n \setminus F_{n+p}) \le \bigvee_{t=1}^{\infty} a_{t,\varphi(t)} \quad \text{for every } n \ge N \text{ and } p \in \mathbb{N}.$$

Fix now $n \in \mathbb{N}$, we have
$m^*(V_n \setminus B) = m^*(V_n \setminus V_{n+p}) + m^*(V_{n+p} \setminus B)$ for every $p \in \mathbb{N}$.

Moreover, $(V_n \setminus V_{n+p})\Delta(F_n \setminus F_{n+p}) \subset (V_n \setminus F_n) \cup (V_{n+p} \setminus F_{n+p})$. So
$V_n \setminus F_n \subset (V_1 \setminus E_1) \cup \ldots \cup (V_n \setminus E_n)$

for every n, and thus $(V_n \setminus V_{n+p})\Delta(F_n \setminus F_{n+p}) \subset (V_1 \setminus E_1) \cup \ldots \cup (V_{n+p} \setminus E_{n+p})$. Hence we deduce

$$m^*(V_n \setminus V_{n+p}) \le m(G) \wedge (m^*(F_n \setminus F_{n+p}) + \sum_{h=1}^{n+p} m^*(V_h \setminus E_h)) \le \bigvee_{t=1}^{\infty} a_{t,\varphi(t)} + \bigvee_{t=1}^{\infty} A_{t,\varphi(t)}$$

for every $n \ge N$ and $p \in \mathbb{N}$. Therefore we get

$$m^*(V_n \setminus B) \le \bigvee_{t=1}^{\infty} a_{t,\varphi(t)} + \bigvee_{t=1}^{\infty} A_{t,\varphi(t)} + m^*(V_{n+p} \setminus B)$$

for each $n \ge N$ and $p \in \mathbb{N}$. Letting p tend to $+\infty$, we get

$$m^*(V_n \setminus B) \le \bigvee_{t=1}^{\infty} b_{t,\varphi(t)} \quad \text{for every } n \ge N.$$

This yields the required density property.

We now prove σ-additivity of μ on $\sigma(\mathcal{L})$. Thanks to Proposition 3.3, it is enough to show that μ is globally (s)-bounded with respect to some regulator $(d_{t,l})_{t,l}$.

As already seen, the regulator $(c_{t,l})_{t,l}$ works for the density property above. Then, thanks to Lemma 2.3, there is a regulator $(C_{t,l})_{t,l}$ such that for every $q \in \mathbb{N}$ and $\varphi \in \mathbb{N}^{\mathbb{N}}$ one has

$$m(G) \wedge (\sum_{s=1}^{q} (\bigvee_{t=1}^{\infty} c_{t,\varphi(t+s)})) \le \bigvee_{t=1}^{\infty} C_{t,\varphi(t)}.$$

Set now $d_{t,l} := 2(C_{t,l} + a_{t,l})$, $t, l \in \mathbb{N}$. We shall show that $(d_{t,l})_{t,l}$ is the requested regulator for global (s)-boundedness of μ. Infact, fix any disjoint sequence $(H_n)_n$ in $\sigma(\mathcal{L})$ and any $\varphi \in \mathbb{N}^{\mathbb{N}}$: by the density property, for each n there is $F_n \in \mathcal{L}$ with $\mu(H_n \Delta F_n) \le \bigvee_{t=1}^{\infty} c_{t,\varphi(t+n)}$. Set now

$F_1^* := F_1, F_n^* := F_n \setminus (\bigcup_{r=1}^{n-1} F_r)$ for every $n \ge 2$. We get $H_n \Delta F_n^* \subset \bigcup_{r=1}^{n} (H_r \Delta F_r)$,

$$\mu(H_n) \le \mu\,(H_n \Delta F_n^*) + \mu\,(F_n^*) \le m(G) \wedge$$

$$\sum_{r=1}^{n} \mu(H_r \Delta F_r) + m(F_n^*) \le \bigvee_{t=1}^{\infty} C_{t,\varphi(t)} + m(F_n^*).$$

Thus for n large enough we get $m(F_n^*) \le \bigvee_{t=1}^{\infty} a_{t,\varphi(t)}$ and $\mu\,(H_n) \le \bigvee_{t=1}^{\infty} d_{t,\varphi(t)}$. □

We now extend Theorem 3.27 to the case of a sequence of globally uniformly (s)-bounded measures $m_n : \mathcal{L} \to R,\ n \in \mathbb{N}$.

Theorem 3.28 *Let $(m_n)_n$ be a sequence of R-valued equibounded, globally uniformly (s)-bounded and globally σ-additive measures, defined on an algebra $\mathcal{L} \subset \mathcal{P}(G)$. Then there exist globally σ-additive measures M_n: $\sigma(\mathcal{L}) \to R$, each of which extending m_n, such that the following density property holds:*

3.28.1) *there exists a regulator $(c_{t,l})_{t,l}$ such that for any $A \in \sigma(\mathcal{L})$ and $\varphi \in \mathbb{N}^{\mathbb{N}}$*

there is $F \in \mathcal{L}$ with $v(M_n)(A \Delta F) \le \bigvee_{t=1}^{\infty} c_{t,\varphi(t)}$ for any $n \in \mathbb{N}$.

Moreover, the measures M_n are globally uniformly (s)-bounded.

Proof: As before, we can and do assume that all the measures involved are positive. Let $u := \bigvee\{m_n(G) : n \in \mathbb{N}\}$ and $(a_{t,l})_{t,l}$ be a regulator, working for global uniform (s)-boundedness of the measures m_n. Moreover, let $(A_{t,l})_{t,l}$ be a regulator, according with the Fremlin lemma 2.3, such that $u \wedge \sum_{n=1}^{q} (\bigvee_{t=1}^{\infty} a_{t,\varphi(t+n)}) \le \bigvee_{t=1}^{\infty} A_{t,\varphi(t)}$ for every $\varphi \in \mathbb{N}^{\mathbb{N}}$ and $q \in \mathbb{N}$.

In this way, setting $c_{t,l} = 4(a_{t,l} + A_{t,l})$, it is possible to show that the regulator $(c_{t,l})_{t,l}$ works for the density property.

Indeed, arguing as in the proof of Theorem 3.27, it is not difficult to check that for each $H \in F_{\sigma,\delta}$ and $\varphi \in \mathbb{N}^{\mathbb{N}}$ there exists a set $F \in \mathcal{L}$, with $M_n\,(H \Delta F) \leq \bigvee_{t=1}^{\infty} c_{t,\varphi(t)}$ for every $n \in \mathbb{N}$.

However, if A is any element of $\sigma(\mathcal{L})$, for each $n \in \mathbb{N}$ there is a set $H_n \in F_{\sigma,\delta}$ with $A \subset H_n$ and $M_n\,(H_n \setminus A) = 0$. So, the problem is, for the set A, to find a *unique* element $H \in F_{\sigma,\delta}$, containing A, and such that $M_n\,(H \setminus A) = 0$ for all n.

We claim that $H := \bigcap_{n=1}^{\infty} H_n$ is the requested set. Indeed, $A \subset H$ by construction, and $M_n\,(H \setminus A) \leq M_n\,(H_n \setminus A) = 0$ for every n. It is not difficult to see that $H \in F_{\sigma,\delta}$. This completes the proof of density of $\mathcal{L}$.

Now, proceeding as in the proof of Theorem 3.27, it is possible to show global σ-additivity of M_n, by means of the regulator $(d_{t,l})_{t,l}$, which is related to $(a_{t,l})_{t,l}$ like in that proof, with the only difference that $m(G)$ is replaced by u, when applying the Fremlin lemma 2.3. The same argument shows that global uniform (s)-boundedness of $(M_n)_n$ follows from that of $(m_n)_n$. □

A consequence of Theorems 3.28 and 2.2 is the following:

Theorem 3.29 *Let R be a Dedekind complete and weakly σ-distributive (ℓ)-group, $m_n : \mathcal{L} \to R$, $n \in \mathbb{N}$, be a sequence of equibounded, globally uniformly (s)-bounded and globally σ-additive measures, with respect to (O)-convergence. Then there exist globally σ-additive extension measures $M_n : \sigma(\mathcal{L}) \to R$ of m_n and an (O)-sequence $(b_k)_k$, such that for each $A \in \sigma(\mathcal{L})$ and $k \in \mathbb{N}$ there is $F \in \mathcal{L}$ with $v\,(M_n)\,(A \Delta F) \leq b_k$ for any $n \in \mathbb{N}$.*

Furthermore, the measures M_n are globally uniformly (s)-bounded.

We now present some results in which the Stone Isomorphism technique is used. This tool allows to study some properties of finitely additive measures by analyzing some corresponding properties of σ-additive measures, which sometimes are easier to handle, even when the measures involved take values in a Dedekind complete (ℓ)-group.

A topological space Q^* is *totally disconnected* iff the only connected subsets of Q^* are the empty set and the one-point sets. It is known (see also Martellotti (2001), Sikorski (1964), Stone (1936, 1938 and 1949)) that, if $\mathcal{L}$ is any algebra of subsets of a nonempty set G, then there is a compact and totally disconnected topological space Q^*, such that $\mathcal{L}$ is lattice isomorphic to the algebra $\mathcal{Q}$ of all open-closed sets of Q^*. This isomorphism $\psi : \mathcal{L} \to \mathcal{Q}$ is called the *Stone Isomorphism*, and Q^* is called the *Stone space*. Observe that every finitely additive measure m defined on $\mathcal{L}$ and with values in a Dedekind complete (ℓ)-group R admits an extension $\overline{m} : \mathcal{Q} \to R$ defined by setting

$$\overline{m}(B) := m(\psi^{-1}(B)) \quad \text{for each } B \in \mathcal{Q}. \tag{2.78}$$

We now show that $\overline{m}$ is globally σ-additive on $\mathcal{Q}$. Let $(C_k)_k$ be a sequence of pairwise disjoint open-closed sets of Q^*, whose union is an open-closed set F. Since Q^* is compact and F is closed, then F is compact, and hence the covering $\{C_k : k \in \mathbb{N}\}$ of F admits a finite subcovering of F, say $\{C_{i_1}, C_{i_2}, \dots, C_{i_q}\}$. We get $F \subset \bigcup_{j=1}^{q} C_{i_j} \subset \bigcup_{k=1}^{\infty} C_k = F$, and hence $\bigcup_{j=1}^{q} C_{i_j} = \bigcup_{k=1}^{\infty} C_k$.

Thus the C_k's are eventually empty, and so every finitely additive R-valued measure turns to be globally σ-additive.

Analogously it is possible to see that every monotone sequence $(H_k)_k$ in Σ is definitely constant, as soon as its limit in the set-theoretic sense belongs to Σ.

If m is positive and R is Dedekind complete and weakly σ-distributive, the Carathéodory extension of $\overline{m}$ to the σ-algebra $\sigma(\mathcal{Q})$ is called the *Stone extension* of m. Note that the Stone extension is unique. In the general case, we consider $m^{(+)}$ and $m^{(-)}$, the positive and negative parts of m, and we deal with their respective Stone extensions.

If R is any Dedekind complete (ℓ)-group, not necessarily weakly σ-distributive, and one deals with (s)-boundedness and σ-additivity not necessarily in the global sense, though in general it is not possible to have always the Carathéodory extension (see also Wright (1971)), we get the Stone-type extension for positive R-valued measures. We now present its construction by means of the tool of the transfinite induction.

Given a positive finitely additive measure $m:\mathcal{L}\to R$, let $\overline{m}$ and $\mathcal{Q}$ be as in (2.78). Following Boccuto (1995b), we construct a σ-additive extension of $\overline{m}$ to the σ-algebra $\sigma(\mathcal{Q})$ generated by $\mathcal{Q}$, which will be the Stone extension. First of all note that, thanks to the Maeda-Ogasawara-Vulikh theorem, there is a nowhere dense set $N\subset\Omega$ such that $m_\omega(\cdot):=m(\cdot)(\omega)$ is a finitely additive positive real-valued measure for each $\omega\in\Omega\setminus N$. By arguing as before and taking $R=\mathbb{R}$, we get that the real-valued measure $\overline{m_\omega}=m_\omega\circ\psi^{-1}$ is σ-additive on $\mathcal{Q}$, and so it admits its Stone extension ν_ω to the σ-algebra $\sigma(\mathcal{Q})$.

We now proceed by transfinite induction (see also Albuquerque (1944), Leblanc and Fox (1956)). Let $\Sigma_0:=\mathcal{Q}$ and $\mu_0:=\overline{m}$. If α is an ordinal of first kind, and $\alpha-1$ is the ordinal preceding α, supposed defined $\Sigma_{\alpha-1}$, we put

$$\Sigma_{\alpha-1,\sigma}=\{F:F=\bigcup_{n=1}^{\infty}F_n, F_n\in\Sigma_{\alpha-1}, (F_n)_n \text{ is increasing}\},$$

$$\Sigma_{\alpha-1,\sigma\delta}=\{F:F=\bigcap_{n=1}^{\infty}F_n, F_n\in\Sigma_{\alpha-1,\sigma}, (F_n)_n \text{ is decreasing}\}, \qquad \textbf{(2.79)}$$

$$\Sigma_\alpha=\{F:F=\lim_n F_n : F_n\in\Sigma_{\alpha-1}\}\subset\Sigma_{\alpha-1,\sigma\delta},$$

where the limit in (2.79) is intended in the set-theoretic sense. If α is an ordinal of second kind, set

$\Sigma_\alpha=\bigcup_{\beta<\alpha}\Sigma_\beta$. Observe that $\sigma(\mathcal{Q})=\Sigma_{\omega_1}$, where ω_1 is the first uncountable ordinal (see Leblanc and Fox (1956)).

If α is an ordinal of first kind, supposed defined $\Sigma_{\alpha-1}$ and $\mu_{\alpha-1}$, let $M_\alpha:\Sigma_{\alpha-1,\sigma}\to R$ by setting $M_\alpha(A)=\bigvee_n \mu_{\alpha-1}(F_n)$ whenever $A=\bigcup_{n=1}^{\infty}F_n$, $F_n\in\Sigma_{\alpha-1}$, $(F_n)_n$ is increasing, and define $\mu_\alpha^*:\Sigma_{\alpha-1,\sigma\delta}\to R$ by

$\mu^*_\alpha(A) = \bigwedge_n M_\alpha(B_n)$ whenever $A = \bigcap_{n=1}^{\infty} F_n$, $F_n \in \Sigma_{\alpha-1,\sigma}$, $(F_n)_n$ is decreasing. Finally, let μ_α be the restriction of μ^*_α to Σ_α.

If α is an ordinal of second kind, define $\mu_\alpha : \Sigma_\alpha \to R$ by $\mu_\alpha(F) = \mu_\beta(F)$, whenever $F \in \Sigma_\beta$ with $\beta < \alpha$. Note that this definition makes sense (see also Boccuto (1995b)). The measure μ_{ω_1} is the requested Stone extension, which is uniquely determined for every Dedekind complete (ℓ)-group R (weakly σ-distributive or not). Indeed the following result holds.

Theorem 3.30 (Boccuto (1995b), Theorem 2.6) *Let R be any Dedekind complete lattice group. Then the measure $\mu_\alpha : \Sigma_\alpha \to R$ is σ-additive for every $\alpha \leq \omega_1$.*

In order to get Theorem 3.30, in Boccuto (1995b) the following lemma, related to the Maeda-Ogasawara-Vulikh Theorem 2.7., is used.

Lemma 3.31 (Boccuto (1995b), Lemma 2.7) *Let α be any ordinal with $\alpha \leq \omega_1$ and Ω, Σ_α, μ_α be as above. Then for each set $B \in \Sigma_\alpha$ there is a meager set L_B with $\nu_\omega(B) = \mu_\alpha(B)(\omega)$ for any $\omega \in \Omega \setminus L_B$.*

2.3.4. Bounded Functions and Limits

We now deal with some other properties of finitely additive measures, taking values in a lattice group R, in particular proving that every bounded R-valued function admits limit with respect to any ultrafilter at least in a "weak" sense (see also Boccuto and Candeloro (2009b)).

Given a nonempty abstract set Λ, a function $f : \Lambda \to R$ and a (Λ)-free filter $\mathcal{F}$ of Λ, we say that f has *$(\mathcal{F})$-limit* in R iff there exist a directed set $(\Xi, \leq)$, an element l in R and a decreasing net $(r_\xi)_{\xi \in \Xi}$ in R, with $\bigwedge_\xi r_\xi = 0$, and such that, for each $\xi \in \Xi$,

$$\{t \in \Lambda : |f(t) - l| \leq r_\xi\} \in \mathcal{F}.$$

In other terms, we have $(\mathcal{F})\lim_{t\in\Lambda} f(t) = l$ if and only if

$$\bigwedge_{F\in\mathcal{F}} \bigvee_{t\in F} f(t) = l = \bigvee_{F\in\mathcal{F}} \bigwedge_{t\in F} f(t).$$

Observe that $(O\mathcal{F})$-convergence and $(D\mathcal{F})$-convergence (when R is also weakly σ-distributive) imply $\mathcal{F}$-convergence and, when R is super Dedekind complete and weakly σ-distributive, these convergences are equivalent (see also Boccuto (2003), Boccuto and Candeloro (2009b)).

A Dedekind complete (ℓ)-group R has the *strong limit property* iff for any abstract nonempty set Λ, any bounded function $f:\Lambda\to R$ and any ultrafilter $\mathcal{F}$ of Λ, the $(\mathcal{F})$-limit of f exists in R. Observe that the space $\mathbb{R}$ has the strong limit property, thanks to Proposition 1.12. From this, it clearly follows that the space $\mathbb{R}^D$, for any nonempty set D, has the strong limit property, when endowed with the natural order and algebraic structures.

We now give a further example. Fix an abstract nonempty set D, and denote by $C(\beta D)$ the space of all continuous functions $\psi:\beta D\to\mathbb{R}$, where βD is the Stone-Čech compactification of D. The space $C(\beta D)$ has a natural ordering, and is closed under arbitrary suprema and infima of bounded subsets.

Given any bounded function $\psi: D\to\mathbb{R}$, we denote by $\overline{\psi}$ the unique continuous extension of ψ to βD. It is not difficult to check that, given any equibounded family $(\psi_h)_{h\in H}$ of functions defined on D and taking values in $\mathbb{R}$, we always have

$$(\bigvee_{h\in H}\overline{\psi_h})(d) = \sup_{h\in H}\psi_h(d), \qquad (\bigwedge_{h\in H}\overline{\psi_h})(d) = \inf_{h\in H}\psi_h(d), \text{ for all } d\in D.$$

Fix an ultrafilter $\mathcal{F}$ of Λ and a bounded function $f:\Lambda\to C(\beta D)$. We get the following

Lemma 3.32 (see Boccuto and Candeloro (2009b), Lemma 3.2) *Let us set*

$$l_1 := \bigwedge_{F\in\mathcal{F}} \bigvee_{t\in F} f(t), \qquad l_2 := \bigvee_{F\in\mathcal{F}} \bigwedge_{t\in F} f(t),$$

where the involved suprema and infima are taken in $C(\beta D)$. *Then* $l_1 = l_2$, *and hence* $C(\beta D)$ *has the strong limit property.*

We now turn to the concept of *weak filter limit*. Given two Dedekind complete (ℓ)-groups R and R^*, we say that R is *embedded* in R^* iff there exist a one-to-one Riesz homomorphism (called *embedding*) $j : R \to R^*$ such that $j(R)$ is a Dedekind complete subgroup of R^*, and a surjective homomorphism $\pi : R^* \to R$ (called *projection*) such that $\pi \circ j$ is the identity function on R. When this happens, we also say that R^* is an *extension* of R.

In general, embeddings reduce the *gap* existing between the two quantities

$$\bigvee_{F \in \mathcal{F}} \bigwedge_{t \in F} f(t), \qquad \bigwedge_{F \in \mathcal{F}} \bigvee_{t \in F} f(t),$$

whenever $f : \Lambda \to R$ is a bounded function, and $\mathcal{F}$ any (Λ)-free filter of Λ. Indeed, we get:

Theorem 3.33 (see Boccuto and Candeloro (2009b), Theorem 3.4) *Assume that* $f : \Lambda \to R$ *is a bounded function, and* $\mathcal{F}$ *is any filter of* Λ. *If* R^* *is any extension of* R, *with embedding* j *and projection* π, *then*

$$\bigvee_{F \in \mathcal{F}} \bigwedge_{t \in F} f(t) \le \pi\Big(\bigvee_{F \in \mathcal{F}} \bigwedge_{t \in F} j(f(t))\Big) \le \pi\Big(\bigwedge_{F \in \mathcal{F}} \bigvee_{t \in F} j(f(t))\Big) \le \bigwedge_{F \in \mathcal{F}} \bigvee_{t \in F} f(t).$$

Let R be any Dedekind complete lattice group and R^* any extension of R. We say that the extension of R into R^* is *maximal* if, for every abstract set Λ, every ultrafilter $\mathcal{F}$ of Λ and any bounded function $f : \Lambda \to R$, we have

$$\pi\Big(\bigwedge_{F \in \mathcal{F}} \bigvee_{t \in F} j(f(t))\Big) = \pi\Big(\bigvee_{F \in \mathcal{F}} \bigwedge_{t \in F} j(f(t))\Big). \qquad \textbf{(2.80)}$$

If this happens, then the common value in (2.80) is called *weak* $\mathcal{F}$ *- limit* of f.

In particular, if R has an extension R^* with the strong limit property, then R^* is a maximal extension. We shall see later that this is the case, for example, if R has an order unit.

More generally, given a lattice group R, we say that a bounded function $f:\Lambda\to R$ has a *weak limit* with respect to a filter $\mathcal{F}$ of Λ (shortly, *weak* $\mathcal{F}$*-limit* or $(w\mathcal{F})\lim$) iff the range of f is contained in a Dedekind complete space R_0 with a maximal extension R_0^*, and

$$\pi(\bigwedge_{F\in\mathcal{F}}\bigvee_{t\in F} j(f(t))) = \pi(\bigvee_{F\in\mathcal{F}}\bigwedge_{t\in F} j(f(t))).$$

In this case the weak $\mathcal{F}$-limit l, namely the common value, is an element l of R_0, and we write $l.=(w\mathcal{F})\lim_{t\in\Lambda} f(t)$.

Thanks to Theorem 3.33, we can see that, if $f:\Lambda\to R$ is a bounded function, then the existence of the $\mathcal{F}$-limit implies that any weak $\mathcal{F}$-limit coincides with it. Moreover, it is not difficult to check that the weak filter limit, when existing, satisfies the usual linearity properties, provided that it is performed always with respect to the same maximal extension.

We now turn to show that a weak $\mathcal{F}$-limit always exists, as soon as $f:\Lambda\to R$ is a bounded function and $\mathcal{F}$ is an ultrafilter of Λ. To this aim, we recall the following well-known result, which is related to the Maeda-Ogasawara-Vulikh representation theorem (see also Chojnacki (1986), Rainwater (1959)).

Theorem 3.34 *Let Ω be any Hausdorff, compact, extremely disconnected topological space, and denote by D the support set of Ω, endowed with the discrete topology. Then, Ω (with its original topology) is a subspace of the Stone-Čech compactification βD and there exists a continuous onto function $r:\beta D\to\Omega$ (retract), such that $r_{|\Omega}$ is the identity function.*

Thanks to Theorems 2.7 and 3.34 we can deduce that, if R is a Dedekind complete vector lattice endowed with an order unit, then R admits a maximal extension. Indeed, R can be viewed as a space of the type $C(\Omega)$, for a suitable Stone space Ω (see also Wright (1969, 1971)), and $C(\Omega)$ is embedded in $C(\beta D)$, according with Theorem 3.34: indeed, for any function $\phi\in C(\Omega)$ we can define $j(\phi):\beta D\to\mathbb{R}$ as $j(\phi)(x)=\phi(r(x))$ for every $x\in\beta D$, and also we can define a projection $\pi:C(\beta D)\to C(\Omega)$ as $\pi(\psi):=\psi_{|\Omega}$ for every $\psi\in C(\beta D)$. By Lemma 3.32, the space $C(\beta D)$ has the strong limit property, and hence the claim is proved.

We now turn to the general case of a bounded R-valued function f.

Theorem 3.35 (see Boccuto and Candeloro (2009b), Theorem 3.8) *Let R be any Dedekind complete Riesz space, $f:\Lambda \to R$ be any bounded function and $\mathcal{F}$ be any ultrafilter of Λ. Then there exists in R a weak $\mathcal{F}$-limit of f.*

Proof: Observe that, if $f:\Lambda \to R$ is bounded and e is the supremum of the range of $|f|$, then the range of f is contained in the vector lattice $V[e] := \{b \in R : \text{there is } \lambda > 0 \text{ with } |b| \leq \lambda e\}$.

Since R is Dedekind complete, $V[e]$ is too, and hence $V[e]$ is a Dedekind complete Riesz space with order unit. Thus $V[e]$ has a maximal extension, and this shows that f has weak $\mathcal{F}$-limit in $V[e]$, and hence in R. □

More precisely, if $f:\Lambda \to R$ is a bounded function and $e \in R$ is such that $|f(t)| \leq e$ for every $t \in \Lambda$, then we can choose the space $C(\beta D)$ as an extension of $V[e]$, and define $f_0(t)(x) = f(t)(r(x))$, $t \in \Lambda$, $x \in \beta D$. Denoting by l the $\mathcal{F}$-limit of f_0, the weak $\mathcal{F}$-limit of f is $\pi(l)$, that is $l_{|\Omega}$, where Ω is the compact Hausdorff Stonian space such that $V[e]$ is isometrically and lattice isomorphic to $C(\Omega)$ (see also Kadison (1951), Theorem 4.1, Wright (1971), §1).

2.3.5. M-Measures and their Extensions

We now deal with some other kinds of extension results, for not necessarily finitely additive lattice group-valued measures, which have several applications in different branches of Mathematics, for instance probability, decision making, quantum logics, fuzzy sets and measures. It is often advisable to consider monotone measures (capacities), continuous and/or compatible with respect to suprema and/or infima (*M-measures,* see also Krachounov (2006)). Some related topics are the so-called *intuitionistic fuzzy events* (shortly *IF-events*, see also Atanassov (1999)). They are pairs of measurable functions $A = (\mu_A, \nu_A)$, where $\mu_A, \nu_A : \Omega \to [0,1]$ have the property that $\mu_A + \nu_A \leq 1$.

If $\mathcal{O}$ be the family of all IF-events, then it is possible to investigate several properties of the set functions defined in $\mathcal{O}$ and with values in [0,1], of the type $P : \mathcal{O} \to [0,1]$, which are compatible with respect to finite suprema and infima (see for instance Riečan (2007)). In quantum logics it is often advisable to replace random variables with *observables*, that is functions defined on the family of all Borel subsets of Ω and with values in $\mathcal{O}$ (see also Krachounov (2006)). Every member of $\mathcal{O}$ can be identified as an element of the type $m(B)=(m^{\#}(B),1-m^{\#}(B))$, where m is a particular observable. For any $g \in G$ the set function $B \mapsto m^{\#}(B)(g)$ is a [0,1]-valued M-measure. We now consider the M-measures with values in lattice groups, extending the extension theorem proved in Riečan (2007) for real-valued measures. Note that in this setting it is possible to consider directly the elements of $\mathcal{O}$, and not necessarily its components.

Given an increasing (resp. decreasing) set sequence $(A_n)_n$, we write $A_n \nearrow A$ (resp. $A_n \searrow A$) when $\bigcup_{n=1}^{\infty} A_n = A$ (resp. $\bigcap_{n=1}^{\infty} A_n = A$).

Let R be a super Dedekind complete and weak σ-distributive lattice group. For a sake of simplicity, if $(a_n)_n$ is a sequence in R, we denote by $\lim_n a_n$ its order limit, which coincides with its (D)-limit. Let W be a ring of subsets of an arbitrary nonempty set G. A bounded set function $\mu : W \to R$ is an *M-measure* iff it fulfils the following conditions:

(*i*) $\mu(\emptyset)=0$;

(*ii*) $\mu(A \cup B)=\mu(A) \vee \mu(B)$, $\mu(A \cap B)=\mu(A) \wedge \mu(B)$ for every A, $B \in W$;

(*iii*) μ is *continuous* from below and from above, namely $\lim_n \mu(A_n)=\mu(A)$ (resp. $\lim_n \mu(B_n)=\mu(B)$) whenever $A_n \nearrow A$, (resp. $B_n \searrow B$), A_n, A (resp. $B_n, B) \in W$, $n \in \mathbb{N}$.

We will prove the following extension theorem for lattice group-valued M-measures. Here we consider (D)-convergence, snce it allows us to treat a series of (D)-sequences like a single regulator, by virtue of the Fremlin lemma 2.3.

Theorem 3.36 (Boccuto and Riečan (2010), Theorem 3.1) *Let R be a super Dedekind complete weakly σ-distributive lattice group. Then every bounded R-valued M-measure μ, defined on W, has a unique extension $\overline{\mu}$ on the σ-ring $\sigma(W)$ generated by W, which is an*

M-measure.

Let W^+ be the class of all sets A with $A:=\bigcup_{n=1}^{\infty} A_n$, $A_n \subset A_{n+1}$, $A_n \in W$ for all $n \in \mathbb{N}$.

As $0 \le \mu(A_n) \le \mu(A_{n+1}) \le \bigvee_{E\in W}\mu(E)$, the limit $\lim_n \mu(A_n)$ exists in R. We now show that this limit is independent of the considered sequence $(A_n)_n$, and we will denote it by $\mu^+(A)$.

Lemma 3.37 (Boccuto and Riečan (2010), Lemma 3.2) *If A_n, $B_q \in W$ ($n, q \in \mathbb{N}$), $A_n \nearrow A$, $B_q \nearrow A$. Then $\lim_n \mu(A_n) = \lim_q \mu(B_q)$.*

Proof: For each $q \in \mathbb{N}$ we have $A_n \cap B_q \nearrow A \cap B_q = B_q$, and so

$$\mu(B_q) = \lim_n \mu(A_n \cap B_q) \le \lim_n \mu(A_n).$$

From arbitrariness of $q \in \mathbb{N}$ we deduce $\lim_q \mu(B_q) \le \lim_n \mu(A_n)$. The converse inequality can be proved similarly. □

Proposition 3.38 (Boccuto and Riečan (2010), Proposition 3.3) *Let $A_k \in W^+$ ($k \in \mathbb{N}$) and $A_k \nearrow A$. Then we get $A \in W^+$ and $\lim_k \mu^+(A_k) = \mu^+(A)$.*

Proof: As $A_k \in W^+$, there are $A_{k,n} \in W$, $n \in \mathbb{N}$, such that $A_{k,n} \nearrow A_k$ $(n \to +\infty)$. For each $n \in \mathbb{N}$, put $B_n := \bigcup_{k=1}^{n} A_{k,n}$. We get $B_n \in W$ and $B_n \nearrow A$, and hence $A \in W^+$. Furthermore we have

$$\mu^+(A) = \lim_k \mu(B_k) \le \lim_k \mu^+(A_k) \le \mu^+(A).$$

Proposition 3.39 (Boccuto and Riečan (2010), Proposition 3.4) *Let W^- be the class of the sets A with $A:=\bigcap_{n=1}^{\infty} A_n$, $A_n \supset A_{n+1}$, $A_n \in W$ for every $n \in \mathbb{N}$, and put*

$\mu^-(A) := \lim_n \mu(A_n)$. *Let* $A_k \in W^-$ $(k \in \mathbb{N})$ *and* $A_k \searrow A$. *Then we get* $A \in W^-$ *and moreover* $\lim_k \mu^-(A_k) = \mu^-(A)$.

Proposition 3.40 (Boccuto and Riečan (2010), Proposition 3.5) *If* $A, B \in W^+$ *and* $C, D \in W^-$, *then* $A \cup B$, $A \cap B \in W^+$; $C \cup D$, $C \cap D \in W^-$, *and*

$$\mu^+(A \cup B) = \mu^+(A) \vee \mu^+(B), \quad \mu^+(A \cap B) = \mu^+(A) \wedge \mu^+(B),$$

$$\mu^-(C \cup D) = \mu^-(C) \vee \mu^-(D), \quad \mu^-(C \cap D) = \mu^-(C) \wedge \mu^-(D).$$

Proposition 3.41 (Boccuto and Riečan (2010), Proposition 3.6) *If* $B \in W^+$, $C \in W^-$, $C \subset B$, *then* $\mu^-(C) \leq \mu^+(B)$.

Proof: If $B_n \in W$, $B_n \nearrow B$, then $C = (C \setminus B_n) \cup (C \cap B_n)$, and thus

$$\mu^-(C) = \mu^-((C \setminus B_n) \cup (C \cap B_n)) = \mu^-(C \setminus B_n) \vee \mu^-(C \cap B_n)$$
$$\leq \mu^-(C \setminus B_n) \vee \mu(B_n) \leq \mu^-(C \setminus B_n) \vee \mu^+(B).$$

Moreover we get $C \setminus B_n \in W^-$, $C \setminus B_n \searrow C \setminus B$ and $C \setminus B = \varnothing$. So we obtain

$$\mu^-(C) \leq (\lim_n \mu^-(C \setminus B_n)) \vee \mu^+(B) = 0 \vee \mu^+(B) = \mu^+(B). \ \square$$

Let W^* be the ideal generated by W^+. For any $A \in W^*$, put $\mu^*(A) = \bigwedge \{\mu^+(B) : B \in W^+, B \supset A\}$,

$\mu_*(A) = \bigvee \{\mu^-(C) : C \in W^-, C \subset A\}$. By Proposition 3.41 we get:

Proposition 3.42 (Boccuto and Riečan (2010), Proposition 3.8) *For every* $A \subset G$, $\mu_*(A) \leq \mu^*(A)$.

Proposition 3.43 (Boccuto and Riečan (2010), Proposition 3.9) *Let* $A_n \nearrow A$ *and* $B_n \searrow B$. *Then* $\mu^*(A_n) \nearrow \mu^*(A)$, $\mu_*(B_n) \searrow \mu_*(B)$.

Proof: As $A_n \subset A$ for every $n \in \mathbb{N}$, then $\mu^*(A_n) \le \mu^*(A)$ for each positive integer n, and thus $\lim_n \mu^*(A_n) \le \mu^*(A)$.

We now prove the converse inequality. Thanks to super Dedekind completeness of R, for each $n \in \mathbb{N}$ a sequence $(B_n^q)_q$ can be found, such that $W^+ \ni B_n^q \supset A_n$ for every $q \in \mathbb{N}$ and $\mu^*(A_n) = \bigwedge_{q=1}^{\infty} \mu^+(B_n^q)$. Set $D_n^q := \bigcap_{i=1}^{q} B_n^i$, $n, q \in \mathbb{N}$: then

$$\mu^*(A_n) \le \bigwedge_{q=1}^{\infty} \mu^+(D_n^q) \le \bigwedge_{q=1}^{\infty} \mu^+(B_n^q) = \mu^*(A_n);$$

$\mu^+(D_n^q) \searrow \mu^*(A_n)$ as q tends to $+\infty$. For every $n \in \mathbb{N}$ there is a (D)-sequence $(a_{t,l}^{(n)})_{t,l}$ such that for each $\varphi \in \mathbb{N}^{\mathbb{N}}$ and $n \in \mathbb{N}$ there exists $k_n \in \mathbb{N}$ with the property that $\mu^+(D_n^{k_n}) - \mu^*(A_n) \le \bigvee_{t=1}^{\infty} a_{t,\varphi(t+n)}^{(n)}$. Set $C_n = D_n^{k_n}$, $n \in \mathbb{N}$: we get $C_n \supset A_n$, $\mu^+(C_n) \le \mu^*(A_n) + \bigvee_{t=1}^{\infty} a_{t,\varphi(t+n)}^{(n)}$, $A = \bigcup_{n=1}^{\infty} A_n \subset \bigcup_{n=1}^{\infty} C_n$, and thus

$$\mu^*(A) \le \mu^+\left(\bigcup_{n=1}^{\infty} C_n\right) = \bigvee_{n=1}^{\infty} \mu^+(C_n) \le \bigvee_{n=1}^{\infty}\left(\mu^*(A_n) + \bigvee_{t=1}^{\infty} a_{t,\varphi(t+n)}^{(n)}\right)$$

$$\le \bigvee_{n=1}^{\infty} \mu^*(A_n) + \sum_{n=1}^{\infty}\left(\bigvee_{t=1}^{\infty} a_{t,\phi(t+n)}^{(n)}\right);$$

$$0 \le \mu^*(A) - \bigvee_{n=1}^{\infty} \mu^*(A_n) \le \mu^*(A).$$

So we obtain $0 \le \mu^*(A) - \bigvee_{n=1}^{\infty} \mu^*(A_n) \le \mu^*(A) \wedge \sum_{n=1}^{\infty}\left(\bigvee_{t=1}^{\infty} a_{t,\phi(t+n)}^{(n)}\right)$.

Thanks to the Fremlin lemma 2.3 there exists a (D)-sequence $(a_{t,l})_{t,l}$ in R with

$$0 \le \mu^*(A) - \bigvee_{n=1}^{\infty} \mu^*(A_n) \le \bigvee_{t=1}^{\infty} a_{t,\varphi(t)} \text{ for each } \varphi \in \mathbb{N}^{\mathbb{N}}.$$

Since R is weakly σ-distributive and $\varphi \in \mathbb{N}^{\mathbb{N}}$ is arbitrary, then

$$0 \le \mu^*(A) - \bigvee_{n=1}^{\infty} \mu^*(A_n) \le \bigwedge_{\varphi \in \mathbb{N}^{\mathbb{N}}} \left(\bigvee_{t=1}^{\infty} a_{t,\varphi(t)} \right) = 0, \text{ namely } \mu^*(A) = \bigvee_{n=1}^{\infty} \mu^*(A_n). \ \square$$

Corollary 3.44 (Boccuto and Riečan (2010), Corollary 3.10)

If $M := \{A \subset G : \mu^*(A) = \mu_*(A)\}$, *then* M *is a monotone class, namely* A, $B \in M$ *whenever* A_n, $B_n \in M$ *with* $A_n \nearrow A$, $B_n \searrow B$.

Proposition 3.45 (Boccuto and Riečan (2010), Proposition 3.11) *The measure* μ^* *extends* μ. *Moreover, we get* $M \supset W$.

Proof: Choose $A \in W$. Then we have

$$\mu(A) = \mu^+(A) = \mu^-(A), \mu^*(A) \le \mu^+(A) = \mu^-(A) \le \mu_*(A), \mu^*(A) \ge \mu_*(A).$$

So we obtain $\mu(A) = \mu^*(A) = \mu_*(A)$. This ends the proof. $\square$

It is not difficult to prove the following

Proposition 3.46 (Boccuto and Riečan (2010), Proposition 3.12) *Let* $\sigma(W)$ *be the* σ*-ring generated by* W. *We get* $\sigma(W) \subset M$.

Proposition 3.47 (see also Boccuto and Riečan (2010), Remark 3.13) *Let* $A, B \in W^*$. *Then*

$$\mu^*(A \cup B) = \mu^*(A) \vee \mu^*(B) \text{ and } \mu^*(A \cap B) = \mu^*(A) \wedge \mu^*(B).$$

Proof: Without loss of generality, it is enough to prove that $\mu^*(A \cup B) = \mu^*(A) \vee \mu^*(B)$, since the proof of the other equality is analogous. Choose $A, B \in \sigma(W)$. As R is super Dedekind complete, there are two decreasing sequences $(C_n)_n$, $(D_n)_n$ belonging to W^+, such that $C_n \supset A$ and $D_n \supset B$ for every $n \in \mathbb{N}$,

$$\mu^*(A) = \bigwedge_{n=1}^{\infty} \mu^*(C_n), \qquad \mu^*(B) = \bigwedge_{n=1}^{\infty} \mu^*(D_n)$$

and hence $\mu^*(A \cup B) \leq \mu^*(C_n \cup D_n) = \mu^*(C_n) \vee \mu^*(D_n)$.

Taking the infimum with respect to *n,* we get that $\mu^*(A \cup B) \leq \mu^*(A) \vee \mu^*(B)$. The opposite inequality is straightforward. □

Proposition 3.48 (Boccuto and Riečan (2010), Proposition 3.14) *If* $\overline{\mu} = \mu^*_{|\sigma(W)}$, *then* $\overline{\mu}$ *is an extension of* μ, *and is also an* M *-measure.*

Proof: Since $\overline{\mu} = \mu^*_{|\sigma(W)}$, $\mu = \mu^*_{|W}$, we get $\mu = \overline{\mu}_{|W}$. Therefore, $\overline{\mu}(\emptyset) = 0$. The boundedness of $\overline{\mu}$ is straightforward. The above and below continuity of $\overline{\mu}$ follow from Corollary 3.44.

By Proposition 3.47, we get $\overline{\mu}(A \cup B) = \overline{\mu}(A) \vee \overline{\mu}(B)$. The proof of the inequality $\overline{\mu}(A \cap B) = \overline{\mu}(A) \wedge \overline{\mu}(B)$ is similar. □

Proof of Theorem 3.36: For the existence of $\overline{\mu}$, see Proposition 3.48. Concerning the uniqueness, pick an M -measure $\nu : \sigma(W) \to R$ such that $\nu_{|W} = \mu$, and define

$$\Pi := \{A \in \sigma(W) : \nu(A) = \overline{\mu}(A)\}.$$

It is easy to check that Π is a monotone class containing W. So Π contains also $\sigma(W)$. □

We now recall an extension result for M -measures with values in *any* Dedekind complete (ℓ)-group, which extends Theorem 3.30 to M -measures (for related results, see also Boccuto (1995b), Khurana (1980-1981), Kusraev and Malyugin (1988), Lipecki (1983, 1988 and 1990), Malyugin (1991) and Wright (1971)).

A ring W of parts of a set G is said to be *perfect* iff for each increasing (resp. decreasing) sequence $(A_n)_n$ in W, whose union (resp. intersection) belongs to W, there exists an index k and an element A_k with $A_n = A_k$ for every $n > k$. Note that the class of the open-closed sets of a Stone space is perfect (see also Boccuto (1995b)).

The following results holds.

Theorem 3.49 (Boccuto and Riečan (2010), Theorem 4.1) *If R is a Dedekind complete lattice group and W is perfect, then every M-measure $\mu : W \to R$ admits a unique extension $\overline{\mu} : \sigma(W) \to R$, which is also an M-measure.*

CHAPTER 3

Classical Limit Theorems in Lattice Groups

Abstract: We consider several versions of limit theorems for lattice group-valued measures, in which both pointwise convergence of the involved measures and the notions of σ-additivity, (s)-boundedness, regularity, are given in the global sense, that is with respect to a common regulator. We present the construction of some kinds of integrals in the vector lattice context and some Vitali and Lebesgue theorems. Successively we prove some other kinds of limit theorems, in which the main properties of the measures are considered in the classical like sense. Finally, we give different types of decomposition theorems for lattice group-valued measures.

Keywords: Axiomatic convergence, Bochner integral, Brooks-Jewett theorem, convergence in L^1, convergence in measure, Dieudonné theorem, dominated convergence theorem, Lattice group, Lebesgue decomposition, Nikodým convergence theorem, optimal integral, Rickart integral, Schur theorem, Sobczyk-Hammer decomposition, Stone Isomorphism technique, ultrafilter measure, uniform integrability, Vitali theorem, Vitali-Hahn-Saks theorem, Yosida-Hewitt decomposition.

In this chapter we present some different types of limit theorems, for measures taking values in lattice groups. We consider both the countably additive and the finitely additive case, and we relate them by means both of the Stone Isomorphism technique and Drewnowski-type theorems, which allow us to investigate countably additive extensions or restrictions of finitely additive measures, respectively, and about which we dealt in Chapter 2. In the context of lattice groups, we first consider the tool of (D)-convergence and study the case in which the notions of (s)-boundedness, σ-additivity, (absolute) continuity, regularity, are given relatively to a single (D)-sequence, as well as the pointwise convergence of the involved measures. Successively we consider also order convergence and the case in which the pointwise convergence of measures is given relatively to a single order sequence, but not necessarily σ-additivity, (s)-boundedness, and so on. In proving limit theorems, often some technical lemmas play a crucial role, by means of which it is possible to demonstrate that uniform (s)-boundedness of a sequence of σ-additive (absolutely) continuous or regular measures implies uniform σ-additivity, uniform (absolute) continuity or uniform regularity respectively. When these concepts are intended with respect to a single regulator or (O)-sequence, it is enough to use techniques analogous to the

Antonio Boccuto & Xenofon Dimitriou

classical case. When we deal with (s)-boundedness, σ-additivity, (absolute) continuity and regularity not necessarily with respect to a same (O)- or (D)-sequence, in order to overcome some technical difficulties we use the Maeda-Ogasawara-Vulikh representation theorem for Archimedean lattice groups, we study these properties for sequences of real-valued measures in the complement of a suitable meager subset of a compact extremely disconnected Hausdorff topological space, and we use a density argument as a consequence of the Baire category theorem. These theorems were proved in Boccuto and Candeloro (2002b and 2004a) and in Candeloro (2002), when (s)-boundedness and the related topics are meant with respect to a common (O)- or (D)-sequence, and in Boccuto and Candeloro (2010 and 2011), Boccuto, Dimitriou and Papanastassiou (2010c and 2011a) where they are meant in the classical sense. Some other version of Schur and Nikodým-type theorems were proved in Boccuto, Dimitriou and Papanastassiou (2012c), Boccuto and Papanastassiou (2007). Some other theorems of this kind in similar contexts were proved in Avallone (2006), Avallone, Rinauro and Vitolo (2007), Barbieri (2009a-c). We give a Vitali-type theorem for a Bochner-type integral for Riesz space-valued functions with respect to a σ-additive positive extended real-valued measure, and we get a similar construction of an abstract integral in the Riesz space setting, in which it is required that the convergences involved satisfy some suitable axioms, which are fulfilled by filter convergence. We also present a construction of an integral with respect to lattice group-valued measures, not necessarily finitely additive, but compatible with respect to supremum, together with some main convergence theorems (see also Boccuto, Dimitriou and Papanastassiou (2010a)). For a related literature see also Benvenuti, Mesiar and Vivona (2002), Boccuto (1993, 1995a and 1997), Boccuto and Candeloro (2002a-c, 2004a-d, 2005, 2008, 2009a-b and 2010-2011), Boccuto, Candeloro and Sambucini (2007 and 2011), Boccuto and Riečan (2006 and 2008-2010), Boccuto, Riečan and Sambucini (2010), Boccuto, Riečan and Vrábelová (2009), Boccuto and Sambucini (1996a-b and 1997), Candeloro and Sambucini (2014b), Choquet (1954), Halmos (1950), Haluška (1993), Haluška and Hutník (2010), Kawabe (2008 and 2012), Mcgill (1975), Riečan and Neubrunn (1997), Riečan and Vrábelová (1988), Vrábelová and Riečan (1996), Wang and Klir (2009) and their bibliographies. Finally, we prove some decomposition theorems for (ℓ)-group-valued measures, by using convergence theorems and, in the finitely additive case, the tool of the Stone Isomorphism technique. For a related literature, see also Avallone, Barbieri and Vitolo (2003, 2008 and 2010), Avallone, Barbieri, Vitolo and Weber (2009), Avallone and Vitolo (2003, 2009 and 2013), Barbieri, Valente and Weber (2012), Brooks (1969a), Brooks and Candeloro (2004), Cavaliere, de Lucia, De Simone and Ventriglia (2013), de Lucia and Pap (1996), Dvurečenskij, de Lucia and Pap (1996), Hammer and Sobczyk (1944), Pavlakos (1978), Rickart (1943), Rüttiman (1994), Schmidt (1982, 1986, 1989 and 1998), Yosida and Hewitt (1952).

3.1. Convergence Theorems in the Global Sense

We present some convergence theorems (Brooks-Jewett, Vitali-Hahn-Saks, Schur, Nikodým, Dieudonné theorems) for finitely or countably additive measures, taking values in (ℓ)-groups, with respect to (D)-convergence (see for instance Boccuto, Riečan and Vrábelová (2009), Riečan and Neubrunn (1997)). In this context, all the fundamental concepts, for example pointwise convergence of measures, (s)-boundedness, σ-additivity, regularity, are intended relatively to a single (D)-sequence or regulator.

We give some relations between global (s)-boundedness and absolute continuity, and in the setting of uniform (s)-bounded measures we present a kind of uniform extension, in the global σ-additive case, and a result on global uniform absolute continuity for a sequence of equibounded, globally uniformly (s)-bounded finitely additive absolutely continuous measures.

3.1.1. Uniform (s)-Boundedness and Related Topics

Concerning the main concepts and properties about lattice group-valued measures, we refer to Chapters 1 and 2. We assume that G is an abstract nonempty set, $\mathcal{E}$, $\mathcal{L}$ and Σ are a lattice, an algebra and a σ-algebra of subsets of G respectively.

Let R be a Dedekind complete and weakly σ-distributive lattice -group. We begin with relating global uniform (s)-boundedness and global absolute continuity of lattice group-valued measures.

Theorem 1.1 (see Boccuto and Candeloro (2002b), Theorem 4.4) *Let $\lambda : \Sigma \to \mathbb{R}$ be a positive σ-additive measure, and $m_j : \Sigma \to R$, $j \in \mathbb{N}$, be a sequence of globally λ-absolutely continuous and globally uniformly (s)-bounded measures. Then the m_j's are globally uniformly λ-absolutely continuous.*

Proof: We can and do suppose that the measures m_j are positive, otherwise we will consider their positive and negative parts, $m_j^{(+)}$ and $m_j^{(-)}$, respectively.

Let $(a_{t,l})_{t,l}$ be a regulator associated with global uniform (s)-boundedness. We prove that it satisfies the thesis of the theorem. Otherwise there are $\varphi \in \mathbb{N}^{\mathbb{N}}$, $(C_k)_k$ in Σ and a strictly increasing sequence $(j_k)_k$ of integers, with $\lambda(C_k) < 2^{-k}$ and

$$m_{j_k}(C_k) \not\leq \bigvee_{t=1}^{\infty} a_{t,\varphi(t)} \tag{3.1}$$

for every k. Let $A_k = \bigcup_{l=k}^{\infty} C_l$, then $\lambda(A_k) \leq 2^{1-k}$ for any k, and thus we get $(D) \lim_k m_j(A_k) = 0$ for each j. By virtue of Proposition 3.14 of Chapter 2 and weak σ-distributivity of R, we find a positive integer k_0 with $\bigvee_{j\in\mathbb{N}} m_j(C_k) \leq \bigvee_{t=1}^{\infty} a_{t,\varphi(t)}$ whenever $k \geq k_0$, getting a contradiction with (3.1). This ends the proof. □

Before giving the main convergence theorems with respect to a single regulator, we prove the following technical results.

Let $m : \mathcal{L} \to R$ be a positive finitely additive measure, and F_σ the family of all countable unions of elements of $\mathcal{L}$. For every $B \in F_\sigma$, $B = \bigcup_{n=1}^{\infty} F_n$, define

$$M(B) = \bigvee_{n\in\mathbb{N}} (\sum_{j=1}^{n} m(F_j)).$$

Note that the construction of M makes sense, thanks to σ-additivity of m (see also Bartle (1966)). Let $\sigma(\mathcal{L})$ be the σ-algebra generated by $\mathcal{L}$. For every $A \in \sigma(\mathcal{L})$, put

$$M(A) = \bigwedge \{M(B) : B \in F_\sigma, A \subset B\}.$$

In a similar way, starting with the non-negative real-valued measure λ, it is possible to construct the measure $P : \sigma(\mathcal{L}) \to \mathbb{R}$, playing an analogous role.

We now prove the following result about σ-additive extensions and a density property of $\mathcal{L}$ in $\sigma(\mathcal{L})$.

Theorem 1.2 (see Boccuto and Candeloro (2002b), Theorem 4.6) *The measure* $M:\sigma(\mathcal{L})\to R$ *is globally* σ*-additive, globally* λ*-absolutely continuous, and its restriction to* $\mathcal{L}$ *is equal to* m. *Furthermore there is a (D)-sequence* $(a_{t,l})_{t,l}$, *such that for every* $A\in\sigma(\mathcal{L})$ *and* $\varphi\in\mathbb{N}^{\mathbb{N}}$ *there is* $F\in\mathcal{L}$, *with* $M(A\Delta F)\leq\bigvee_{t=1}^{\infty}a_{t,\varphi(t)}$.

Proof: First of all, observe that $M(F)=m(F)$ for each $F\in\mathcal{L}$. Now, since m is globally λ-absolutely continuous, there exists a (*D*)-sequence $(a_{t,l})_{t,l}$ with the property that for every $\varphi\in\mathbb{N}^{\mathbb{N}}$ there is $\delta>0$ with $m(F)\leq\bigvee_{t=1}^{\infty}a_{t,\varphi(t)}$ for each $F\in\mathcal{L}$ such that $\lambda(F)<\delta$. Choose $A\in\sigma(\mathcal{L})$ such that $P(A)<\delta/2$ and $B\in F_{\sigma}$ with $A\subset B$ and $P(B)<\delta$. Then $B=\bigcup_{n=1}^{\infty}F_n$, where $(F_n)_n$ is a suitable increasing sequence of elements of $\mathcal{L}$. We get $P(F_n)<\delta$, and thus $M(F_n)\leq\bigvee_{t=1}^{\infty}a_{t,\varphi(t)}$, for each n. So we obtain $M(B)\leq\bigvee_{t=1}^{\infty}a_{t,\varphi(t)}$ and $M(A)\leq\bigvee_{t=1}^{\infty}a_{t,\varphi(t)}$. Hence, *M* is globally absolutely continuous with respect to *P*.

Choose arbitrarily A in $\sigma(\mathcal{L})$ and $\varphi\in\mathbb{N}^{\mathbb{N}}$, and let δ be associated with global *P*-absolute continuity of *M*. There is $F\in\mathcal{L}$, with $P(F\Delta A)<\delta$ (see also Maharam (1987), Theorem 2, (1)). By global absolute continuity, we get $M(F\Delta A)\leq\bigvee_{t=1}^{\infty}a_{t,\varphi(t)}$. We now prove global σ-additivity of *M:* this will follow from global absolute continuity and finite additivity of *M*. Choose arbitrarily two disjoint sets A_1, A_2 in $\sigma(\mathcal{L})$, and let $(a_{t,l})_{t,l}$ be a (*D*)-sequence, associated with global P-absolute continuity of *M*. Choose $\varphi\in\mathbb{N}^{\mathbb{N}}$, and pick a $\delta>0$ be associated with global absolute continuity. There exist two disjoint elements F_1, $F_2\in\mathcal{L}$, with $P(F_l\Delta A_l)<\delta/2$, $l=1,2$. Let $F=F_1\cup F_2$, $A=A_1\cup A_2$, then $P(F\Delta A)<\delta$, and thus $M(F\Delta A)\leq\bigvee_{t=1}^{\infty}a_{t,\varphi(t)}$. So we obtain

$$|M(A)-(M(A_1)+M(A_2))|\leq 3\bigvee_{t=1}^{\infty}a_{t,\varphi(t)}.$$

From this and weak σ-distributivity of R we get additivity of M. This ends the proof. □

We now recall the following result about a uniform density property.

Theorem 1.3 (see Boccuto and Candeloro (2002b), Theorem 4.7) *Suppose that* $(m_n)_n$ *is an equibounded and globally uniformly* (s)*-bounded sequence of global* σ*-additive measures defined on an algebra* $\mathcal{L}$*, with values in* R *and globally absolutely continuous with respect to a non-negative and* σ*-additive measure* $\lambda : \mathcal{L} \to \mathbb{R}$ *. Then a sequence* $(\pi_n)_n$ *of globally* σ *-additive and globally uniformly P-absolutely continuous measures can be found, which extends* m_n *to the whole of on* $\sigma(\mathcal{L})$*. Moreover there is a (D)-sequence* $(\beta_{t,l})_{t,l}$ *with the property that to each* $A \in \sigma(\mathcal{L})$ *and* $\varphi \in \mathbb{N}^{\mathbb{N}}$ *a set* $F \in \mathcal{L}$ *can be associated, with*

$$v(\pi_n)(F \Delta A) \leq \bigvee_{t=1}^{\infty} \beta_{t,\varphi(t)} \quad \textit{for every } n \in \mathbb{N}.$$

Proof: Without restriction, suppose that the measures m_n are positive. Set $\pi_n = M_n$, where M_n is as in Theorem 1.2. By virtue of Theorem 1.2 and the Fremlin lemma, we find a (*D*)-sequence $(a_{t,l})_{t,l}$ satisfying the following properties:

(*i*) For each $\varphi \in \mathbb{N}^{\mathbb{N}}$ and $n \in \mathbb{N}$ there exists a $\delta > 0$ with

$$P(A) < \delta \Rightarrow \pi_n(A) \leq \bigvee_{t=1}^{\infty} a_{t,\varphi(t)}$$

for each $A \in \sigma(\mathcal{L})$;

(*ii*) If $(F_k)_k$ is a disjoint sequence in $\mathcal{L}$ and $\varphi \in \mathbb{N}^{\mathbb{N}}$, then there is a natural number k_0 with

$$m_n(F_k) \leq \bigvee_{t=1}^{\infty} a_{t,\varphi(t)}$$

for all $k \geq k_0$ and $n \in \mathbb{N}$.

By the Fremlin lemma, we find another regulator $(b_{t,l})_{t,l}$ with

$$u \wedge \sum_{k=1}^{q} (\bigvee_{t=1}^{\infty} a_{t,\varphi(t+k)}) \leq \bigvee_{t=1}^{\infty} b_{t,\varphi(t)}, \qquad \textbf{(3.2)}$$

for each $q \in \mathbb{N}$ and $\varphi \in \mathbb{N}^{\mathbb{N}}$ (here, $u := \bigvee_{n \in \mathbb{N}} m_n(G)$).

We prove that the regulator $(2b_{t,l})_{t,l}$ satisfies global uniform (s)-boundedness for the π_n's. Otherwise we find a disjoint sequence $(A_k)_k$ of elements of $\sigma(\mathcal{L})$, an increasing sequence $(n_k)_k$ in $\mathbb{N}$ and a $\varphi \in \mathbb{N}^{\mathbb{N}}$ with

$$\pi_{n_k}(A_k) \not\leq 2 \bigvee_{t=1}^{\infty} b_{t,\varphi(t)} \quad \text{for every } k \in \mathbb{N}. \qquad \textbf{(3.3)}$$

By virtue of Theorem 1.2, for every k there is an $F_k \in \mathcal{L}$, with

$$\pi_{n_r}(F_k \Delta A_k) \leq \bigvee_{t=1}^{\infty} a_{t,\varphi(t+k)} \text{ for each } r = 1,2,\ldots,k.$$

Put

$$F_1^* = F_1, \; F_k^* = F_k \setminus (\bigcup_{r=1}^{k-1} F_r), \; k \geq 2.$$

Note that $F_k^* \in \mathcal{L}$ for every k, the F_k^*'s are disjoint and that

$$A_k \Delta F_k^* \subset \bigcup_{r=1}^{k} (A_r \Delta F_r) \quad \text{for each } k. \qquad \textbf{(3.4)}$$

From (3.4) we obtain

$$\pi_{n_k}(A_k \Delta F_k^*) = u \wedge \pi_{n_k}(A_k \Delta F_k^*) \leq u \wedge \sum_{r=1}^{k} (\bigvee_{t=1}^{\infty} a_{t,\varphi(t+r)}) \text{ for all } k$$

and by (3.2) we have

$$\pi_{n_k}(A_k) \le m_{n_k}(F_k^*) + \pi_{n_k}(A_k \Delta F_k^*) \le m_{n_k}(F_k^*) + \bigvee_{t=1}^{\infty} b_{t,\varphi(t)} \text{ for all } k. \quad \mathbf{(3.5)}$$

Since the m_n 's are globally uniformly (s)-bounded, there is a natural number k_0 with

$$m_n(F_k^*) \le \bigvee_{t=1}^{\infty} a_{t,\varphi(t)} \le \bigvee_{t=1}^{\infty} b_{t,\varphi(t)}$$

for every $k \ge k_0$ and $n \in \mathbb{N}$. From (3.5) we get $\pi_{n_k}(A_k) \le 2\bigvee_{t=1}^{\infty} b_{t,\varphi(t)}$ for each $k \ge k_0$, which is absurd, thanks to (3.3). Thus, the π_n's are globally uniformly (s)-bounded on $\sigma(\mathcal{L})$. By virtue of Theorem 1.1 applied to the π_n's, we get that they are globally *P*-uniformly absolutely continuous. From this we get the last assertion, taking the (*D*)-sequence $\beta_{t,l} = 2b_{t,l}$, t, $l \in \mathbb{N}$. □

In the finitely additive case we have the following

Theorem 1.4 (see Boccuto and Candeloro (2002b), Theorem 4.8) *Let* $(m_j)_j$ *be an equibounded sequence of globally uniformly* (s)*-bounded finitely additive measures, defined on an algebra* $\mathcal{L}$*, taking values in R and globally absolutely continuous with respect to a single finitely additive non-negative measure* $\lambda : \mathcal{L} \to \mathbb{R}$*. Then the* m_j*'s are globally uniformly* λ*-absolutely continuous.*

Proof: Let Q^* be the Stone space associated with $\mathcal{L}$, $\mathcal{Q}$ be the algebra of all open-closed subsets of Q^*, $\psi : \mathcal{L} \to \mathcal{Q}$ be the Stone isomorphism, and put

$$(\psi\lambda)(B) = \lambda(\psi^{-1}(B)), \quad (\psi m_j)(B) = m_j(\psi^{-1}(B)), \ B \in \mathcal{Q}.$$

Note that, for every *j*, ψm_j is globally absolutely continuous with respect to $\psi\lambda$ and globally uniformly (s)-bounded. Furthermore, $\psi\lambda$ and ψm_j are globally σ-additive for each *j*. Thanks to Theorem 1.3, the ψm_j's are globally uniformly absolutely continuous with respect to $\psi\lambda$ for every *j*. From this it follows that m_j is globally uniformly λ-absolutely continuous for each *j*. □

We now present a Vitali-Hahn-Saks-Nikodým-type theorem, in the global σ-additive case and assuming pointwise convergence of the involved measures with respect to a single regulator.

Theorem 1.5 (see Boccuto and Candeloro (2002b), Theorem 5.4) *Let $(m_j)_j$ be an equibounded sequence of R-valued globally σ-additive measures, defined on a σ-algebra Σ, $\mathcal{G} \subset \Sigma$ be a lattice closed under countable disjoint unions, and suppose that $(m_j)_j$ (RD)-converges on $\mathcal{G}$. Then the m_j's are globally uniformly (s)-bounded on $\mathcal{G}$.*

Proof: By virtue of the Fremlin lemma and global σ-additivity of the measures m_j, there exists a regulator $(a_{t,l})_{t,l}$ with the property that, for each decreasing sequence $(B_h)_h$ of elements of Σ with $\bigcap_{h=1}^{\infty} B_h = \emptyset$, for any $j \in \mathbb{N}$ and $\varphi \in \mathbb{N}^{\mathbb{N}}$ there is an integer h_0 with

$$v(m_j)(B_{h_0}) \leq \bigvee_{t=1}^{\infty} a_{t,\varphi(t)} \tag{3.6}$$

(when no confusion can arise, $v = v_{\Sigma}$). By (RD)-convergence, we can find a (D)-sequence $(b_{t,l})_{t,l}$ such that for each $A \in G$ and $\varphi \in \mathbb{N}^{\mathbb{N}}$ there is a natural number j_0 with

$$| m_j(A) - m_{j+p}(A) | \leq \bigvee_{t=1}^{\infty} b_{t,\varphi(t)} \tag{3.7}$$

whenever $j \geq j_0$ and $p \in \mathbb{N}$.

We now show global uniform (s)-boundedness. Let $c_{t,l} = a_{t,l} \vee b_{t,l}$, t, $l \in \mathbb{N}$: so $(6c_{t,l})_{t,l}$ is a (D)-sequence. We prove that for every disjoint sequence $(H_k)_k$ of elements of $\mathcal{G}$ and $\varphi \in \mathbb{N}^{\mathbb{N}}$ there is $k_0 \in \mathbb{N}$ with

$$\bigvee_{j \in \mathbb{N}} v_{\mathcal{G}}(m_j)(H_k) \leq 6 \bigvee_{i=1}^{\infty} c_{i,\varphi(i)}$$

for every $k \ge k_0$.

If not, then there are a disjoint sequence $(H_r)_r$ of elements of $\mathcal{G}$ and $\varphi \in \mathbb{N}^{\mathbb{N}}$ with

$$\bigvee_{j\in\mathbb{N}} v_{\mathcal{G}}(m_j)(H_r) \not\le 6 \bigvee_{t=1}^{\infty} c_{t,\varphi(t)} \tag{3.8}$$

for each $r \in \mathbb{N}$. Let $c := \bigvee_{t=1}^{\infty} c_{t,\varphi(t)}$. Since (3.8) is true for every r, we find a natural number n_1, with $v_{\mathcal{G}}(m_{n_1})(H_1) \not\le 6c$, and so there is a set $A_1 \in \mathcal{G}$, $A_1 \subset H_1$, with

$$| m_{n_1}(A_1) | \not\le 6c.$$

By virtue of (RD)-convergence, there exists $n_1' > n_1$ with

$$| m_n(A_1) - m_{n+p}(A_1) | \le \bigvee_{t=1}^{\infty} b_{t,\varphi(t)}$$

whenever $n \ge n_1'$ and $p \in \mathbb{N}$. Thanks to global σ-additivity of $m_1, m_2, \ldots, m_{n_1'}$, there is $r_1 > 1$ with

$$[v(m_1) \vee v(m_2) \vee \ldots \vee v(m_{n_1'})](\bigcup_{r \ge r_1} H_r) \le \bigvee_{t=1}^{\infty} a_{t,\varphi(t)}.$$

Since H_{r_1} fulfils (3.8), we find $n_2 > n_1'$ with $v_{\mathcal{G}}(m_{n_2})(H_{r_1}) \not\le 6c$. There is a set $A_2 \in \mathcal{G}$, $A_2 \subset H_{r_1}$, with

$$| m_{n_2}(A_2) | \not\le 6c.$$

Set $\mathcal{G}_2 = \{A_1, A_2, A_1 \cup A_2\}$. Thanks to (RD)-convergence, there is $n_2' > n_2$ with

$$| m_n(E) - m_{n+p}(E) | \le \bigvee_{t=1}^{\infty} b_{t,\varphi(t)}$$

whenever $n \geq n_2'$, $p \in \mathbb{N}$ and $E \in G_2$. Thanks to σ-additivity of $m_1, m_2, \ldots, m_{n_2'}$, there is $r_2 > r_1$ with

$$[v(m_1) \vee v(m_2) \vee \ldots \vee v(m_{n_2'})](\bigcup_{r \geq r_2} H_r) \leq \bigvee_{t=1}^{\infty} a_{t,\varphi(t)}.$$

Since H_{r_2} fulfils (3.8), we find an $n_3 > n_2'$ with $v_{\mathcal{G}}(m_{n_3})(H_{r_2}) \not\leq 6c$, and a set $A_3 \in \mathcal{G}$, $A_3 \subset H_{r_2}$, with

$$| m_{n_3}(A_3) | \not\leq 6c.$$

By induction, we get the existence of sequences $(r_l)_l$, $(n_l)_l$, $(n_l')_l$, $(A_l)_l$ with the property that for every $l \in \mathbb{N}$ we get:

$$n_{l+1} > n_l' > n_l, \; r_{l+1} > r_l, \; A_{l+1} \subset H_{r_l},$$

$$| m_{n_l}(A_l) | \not\leq 6c, \tag{3.9}$$

$$[v(m_1) \vee v(m_2) \vee \ldots \vee v(m_{n_l'})](\bigcup_{r \geq r_l} H_r) \leq \bigvee_{t=1}^{\infty} a_{t,\varphi(t)}, \tag{3.10}$$

$$| m_n(E) - m_{n+p}(E) | \leq \bigvee_{i=1}^{\infty} b_{t,\varphi(t)} \tag{3.11}$$

for all $n \geq n_l'$, $p \in \mathbb{N}$, $E \in \mathcal{G}_l$ (here $\mathcal{G}_l$ is the class of all sets, which are suitable finite unions of elements from the set $\{A_1, \ldots, A_l\}$. Put $A = \bigcup_{l=1}^{\infty} A_l$.

We will prove that the relation (3.7) applied to A yields a contradiction. For each $l \in \mathbb{N}$ we have

$$m_{n_l'}(A) - m_{n_{l+1}}(A) = [m_{n_l'}(A_1 \cup A_2 \cup \ldots \cup A_l) - m_{n_{l+1}}(A_1 \cup A_2 \cup \ldots \cup A_l)] +$$

$$+[m_{n_l'}(A_{l+1})-m_{n_{l+1}}(A_{l+1})]+[m_{n_l'}(\bigcup_{s=l+2}^{\infty}A_s)-m_{n_{l+1}}(\bigcup_{s=l+2}^{\infty}A_s)]. \quad \textbf{(3.12)}$$

Now we get:

$$|m_{n_l'}(A_1\cup A_2\cup\ldots\cup A_l)-m_{n_{l+1}}(A_1\cup A_2\cup\ldots\cup A_l)|\le\bigvee_{t=1}^{\infty}b_{t,\varphi(t)}$$

Thanks to (3.11). By (3.10), we have also

$$|m_{n_l'}(A_{l+1})|\le v(m_{n_l'})(\bigcup_{r\ge r_l}H_r)\le\bigvee_{t=1}^{\infty}a_{t,\varphi(t)},$$

$$|m_{n_l'}(\bigcup_{s=l+2}^{\infty}A_s)|\le\bigvee_{t=1}^{\infty}a_{t,\varphi(t)},$$

$$|m_{n_{l+1}}(\bigcup_{s=l+2}^{\infty}A_s)|\le\bigvee_{t=1}^{\infty}a_{t,\varphi(t)}.$$

By (3.7) applied to A, there is $l_0\in\mathbb{N}$ with

$$|m_{n_l'}(A)-m_{n_{l+1}}(A)|\le c$$

whenever $l\ge l_0$. From (3.12) we obtain

$$|m_{n_{l+1}}(A_{l+1})|\le 5c$$

for each $l\ge l_0$, getting a contradiction with (3.9). This ends the proof. □

Corollary 1.6 (see Boccuto and Candeloro (2002b), Corollary 5.5) *Let $(m_j)_j$ be an equibounded sequence of globally σ-additive R-valued measures, defined on a σ-algebra Σ and (RD)-convergent. Then the m_j's are globally uniformly σ-additive on Σ, and the limit measure is globally σ-additive.*

Proof: Global uniform (s)-boundedness is a consequence of Theorem 1.5 applied to $\mathcal{G}$. Global uniform σ-additivity follows from Corollary 3.15 of Chapter 2.

From this, since R is weakly σ-distributive, we get global σ-additivity of the limit measure. □

The next step is a Schur-type theorem. We first state a technical lemma.

Lemma 1.7 (see Boccuto, Dimitriou and Papanastassiou (2012c), Lemma 3.7) *Let $m_j : \mathcal{P}(\mathbb{N}) \to R$, $j \in \mathbb{N}$, be an equibounded sequence of σ-additive measures, and $m_0 : \mathcal{P}(\mathbb{N}) \to R$ be a set function, with (RD) $\lim_j m_j = m_0$. Then there exists a regulator $(c_{t,l})_{t,l}$ in R such that, for every $\varphi \in \mathbb{N}^{\mathbb{N}}$ and for all sequences $(j_s)_s$, $(p_s)_s$ in $\mathbb{N}$ with $j_s \geq s$ for every $s \in \mathbb{N}$, there is $\overline{s} \in \mathbb{N}$ with*

$$| m_{j_s}(A) - m_{j_s+p_s}(A) | \leq \bigvee_{t=1}^{\infty} c_{t,\varphi(t)} \tag{3.13}$$

for any $s \geq \overline{s}$ and $A \subset \mathbb{N}$.

Proof: Let

$$u = \bigvee_{j \in \mathbb{N}, A \subset \mathbb{N}} | m_j(A) |, \tag{3.14}$$

and for each $j \in \mathbb{N}$ let $(a_{t,l}^{(j)})_{t,l}$ be a (D)-sequence in connection with σ-additivity of m_j. For every $\varphi \in \mathbb{N}^{\mathbb{N}}$ and $j \in \mathbb{N}$, let us define $\xi_j : \mathbb{N} \to \mathbb{N}$ as follows: $\xi_j(n) = \varphi(n+j)$, for all $n \in \mathbb{N}$. So, for every $j \in \mathbb{N}$, the regulator $(a_{t,l}^{(j)})_{t,l}$ is such that in correspondence with ξ_j there exists $\overline{n} \in \mathbb{N}$ with

$$| m_j(A) | \leq \bigvee_{t=1}^{\infty} a_{t,\xi_j(t)}^{(j)} = \bigvee_{t=1}^{\infty} a_{t,\varphi(t+j)}^{(j)} \tag{3.15}$$

for all $A \subset [\overline{n}, +\infty[$ (here, the halflines and intervals are meant in $\mathbb{N}$). By (3.14) and (3.15) it follows that

$$| m_j(A) | \leq u \wedge \Big(\sum_{j=1}^{q} \big(\bigvee_{t=1}^{\infty} a_{t,\varphi(t+j)}^{(j)}\big)\Big) \tag{3.16}$$

for each $A \subset [\overline{n}, +\infty[$ and $q \in \mathbb{N}$.

By virtue of (3.16) and the Fremlin lemma there exists a (D)-sequence $(a_{t,l})_{t,l}$ such that for every $\varphi \in \mathbb{N}^{\mathbb{N}}$ and $j \in \mathbb{N}$ there is $\overline{n} \in \mathbb{N}$ with

$$| m_j(A) | \leq \bigvee_{t=1}^{\infty} a_{t,\varphi(t)} \tag{3.17}$$

for each $A \subset [\overline{n}, +\infty[$, that is

$$v_{\mathcal{P}(\mathbb{N})}(m_j)\,(\,[\overline{n}, +\infty\,[) \leq \bigvee_{t=1}^{\infty} a_{t,\varphi(t)}. \tag{3.18}$$

Hence the measures m_j are globally σ-additive with respect to the regulator $(a_{t,l})_{t,l}$.

Let $(b_{t,l})_{t,l}$ satisfy the condition of (RD)-convergence of the m_j's to m_0 on $\mathcal{P}(\mathbb{N})$. For each $\varphi \in \mathbb{N}^{\mathbb{N}}$, $n \in \mathbb{N}$ and for all sequences $(j_s)_s, (p_s)_s$ in $\mathbb{N}$ with $j_s \geq s$ for every s, there is $s_0 \in \mathbb{N}$ with

$$\bigvee_{A \subset [1,n]} | \lambda_s(A) | \leq 2 \bigvee_{t=1}^{\infty} b_{t,\varphi(t)} \tag{3.19}$$

for all $s \geq s_0$, where $\lambda_s = m_{j_s} - m_{j_s + p_s}$. From (3.19) and Lemma 3.4 of Chapter 2 it follows that

$$| \lambda_s(\{1\}) | + \ldots + | \lambda_s(\{n\}) | \leq 2 \bigvee_{t=1}^{\infty} b_{t,\varphi(t)} \tag{3.20}$$

for all $s \geq s_0$. Set now

$$c_{t,l} = 8 a_{t,l} + 12 b_{t,l} \tag{3.21}$$

for all t, $l \in \mathbb{N}$. We prove that the (D)-sequence $(c_{t,l})_{t,l}$ defined in (3.21) satisfies condition (3.13). Otherwise there are an element $\varphi \in \mathbb{N}^{\mathbb{N}}$ and two sequences $(j_s)_s$, $(p_s)_s$ in $\mathbb{N}$, with $j_s \ge s$ for each $s \in \mathbb{N}$, and such that for every $s \in \mathbb{N}$ there is a set $A_s \subset \mathbb{N}$ with

$$| \lambda_s(A_s) | \not\le \bigvee_{t=1}^{\infty} c_{t,\varphi(t)}. \tag{3.22}$$

Arguing analogously as in (3.15-3.18), since the λ_s's are equibounded and globally σ-additive with respect to the regulator $(2a_{t,l})_{t,l}$ and by virtue of the Fremlin lemma, in correspondence with the function φ satisfying (3.22) and λ_1 it is possible to find a natural number n_1 such that

$$v_{\mathcal{P}(\mathbb{N})}(\lambda_1)([n_1+1,+\infty[) \le 2\bigvee_{t=1}^{\infty} a_{t,\varphi(t)}. \tag{3.23}$$

By (RD)-convergence of the λ_j's to 0 with respect to the regulator $(2b_{t,l})_{t,l}$, which is an easy consequence of (RD)-convergence of $(m_j)_j$ to m_0 with respect to $(b_{t,l})_{t,l}$, proceeding analogously as in (3.19) and (3.20), in correspondence with n_1 there exists $s_1 \in \mathbb{N}$ such that

$$| \lambda_s(\{1\}) | + \ldots + | \lambda_s(\{n_1\}) | \le 2\bigvee_{t=1}^{\infty} b_{t,\varphi(t)} \tag{3.24}$$

for all $s \ge s_1$. Proceeding similarly as above, it is possible to associate to λ_{s_1} a natural number $n_2 > n_1$ such that

$$v_{\mathcal{P}(\mathbb{N})}(\lambda_{s_1})([n_2+1,+\infty[) \le 2\bigvee_{t=1}^{\infty} a_{t,\varphi(t)}, \tag{3.25}$$

and to find $s_2 > s_1$ with

$$| \lambda_s(\{1\}) | + \ldots + | \lambda_s(\{n_2\}) | \le 2\bigvee_{t=1}^{\infty} b_{t,\varphi(t)} \tag{3.26}$$

for all $s \ge s_2$. Proceeding by induction, we get the existence of two strictly increasing sequences $(n_h)_h$ and $(s_h)_h$ in $\mathbb{N}$ such that for all $h \in \mathbb{N}$ we have

$$v_{\mathcal{P}(\mathbb{N})}(\lambda_{s_h})([n_h+1,+\infty[) \leq 2\bigvee_{t=1}^{\infty} a_{t,\varphi(t)} \tag{3.27}$$

and

$$|\lambda_s(\{1\})|+\ldots+|\lambda_s(\{n_h\})| \leq 2\bigvee_{t=1}^{\infty} b_{t,\varphi(t)} \tag{3.28}$$

for every $s \geq s_h$. Set now $s_0 = 1$, $n_0 = 0$ and

$$A = \bigcup_h (A_{s_h} \cap [n_h+1, n_{h+1}]).$$

Note that $A \cap [1,n_1] = A_1 \cap [1,n_1]$. From this and taking into account the finite additivity of λ_1 we have:

$$\lambda_1(A) = \lambda_1(A_1 \cap [1,n_1]) + \lambda_1(A \cap [n_1+1,+\infty[),$$

$$\lambda_1(A_1) = \lambda_1(A_1 \cap [1,n_1]) + \lambda_1(A_1 \cap [n_1+1,+\infty[).$$

Observe now that $A \cap [n_h+1, n_{h+1}] = A_{s_h} \cap [n_h+1, n_{h+1}]$ for all $h \in \mathbb{N}$. From this and the finite additivity of λ_{s_h} we get:

$$\lambda_{s_h}(A) = \lambda_{s_h}(A \cap [1,n_h]) + \lambda_{s_h}(A_{s_h} \cap [n_h+1, n_{h+1}]) + \lambda_{s_h}(A \cap [n_{h+1}+1,+\infty[), \tag{3.29}$$

$$\lambda_{s_h}(A_{s_h}) = \lambda_{s_h}(A_{s_h} \cap [1,n_h]) + \lambda_{s_h}(A_{s_h} \cap [n_h+1, n_{h+1}]) + \lambda_{s_h}(A_{s_h} \cap [n_{h+1}+1,+\infty[) \tag{3.30}$$

for every $h \in \mathbb{N}$. From (3.27), (3.28), (3.29) and (3.30), for every $h \in \mathbb{N}$ we obtain:

$$|\lambda_{s_h}(A) - \lambda_{s_h}(A_{s_h})| \leq 4\bigvee_{t=1}^{\infty} a_{t,\varphi(t)} + 4\bigvee_{t=1}^{\infty} b_{t,\varphi(t)}. \tag{3.31}$$

By (RD)-convergence of the λ_j's to 0 with respect to the regulator $(2b_{t,l})_{t,l}$, in correspondence with A there exists a positive integer h_0 such that for every $h \geq h_0$ we get:

$$| \lambda_{s_h}(A) | \leq \bigvee_{t=1}^{\infty} 2 b_{t,\varphi(t)}. \quad \mathbf{(3.32)}$$

From (3.31) and (3.32), for all $h \geq h_0$ we have:

$$| \lambda_{s_h}(A_{s_h}) | \leq | \lambda_{s_h}(A) | + | \lambda_{s_h}(A) - \lambda_{s_h}(A_{s_h}) | \leq 8 \bigvee_{t=1}^{\infty} a_{t,\varphi(t)} + 12 \bigvee_{t=1}^{\infty} b_{t,\varphi(t)} = \bigvee_{t=1}^{\infty} c_{t,\varphi(t)}.$$

This is a contradiction with (3.22) and proves the lemma. □

We now turn to the Schur theorem (Boccuto, Dimitriou and Papanastassiou (2012c), Theorem 3.1, see also Boccuto and Candeloro (2002b), Corollary 5.6).

Theorem 1.8 *Let* $m_j : \mathcal{P}(\mathbb{N}) \to R$, $j \in \mathbb{N}$, *be an equibounded sequence of* σ*-additive measures, and let* $(RD) \lim_j m_j = m_0$. *Then*

$$(D) \lim_j \left(\sum_{n=1}^{\infty} | m_j(\{n\}) - m_0(\{n\}) | \right) = 0. \quad \mathbf{(3.33)}$$

Moreover m_0 *is globally* σ*-additive,* $(UD) \lim_j m_j = m_0$ *and the measures* m_j *are globally uniformly* σ*-additive.*

Proof: Proceeding analogously as at the beginning of Lemma 1.7, it is possible to check that the measures m_j, $j \in \mathbb{N}$, are globally σ-additive. Thus, thanks to Corollary 1.6, they are globally uniformly σ-additive, and m_0 is globally σ-additive. So we find a (D)-sequence $(a_{t,l})_{t,l}$ with the property that for each disjoint sequence $(H_k)_k$ in $\mathcal{P}(\mathbb{N})$ and $\varphi \in \mathbb{N}^{\mathbb{N}}$ there is $k_0 \in \mathbb{N}$ with

$$v(m_j)\Big(\bigcup_{k \geq k_0} H_k\Big) \leq \bigvee_{t=1}^{\infty} a_{t,\varphi(t)}$$

for all $j \in \mathbb{N} \cup \{0\}$.

By Lemma 3.4 of Chapter 2, for all j, $h \in \mathbb{N}$ we get:

$$\sum_{q=1}^{h} | m_j(\{q\}) - m_0(\{q\}) | \le \sum_{q=1}^{h} | m_j(\{q\}) | + \sum_{q=1}^{h} | m_0(\{q\}) | \tag{3.34}$$

$$\le 2 \bigvee_{A \cap [1,h]} | m_j(A) | + 2 \bigvee_{A \cap [1,h]} | m_0(A) | .$$

Since the measures m_j are equibounded, from (3.34) it follows that there is $v \in R$, $v \ge 0$, with

$$\sum_{q=1}^{h} | m_j(\{q\}) - m_0(\{q\}) | \le v \quad \text{for all } j, h \in \mathbb{N}. \tag{3.35}$$

By (RD)-convergence of $(m_j)_j$ to m_0 there exists a regulator $(\alpha_{t,l})_{t,l}$ such that, for every $q \in \mathbb{N}$ and $\varphi \in \mathbb{N}^{\mathbb{N}}$ there is a natural number j_0 with

$$| m_j(\{q\}) - m_0(\{q\}) | \le \bigvee_{t=1}^{\infty} \alpha_{t,\varphi(t+q)} \quad \text{for all } j \ge j_0. \tag{3.36}$$

From (3.35), (3.36) and the Fremlin lemma we get the existence of a (D)-sequence $(b_{t,l})_{t,l}$ with

$$\sum_{q=1}^{h} | m_j(\{q\}) - m_0(\{q\}) | \le v \wedge \left(\sum_{q=1}^{h} \left(\bigvee_{t=1}^{\infty} \alpha_{t,\varphi(t+q)}\right)\right) \le \bigvee_{t=1}^{\infty} b_{t,\varphi(t)} \tag{3.37}$$

for all $j \ge j_0$ and $h \in \mathbb{N}$. Thus (3.33) follows from (3.37).

We now prove (UD)-convergence. Let $A_{t,l} = a_{t,l} \vee b_{t,l}$, t, $l \in \mathbb{N}$. We claim that for every $\varphi \in \mathbb{N}^{\mathbb{N}}$ there is $j^* \in \mathbb{N}$ with

$$| m_j(F) - m_0(F) | \le 3 \bigvee_{t=1}^{\infty} A_{t,\varphi(t)}$$

for all $F \subset \mathbb{N}$ and $j \ge j^*$. Indeed, fix $\varphi \in \mathbb{N}^{\mathbb{N}}$, and take $H_k = \{k\}$ for each $k \in \mathbb{N}$. By global uniform σ-additivity, there is an integer k_0 with

$$v(m_j)([k_0,+\infty[) \le \bigvee_{t=1}^{\infty} a_{t,\varphi(t)}$$

for any $j \in \mathbb{N} \cup \{0\}$. Now there is an integer j^* with

$$\sum_{q=1}^{k_0} | m_j(\{q\}) - m_0(\{q\}) | \le \bigvee_{t=1}^{\infty} b_{t,\varphi(t)}$$

for all $j \ge j^*$. Thus, fixed arbitrarily $F \subset \mathbb{N}$, we have, for each $j \ge j^*$:

$$| m_j(F) - m_0(F) | \le | m_j(F \cap [1,k_0]) - m_0(F \cap [1,k_0]) | +$$

$$+ v(m_j)(F \cap [k_0+1,+\infty[) + v(m_0)(F \cap [k_0+1,+\infty[) \le$$

$$\bigvee_{t=1}^{\infty} b_{t,\varphi(t)} + 2\bigvee_{t=1}^{\infty} a_{t,\varphi(t)} \le 3\bigvee_{t=1}^{\infty} A_{t,\varphi(t)}.$$

This concludes the proof. □

We now give another proof of Theorem 1.8, which follows directly from Lemma 1.7 and is independent of Corollary 1.6., but uses Proposition 1.8 of Chapter 4.

Proof: Thanks to Lemma 1.7, there exists a regulator $(c_{t,l})_{t,l}$ such that, for each $\varphi \in \mathbb{N}^{\mathbb{N}}$ and for all sequences $(j_s)_s$, $(p_s)_s$ in $\mathbb{N}$ with $j_s \ge s$ for any s, there is $\overline{s} \in \mathbb{N}$ with

$$| m_{j_s}(A) - m_{j_s+p_s}(A) | \le \bigvee_{t=1}^{\infty} c_{t,\varphi(t)} \qquad \textbf{(3.38)}$$

for any $s \ge \overline{s}$ and $A \subset \mathbb{N}$. From (3.38) and Lemma 3.4 of Chapter 2 applied to the set functions $m_{j_s} - m_{j_s+p_s}$, $s \ge \overline{s}$, it follows that for all $\varphi \in \mathbb{N}^{\mathbb{N}}$ and for all sequences $(j_s)_s$, $(p_s)_s$ in $\mathbb{N}$ with $j_s \ge s$ for each s, there exists $\overline{s} \in \mathbb{N}$ such that

$$\sum_{n=1}^{q} | m_{j_s}(\{n\}) - m_{j_s+p_s}(\{n\}) | \le 2 \bigvee_{S \subset [1,q]} | m_{j_s}(S) - m_{j_s+p_s}(S) | \le 2\bigvee_{t=1}^{\infty} c_{t,\varphi(t)} \qquad \textbf{(3.39)}$$

for every $s \geq \overline{s}$ and $q \in \mathbb{N}$. Taking in (3.39) the supremum as q varies in $\mathbb{N}$, we obtain that for any $\varphi \in \mathbb{N}^{\mathbb{N}}$, for all sequences of natural numbers $(j_s)_s$, $(p_s)_s$ with $j_s \geq s$ for every s, there is $\overline{s} \in \mathbb{N}$ such that, whenever $s \geq \overline{s}$,

$$\sum_{n=1}^{\infty} | m_{j_s}(\{n\}) - m_{j_s+p_s}(\{n\}) | \leq 2 \bigvee_{t=1}^{\infty} c_{t,\varphi(t)}. \tag{3.40}$$

From (3.40) it follows that for any $\varphi \in \mathbb{N}^{\mathbb{N}}$ there exists $\overline{j}$ with the property that, for each $j \geq \overline{j}$ and $p \in \mathbb{N}$,

$$\sum_{n=1}^{\infty} | m_j(\{n\}) - m_{j+p}(\{n\}) | \leq 2 \bigvee_{t=1}^{\infty} c_{t,\varphi(t)}. \tag{3.41}$$

Otherwise, there is $\varphi \in \mathbb{N}^{\mathbb{N}}$ such that for any $s \in \mathbb{N}$ there are j_s, $p_s \in \mathbb{N}$ with $j_s \geq s$ and such that

$$\sum_{n=1}^{\infty} | m_{j_s}(\{n\}) - m_{j_s+p_s}(\{n\}) | \not\leq 2 \bigvee_{t=1}^{\infty} c_{t,\varphi(t)},$$

that is there exist $\varphi \in \mathbb{N}^{\mathbb{N}}$ and two sequences $(j_s)_s$, $(p_s)_s$ in $\mathbb{N}$ with $j_s \geq s$ for each $s \in \mathbb{N}$ and such that

$$\sum_{n=1}^{\infty} | m_{j_s}(\{n\}) - m_{j_s+p_s}(\{n\}) | \not\leq 2 \bigvee_{t=1}^{\infty} c_{t,\varphi(t)}$$

whenever $s \in \mathbb{N}$. This contradicts (3.40), and thus (3.41) is proved.

Set now, for all j, $n \in \mathbb{N}$, $a_n^{(j)} = m_j(\{n\})$, and for every $j \in \mathbb{N}$, put $a^{(j)} = (a_n^{(j)})_n$.

From (3.41) we get that the sequence $(a^{(j)})_j$ is Cauchy in l_R^1. As the space l_R^1 is complete, $(a^{(j)})_j$ is convergent in l_R^1, and thus there exist an element $a \in l_R^1$, $a = (a_n^{(0)})_n$, and a regulator $(\beta_{t,l})_{t,l}$ such that for every $\varphi \in \mathbb{N}^{\mathbb{N}}$ there is $\overline{j} \in \mathbb{N}$ with

$$\sum_{n=1}^{\infty} | m_j(\{n\}) - a_n^{(0)} | \leq \bigvee_{l=1}^{\infty} \beta_{t,\varphi(t)} \tag{3.42}$$

for every $j \geq \overline{j}$. Note that convergence of $(a^{(j)})_j$ in l_R^1 implies (D)-convergence of $(m_j(\{n\}))_j$ to $a_n^{(0)}$ for all $n \in \mathbb{N}$, and this limit is unique, thanks to weak σ-distributivity of R. Thus we get $a_n^{(0)} = m_0(\{n\})$ for every $n \in \mathbb{N}$.

We now prove that m_0 is σ-additive. For all $A \subset \mathbb{N}$ set

$$m^*(A) = (D)\lim_q \Big(\sum_{n \in A \cap [1,q]} m_0(\{n\}) \Big) = \sum_{n \in A} m_0(\{n\}). \tag{3.43}$$

By Proposition 2.1 of Chapter 2, the limit in (3.43) exists in R uniformly with respect to $A \subset \mathbb{N}$. This also implies that

$$\sum_{n \in A, n \geq q} m_0(\{n\}) = (D)\lim_l \Big(\sum_{n \in A \cap [q,l]} m_0(\{n\}) \Big)$$

exists in R for every $q \in \mathbb{N}$ and that

$$(D)\lim_q \sum_{n \in A, n \geq q} m_0(\{n\}) = 0 \tag{3.44}$$

uniformly with respect to $A \subset \mathbb{N}$. We now claim that

$$m^*(A) = m_0(A) \text{ for every } A \subset \mathbb{N}. \tag{3.45}$$

Indeed, we have:

$$m_0(A) - m^*(A) = (D)\lim_j [m_j(A) - m^*(A)] \tag{3.46}$$

$$= (D)\lim_j [(D)\lim_q \Big(\sum_{n \in A \cap [1,q]} (m_j(\{n\}) - m_0(\{n\})) \Big)]$$

for each $A \subset \mathbb{N}$; moreover from (3.42) it follows that to every $\varphi \in \mathbb{N}^{\mathbb{N}}$ a natural number $\overline{j}$ can be associated, with the property that, for every $j \geq \overline{j}$ and $A \subset \mathbb{N}$,

$$|(D)\lim_q \Big(\sum_{n\in A\cap[1,q]} (m_j(\{n\})-m_0(\{n\}))\Big)| \le \sum_{n\in A} |m_j(\{n\})-m_0(\{n\})| \tag{3.47}$$

$$\le \sum_{n=1}^{\infty} |m_j(\{n\})-m_0(\{n\})| \le \bigvee_{t=1}^{\infty} \beta_{t,\varphi(t)}.$$

From (3.46) and (3.47), since R is weakly σ-distributive, we get:

$$0 \le |m_0(A)-m^*(A)| \le \bigwedge_{\varphi\in\mathbb{N}^{\mathbb{N}}} \Big(\bigvee_{t=1}^{\infty} \beta_{t,\varphi(t)}\Big) = 0$$

for all $A \subset \mathbb{N}$; thus we get (3.45). From (3.43), (3.44) and (3.45) we have, uniformly with respect to $A \subset \mathbb{N}$:

$$(D)\lim_q |m^*(A\cap[q,+\infty[)| = (D)\lim_q |\sum_{n\in A, n\ge q} m^*(\{n\})| \tag{3.48}$$

$$= (D)\lim_q |\sum_{n\in A, n\ge q} m_0(\{n\})| = 0.$$

From (3.48) it follows that $(D)\lim_q (\bigvee\{|m^*(B)| : B \subset [q,+\infty[\}) = 0$, namely

$$(D)\lim_q v_{\mathcal{P}(\mathbb{N})}(m^*)([q,+\infty[) = 0. \tag{3.49}$$

From (3.49), by applying Proposition 1.8 of Chapter 4, it will follow that m^* is σ-additive on $\mathcal{P}(\mathbb{N})$, and hence m_0 is σ-additive too, since m_0 coincides with m^*.

From σ-additivity of m_0 and (3.47) it follows that to every $\varphi\in\mathbb{N}^{\mathbb{N}}$ there corresponds $\overline{j}\in\mathbb{N}$ such that, for any $j \ge \overline{j}$ and $A \subset \mathbb{N}$, we have

$$|m_j(A)-m_0(A)| = |(D)\lim_q \Big(\sum_{n\in A\cap[1,q]} (m_j(\{n\})-m_0(\{n\}))\Big)| \le \bigvee_{t=1}^{\infty} \beta_{t,\varphi(t)},$$

and hence

$$(UD)\lim_j m_j = m_0. \tag{3.50}$$

By virtue of (3.50) there exists a regulator $(h_{t,l})_{t,l}$ such that to every $\varphi \in \mathbb{N}^{\mathbb{N}}$ a positive integer $\overline{j}$ can be associated, with

$$| m_j(A) - m_0(A) | \le \bigvee_{t=1}^{\infty} h_{t,\varphi(t)} \tag{3.51}$$

whenever $j \ge \overline{j}$ and $A \subset \mathbb{N}$. Moreover, by virtue of σ-additivity of the measures m_0 and m_j, their equiboundedness and the Fremlin lemma, arguing analogously as in (3.15-3.17), we get the existence of a (D)-sequence $(f_{t,l})_{t,l}$ such that, for each $\varphi \in \mathbb{N}^{\mathbb{N}}$ and j there is $\overline{n} = \overline{n}(\varphi, j) \in \mathbb{N}$ with

$$| m_j(A) | \le \bigvee_{t=1}^{\infty} f_{t,\varphi(t)} \text{ for every } A \subset [\overline{n}, +\infty[. \tag{3.52}$$

Fix arbitrarily $\varphi \in \mathbb{N}^{\mathbb{N}}$, and let $\overline{j}$ be as in (3.51). In correspondence with φ and $j = 0, 1, \dots, \overline{j} - 1$, there exist $\overline{n}_0, \overline{n}_1, \dots, \overline{n}_{\overline{j}-1}$ as in (3.52). Set $n^* = \max\{\overline{n}_0, \overline{n}_1, \dots, \overline{n}_{\overline{j}-1}\}$: we have

$$| m_j(A) | \le \bigvee_{t=1}^{\infty} f_{t,\varphi(t)} \quad \text{for each } A \subset [n^*, +\infty[. \tag{3.53}$$

Moreover, for every $j \ge \overline{j}$ and $A \subset [n^*, +\infty[$, we get

$$| m_j(A) | \le | m_j(A) - m_0(A) | + | m_0(A) | \le \bigvee_{t=1}^{\infty} h_{t,\varphi(t)} + \bigvee_{t=1}^{\infty} f_{t,\varphi(t)}. \tag{3.54}$$

Global uniform σ-additivity of the m_j's and global σ-additivity of m_0 follow from (3.53) and (3.54). □

We now give a finitely additive version of Theorem 1.5.

Corollary 1.9 (see Boccuto and Candeloro (2002b), Corollary 5.7) *Let* $(m_j)_j$ *be an equibounded sequence of finitely additive R-valued measures, defined on a* σ*-algebra* Σ*, and* (RD)*-convergent to* m_0*. If the* m_j*'s are globally absolutely continuous relatively to a finitely additive non-negative measure* $\lambda:\Sigma\to\mathbb{R}$*, then the* m_j*'s are globally uniformly* (s)*-bounded and globally uniformly* λ*-absolutely continuous.*

Proof: Thanks to Theorem 1.4, it will be enough to prove global uniform (s)-boundedness. By the Fremlin lemma there exists a regulator $(a_{t,l})_{t,l}$ with the property that, for each disjoint sequence $(H_k)_k$ from Σ, $\varphi\in\mathbb{N}^{\mathbb{N}}$ and $j\in\mathbb{N}$ there is $k_0\in\mathbb{N}$ with

$$v(m_j)(H_k)\le\bigvee_{t=1}^{\infty}a_{t,\varphi(t)}$$

whenever $k\ge k_0$. By virtue of (RD)-convergence, we find a (*D*)-sequence $(b_{t,l})_{t,l}$ with the property that for each $A\in\Sigma$ and $\varphi\in\mathbb{N}^{\mathbb{N}}$ there exists $j_0\in\mathbb{N}$ with

$$|\,m_j(A)-m_0(A)\,|\le\bigvee_{t=1}^{\infty}b_{t,\varphi(t)}\ \text{for any}\ j\ge j_0.$$

We now prove global uniform (s)-boundedness. Let $c_{t,l}=a_{t,l}\vee b_{t,l}$, t, $l\in\mathbb{N}$. We prove that the (*D*)-sequence $(6c_{t,l})_{t,l}$ satisfies the condition of uniform (s)-boundedness. Otherwise there exist a disjoint sequence $(H_k)_k$ of sets from Σ, an element $\varphi\in\mathbb{N}^{\mathbb{N}}$, and a strictly increasing sequence $(j_k)_k$ in $\mathbb{N}$, with

$$|\,m_{j_k}(H_k)\,|\not\le 6\bigvee_{t=1}^{\infty}c_{t,\varphi(t)} \tag{3.55}$$

whenever $k\in\mathbb{N}$. We find a subsequence $(H_{k_r})_r$ such that λ is σ-additive on the σ-algebra $\mathcal{A}$ generated by the H_{k_r}'s (see also Drewnowski (1972a) and Theorem 3.24 of Chapter 2). Thus the measures m_j are σ-additive on $\mathcal{A}$. Thanks to Corollary 1.6, the double sequence $(6c_{t,l})_{t,l}$ is a regulator associated with uniform (s)-boundedness of the m_j's on $\mathcal{A}$, getting a contradiction with (3.55). This concludes the proof. □

A consequence of the previous results is the following

Corollary 1.10 *Let $\lambda : \Sigma \to [0,+\infty]$ be a positive globally σ-additive measure, defined on a σ-algebra $\Sigma \subset \mathcal{P}(G)$. If $m_j : \Sigma \to R$, $j \in \mathbb{N}$, is a sequence of globally uniformly σ-additive and globally λ-absolutely continuous measures, then the m_j's are globally uniformly λ-absolutely continuous. Moreover, if the m_j's are globally uniformly λ-absolutely continuous, then they are globally uniformly σ-additive.*

We now prove a Brooks-Jewett-type theorem for globally (s)-bounded lattice-group valued measures.

Theorem 1.11 (see Boccuto and Candeloro (2004a), Theorem 6.8) *Let R be a super Dedekind complete and weakly σ-distributive (ℓ)-group, and $(m_j)_j$ be any equibounded sequence of globally (s)-bounded R-valued measures, defined on a σ-algebra Σ. Assume that there exists (RD) $\lim_j m_j = m_0$ on Σ. Then the measures m_j are globally uniformly (s)-bounded on Σ, and m_0 is globally (s)-bounded on Σ.*

Proof: Thanks to global (s)-boundedness and equiboundedness of the m_j's, and by the Fremlin Lemma, there exists a regulator $(a_{t,l})_{t,l}$ such that, for each disjoint sequence $(C_k)_k$ in Σ, every $\varphi \in \mathbb{N}^{\mathbb{N}}$ and $n \in \mathbb{N}$, there is a positive integer k_0, with $v(m_n)(C_k) \leq \bigvee_{t=1}^{\infty} a_{t,\varphi(t)}$ whenever $k \geq k_0$. Moreover, thanks to (RD)-convergence, there exists a regulator $(b_{t,l})_{t,l}$ such that, for every $A \in \Sigma$ and $\varphi \in \mathbb{N}^{\mathbb{N}}$, there exists an integer j_0 with $| m_j(A) - m_0(A) | \leq \bigvee_{t=1}^{\infty} b_{t,\varphi(t)}$ for every $j \geq j_0$.

We prove global uniform (s)-boundedness. Setting $c_{t,l} := a_{t,l} \vee b_{t,l}$, t, $l \in \mathbb{N}$, then $(6c_{t,l})_{t,l}$ is a regulator. We prove that it satisfies the condition of global uniform (s)-boundedness. Otherwise there are a disjoint sequence $(H_k)_k$ in Σ, a $\varphi \in \mathbb{N}^{\mathbb{N}}$, and a subsequence $(m_{n_k})_k$, with

$$| m_{n_k}(H_k) | \not\leq 6 \bigvee_{t=1}^{\infty} c_{t,\varphi(t)} \tag{3.56}$$

for every k. Using Corollary 3.26 of Chapter 2 and taking into account that, by super Dedekind completeness and weak σ-distributivity of R, (D)- and (O)-convergence are equivalent in this context, we find a subsequence $(H_{k_r})_r$ such that every m_j is σ-additive in the σ-algebra $\mathcal{A}$ generated by the H_{k_r}'s. Now, proceeding analogously as in Corollary 1.6, we see that the regulator $(6c_{t,l})_{t,l}$ satisfies the condition of global uniform (s)-boundedness of the measures m_n in $\mathcal{A}$, obtaining a contradiction with (3.56).

Global (s)-boundedness of m_0 follows easily from global uniform (s)-boundedness of the m_j's and weak σ-distributivity of R. This ends the proof. □

A consequence of the Stone Isomorphism technique and Theorems 1.10, 2.2 of Chapter 2 and 3.29 of Chapter 2 is the following

Theorem 1.12 *Let R be a super Dedekind complete and weakly σ-distributive (ℓ)-group, and $m_n : \Sigma \to R$, $n \in \mathbb{N}$, be an equibounded sequence of globally (s)-bounded finitely additive measures, (RO)-convergent to a set function m. Then the sequence $(M_n(B))_n$ of the Stone extensions of the m_n's is (RO)-convergent to $M(B)$ for each $B \in \sigma(\mathcal{Q})$, where $\mathcal{Q}$ is the Stone algebra associated with Σ, $\sigma(\mathcal{Q})$ is the σ-algebra generated by $\mathcal{Q}$ and M is the Stone extension of m defined on $\sigma(\mathcal{Q})$.*

3.1.2. The Dieudonné Theorem

Now we give some versions of Dieudonné theorems. Let $\mathcal{G}$, $\mathcal{H} \subset \Sigma$ be two lattices of subsets of G, such that the complement of each element of $\mathcal{H}$ belongs to $\mathcal{G}$, and $\mathcal{G}$ is closed under countable disjoint unions. We begin with the following result.

Proposition 1.13 (see also Boccuto and Candeloro (2002a), Proposition 2.6) *Let $m_j : \Sigma \to R$, $j \in \mathbb{N}$, be a sequence of equibounded globally regular measures. Then there exists a regulator $(p_{t,l})_{t,l}$ such that for every $A \in \Sigma$ there exist two sequences $(F_n)_n$ in $\mathcal{H}$, $(G_n)_n$ in $\mathcal{G}$, with $F_n \subset F_{n+1} \subset A \subset G_{n+1} \subset G_n$ for every $n \in \mathbb{N}$, and the double sequence $v(m_j)(G_n \setminus F_n)$, n, $j \in \mathbb{N}$, (RD)-converges to 0 with respect to $(p_{t,l})_{t,l}$.*

Proof: Set

$$u := \bigvee_j [\bigvee_{A \in \Sigma} |m_j(A)|]. \quad \textbf{(3.57)}$$

By hypothesis, for every $A \in \Sigma$ and $j \in \mathbb{N}$ there exist a regulator $(\gamma_{t,l}^{(j)})_{t,l}$ and two sequences $(G_n^{(j)})_n$, $(F_n^{(j)})_n$ such that $F_n^{(j)} \in \mathcal{H}$, $G_n^{(j)} \in \mathcal{G}$ for all j, $n \in \mathbb{N}$, and

$$F_n^{(j)} \subset F_{n+1}^{(j)} \subset A \subset G_{n+1}^{(j)} \subset G_n^{(j)} \text{ for all } j,\ n \in \mathbb{N}, \quad \textbf{(3.58)}$$

and for every $\varphi \in \mathbb{N}^{\mathbb{N}}$ and $j \in \mathbb{N}$ we find $n_0 = n_0(\varphi, j) \in \mathbb{N}$, with

$$v(m_j)(G_n^{(j)} \setminus F_n^{(j)}) \leq \bigvee_{t=1}^{\infty} \gamma_{t,\varphi(t)}^{(j)} \quad \textbf{(3.59)}$$

for every $n \geq n_0$. Moreover, thanks to the Fremlin lemma, there exists a (D)-sequence $(p_{t,l})_{t,l}$ with

$$u \wedge (\sum_{j=1}^{q} (\bigvee_{t=1}^{\infty} \gamma_{t,\varphi(t+j)}^{(j)})) \leq \bigvee_{t=1}^{\infty} p_{t,\varphi(t)} \quad \text{for all } q \in \mathbb{N} \text{ and } \varphi \in \mathbb{N}^{\mathbb{N}}, \quad \textbf{(3.60)}$$

where u is as in (3.57).

For every $n \in \mathbb{N}$, set $G_n = \bigcap_{j \le n} G_n^{(j)}$; $F_n = \bigcup_{j \le n} F_n^{(j)}$: then $G_n \in \mathcal{G}$, $F_n \in \mathcal{H}$, and $A \subset G_n$, $A \supset F_n$ for all $n \in \mathbb{N}$. Note that $G_{n+1} \subset G_n$ and $F_{n+1} \supset F_n$ for every n. Since $G_n \setminus F_n \subset G_n^{(j)} \setminus F_n^{(j)}$ for all j, $n \in \mathbb{N}$, then for every $\varphi \in \mathbb{N}^{\mathbb{N}}$ and j there exists $n_0 = n_0(\varphi, j) \in \mathbb{N}$, with $v(m_j)(G_n \setminus F_n) \le \bigvee_{t=1}^{\infty} p_{t,\varphi(t)}$ for all $n \ge n_0$. This ends the proof. □

We will give a Dieudonné-type theorem for globally regular lattice group-valued measures. We begin with the following

Lemma 1.14 (see Boccuto and Candeloro (2002a), Lemma 3.1) *Let R, G, Σ, $\mathcal{G}$, $\mathcal{H}$ be as in Proposition 1.13, and $m : \Sigma \to R$ be a globally regular bounded measure. Then for each $A \in \Sigma$ and $V \in \mathcal{G}$ we get*

$$v_{\Sigma}(m)(A) = v_{\mathcal{H}}(m)(A), \tag{3.61}$$

$$v_{\Sigma}(m)(V) = v_{\mathcal{G}}(m)(V). \tag{3.62}$$

Proof: The equality (3.61) follows from global regularity and weak σ-distributivity of R.

So, let $V \in \mathcal{G}$, $(\gamma_{t,l})_{t,l}$ be a (D)-sequence in connection with global regularity, $B \in \Sigma$, $B \subset V$, and choose $\varphi \in \mathbb{N}^{\mathbb{N}}$. Then there is a set $G \in \mathcal{G}$, $G \supset B$ with $v(m)(G \setminus B) \le \bigvee_{t=1}^{\infty} \gamma_{t,\varphi(t)}$, and hence

$$| m(B) | \le | m(G) | + \bigvee_{t=1}^{\infty} \gamma_{t,\varphi(t)}.$$

Without restriction, we assume $G \subset V$. Thus we obtain

$| m(B) | \le v_{\mathcal{G}}(m)(V) + \bigvee_{t=1}^{\infty} \gamma_{t,\varphi(t)}$. By arbitrariness of B, we deduce

$$v_{\Sigma}(m)(V) \le v_{\mathcal{G}}(m)(V) + \bigvee_{t=1}^{\infty} \gamma_{t,\varphi(t)}.$$

Finally, since R is weakly σ-distributive, we get $v_\Sigma(m)(V) \le v_{\mathcal{G}}(m)(V)$, and so the two elements coincide. Thus (3.62) is proved. □

Lemma 1.15 (see also Boccuto and Candeloro (2002a), Lemma 3.2) *Let* $m_j : \Sigma \to R$, $j \in \mathbb{N}$, *be an equibounded sequence of measures, globally regular and globally uniformly* (s)*-bounded on* $\mathcal{G}$ *(relatively to a* (D)*-sequence* $(b_{t,l})_{t,l}$*).*

Then the m_j*'s are globally uniformly regular and globally uniformly* (s)*-bounded on* Σ.

Proof: Let $(K_n)_n$ be any disjoint sequence from Σ. Observe that the conditions of Proposition 1.13 are fulfilled. Let $(p_{t,l})_{t,l}$ be the same regulator as in that proposition, define u as in (3.57), and let $(d_{t,l})_{t,l}$ be a (D)-sequence such that

$$u \wedge \left(\sum_{h=1}^{q}\left(\bigvee_{t=1}^{\infty} p_{t,\varphi(t+h)}\right)\right) \le \bigvee_{t=1}^{\infty} d_{t,\varphi(t)} \qquad \text{for all } q \in \mathbb{N} \text{ and } \varphi \in \mathbb{N}^{\mathbb{N}}.$$

Set $e_{t,l} = 2(b_{t,l} + d_{t,l})$, $t, l \in \mathbb{N}$. We prove that $(D)\lim_n (\bigvee_j v(m_j)\ (K_n)) = 0$ with respect to the regulator $(e_{t,l})_{t,l}$. Otherwise there exists $\varphi \in \mathbb{N}^{\mathbb{N}}$ such that for every $k \in \mathbb{N}$ there exist $n_k \ge k$, $j_k \in \mathbb{N}$, $A_k \in \Sigma$ with $A_k \subset K_{n_k}$ and

$$| m_{j_k}(A_k) | \not\le \bigvee_{t=1}^{\infty} e_{t,\varphi(t)}. \qquad \textbf{(3.63)}$$

By (3.61), we may suppose that $A_k \in H$ for each $k \in \mathbb{N}$.

Fix $k \in \mathbb{N}$, and let $b = \bigvee_{t=1}^{\infty} b_{t,\varphi(t)}$, $e = \bigvee_{t=1}^{\infty} e_{t,\varphi(t)}$. Note that, by global regularity of the m_j's, there are $G_k \in \mathcal{G}$, $F_k \in \mathcal{H}$ with $A_k \subset G_k \subset F_k$ and

$$[v(m_1) \vee \ldots \vee v(m_{j_k})]\,(F_k \setminus A_k) \le \bigvee_{t=1}^{\infty} p_{t,\varphi(t+k)}.$$

Put $G_1^* = G_1$, $G_k^* = G_k \setminus (\bigcup_{h=1}^{k-1} F_h)$, $k \geq 2$. The G_k^*'s are disjoint elements of $\mathcal{G}$, and so there is $k_0 \in \mathbb{N}$ with $\bigvee_j v(m_j)(G_k^*) \leq b$ for all $k \geq k_0$. Now, as $A_{k+1} \setminus G_{k+1}^* \subset \bigcup_{h=1}^{k} (F_h \setminus A_h)$ for every k, we get:

$$|m_{j_k}(A_k)| \leq |m_{j_k}(A_k \cap G_k^*)| + |m_{j_k}(A_k \setminus G_k^*)|$$

$$\leq b + u \wedge (\sum_{h=1}^{k} (\bigvee_{t=1}^{\infty} p_{t,\varphi(t+h)})) \leq e \quad \text{for all } k \geq k_0.$$

This contradicts (3.63). So, the m_j's are globally uniformly (s)-bounded on Σ.

Now we prove global uniform regularity. For any $A \in \Sigma$ there are two sequences $(F_n)_n$ in $\mathcal{H}$ and $(G_n)_n$ in $\mathcal{G}$, fulfilling global regularity relatively to $(p_{t,l})_{t,l}$. Since $(G_n \setminus F_n)_n$ is decreasing, by Propositions 1.13 and 3.14 of Chapter 2 and weak σ-distributivity of R, the sequence

$$\bigvee_j v(m_j)(G_n \setminus F_n), \; n \in \mathbb{N}, \tag{3.64}$$

(D)-converges to 0. This ends the proof. □

We now turn to the Dieudonné theorem, in both σ-additive and finitely additive cases.

Theorem 1.16 (see also Boccuto and Candeloro (2002a), Theorem 3.3) *Let* G, R, Σ, $\mathcal{G}$, $\mathcal{H}$ *be as above, and suppose that* $m_j : \Sigma \to R$, $j \in \mathbb{N}$, *are equibounded, globally regular and globally* σ*-additive measures on* Σ*, such that there is in* R *the limit* $m_0 = (RD) \lim_j m_j$ *on* $\mathcal{G}$*. Then we get:*

(a) *The measures* m_j, $j \in \mathbb{N}$, *are globally uniformly regular and globally uniformly* (s)*-bounded on* Σ.

(b) *The limit* $m_0 = (RD)\lim_j m_j$ *exists in R on the whole of* Σ.

(c) *The measures* m_j *are globally uniformly* σ*-additive on* Σ.

(d) *The measure* m_0 *is globally regular and globally* σ*-additive on* Σ.

Proof: (a) By Theorem 1.5, the measures m_j are globally uniformly (s)-bounded on $\mathcal{G}$; so, by Lemma 1.15 we get global uniform regularity and global uniform (s)-boundedness of the measures m_j on Σ.

(b) Fix $A \in \Sigma$, and let $(y_{t,l})_{t,l}$ be a (D)-sequence associated with global uniform regularity. For any $\varphi \in \mathbb{N}^{\mathbb{N}}$ there exists $G \in \mathcal{G}$ with $A \subset G$ and $v(m_j)(G \setminus A) \le \bigvee_{t=1}^{\infty} y_{t,\varphi(t)}$ for all j.

In correspondence with G we find a $j_0 \in \mathbb{N}$ with

$$| m_j(G) - m_{j+p}(G) | \le \bigvee_{t=1}^{\infty} \alpha_{t,\varphi(t)} \quad \text{for all } j \ge j_0 \text{ and } p \in \mathbb{N},$$

where $(\alpha_{t,l})_{t,l}$ is a regulator associated with (RD)-convergence on $\mathcal{G}$. So we get

$$| m_j(A) - m_{j+p}(A) | \le 2\bigvee_{t=1}^{\infty} y_{t,\varphi(t)} + \bigvee_{t=1}^{\infty} \alpha_{t,\varphi(t)} \quad \text{for all } j \ge j_0 \text{ and } p \in \mathbb{N}. \quad \textbf{(3.65)}$$

From (3.65) it follows that the sequence $(m_j(A))_j$ is a (D)-Cauchy sequence in R. By Dedekind completeness of R, and since Dedekind completeness implies (D)-completeness (see also Chapter 2), then the sequence $(m_j(A))_j$ is (D)-convergent. Thus (b) is proved. The statement (c) follows from (b) and Corollary 1.6, while (d) is a consequence of (a), (b), (c) and weak σ-distributivity of R. □

Theorem 1.17 (see also Boccuto and Candeloro (2002a), Theorem 3.4) *Let* G, R, Σ, $\mathcal{G}$, $\mathcal{H}$ *be as in Proposition* 1.13, *and* $m_j : \Sigma \to R$, $j \in \mathbb{N}$, *be equibounded globally regular measures, globally absolutely continuous relatively to a real-*

valued, non-negative and finitely additive measure λ *on* Σ. *Let the limit* $m_0 = (RD) \lim_j m_j$ *exist in* R *on* $\mathcal{G}$.

Then we get:

1.17.1) *The measures* m_j, $j \in \mathbb{N}$, *are globally uniformly regular, globally uniformly* λ*-absolutely continuous and globally uniformly* (s)*-bounded on* Σ.

1.17.2) *The limit* $m_0 = (RD) \lim_j m_j$ *exists in* R *on* Σ.

1.17.3) m_0 *is globally* (s)*-bounded, globally regular and globally* λ*-absolutely continuous on* Σ.

Proof: 1.17.1) Let $(\alpha_{t,l})_{t,l}$ be a (D)-sequence associated with (RD)-convergence in $\mathcal{G}$, and let $(\beta_{t,l})_{t,l}$ be a (D)-sequence (whose existence is guaranteed by global absolute continuity and the Fremlin Lemma) with the property that, for any disjoint sequence $(H_k)_k$ in Σ, $j \in \mathbb{N}$ and $\varphi \in \mathbb{N}^{\mathbb{N}}$, there exists $k_0 \in \mathbb{N}$ with $v(m_j)(H_k) \le \bigvee_{t=1}^{\infty} \beta_{t,\varphi(t)}$ whenever $k \ge k_0$.

Let now $c_{t,l} = \alpha_{t,l} \vee \beta_{t,l}$, t, $l \in \mathbb{N}$. We prove that $(6c_{t,l})_{t,l}$ satisfies the condition of global uniform (s)-boundedness of the measures m_j on $\mathcal{G}$. Otherwise we find a disjoint sequence $(G_k)_k$ in $\mathcal{G}$, a $\varphi \in \mathbb{N}^{\mathbb{N}}$ and a subsequence $(j_k)_k$ in $\mathbb{N}$ with

$$| m_{j_k}(G_k) | \not\le 6 \bigvee_{t=1}^{\infty} c_{t,\varphi(t)} \tag{3.66}$$

for each $k \in \mathbb{N}$. Let V be the union of the G_k's and $\mathcal{A}$ be the σ-algebra generated by the G_k's in V: the sequence $m_j : \mathcal{A} \to R$, $j \in \mathbb{N}$, (RD)-converges to m_0. From Corollary 1.9 we deduce global $\mathcal{A}$-uniform (s)-boundedness of the measures m_j relatively to $(6c_{t,l})_{t,l}$, and this contradicts (3.66). Thus, the measures m_j are globally uniformly (s)-bounded on $\mathcal{G}$, and so they are globally uniformly (s)-bounded on Σ and globally uniformly regular, by Lemma 1.15, and globally uniformly λ-absolutely continuous, thanks to Theorem 1.4.

The proof of 1.17.2) is analogous to the argument in Theorem 1.16.

1.17.3) Global (s)-boundedness, global regularity and global absolute continuity of m_0 on Σ are easy consequences of 1.17.1), 1.17.2) and weak σ-distributivity of R. □

3.2. Construction of Integrals

3.2.1. Bochner-Type Integrals

We now present a Bochner-type integral for Riesz space-valued functions, in which it is possible to apply the results obtained in the previous section, because the defined integrals turn to be globally σ-additive measures. We study the main properties of this integral, which is defined with respect to extended real-valued measures, not necessarily finite, and give some Vitali-type theorems.

Under the same above hypotheses and notations, let $\lambda : \Sigma \to [0,+\infty]$ denote a σ-additive measure, and assume that R is a Dedekind complete and weakly σ-distributive Riesz space.

We say that a function $f : G \to R$ is *integrable simple* iff it is of the type $f = \sum_{i=1}^{n} a_i \chi_{A_i}$, where $n \in \mathbb{N}$, $A_i \in \Sigma$ and $\lambda(A_i) < +\infty$, $i = 1,2,\ldots,n$.

If f is an integrable simple function, $f = \sum_{i=1}^{n} a_i \chi_{A_i}$, then, as in the classical setting, put

$$\int_E f\,d\lambda = \sum_{i=1}^{n} a_i\,\lambda(A_i \cap E), \quad \text{for all } E \in \Sigma.$$

The integral is well-defined and the function $\int_{\cdot} |f|\,d\lambda$ is globally λ-absolutely continuous.

A function sequence $(f_n)_n$ in R^G *converges in measure* to $f \in R^G$ (with respect to λ) iff there is a regulator $(a_{t,l})_{t,l}$ in R such that, for all $\varphi \in \mathbb{N}^{\mathbb{N}}$, there exists a sequence $(C_n)_n$ in Σ, with

$$\{g \in G : | f_n(g) - f(g) | \not\leq \bigvee_{t=1}^{\infty} a_{t,\varphi(t)}\} \subset C_n \quad \text{for all } n \in \mathbb{N} \tag{3.67}$$

and

$$\lim_n \lambda(C_n) = 0. \tag{3.68}$$

A sequence $(f_n)_n$ of integrable simple functions is said to *converge in* L^1 to f_0 iff $(D)\lim_n \int_G | f_n - f_0 | d\lambda = 0$, and *Cauchy in* L^1 iff $(D)\lim_n \int_G | f_n - f_{n+p} | d\lambda = 0$ uniformly with respect to $p \in \mathbb{N}$. A sequence $(f_n)_n$ is *uniformly integrable* iff there is a regulator $(z_{t,l})_{t,l}$ such that for every $\varphi \in \mathbb{N}^{\mathbb{N}}$ there exists $\delta_0 \in \mathbb{R}^+$ with the property that: for each $E \in \Sigma$ with $\lambda(E) \leq \delta_0$, we have $\int_E | f_n | d\lambda \leq \bigvee_{t=1}^{\infty} z_{t,\varphi(t)}$, that is iff the integrals $\int_{\cdot} | f_n | d\lambda$ are globally uniformly λ-absolutely continuous. We now prove the following technical results.

Proposition 2.1 (see Boccuto and Candeloro (2002c), Proposition 3.1) *If $(f_n)_n$ is a Cauchy in L^1 sequence of integrable simple functions, then the integrals $\int_{\cdot} | f_n | d\lambda$ are equibounded and globally uniformly λ-absolutely continuous.*

Proof: The sequence $(\int_G | f_n | d\lambda)_n$ is bounded: indeed it is Cauchy, since $(f_n)_n$ is Cauchy in L^1 and

$$| \int_G (| f_n | - | f_q |) d\lambda | \leq \int_G | f_n - f_q | d\lambda \tag{3.69}$$

for each n, $q \in \mathbb{N}$. Furthermore, for any $n \in \mathbb{N}$ the measure $\int_{\cdot} | f_n | d\lambda$ is globally λ-absolutely continuous, and hence globally σ-additive. By the Cauchy condition, we get (RD)-convergence of $(\int_A f_n d\lambda)_n$ for any $A \in \Sigma$. By Theorem

1.5, the measures $\int f_n \, d\lambda$ are uniformly (s)-bounded, and so thanks to Corollary 1.10 we get the assertion. □

Corollary 2.2 (see Boccuto and Candeloro (2002c), Corollary 3.2) *Let $(f_n)_n$ be a Cauchy in L^1 sequence of integrable simple functions, and set $\lambda(E) = (D)\lim_n \int_E |f_n| d\lambda$, $E \in \Sigma$. Then λ is globally σ-additive.*

Proof: Since the measures $(\int f_n \, d\lambda)_n$ are globally σ-additive and globally uniformly (s)-bounded, then they are globally uniformly σ-additive. The assertion follows from this and weak σ-distributivity of R. □

Definition 2.3 (see Boccuto and Candeloro (2002c), Definition 3.3) A function $f_0 \in R^G$ is *integrable* (with respect to λ) iff there exists a *defining sequence* for f_0, that is a Cauchy in L^1 sequence $(f_n)_n$ of integrable simple functions, which converges in measure to f_0. We set:

$$\int_E f_0 \, d\lambda = (D)\lim_n \int_E f_n \, d\lambda \quad \text{for all } E \in \Sigma. \tag{3.70}$$

Note that the limit in (3.70) exists uniformly with respect to $E \in \Sigma$, because Dedekind completeness implies (D)-completeness uniformly with respect to a parameter (see also Boccuto (1997), Theorem 2.16, Boccuto, Riečan and Vrábelová (2009), Duchoň and Riečan (1996), Luxemburg and Zaanen (1971), Vulikh (1967)).

If f_0 is integrable, then also $|f_0|$ is integrable, because if $(f_n)_n$ is a defining sequence for f_0, then $(|f_n|)_n$ is a defining sequence for $|f_0|$. We see that the integral in (3.70) is well-defined.

Theorem 2.4 (see Boccuto and Candeloro (2002c), Theorem 3.4) *Let $f_0 \in R^G$ be an integrable function, and $(f_n)_n$ and $(h_n)_n$ be two defining sequences for f_0. Then*

$$(D)\lim_n \int_E f_n \, d\lambda = (D)\lim_n \int_E h_n \, d\lambda \text{ for all } E \in \Sigma.$$

Proof: Observe that $(f_n - h_n)_n$ is Cauchy in L^1 and convergent in measure to 0, and by Proposition 2.1, $(|f_n - h_n|)_n$ is uniformly integrable. We prove that $(D)\lim_n \int_G |f_n - h_n| \, d\lambda = 0$. Let $g_n = f_n - h_n$, $(z_{t,l})_{t,l}$ and $(a_{t,l})_{t,l}$ be two regulators, associated with uniform integrability of $|g_n|$ and convergence in measure of $(g_n)_n$ to 0, respectively. Pick $\varphi \in \mathbb{N}^{\mathbb{N}}$, and let $(C_n)_n$ be as in (3.67) in connection with convergence in measure of $(g_n)_n$ to 0. For any $n \in \mathbb{N}$ and $x \in G \setminus C_n$ we get:

$$|g_n(x)| \le \bigvee_{t=1}^{\infty} a_{t,\varphi(t)}. \tag{3.71}$$

Fix arbitrarily $H \in \Sigma$ with $\lambda(H) < +\infty$. For any $\varphi \in \mathbb{N}^{\mathbb{N}}$ there is $n_0 \in \mathbb{N}$ such that, for all $n \ge n_0$,

$$\int_H |g_n| \, d\lambda \le \int_{H \setminus C_n} |g_n| \, d\lambda + \int_{H \cap C_n} |g_n| \, d\lambda \le \left(\bigvee_{t=1}^{\infty} a_{t,\varphi(t)}\right) \cdot \lambda(H) + \bigvee_{t=1}^{\infty} z_{t,\varphi(t)},$$

and hence,

$$(D)\lim_n \int_H |g_n| \, d\lambda = 0. \tag{3.72}$$

By Corollary 2.2, the measure $(D)\lim_n \int_\cdot |g_n| \, d\lambda$ are globally σ-additive. Thus, we have:

$$(D)\lim_n \int_F |g_n| \, d\lambda = 0 \tag{3.73}$$

for every $F \in \Sigma$ with σ-finite measure λ.

Since the functions g_n are integrable simple, we find a set $E_0 \in \Sigma$ of σ-finite measure λ with

$$\int_{G\setminus E_0} | g_n | d\lambda = 0, \tag{3.74}$$

and thus, thanks to (3.73) and (3.74), we have
$(D)\lim_n \int_G | g_n | d\lambda = (D)\lim_n \int_{E_0} | g_n | d\lambda = 0$, since E_0 has σ-finite measure λ.
This ends the proof. □

We can give the notions of convergence in L^1 and Cauchy in L^1 for sequences of integrable functions analogously as for integrable simple functions. Note that the sum of two integrable functions f and h is integrable, and

$$\int_E (f+h)d\lambda = \int_E f\,d\lambda + \int_E h\,d\lambda \quad \text{for all } E \in \Sigma.$$

Moreover observe that, by virtue of Proposition 2.2 and Theorem 2.4, the measure $\int_{\cdot} f\,d\lambda$ is globally σ-additive for any integrable function f. We now prove the following:

Lemma 2.5 (see Boccuto and Candeloro (2002c), Lemma 3.6) *If f is integrable, then he measure $\int_{\cdot} | f | d\lambda$ is globally λ-absolutely continuous.*

Proof: Let $(f_n)_n$ be a defining sequence for f, then $(| f_n |)_n$ is a defining sequence for $| f |$. If $E \in \Sigma$ and $\lambda(E) = 0$, then by global λ-absolute continuity of the integrals $\int_{\cdot} | f_n | \, d\lambda$ we get:

$$\int_E | f_n | d\lambda = 0 \quad \text{for all } n \in \mathbb{N}. \tag{3.75}$$

Taking the (D)-limit in (3.75) with respect to n and using weak σ-distributivity of R, we obtain $\int_E | f | d\lambda = 0$, which ends the proof. □

We now give a Vitali-type theorem.

Theorem 2.6 (see Boccuto and Candeloro (2002c), Theorem 3.7) *If $\lambda : \Sigma \to \widetilde{\mathbb{R}}$ is a σ-additive and σ-finite positive measure, $(f_n)_n$ is a sequence of integrable* *functions in* R^G*, convergent in measure to an integrable function* $f_0 \in R^G$*, and the integrals* $\int_E |f_n| \, d\lambda$ *are globally uniformly* (s)*-bounded, then* $(f_n)_n$ *converges in* L^1 *to* f_0*.*

Conversely, if $(f_n)_n$ *is a sequence of integrable functions, convergent in measure and in* L^1 *to an integrable function* f_0*, then the integrals* $\int_E |f_n| \, d\lambda$ *are globally uniformly* (s)*-bounded, globally uniformly* σ*-additive and globally uniformly* λ*-absolutely continuous.*

Proof: We begin with the first part. Since f_0 is integrable, without restriction we can suppose $f_0 = 0$. Proceeding similarly as in Proposition 2.1, we can prove that the measure $E \mapsto \int_E |f| d\lambda$ are uniformly bounded. Since they are by hypothesis globally uniformly (s)-bounded, and globally λ-absolutely continuous thanks to Lemma 2.5, by virtue of Corollary 1.10 they are globally uniformly λ-absolutely continuous and globally uniformly σ-additive.

By σ-finiteness of λ there exists an increasing sequence $(E_s)_s$ in Σ with $\bigcup_{s=1}^{\infty} E_s = G$ and $\lambda(E_s) < +\infty$ for all $s \in \mathbb{N}$. Set $F_s := G \setminus E_s$, $s \in \mathbb{N}$: of course the sequence $(F_s)_s$ is decreasing, and their intersection is the empty set. By global uniform σ-additivity of the measures $E \mapsto \int_E |f_n| \, d\lambda$, $n \in \mathbb{N}$, there exists a regulator $(z_{t,l})_{t,l}$ such that for any $\varphi \in \mathbb{N}^{\mathbb{N}}$ there exists $s_0 \in \mathbb{N}$ with

$$\int_{F_{s_0}} |f_n| d\lambda \leq \bigvee_{t=1}^{\infty} z_{t,\varphi(t)} \quad \text{for all } n \in \mathbb{N}. \tag{3.76}$$

By convergence in measure to 0 of $(f_n)_n$, there exists a (*D*)-sequence $(a_{t,l})_{t,l}$ with the property that for any $\varphi \in \mathbb{N}^{\mathbb{N}}$ there is a sequence $(C_n)_n$ in Σ which fulfils (3.67) and with $\lim_n \lambda(C_n) = 0$. Fix arbitrarily $\varphi \in \mathbb{N}^{\mathbb{N}}$, and let $(C_n)_n$ be as above. For every n and $s \in \mathbb{N}$, we get:

$$\int_{E_s} |f_n| d\lambda \leq \int_{E_s \cap C_n} |f_n| d\lambda + \int_{E_s \setminus C_n} |f_n| d\lambda \leq \int_{C_n} |f_n| d\lambda + \lambda(E_s) \bigvee_{t=1}^{\infty} a_{t,\varphi(t)}.$$

From global uniform absolute continuity of the measures $E \mapsto \int_E |f_n| \, d\lambda$, $n \in \mathbb{N}$, it follows the existence of a (D)-sequence $(u_{t,l})_{t,l}$ with the property that for any $\varphi \in \mathbb{N}^{\mathbb{N}}$ there is $n_1 \in \mathbb{N}$ with $\bigvee_{q \in \mathbb{N}} (\int_{C_n} |f_q| d\lambda) \leq \bigvee_{t=1}^{\infty} u_{t,\varphi(t)}$ for each $n \geq n_1$. Thus for any $n \geq n_1$ and $s \in \mathbb{N}$ we get:

$$\int_{E_s} |f_n| d\lambda \leq \bigvee_{t=1}^{\infty} u_{t,\varphi(t)} + \lambda(E_s) \bigvee_{t=1}^{\infty} a_{t,\varphi(t)}. \tag{3.77}$$

From (3.77) it follows that $(f_n)_n$ converges in L^1 to 0 on every set E_s, namely for any $s \in \mathbb{N}$ there is a (D)-sequence $(A^s_{t,l})_{t,l}$ with the property that, for every fixed $\varphi \in \mathbb{N}^{\mathbb{N}}$, there exists $\overline{n}$ with $\int_{E_s} |f_n| d\lambda \leq \bigvee_{t=1}^{\infty} A^s_{t,\varphi(t+s)}$ for each $n \geq \overline{n}$. By equiboundedness of the integrals, the (D)-sequences $(A^s_{t,l})_{t,l}$ can be taken bounded by some fixed element $u \in R$, and therefore, by applying the Fremlin lemma, we get a single (D)-sequence $(A_{t,l})_{t,l}$ such that for any $s \in \mathbb{N}$ and $\varphi \in \mathbb{N}^{\mathbb{N}}$ there is a positive integer $n_0(s, \varphi)$ with

$$\int_{E_s} |f_n| d\lambda \leq \bigvee_{t=1}^{\infty} A_{t,\varphi(t)} \tag{3.78}$$

for every $n \geq n_0$. Take now $(b_{t,l})_{t,l} = (A_{t,l} + z_{t,l})_{t,l}$, and fix $\varphi \in \mathbb{N}^{\mathbb{N}}$. Then there is an integer s_0 such that (3.76) hold, for all n. There is a natural number $n_0(s_0, \varphi)$ satisfying (3.78) for all $n \geq n_0$. So we get

$$\int_G |f_n| \, d\lambda \leq \int_{F_{s_0}} |f_n| \, d\lambda + \int_{E_{s_0}} |f_n| \, d\lambda \leq \bigvee_{t=1}^{\infty} b_{t,\varphi(t)}$$

for any $n \geq n_0$, and thus the first part is proved.

Now we prove the last part. If $(f_n)_n$ tends to f_0 in L^1, we see that the sequence $(\int_G |f_n| d\lambda)_n$ is bounded. Thanks to global σ-additivity of the measures $E \mapsto \int_E |f_n| \, d\lambda, n \in \mathbb{N}$ and the Fremlin lemma, we find a (D)-sequence $(\gamma_{t,l})_{t,l}$ with the property that for any disjoint sequence $(H_k)_k$ in Σ, $\varphi \in \mathbb{N}^{\mathbb{N}}$ and $n \in \mathbb{N}$ there is $k_0 \in \mathbb{N}$ with

$$\int_{U_{k_0}} |f_n| \, d\lambda \leq \bigvee_{t=1}^{\infty} \gamma_{t,\varphi(t)}, \tag{3.79}$$

where U_{k_0} is the union of those H_k with $k \geq k_0$. By convergence in L^1, we find a regulator $(\beta_{t,l})_{t,l}$ such that for any $\varphi \in \mathbb{N}^{\mathbb{N}}$ there exists $n_0 \in \mathbb{N}$, satisfying

$$\int_G |f_n - f_0| d\lambda \leq \bigvee_{t=1}^{\infty} \beta_{t,\varphi(t)} \tag{3.80}$$

for any $n \geq n_0$. Set $a_{t,l} = 2(\gamma_{t,l} + \beta_{t,l})$, and pick a disjoint sequence $(H_k)_k$ in Σ. For each $\varphi \in \mathbb{N}^{\mathbb{N}}$ there is n_0 satisfying (3.80). In correspondence with n_0 and φ, we find a k_0 fulfilling (3.79) for $n = 0, 1, \ldots, n_0 - 1$. If $n \leq n_0 - 1$, then $\int_{U_{k_0}} |f_n| d\lambda \leq \bigvee_{t=1}^{\infty} \gamma_{t,\varphi(t)} \leq \bigvee_{t=1}^{\infty} a_{t,\varphi(t)}$. If $n \geq n_0$, then

$$\int_{U_{k_0}} |f_n| d\lambda \leq \int_{U_{k_0}} |f_n - f_0| d\lambda + \int_{U_{k_0}} |f_0| d\lambda \leq \bigvee_{t=1}^{\infty} \gamma_{t,\varphi(t)} + \bigvee_{t=1}^{\infty} \beta_{t,\varphi(t)} \leq \bigvee_{t=1}^{\infty} a_{t,\varphi(t)}.$$

So we obtain global uniform σ-additivity, and also global uniform (s)-boundedness and global uniform absolute continuity. This ends the proof. □

In general, convergence in measure is not a consequence of convergence in L^1 (see also Boccuto (1995a), Example 3.14; Boccuto (1997), Example 3.23.1). Indeed, let $R_1 = R_3 = \{f : D \to \mathbb{R},\ f \text{ is bounded}\}$, where $D = [0,1]$ endowed with the discrete topology, $R_2 = \mathbb{R}$, $G = [0,1]$, $\nu : \mathcal{P}(G) \to [0,1]$ be a finitely additive extension of the Lebesgue measure. Observe that R_1 is weakly σ-distributive (see also Boccuto (2003)), and hence, in the space R_1, (D)-

convergence coincides with order convergence, which is pointwise convergence dominated by an element of R_1, or equivalently by a real constant (see also Luxemburg and Zaanen (1971)). Set $L=\{f:G\to R_1,\ f \text{ is bounded }\}$, and for each $f\in L$ let $(\int_G f(g)d\nu(g))(d)=\int_G f(g)(d)d\nu(g)$ be the "componentwise" Lebesgue integral, which is a linear functional, defined on L and with values in $\mathbb{R}^D$. Let $\oplus$ the sum mod 1 of two elements of [0,1], and put $f_n(g)(d)=\sqrt{n}$ if $g\in[d,d\oplus\frac{1}{n}]$ and 0 otherwise, $n\in\mathbb{N}$, $g\in G$, $d\in D$.

For every $n\in\mathbb{N}$ and $g\in G$, we have $\int_G f_n(g)\,d\nu(g)=\underline{\frac{1}{\sqrt{n}}}$, where $\underline{\frac{1}{\sqrt{n}}}$ is the function which associates the real constant $\frac{1}{\sqrt{n}}$ with every $d\in D$. Thus, convergence in L^1 to 0 holds. We prove that $(f_n)_n$ does not converge in measure to 0. Let $(p_n)_n$ and $(q_n)_n$ be two (O)-sequences in R_1 and $\mathbb{R}$ respectively. There is an integer $\overline{n}$ such that, for every $n\geq\overline{n}$, $\sqrt{n}\geq m_1$, where $m_1=\sup_{d\in D} p_1(d)$. Thus $f_n(g)\not\leq p_1$, and a fortiori $f_n(g)\not\leq p_n$, whenever $n\geq\overline{n}$ and $g\in G$. Set $A_n=\{g\in G: f_n(g)\not\leq p_n\}$. We have $A_n=G, \mu(A_n)=1$ and thus $\mu(A_{\overline{n}})\not\leq q_{\overline{n}}$ for each $n\geq\overline{n}$. □

Furthermore, observe that, in our context, the monotone convergence theorem in general does not hold, even when the involved measure μ is σ-additive (see also Boccuto (1997), Example 3.23.2). Indeed, let R_j, $j=1,2,3$, G, D be as above, and let us add to R_1 an extra element, $+\infty$, satisfying the analogous properties as in the classical setting. Furthermore, let $\mathcal{M}\subset\mathcal{P}([0,1])$ be the σ-algebra of all measurable subsets of $[0,1]$, $\nu:\mathcal{M}\to[0,1]$ be the Lebesgue measure, and put

$$h_n(g)(d)=1 \quad \text{if } g\in[d,d\oplus\frac{1}{2^n}] \text{ and 0 otherwise, } n\in\mathbb{N},\ g\in G, d\in D.$$

Set $k_n(g)(d)=\sum_{j=1}^n h_j(g)(d)$. For every n and d we get

$$\int_G k_n(g)(d)\, d\nu(g) = \sum_{j=1}^{n} \int_G h_j(g)(d)\, d\nu(g) = \sum_{j=1}^{n} \frac{1}{2^j}.$$

So we have $\int_G k_n(g)\, d\nu(g) = \sum_{j=1}^{n} \underline{\frac{1}{2^j}}$, where $\underline{\frac{1}{2^j}}$ is the function which associates the constant $\frac{1}{2^j}$ with every $d \in D$. Therefore we obtain that

$\bigvee_n \int_G k_n(g)\, d\nu(g) = \underline{1}$. Now, we prove that

$$\bigvee_n k_n(g) = +\infty \quad \text{for all } g \in G.$$

Fix arbitrarily $g \in G$. If the set $\{k_n(g) : n \in \mathbb{N}\}$ is bounded from above, then there is a positive real number $M > 0$ with $k_n(g)(d) \le M$ for all $n \in \mathbb{N}$ and $d \in D$. If $n_0 > M$, then $k_{n_0}(g)(g) = n_0$, obtaining a contradiction. □

We now give the following lemma (for similar results existing in the literature in the case $R = \mathbb{R}$, see also Diestel (1984)).

Lemma 2.7 (see Boccuto and Candeloro (2002c), Lemma 3.8) *Let $f : G \to R$ be an integrable function, and define $m(A) = \int_A f\, d\lambda$ for every $A \in \Sigma$. Then*

$$\int_A |f|\, d\lambda \le 2\nu(m)(A) \text{ for each } A \in \Sigma. \tag{3.81}$$

Proof: We first consider the case $f = \sum_{i=1}^{n} a_i\, \chi_{A_i}$, where $n \in \mathbb{N}$, $a_i \in R$ and $A_i \in \Sigma$ for any $i \in [1, n]$. By the Maeda-Ogasawara-Vulikh representation theorem, we find a meager set $N \subset \Omega$ with $f(x)(\omega) = \sum_{i=1}^{n} a_i(\omega)\chi_{A_i}(x)$ for any $x \in G$ and $\omega \in \Omega \setminus N$. For such x and ω, set $f_\omega(x) = f(x)\ (\omega)$. Note that f_ω is an integrable simple real-valued function, and

$$\int_A f_\omega\, d\lambda = \left(\int_A f d\lambda\right)(\omega) = m(A)(\omega) \quad \text{for all } A \in \Sigma.$$

Observe that, for real-valued functions, (3.81) is satisfied, and thus, in the complement of a suitable a meager set N' greater than N, we get

$$\int_A |f_\omega| \, d\lambda \le 2v(m_\omega)(A),$$

where m_ω is the real-valued measure $A \mapsto m(A)\,(\omega)$. Thanks to the Maeda-Ogasawara-Vulikh theorem, we get (3.81). Now, for general f, let $(f_n)_n$ be a defining sequence for f (see Definition 2.3). Then we get

$$\int_A |f| \, d\lambda = \lim_n \int_A |f_n| \, d\lambda \le 2 \ (O)\limsup_n v(m_n)(A), \tag{3.82}$$

where $m_n(A) = \int_A f_n \, d\lambda$ for every $A \in \Sigma$. Now we have

$$v(m_n)(A) = \bigvee_{B \subset A} |\int_B f_n d\lambda| \le \bigvee_{B \subset A} |\int_B (f_n - f) \, d\lambda + \int_B f \, d\lambda| \le$$

$$\le \bigvee_{B \subset A} \{\int_B |f_n - f| \, d\lambda + |\int_B f \, d\lambda|\} \le \int_G |f_n - f| \, d\lambda + v(m)(A).$$

So, by (3.82), we obtain $\int_A |f| \, d\lambda \le 2 \ (O)\limsup_n \int_G |f_n - f| \, d\lambda + 2v(m)(A)$.

Since $(f_n)_n$ converges in L^1 to f and taking into account weak σ-distributivity of R, we get (3.81). □

Proposition 2.8 (see Boccuto and Candeloro (2002c), Proposition 3.9) *A sequence $(f_n)_n$ of integrable functions converges to f_0 in L^1 if and only if*

$$\lim_n \int_A f_n \, d\lambda = \int_A f_0 \, d\lambda \text{ uniformly with respect to } A \in \Sigma.$$

Proof: We have already proved the "only if" part. We now turn to the "if" part: without restriction, we can suppose $f_0 = 0$ and $\lim_n |\int_A f_n \, d\lambda| = 0$ uniformly with respect to A. By Lemma 2.7, we obtain $\lim_n \int_G |f_n| \, d\lambda \le 2 \lim_n v(m_n)(G) = 0$, that is the assertion. □

As a consequence, we get the following characterization of convergence in L^1.

Theorem 2.9 (see Boccuto and Candeloro (2002c), Theorem 3.11) *If* $(f_n)_n$ *is a sequence of integrable functions, convergent in measure to an integrable function* $f_0 \in R^G$*, and if the set* $\{\int_G |f_n|\, d\lambda : n \in \mathbb{N}\}$ *is bounded in R, then* $(f_n)_n$ *converges in* L^1 *to* f_0 *if and only if for each* $A \in \Sigma$ *we get* $(RD)\lim_n \int_A f_n \, d\lambda = \int_A f_0 \, d\lambda$.

Proof: Thanks to Proposition 2.8, we will prove only the "if" part. As the integrals of the functions f_n, $n \in \mathbb{N}$, form an (RD)-convergent sequence of globally σ-additive measures, then, by Theorem 1.5, these integrals are globally uniformly (s)-bounded. By Theorem 2.6, $(f_n)_n$ converges to f_0 in L^1. □

We now observe that it is possible to do a similar construction of an abstract integral involving Riesz space-valued functions and vector lattice-valued measures, with respect to suitable convergences, satisfying certain axioms, which are fulfilled by $\mathcal{F}$-, $(O\mathcal{F})$- and $(D\mathcal{F})$-convergence (see also Boccuto and Candeloro (2009a-b)), by the classical (O)- and (D)-convergences (see for instance Boccuto (1997)) and by almost convergence. We now give an axiomatic approach of convergence of sequences, taking values in a Dedekind complete vector lattice R (see also Boccuto and Candeloro (2004b-c, 2009a), Kuratowski (1966 and 1968)).

A *convergence* is a pair $(\mathcal{S}, \mathcal{N})$, where S is a linear subspace of $R^{\mathbb{N}}$ and $\mathcal{N} : \mathcal{S} \to R$ is a function, satisfying the following axioms:

a) $\mathcal{N}((\zeta_1 x_n + \zeta_2 x_n)_n) = \zeta_1 \mathcal{N}((x_n)_n) + \zeta_2 \mathcal{N}((y_n)_n)$ for every pair of sequences $(x_n)_n$, $(y_n)_n \in S$ and ζ_1, $\zeta_2 \in \mathbb{R}$; if $(x_n)_n$, $(y_n)_n \in S$ and $x_n \le y_n$ for each n with the exception of a finite number of indexes, then $\mathcal{N}((x_n)_n) \le \mathcal{N}((y_n)_n)$.

b) If $(x_n)_n$ is such that the set $\{n \in \mathbb{N} : x_n \neq l\}$ is empty or finite, then $(x_n)_n \in \mathcal{S}$ and $\mathcal{N}((x_n)_n) = l$; if $(x_n)_n$, $(y_n)_n$ satisfy the property that the set $\{n \in \mathbb{N} : x_n \neq y_n\}$ is empty or finite and $(x_n)_n \in S$, then also $(y_n)_n \in S$, and $\mathcal{N}((y_n)_n) = \mathcal{N}((x_n)_n)$.

c) If $(x_n)_n \in S$, then $(|x_n|)_n \in S$ and $\mathcal{N}((|x_n|)_n) = |\mathcal{N}((x_n)_n)|$.

d) If $(x_n)_n$, $(z_n)_n \in S$, $\mathcal{N}((x_n)_n) = \mathcal{N}((z_n)_n)$, and $x_n \le y_n \le z_n$ for n greater than a suitable integer n_0, then $(y_n)_n \in \mathcal{S}$.

e) If $u \in R$, $u \ge 0$, then $(\frac{1}{n}u)_n$ is an element of $\mathcal{S}$, and $\mathcal{N}((\frac{1}{n}u)_n) = 0$.

Let R_1, R_2, R be three Dedekind complete Riesz spaces, endowed with three convergence structures, fulfilling axioms a), …, e) above, which we denote by $(\mathcal{S}, \mathcal{N})$.

Assumptions 2.10 We say that (R_1, R_2, R) is a *product triple* iff a function $\cdot : R_1 \times R_2 \to R$ (*product*) is defined, with

i) $(r_1 + s_1) \cdot r_2 = r_1 \cdot r_2 + s_1 \cdot r_2$,

ii) $r_1 \cdot (r_2 + s_2) = r_1 \cdot r_2 + r_1 \cdot s_2$,

iii) $[r_1 \ge s_1, r_2 \ge 0] \Rightarrow [r_1 \cdot r_2 \ge s_1 \cdot r_2]$,

iv) $[r_1 \ge 0, r_2 \ge s_2] \Rightarrow [r_1 \cdot r_2 \ge r_1 \cdot s_2]$,

v) $a(r_1 \cdot r_2) = (ar_1) \cdot r_2 = r_1 \cdot (ar_2)$ for all r_j, $s_j \in R_j$, $j = 1,2$, and $a \in \mathbb{R}$;

vi) if $(a_n)_n$ is any sequence in R_2 and $b \in R_1$, then $[\mathcal{N}((a_n)_n) = 0] \Rightarrow [\mathcal{N}((b \cdot a_n)_n) = 0]$;

vii) if $(a_n)_n$ is any sequence in R_1 and $b \in R_2$, then $[\mathcal{N}((a_n)_n) = 0] \Rightarrow [\mathcal{N}((a_n \cdot b)_n) = 0]$.

We now consider an integral for functions defined on any nonempty set G and taking values in a Riesz space R_1, with respect to a finitely additive and positive measure $m : \Sigma \to R_2$, where $\Sigma \subset \mathcal{P}(G)$ is a fixed σ-algebra (for similar constructions in classical cases, see also Dunford and Schwartz (1958), Halmos (1950)). The integral of simple functions is defined in the usual way; however, it is possible to give a more abstract definition, by replacing the "basic" space of simple functions with another suitable linear space L, for example

$L:=\{f \in R_1^G, f$ is bounded$\}$. So, from now on, we assume the existence of a Riesz subspace L of bounded functions, containing all simple functions and closed with respect to the function product, and a linear monotone R-valued functional defined on L (which we denote by $\int_G (\cdot) dm$) and satisfying $\int_G f(g) dm(g) = cm(A)$ for each function f such that $f(g) = c$ when $g \in A$ and $f(g) = 0$ if $g \notin A$. Furthermore, set $\int_A f\, dm = \int_G f \chi_A\, dm$, $A \in \Sigma$.

A function sequence $(f_n)_n$ of R_1^G *converges uniformly* to $f \in R_1^G$ iff

$$\mathcal{N}((\bigvee_{g \in G} | f_n(g) - f(g) |)_n) = 0.$$

A sequence $(f_n)_n$ in R_1^G *converges in measure* to $f \in R_1^G$ iff a sequence $(A_n)_n$ can be found in Σ, such that

$$\mathcal{N}((m(A_n))_n) = 0 \text{ and } \mathcal{N}((\bigvee_{g \in G \setminus A_n} | f_n(g) - f(g) |)_n) = 0.$$

Note that, if $R_1 = R_2 = \mathbb{R}$ endowed with the ordinary convergence, then the definition of convergence in measure is equivalent to the classical one (see also Boccuto and Sambucini (1997), Theorem 4.4). We now recall the following:

Proposition 2.11 (see Boccuto and Candeloro (2009a), Proposition 3.4) *If* $(f_n)_n$, $(h_n)_n \in R_1^G$ *converge in measure to* f, h *and* $a \in \mathbb{R}$. *Then* $(f_n + h_n)_n$, $(f_n \vee h_n)_n$, $(f_n \wedge h_n)_n$, $(a f_n)_n$, $(| f_n |)_n$ *converge in measure to* $f + h$, $f \vee h$, $f \wedge h$, af, $| f |$, *respectively.*

Proof: We consider only $(f_n \vee h_n)_n$. Note that, thanks to the Birkhoff inequality (see Luxemburg and Zaanen (1971)), we get:

$$| f_n(g) \vee h_n(g) - f(g) \vee h(g) | \leq$$

$$| f_n(g) \vee h_n(g) - f_n(g) \vee h(g) | + | f_n(g) \vee h(g) - f(g) \vee h(g) | \leq$$

$$| h_n(g) - h(g) | + | f_n(g) - f(g) |, \; n \in \mathbb{N}, \; g \in G.$$

From axioms a), c), d) of convergence and convergence in measure of $(f_n)_n$, $(h_n)_n$ to f, h we get the assertion. Analogously it is possible to prove the other results. □

We say that $(f_n)_n$ (in L) *converges in* L^1 to $f \in R_1^G$ iff $\mathcal{N}((\int_G |f_n(g) - f(g)| dm(g))_n) = 0$, and that $(f_n)_n$ is *uniformly integrable* iff $\mathcal{N}(m(A_n))_n = 0$ implies $\mathcal{N}((\int_{A_n} |f_n(g)| dm(g))_n) = 0$.

Observe that, for $R_1 = R_2 = \mathbb{R}$ and when the involved convergence is the ordinary one, by virtue of Boccuto and Sambucini (1997), Theorem 4.16, this concept and the corresponding classical one are equivalent. The following result holds.

Theorem 2.12 (see Boccuto and Candeloro (2009a), Theorem 3.7) *If $(f_n)_n$ is a sequence in L, convergent in measure to $f \in L$ and is uniformly integrable, then $(f_n)_n$ converges to f in L^1.*

Proof: First of all note that f is bounded. Let $v \in R_1$, $v \geq 0$, be with $|f(g)| \leq v$ for every $g \in G$. We get

$$\int_A |f(g)| \, dm(g) \leq \int_A v dm(g) = v \cdot m(A) \text{ for all } A \in \Sigma.$$

Let $(A_n)_n$ be a sequence of sets associated with convergence in measure. For each $n \in \mathbb{N}$ we have

$$\int_G |f_n(g) - f(g)| dm(g) = \int_{G \setminus A_n} |f_n(g) - f(g)| dm(g) + \int_{A_n} |f_n(g) - f(g)| dm(g)$$

$$\leq \Big(\bigvee_{g \in G \setminus A_n} |f_n(g) - f(g)| \Big) \cdot m(G) + \int_{A_n} |f_n(g) - f(g)| dm(g) \leq$$

$$\leq \Big(\bigvee_{g \in G \setminus A_n} |f_n(g) - f(g)| \Big) \cdot m(G) + \int_{A_n} |f_n(g)| dm(g) + \int_{A_n} |f(g)| dm(g).$$

The assertion follows from axioms of convergence and properties of the product, taking the limits. □

A function $f \in R_1^G$ is *measurable* iff one may find a sequence $(f_n)_n$ in L, converging in measure to f. A sequence $(f_n)_n$ is *defining* for a function $f \in R_1^G$ iff it converges in measure to f and is uniformly integrable. A positive function $f \in R_1^G$ is *integrable* iff there are a defining sequence $(f_n)_n$ and a set function $l : \Sigma \to R$, with

$$((\bigvee_{A \in \Sigma} | \int_A f_n(g) dm(g) - l(A) |)_n) = 0. \tag{3.83}$$

The following result holds.

Proposition 2.13 (see Boccuto and Candeloro (2009a), Proposition 3.11) *For every* $A \in \Sigma$ *we have*

$$l(A) = \mathcal{N}((\int_A f_n(g) dm(g))_n), \tag{3.84}$$

where $l(A)$ *is independent of the chosen defining sequence.*

Proof: The relation (3.84) is a consequence of axiom d) of convergence and (3.83).

Let $(f_n^1)_n$ and $(f_n^2)_n$ be two defining sequences for the same function f, and define

$$l_j(A) := \mathcal{N}((\int_A f_n^j(g) dm(g))_n), \ A \in \Sigma,$$

$p_n(g) = | f_n^1(g) - f_n^2(g) |$, $g \in G$, $n \in \mathbb{N}$. Uniform integrability (resp. convergence in measure to 0) of $(p_n)_n$ follows from uniform integrability (resp. convergence in measure) of $(f_n^1)_n$ and $(f_n^2)_n$. Thanks to Theorem 2.12, $(p_n)_n$ converges to 0 in L^1. We get

$$| l_1(A) - l_2(A) | \leq | \int_A f_n^1(g) dm(g) - l_1(A) | + | l_2(A) - \int_A f_n^2(g) dm(g) | +$$

$$+|\int_A f_n^1(g)dm(g)-\int_A f_n^2(g)dm(g)|\le$$

$$\le|\int_A f_n^1(g)dm(g)-l_1(A)|+|l_2(A)-\int_A f_n^2(g)dm(g)|+\int_G p_n(g)dm(g).$$

Thus we obtain $|l_1(A)-l_2(A)|=0$, that is $l_1(A)=l_2(A)$ for every $A\in\Sigma$, thanks to the properties of convergence. □

We set $\int_A f(g)dm(g):=l(A)$, $A\in\Sigma$.

A function $f\in R_1^G$ is said to be *integrable* iff f^+ and f^- are integrable, and we put

$$\int_A f(g)dm(g)=\int_A f^+(g)dm(g)-\int_A f^-(g)dm(g),\ \ A\in\Sigma.$$

Observe that, if f is integrable, then also $|f|$ is integrable, and

$$\int_A |f(g)|dm(g)=\int_A f^+(g)dm(g)+\int_A f^-(g)dm(g),\ A\in\Sigma.$$

It is not difficult to see that this integral is linear. We now show that our integral is absolutely continuous.

Theorem 2.14 (see Boccuto and Candeloro (2009a), Theorem 3.14) *If* $f\in R_1^G$ *is an integrable function and* $\mathcal{N}((m(A_n))_n)=0$, *then*

$$\mathcal{N}((\int_{A_n}|f(g)|\,dm(g))_n)=0.$$

Proof: We suppose $f\ge 0$, without restriction. For every sequence $(A_n)_n$ of elements of Σ with $\mathcal{N}((m(A_n))_n)=0$ and $n\in\mathbb{N}$, we have

$$\int_{A_n}|f(g)|dm(g)\le|\int_{A_n}|f(g)|dm(g)-\int_{A_n}|f_n(g)|dm(g)|+\int_{A_n}|f_n(g)|dm(g)\le$$

$$\le\bigvee_{A\in\Sigma}|\int_A|f(g)|dm(g)-\int_A|f_n(g)|dm(g)|+\int_{A_n}|f_n(g)|dm(g).$$

Thanks to uniform integrability, we obtain

$$\mathcal{N}((\int_{A_n} | f(g) | dm(g))_n) = 0. \quad \square$$

Analogously as above, it is possible to formulate the concepts of convergence in L^1 and uniform integrability for integrable functions.

By virtue of Theorem 2.14 and proceeding as in Theorem 2.12, we can give the following version of the Vitali theorem.

Theorem 2.15 (see Boccuto and Candeloro (2009a), Theorem 3.15) *If $(f_n)_n$ is a sequence of integrable functions, converging in measure to an integrable function f and uniformly integrable, then $(f_n)_n$ converges in L^1 to f.*

Observe that, if $(f_n)_n$ converges in L^1 to f, then $(f_n)_n$ is uniformly integrable. Indeed, if $\mathcal{N}((m(A_n))_n) = 0$, then we get

$$\int_{A_n} | f_n(g) | dm(g) \le \int_G | f_n(g) - f(g) | dm(g) + \int_{A_n} | f(g) | dm(g), \quad n \in \mathbb{N},$$

and hence uniform integrability, taking into account absolute continuity of the integral.

We now recall the following property of the integral.

Lemma 2.16 (see also Boccuto and Candeloro (2009a), Lemma 3.17) *Suppose that $f \in R_1^G$ is integrable, and let $h \in R_1^G$ be such that $h = f$ in the complement of a m-null set $O \in \Sigma$. Then h is integrable and*

$$\int_A h(g)\, dm(g) = \int_A f(g)\, dm(g) \text{ for every } A \in \Sigma.$$

Proof: Let $(f_n)_n$ be a defining sequence for f, and for each $n \in \mathbb{N}$ set $h_n(g) = f_n(g)$ if $g \notin O$ and $h_n(g) = 0$ if $g \in O$. It is easy to check that, for every $n \in \mathbb{N}$, $h_n \in L$, $(h_n)_n$ is uniformly integrable and

$$\int_A h_n(g)\, dm(g) = \int_A f_n(g)\, dm(g) \text{ for each } A \in \Sigma.$$

Since $m(\{g \in G : |h_n(g) - h(g)| \neq |f_n(g) - f(g)|\}) = 0$, convergence in measure of $(h_n)_n$ follows from that of $(f_n)_n$. □

Proposition 2.17 *If* $f, h \in R_1^G$, $|f(g)| \le |h(g)|$ *for all* $g \in G \setminus O$ *where* $m(O) = 0$, h *is integrable and* f *is measurable, then* f *is integrable.*

Proof: Without loss of generality, by Lemma 2.16, we can assume that $|f(g)| \le |h(g)|$ for any $g \in G$.

We get $(f(g))^+ \le |h(g)|$, $(f(g))^- \le |h(g)|$ for all $g \in G$; moreover, there are a function sequence $(h_n(g))_n$, convergent in measure to h and uniformly integrable, and a sequence $(f_n(g))_n$, convergent in measure to f. Thanks to Proposition 2.11, $((f_n(g))^+)_n$(resp. $((f_n(g))^-)_n$, $(|h_n(g)|)_n$) converges in measure to $(f(g))^+$ (resp. $(f(g))^-$, $|h(g)|$); furthermore, the sequences

$$(l_n(g) := (f_n(g))^+ \wedge |h_n(g)|)_n, \quad (m_n(g) := (f_n(g))^- \wedge |h_n(g)|)_n \qquad \textbf{(3.85)}$$

are uniformly integrable and converge in measure to $|h(g)| \wedge (f(g))^+ = (f(g))^+$ and $|h(g)| \wedge (f(g))^- = (f(g))^-$, respectively. So, $(l_n(g) - m_n(g))_n$ is a defining sequence for f, and hence the assertion follows. □

Observe that, if L is the space of all bounded functions, Proposition 2.17 holds without assuming measurability on f: indeed, by proceeding analogously as in the proof of Proposition 2.17 and defining l_n, m_n, $n \in \mathbb{N}$, as in (3.85), it is enough to note that the boundedness of the functions h_n, $n \in \mathbb{N}$, is sufficient to get l_n, $m_n \in L$.

We now prove the Lebesgue dominated convergence theorem.

Theorem 2.18 (see Boccuto and Candeloro (2009a), Theorem 3.18) *If* f, h, $f_n \in R_1^G$, $n \in \mathbb{N}$, *are integrable functions, such that* $(f_n)_n$ *converges in measure to* f *and* $|f_n(g)| \le |h(g)|$ *for each* $n \in \mathbb{N}$ *and* m*-almost everywhere, then* $(f_n)_n$ *converges in* L^1 *to* f.

Proof: By Proposition 2.17, the functions f_n, $n \in \mathbb{N}$, are integrable. Moreover, by hypothesis, they are uniformly integrable. Thus Theorem 2.18 follows from Theorem 2.15. □

If L is the space of all bounded functions, then in Theorem 2.18 the hypothesis of measurability of the functions f_n, $n \in \mathbb{N}$, can be dropped (see also Boccuto (1993 and 1995a), Boccuto and Candeloro (2009a)).

We now turn to a Scheffé-type theorem, which is a consequence of Theorem 2.18.

Theorem 2.19 (see Boccuto and Candeloro (2009a), Theorem 3.19) *If f, $f_n \in R_1^G$, $n \in \mathbb{N}$, are positive and integrable functions, such that $(f_n)_n$ converges in measure to f and*

$$\mathcal{N}((\int_G f_n(g)dm(g))_n) = \int_G f(g)dm(g),$$

then $(f_n)_n$ converges to f in L^1.

Proof: Put $h_n(g) = f_n(g) - f(g)$, $n \in \mathbb{N}$, $g \in G$. As $f_n(g) \geq 0$, $f(g) \geq 0$ for all g and n, then $|h_n(g)| \leq f_n(g) + f(g)$ for each $g \in G$ and $n \in \mathbb{N}$. Thus,

$$0 \leq (h_n(g))^- = \frac{|h_n(g)| - h_n(g)}{2} \leq \frac{f_n(g) + f(g) - f_n(g) + f(g)}{2} = f(g)$$

for all g and n.

Set $H_n(g) = (h_n(g))^-$, $g \in G$, $n \in \mathbb{N}$. Note that $(H_n)_n$ converges in measure to zero, and that f and H_n, $n \in \mathbb{N}$, are integrable. Thanks to Theorem 2.18, we get

$$\mathcal{N}((\int_G (h_n(g))^- dm(g))_n) = 0$$

and thus

$$\mathcal{N}((\int_G (h_n(g))^+ dm(g))_n) = \mathcal{N}((\int_G h_n(g)dm(g))_n) = 0.$$

So we obtain

$$\mathcal{N}((\int_G |h_n(g)|dm(g))_n) = \mathcal{N}((\int_G (h_n(g))^+ dm(g))_n) + \mathcal{N}((\int_G (h_n(g))^- dm(g))_n) = 0.$$

This ends the proof. □

3.2.2. Integrals with Respect to Optimal Measures

We now deal with the integral for real-valued functions with respect to *optimal measures*, that is measures compatible with respect to the supremum, with values in a lattice group R (for the case $R = \mathbb{R}$, see also Agbeko (1994, 1995, 2000 and 2009), Shilkret (1971)). These kinds of measures have several applications in quantum structures and observables as well as in intuitionistic fuzzy sets (see also Atanassov (1999), Riečan and Neubrunn (1997)).

Let R be a Dedekind complete Riesz space, G be any abstract nonempty set and $\Sigma \subset \mathcal{P}(G)$ be a σ-algebra. Let us add to R two extra elements, $+\infty$ and $-\infty$, obeying to the usual rules as in the classical case, and set $\overline{R} := R \cup \{+\infty\} \cup \{-\infty\}$.

A set function $P : \Sigma \to R$ is called an *optimal measure* iff $P(A \cup B) = P(A) \vee P(B)$ whenever $A, B \in \Sigma$ and P is *continuous from above,* that is $P(A) = \bigwedge_n P(A_n)$ whenever $(A_n)_n$ is a decreasing sequence in Σ, such that $A = \bigcap_{n=1}^{\infty} A_n$. The following result holds.

Proposition 2.20 (Boccuto, Dimitriou and Papanastassiou (2010a), Proposition 1) *Let $(B_n)_n$ be any increasing sequence in Σ, and $B := \bigcup_{n=1}^{\infty} B_n$. Assume that P is an optimal measure. Then, $P(B) = (O) \lim_n P(B_n) = \bigvee_n P(B_n)$.*

Proof: Observe that $(B \setminus B_n)_n$ is a decreasing sequence in Σ, and thus

$$\bigwedge_n P(B \setminus B_n) = (O) \lim_n P(B \setminus B_n) = P(\emptyset) = 0.$$

Furthermore note that $B_n \subset B$, and so $B = B_n \cup (B \setminus B_n)$ for all $n \in \mathbb{N}$. Hence $P(B) = P(B_n) \vee P(B \setminus B_n)$ for any $n \in \mathbb{N}$. From this and by the Maeda-Ogasawara-Vulikh representation theorem we get a meager set N with $P(\emptyset)(\omega)$,

$P(B)(\omega)$, $P(B_n)(\omega) \in \mathbb{R}$ for any $\omega \in \Omega \setminus N$ and $n \in \mathbb{N}$, and $\inf_n P(B \setminus B_n)(\omega) = \lim_n P(B \setminus B_n)(\omega) = 0$ for all $\omega \in \Omega \setminus N$. We now claim that, for such ω's,

$$P(B)(\omega) = P(B_{n_0})\,(\omega) \tag{3.86}$$

for at least one index $n_0 = n_0(\omega)$. From this, by monotonicity of P, we will get $P(B)(\omega) = P(B_n)(\omega)$ for all $n \geq n_0(\omega)$, and hence

$$\lim_n P(B_n)\,(\omega) = \sup_n P(B_n)(\omega) = P(B)(\omega). \tag{3.87}$$

By virtue of arbitrariness of $\omega \notin N$ and by means of a density argument, from (3.87) we obtain $P(B) = (O)\lim_n P(B_n) = \bigvee_n P(B_n)$, that is the assertion.

So, to complete the proof it remains to prove (3.86). If (3.86) should not hold, then for at least one $\omega_0 \in \Omega \setminus N$ we should get $P(B)(\omega_0) \neq P(B_n)(\omega_0)$, and hence

$$P(B)(\omega_0) > P(B_n)(\omega_0) \tag{3.88}$$

for every $n \in \mathbb{N}$. So we have $P(B)(\omega_0) = P(B \setminus B_n)(\omega_0)$ for all $n \in \mathbb{N}$. Taking the infimum as n varies in $\mathbb{N}$, we should get $P(B)(\omega_0) = 0$. By monotonicity, $P(B_n)(\omega_0) = P(B)(\omega_0) = 0$ for all $n \in \mathbb{N}$, which contradicts (3.88). This concludes the proof. □

Let $s := \sum_{i=1}^n a_i \chi(A_i)$ be any nonnegative simple function, where $A_1, \ldots, A_n \in \Sigma$ and $\bigcup_{i=1}^n A_i = G$. If $A \in \Sigma$, we call *integral of s with respect to P on A* the quantity $\bigvee_{i=1}^n a_i P(A_i \cap A)$, and we denote it by the symbols $\int_A s\,dP$ or $I_A(s)$. If $A = G$, set $\int s\,dP = \int_G s\,dP = I(s) = I_G(s)$.

We now prove that our integral is well-defined, that is it does not depend on the choice of the representation of s. Indeed, if $s := \sum_{i=1}^n a_i \chi(A_i) = \sum_{k=1}^m b_k \chi(B_k)$, then for all $i = 1, \ldots, n$, $k = 1, \ldots, m$ we get $A_i = \bigcup_{k=1}^m (A_i \cap B_k)$,

$B_k = \bigcup_{i=1}^{n}(A_i \cap B_k)$, and hence $P(A_i) = \bigvee_{k=1}^{m} P(A_i \cap B_k)$, $P(B_k) = \bigvee_{i=1}^{n} P(A_i \cap B_k)$. From this it follows that

$$\bigvee_{i=1}^{n} a_i P(A_i) = \bigvee_{i=1}^{n}\left[\bigvee_{k=1}^{m} a_i P(A_i \cap B_k)\right] = \bigvee_{i,k} a_i P(A_i \cap B_k), \tag{3.89}$$

$$\bigvee_{k=1}^{m} b_k P(B_k) = \bigvee_{k=1}^{m}\left[\bigvee_{i=1}^{n} b_k P(A_i \cap B_k)\right] = \bigvee_{i,k} b_k P(A_i \cap B_k). \tag{3.90}$$

The assertion follows from (3.89) and (3.90) taking into account that $b_i = c_k$ whenever $A_i \cap B_k \neq \emptyset$ and $P(A_i \cap B_k) = 0$ whenever $A_i \cap B_k = \emptyset$. □

Proposition 2.21 (Boccuto, Dimitriou and Papanastassiou (2010a), Proposition 2) *Let* s *be any simple function and* $(B_n)_n$ *be any increasing sequence in* Σ, *with* $\bigcup_{n=1}^{\infty} B_n = B$. *Then,* $I_B(s) = \bigvee_n I_{B_n}(s) = (O)\lim_n I_{B_n}(s)$.

Proof: Let $s := \sum_{i=1}^{m} a_i \chi_{A_i}$. Then $I_B(s) = \bigvee_{i=1}^{m} a_i P(A_i \cup B)$, $I_{B_n}(s) = \bigvee_{i=1}^{m} a_i P(A_i \cup B_n)$. For each $i = 1, \dots, m$ we have $A_i \cap B_n \subset A_i \cap B \in \Sigma$ and $\bigcup_{n=1}^{\infty}(A_i \cap B_n) = A_i \cap B$. By Proposition 2.20 we get

$$P(A_i \cap B) = (O)\lim_n P(A_i \cap B_n) = \bigvee_n P(A_i \cap B_n). \tag{3.91}$$

Since the a_i's, $i = 1, \dots, m$, are positive, from (3.91) we get:

$$I_B(s) = \bigvee_{i=1}^{m} a_i P(A_i \cup B) = \bigvee_{i=1}^{m}\left[\bigvee_n a_i P(A_i \cup B_n)\right] = \tag{3.92}$$

$$= \bigvee_n \left[\bigvee_{i=1}^{m} a_i P(A_i \cup B_n)\right] = \bigvee_n I_{B_n}(s) = (O)\lim_n I_{B_n}(s).$$

This concludes the proof. □

Note that $I(s_1+s_2)\le I(s_1)+I(s_2)$ whenever s_1, s_2 are two simple non-negative functions and $I(s_1)\le I(s_2)$ whenever $s_1\le s_2$. Indeed, if (without loss of generality) $s_1=\sum_{i=1}^n a_i\chi(A_i)$, $s_2=\sum_{i=1}^n b_i\chi(A_i)$, then

$$I(s_1+s_2)=\bigvee_{i=1}^n (a_i+b_i)P(A_i)\le \bigvee_{i=1}^n a_i P(A_i)+\bigvee_{i=1}^n b_i P(A_i)=I(s_1)+I(s_2);$$

$$I(s_1)=\bigvee_{i=1}^n a_i P(A_i)\le \bigvee_{i=1}^n b_i P(A_i)=I(s_2) \textbf{ if } s_1\le s_2. \qquad \textbf{(3.93)}$$

Let $f\ge 0$ be a measurable function, $S_f:=\{s \text{ simple: } 0\le s\le f\}$ and $A\in\Sigma$. The quantity $I_A(f):=\bigvee_{s\in S_f} I_A(s)\in \overline{R}$ is called the *integral of* f *on* A *with respect to* P, which we denote also by the symbol $\int_A f\,dP$. Moreover we indicate with $\int_G f\,dP$ or $I(f)$ the quantity $I_G(f)$. We say that $f\ge 0$ is *integrable* (on G) if $I(f)\in R$.

It is not difficult to check that our integral satisfies the classical elementary properties: in particular we get $I_A(af)=aI_A(f)$ whenever $A\in\Sigma$, $a\ge 0$ and $f\ge 0$ is a measurable real-valued function, and from (3.93) it follows easily that, if $0\le f_1\le f_2$ and $A\in\Sigma$, then $I_A(f_1)\le I_A(f_2)$. Furthermore it is readily seen that $I_A(f)\le I_B(f)$ whenever $A\subset B$, $A,B\in\Sigma$ and $f\ge 0$ is measurable.

Proposition 2.22 (Absolute continuity, see Boccuto, Dimitriou and Papanastassiou (2010a), Proposition 4) *Let* $f\ge 0$ *be any measurable bounded function. If* $(O)\lim_n P(E_n)=0$ *for all* $n\in\mathbb{N}$, *then* $(O)\lim_n I_{E_n}(f)=0$.

Proof: Let $a>0$ be such that $0\le f\le a$. It is easy to see that $0\le I_{E_n}(f)\le a\,P(E_n)$ for all $n\in\mathbb{N}$. The assertion follows directly from this and the hypotheses. □

Observe that, even $R=\mathbb{R}$, Proposition 2.22 is in general not true without assuming boundedness of f (see for instance Agbeko (1994), Example 3.2).

We now prove some convergence theorems.

Theorem 2.23 (Boccuto, Dimitriou and Papanastassiou (2010a), Theorem 5) *Let $(f_n)_n$ be an increasing sequence of measurable non-negative functions, and $f := \lim_n f_n$ be the pointwise limit of $(f_n)_n$. Then $I(f) = (O)\lim_n I(f_n)$.*

Proof: First of all let us prove the inequality

$$I(f) \leq \bigvee_n I(f_n) = (O)\lim_n I(f_n). \qquad \textbf{(3.94)}$$

If the sequence $(I(f_n))_n$ is not bounded in R, then the right member in (3.94) is equal to $+\infty$ and so it is nothing to prove. So, let us suppose that $(I(f_n))_n$ is bounded in R: then, by Dedekind completeness of R, the quantity $\bigvee_n I(f_n) = (O)\lim_n I(f_n)$ exists in R.

Fix arbitrarily $\alpha \in (0,1)$ and $s \in S_f$. For each $n \in \mathbb{N}$, set $E_n := \{x \in X : f_n(x) \geq \alpha s(x)\}$. It is easy to check that $(E_n)_n$ is an increasing sequence in Σ and $G = \bigcup_{n=1}^{\infty} E_n$. By virtue of monotonicity of the integral, we get

$$I_{E_n}(\alpha s) \leq I_{E_n}(f_n) \leq I_G(f_n) = I(f_n). \qquad \textbf{(3.95)}$$

Since it is readily seen that $I(\alpha s) = \alpha I(s)$, from this, (3.95) and Proposition 2.21 it follows that

$$\alpha I(s) = I(\alpha s) = I_G(\alpha s) = (O)\lim_n I_{E_n}(\alpha s) = \bigvee_n I_{E_n}(\alpha s) \leq \bigvee_n I(f_n) = (O)\lim_n I(f_n). \qquad \textbf{(3.96)}$$

By arbitrariness of $\alpha \in (0,1)$, from (3.96) we get:

$$I(s) \leq \bigvee_n I(f_n) = (O)\lim_n I(f_n). \qquad \textbf{(3.97)}$$

By arbitrariness of $s \in S_f$, from (3.97) we obtain $I(f) = \bigvee_{s \in S_f} I(s) \leq \bigvee_n I(f_n)$ $= (O)\lim_n I(f_n)$, that is precisely (3.94). Concerning the opposite inequality, we have: $I(f_n) \leq I(f_{n+1}) \leq I(f)$, and so

$\bigvee_n I(f_n) = (O)\lim_n I(f_n) \le I(f)$. This concludes the proof. □

We now state the following monotone convergence theorem for decreasing sequences.

Theorem 2.24 (Boccuto, Dimitriou and Papanastassiou (2010a), Theorem 6) *Let $(k_n)_n$ be a decreasing sequence of measurable functions, and suppose that there is a positive real number a with $0 \le k_n \le a$ for all $n \in \mathbb{N}$, and set $k := \lim_n k_n$. Then $I(k) = (O)\lim_n I(k_n) = \bigwedge_n I(k_n)$.*

Proof: First of all, let us show that

$$I(f+h) \le I(f) + I(h) \tag{3.98}$$

for each measurable non-negative function f and for any $0 \le h \in S$, where S is the class of all simple functions. We have:

$$I(f+h) = \bigvee_{s \in S, 0 \le s \le f+h} I(s) = \bigvee_{s \in S, h \le s \le f+h} I(s) =$$

$$= \bigvee_{s \in S, 0 \le s \le f} I(s+h) \le \bigvee_{s \in S, 0 \le s \le f} [I(s) + I(h)] \le [\bigvee_{s \in S, 0 \le s \le f} I(s)] + I(h) = I(f) + I(h),$$

getting the claim.

Now, fix arbitrarily $\varepsilon > 0$, and let $B_n := \{x \in X : k_n(x) \le k(x) + \varepsilon\}$. It is easy to see that the sequence $(B_n)_n$ is increasing in Σ and $G = \bigcup_{n=1}^{\infty} B_n$. Thus $\emptyset = \bigcap_{n=1}^{\infty} (G \setminus B_n)$. By the continuity from above we get:

$$(O)\lim_n P(G \setminus B_n) = \bigwedge_n P(G \setminus B_n) = 0. \tag{3.99}$$

For each $n \in \mathbb{N}$ we get $k_n \le (k+\varepsilon) \vee a\chi_{(G \setminus B_n)} \le (k+\varepsilon) + a\chi_{(G \setminus B_n)}$. By (3.98) and monotonicity of the integral, we obtain:

$$I(k_n) \le I(k+\varepsilon) + aP(G \setminus B_n) \le I(k) + \varepsilon P(G) + aP(G \setminus B_n).$$

Hence,

$$(O)\limsup_n I(k_n) \le I(k) + \varepsilon P(G) + a\,(O)\limsup_n P(G \setminus B_n) = I(k) + \varepsilon P(G).$$

By arbitrariness of $\varepsilon > 0$ we get: $\bigwedge_n I(k_n) = (O)\lim_n I(k_n) = (O)\limsup_n I(k_n) \le I(k)$.

Concerning the converse inequality, note that, by monotonicity of the integral, $I(k) \le I(k_n)$ for all $n \in \mathbb{N}$, and hence $I(k) \le \bigwedge_n I(k_n) = (O)\lim_n I(k_n) = (O)\limsup_n I(k_n)$. □

Observe that, without assuming equiboundedness of the sequence $(k_n)_n$, in general Theorem 2.24 fails to hold, even when $R = \mathbb{R}$ (see Agbeko (1994), Example 3.1).

We now prove our version of the Fatou Lemma.

Lemma 2.25 (Boccuto, Dimitriou and Papanastassiou (2010a), Lemma 7) *If $(f_n)_n$ is a sequence of measurable non-negative functions, then $I(\liminf_n f_n) \le (O)\liminf_n I(f_n)$. Moreover, if the functions f_n, $n \in \mathbb{N}$, are equibounded, then we get also $I(\limsup_n f_n) \ge (O)\limsup_n I(f_n)$.*

Proof: We prove only the first inequality, since the proof of the second one is analogous. For every $n \in \mathbb{N}$, set $h_n := \bigwedge_{k \ge n} f_k$. The sequence $(h_n)_n$ is increasing, and so by the monotone convergence theorem we get:

$$I(\liminf_n f_n) = I(\bigvee_n h_n) = (O)\lim_n I(h_n) = (O)\liminf_n I(h_n) \le (O)\liminf_n I(f_n). \ \square$$

We now turn to the Lebesgue dominated convergence theorem.

Theorem 2.26 (Boccuto, Dimitriou and Papanastassiou (2010a), Theorem 8) *Let $(f_n)_n$ be an equibounded sequence of measurable non-negative functions, and suppose that $f := \lim_n f_n$ pointwise. Then $I(f) = (O)\lim_n I(f_n)$.*

Proof: Since $f = \limsup_n f_n = \liminf_n f_n$, by virtue of Lemma 2.25 we get

$$I(f) \le (O)\liminf_n I(f_n) \le (O)\limsup_n I(f_n) \le I(f). \qquad \textbf{(3.100)}$$

So all the inequalities in (3.100) are equalities, and hence the theorem is proved. □

3.2.3. Ultrafilter Measures and Integrals

We now give some other property of measures and integrals, and we deal with the $\mathcal{F}$-limit in the ultrafilter setting, to extend vector lattice-valued measures.

Given an abstract nonempty set T and an algebra $\mathcal{L}$ of subsets of T, let us denote by $\mathcal{D}$ the family of all finite tagged partitions Π of T, $\Pi = \{(A_1,t_1),(A_2,t_2),...,(A_k,t_k)\}$, where the *intervals* $A_1,\dots,A_k$ belong to $\mathcal{L}$ and $t_i \in A_i$ for every $i = 1,\dots,k$.

Given two tagged partitions Π_1 and Π_2, we say that Π_2 *refines* Π_1 iff every element of Π_2 is a subset of an element of Π_1. Given a partition $\Pi \in \mathcal{D}$, let $\mathcal{D}(\Pi)$ the class of all partitions which refine Π.

The classes $\mathcal{D}(\Pi)$ form, as Π varies in $\mathcal{D}$, a *filter basis* in $\mathcal{D}$, that is all class of subsets of $\mathcal{D}$ containing some family $\mathcal{D}(\Pi)$, form a filter in $\mathcal{D}$, called the *refinement filter* of $\mathcal{D}$.

Let $m: \mathcal{L} \to R$ be a finitely additive bounded measure. Then we have the following

Proposition 2.27 (see Boccuto and Candeloro (2009b), Proposition 4.2) *There exists a finitely additive measure* $M: \mathcal{P}(T) \to R$ *with* $M_{|\mathcal{L}} = m$.

Proof: For each $\Pi \in \mathcal{D}$, $\Pi = \{(A_1,t_1),(A_2,t_2),...,(A_k,t_k)\}$, and for every bounded function $f: T \to \mathbb{R}$, set

$$\Sigma(f,\Pi) = \sum_{j=1}^{k} f(t_j)m(A_j),$$

and pick an ultrafilter $\mathcal{F}$ of $\mathcal{D}$ which contains the refinement filter. Let e be an order unit in R with the property that the range of m is contained in $V[e]$. Then $\Sigma(f,\cdot)$ is a bounded function defined on $\mathcal{D}$ and taking values in $V[e]$, and so it has a weak $\mathcal{F}$-limit. Put

$$\mathcal{J}(f) := (w\mathcal{F})\lim_{\Pi\in\mathcal{D}} \Sigma(f,\Pi),$$

for each bounded function $f : T \to \mathbb{R}$, and in connection with the *same* maximal extension of $V[e]$. The function $M(A) := \mathcal{J}(\chi_A)$, defined on the class of all parts of T, extends m. It is not difficult to see that $\mathcal{J}$ is linear.

We now prove that $\mathcal{J}(\chi_A) = m(A)$ for every $A \in \mathcal{L}$. Fix $A \in \mathcal{L}$ and put $f = \chi_A$. Let Π_A be the partition formed by A and $T \setminus A$. For each $\Pi \in \mathcal{D}(\Pi_A)$ we get $\Sigma(f,\Pi) = m(A)$, and so $\mathcal{J}(f) = m(A)$, that is the claim. This ends the proof. □

Corollary 2.28 (see Boccuto and Candeloro (2009b), Corollary 4.3) *Let $P : \mathcal{P}(T) \to \mathbb{R}$ be any positive finitely additive measure, and let $f : T \to \mathbb{R}$ be any bounded function. Let $\mathcal{F}$ be an ultrafilter which contains the refinement filter, with $\mathcal{L} = \mathcal{P}(T)$. For each bounded function $f : T \to \mathbb{R}$, let us define*

$$\int_T f\,dP := (w\mathcal{F})\lim \Sigma(f,\Pi),$$

where $(w\mathcal{F})\lim$ is the weak $\mathcal{F}$-limit as in (2.80).

Then, $f \mapsto \int_T f\,dm$ is a linear monotone functional, which is equal to the usual integral, when f has finite range.

We now see that, given any nonempty set T and any function $\tau : T \to T$, it is always possible to construct a finitely additive τ-invariant measure $m : \mathcal{P}(T) \to R$.

Theorem 2.29 (see Boccuto and Candeloro (2009b), Theorem 4.4) *Let $\tau : T \to T$ be any function, $m : \mathcal{L} \to R$ be a τ-invariant bounded finitely additive measure, defined in an algebra $\mathcal{L}$ of subsets of T, that is $\tau^{-1}(A) \in \mathcal{L}$ and $m(A) = m(\tau^{-1}(A))$ for every $A \in \mathcal{L}$. Then there exists a τ-invariant finitely* *additive bounded extension* $\mu : \mathcal{P}(T) \to R$, *that is* $\mu(A) = \mu(\tau^{-1}(A))$ *for each* $A \subset T$.

Proof: Thanks to Proposition 2.27, we find a bounded finitely additive extension m_0 of m to the whole of $\mathcal{P}(T)$. Then let us define, by induction, $m_n(A) = m_{n-1}(\tau^{-1}(A))$ for any $A \subset T$ and $n \in \mathbb{N}$. Note that every m_n coincides with m on $\mathcal{L}$. Set $\mu_n := \frac{1}{n}\sum_{i=1}^{n} m_i$ for any $n \in \mathbb{N}$. If e is any upper bound for the range of $|m_0(\cdot)|$, then the measures m_n have the same upper bound, and

$$| m_n(A) - m_n(\tau^{-1}(A)) | \leq \frac{2e}{n} \text{ for every } A \subset T.$$

Let $\mathcal{F}$ be an ultrafilter of $\mathbb{N}$, pick any maximal extension of $V[e]$ and consider a pointwise weak $\mathcal{F}$-limit μ_0 of the sequence $(\mu_n)_n$. Then μ_0 is the required extension of m: indeed, it is easy to see that it is additive. Invariance follows from the fact that for each $A \in \mathcal{L}$ we get:

$$| \mu(A) - \mu(\tau^{-1}(A)) | = | (w\mathcal{F})\lim_n \mu_n(A) - (w\mathcal{F})\lim_n \mu_n(\tau^{-1}(A)) | \leq \lim_n \frac{2e}{n} = 0.$$

This ends the proof. □

3.3. Further Limit Theorems

In this section we give some convergence theorems, in which σ-additivity, (s)-boundedness and the related concepts are intended in the classical sense and not relatively to a single regulator or order sequence. In this context, a fundamental role in some crucial technical lemmas is played by the Maeda-Ogasawara-Vulikh representation theorem.

3.3.1. Brooks-Jewett Theorem

Theorem 3.1 (see Boccuto and Candeloro (2010), Theorem 3.1) *Let* $\Sigma \subset \mathcal{P}(G)$ *be a* σ*-algebra,* $\mathcal{G} \subset \Sigma$ *be a lattice, closed under countable disjoint unions,* R *be a Dedekind complete lattice group,* Ω *be as in the Maeda-Ogasawara-Vulikh representation theorem. Assume that* $m_j : \Sigma \to R$, $j \in \mathbb{N}$, *is an equibounded* *sequence of finitely additive measures. Suppose that* $(m_j)_j$ (RO)*-converges to a set function* m_0 *on* $\mathcal{G}$.

Then the measures $m_j(\cdot)(\omega)$, $j \in \mathbb{N}$, *are uniformly* (s)*-bounded on* $\mathcal{G}$ *for all* ω *belonging to the complement of a meager subset of* Ω, *and the measures* m_j *are uniformly* (s)*-bounded on* $\mathcal{G}$.

Proof: Observe that, since the m_j's are equibounded, then there exists a nowhere dense set $N_0 \subset \Omega$ such that for all $\omega \in \Omega \setminus N_0$ the functions $m_j(\cdot)(\omega)$, $j \in \mathbb{N}$, are real-valued, finitely additive and bounded on $\mathcal{G}$, and hence (s)-bounded on $\mathcal{G}$, by virtue of Theorem 3.12 of Chapter 2. Moreover, by (RO)-convergence, there is an (O)-sequence $(p_l)_l$ such that for every $l \in \mathbb{N}$ and $A \in \mathcal{G}$ there exists j_0 with

$$| m_j(A) - m_0(A) | \leq p_l \quad \text{for all} \quad j \geq j_0. \tag{3.101}$$

By the Maeda-Ogasawara-Vulikh representation theorem, a meager set $N \subset \Omega$ can be found, without loss of generality $N \supset N_0$, such that the sequence $(p_l(\omega))_l$ is a real-valued (O)-sequence, whenever $\omega \notin N$. Thus for every $l \in \mathbb{N}$ and $A \in \mathcal{G}$ there is $j_0 \in \mathbb{N}$ such that for all $\omega \in \Omega \setminus N$ and $j \geq j_0$ we get:

$$| m_j(A)(\omega) - m_0(A)(\omega) | \leq p_l(\omega). \tag{3.102}$$

This implies that $\lim_j m_j(A)(\omega) = m_0(A)(\omega)$ for each $A \in \mathcal{G}$ and $\omega \in \Omega \setminus N$. Thus for such ω's the real-valued measures $m_j(\cdot)(\omega)$ satisfy the hypotheses of the classical version of the Brooks-Jewett theorem (see Brooks and Jewett (1970), Theorem 2), and so they are uniformly (s)-bounded on $\mathcal{G}$. This concludes the first part of the assertion.

We now prove that the measures m_j, $j \in \mathbb{N}$, are uniformly (s)-bounded on $\mathcal{G}$. Fix arbitrarily any disjoint sequence $(C_k)_k$ in $\mathcal{G}$ and let us check that

$$\bigwedge_s [\bigvee_{k \geq s} (\bigvee_j [\bigvee_{B \in \mathcal{G}, B \subset C_k} | m_j(B) |])] = 0. \tag{3.103}$$

Since the measures $m_j(\cdot)(\omega)$ are uniformly (s)-bounded on $\mathcal{G}$ for every $\omega \in \Omega \setminus N$, where N is as in (3.102), then

$$\inf_s [\sup_{k \geq s} \{\sup_j [v_{\mathcal{G}}(m_j(\cdot)(\omega))(C_k)]\}] = \lim_k \{\sup_j [v_{\mathcal{G}}(m_j(\cdot)(\omega))(C_k)]\} = 0 \quad \textbf{(3.104)}$$

for all $\omega \notin N$. So, in the complement of a suitable meager set (containing N), for all $k \in \mathbb{N}$ we get:

$$\sup_j [\sup_{B \in \mathcal{G}, B \subset C_k} |m_j(B)(\omega)|] = \{\bigvee_j [\bigvee_{B \in \mathcal{G}, B \subset C_k} |m_j(B)|]\}(\omega). \quad \textbf{(3.105)}$$

From (3.104) and (3.105) it follows that, again in the complement of meager sets,

$$\bigwedge_s [\bigvee_{k \geq s} (\bigvee_j [\bigvee_{B \in \mathcal{G}, B \subset C_k} |m_j(B)|])](\omega) = 0. \quad \textbf{(3.106)}$$

By a density argument we get (3.103).

Thus $(O)\limsup_k (\bigvee_j [\bigvee_{B \in \mathcal{G}, B \subset C_k} |m_j(B)|]) = 0$, namely $(O)\limsup_k (\bigvee_j v_{\mathcal{G}}(m_j)(C_k)) = 0$. Thanks to arbitrariness of the chosen sequence $(C_k)_k$, we get uniform (s)-boundedness of the m_j's on $\mathcal{G}$. □

3.3.2. Dieudonné Theorem

We now prove a technical lemma, which uses the Maeda-Ogasawara-Vulikh theorem and is crucial, to prove different types of convergence theorems.

Lemma 3.2 (see Boccuto, Dimitriou and Papanastassiou, (2010c)) *Let $\mathcal{G} \subset \Sigma$ be a lattice, closed under countable disjoint unions, and let there exist a meager set $N \subset \Omega$ with the property that the measures $m_j(\cdot)(\omega)$, $j \in \mathbb{N}$, are uniformly (s)-bounded on Σ for all $\omega \notin N$. Fix $W \in \Sigma$, and assume that $(H_n)_n$ is a decreasing sequence in $\mathcal{G}$, such that $W \subset H_n$ for every $n \in \mathbb{N}$. If*

$$\lim_n [\sup_{A \in \Sigma, A \subset H_n \setminus W} |m_j(A)(\omega)|] = 0 \quad \textit{for every } j \in \mathbb{N} \quad \textbf{(3.107)}$$

in the complement of a meager set $N_W \subset \Omega$, *then*

$$\lim_n (\sup_j [\sup_{A \in \Sigma, A \subset H_n \setminus W} |m_j(A)(\omega)|]) = 0 \tag{3.108}$$

whenever $\omega \in \Omega \setminus (N \cup N_W)$.

Proof: Fix arbitrarily $\omega \in \Omega \setminus (N \cup N_W)$, and let $\mathcal{W} := \{A \in \Sigma : A \cap W = \emptyset\}$. For every $A \in \mathcal{W}$ and $j, q \in \mathbb{N}$ we have

$$m_j(A)(\omega) - m_j(A \setminus H_q)(\omega) = m_j(A \cap H_q)(\omega). \tag{3.109}$$

Since $A \cap H_q \subset H_{q-1} \setminus W$ for all $q \in \mathbb{N}$, from (3.107) and (3.109) for every $j \in \mathbb{N}$ we get:

$$m_j(A)(\omega) = \lim_q m_j(A \setminus H_q)(\omega) \tag{3.110}$$

uniformly with respect to $A \in \mathcal{W}$.

Suppose by contradiction that there exists $\varepsilon > 0$ with the property that for every $p \in \mathbb{N}$ there are $n \in \mathbb{N}$, $n > p$, $j \in \mathbb{N}$ and $A \in \Sigma$ such that $A \subset H_n \setminus W$, $|m_j(A)(\omega)| > \varepsilon$, and hence, thanks to (3.110),

$$|m_j(A \setminus H_q)(\omega)| > \varepsilon \tag{3.111}$$

for q large enough.

In correspondence with $p = 1$, there exist a set $A_1 \in \Sigma$ and three integers $n_1 \in \mathbb{N} \setminus \{1\}$, $j_1 \in \mathbb{N}$ and $q_1 > \max\{n_1, j_1\}$, with $A_1 \subset H_{n_1} \setminus W$, $|m_{j_1}(A_1)(\omega)| > \varepsilon$ and $|m_{j_1}(A_1 \setminus H_{q_1})(\omega)| > \varepsilon$. From (3.107), in correspondence with $j = 1, 2, \ldots, j_1$ we find an integer $h_1 > q_1$ with

$$|m_j(A)(\omega)| \le \varepsilon \tag{3.112}$$

whenever $n \ge h_1$ and $A \subset H_n \setminus W$.

Now we find a set $A_2 \in \Sigma$ and three integers $n_2 > h_1$, $j_2 \in \mathbb{N}$ and $q_2 > \max\{n_2, j_2\}$, with $A_2 \subset H_{n_2} \setminus W$ and

$$| m_{j_2}(A_2)(\omega) | > \varepsilon; \; | m_{j_2}(A_2 \setminus H_{q_2})(\omega) | > \varepsilon. \quad \textbf{(3.113)}$$

From (3.112) and (3.113) it follows that $j_2 > j_1$.

Proceeding by induction, it is possible to construct a sequence $(A_k)_k$ in Σ and three strictly increasing sequences in $\mathbb{N}$, $(n_k)_k$, $(j_k)_k$, $(q_k)_k$, with $q_k > n_k > q_{k-1}$, $q_k > j_k$, $A_k \subset H_{n_k} \setminus W$, $| m_{j_k}(A_k \setminus H_{q_k})(\omega) | > \varepsilon$ for all $k \geq 2$. But this is absurd, since the sets $A_k \setminus H_{q_k}$, $k \in \mathbb{N}$, are disjoint elements of Σ, $\omega \in \Omega \setminus (N \cup N_W)$, and the measures $m_j(\cdot)(\omega)$, $j \in \mathbb{N}$ are uniformly (s)-bounded on Σ for each fixed $\omega \in \Omega \setminus N$. This concludes the proof. □

A similar version of Lemma 3.2, related with global σ-additivity, global τ-continuity and global regularity of measures, is the following:

Lemma 3.3 (see also Boccuto and Dimitriou (2013c), Lemma 2.6) *With the same notations as in Lemma 3.2, let $(\sigma_p)_p$ be an (O)-sequence in R and $m_n : \Sigma \to R$, $n \in \mathbb{N}$, be finitely additive and globally uniformly (s)-bounded measures. Let $W \in \Sigma$ and pick a decreasing sequence $(H_k)_k$ in $\mathcal{G}$, such that $W \subset H_k$ for each $k \in \mathbb{N}$. If*

$$(O)\lim_k v(m_n)(H_k \setminus W) = 0 \text{ for each } n \in \mathbb{N} \quad \textbf{(3.114)}$$

with respect to the same (O)-sequence $(\sigma_p)_p$, then

$$(O)\lim_k \Big(\bigvee_n v(m_n)(H_k \setminus W)\Big) = 0$$

with respect to $(\sigma_p)_p$.

Proof: Let $\mathcal{W}$ be as in Lemma 3.2. For every $A \in \mathcal{W}$ and n, $q \in \mathbb{N}$, we have

$$m_n(A) - m_n(A \setminus H_q) = m_n(A \cap H_q). \quad \textbf{(3.115)}$$

Since $A \cap H_q \subset H_{q-1} \setminus W$ for any $q \in \mathbb{N}$, from (3.114) and (3.115), for each $n \in \mathbb{N}$ we get:

$$m_n(A) = (O)\lim_q m_n(A \setminus H_q) \quad \text{for all } A \in \mathcal{W}. \tag{3.116}$$

By contradiction, suppose that there is $p \in \mathbb{N}$ such that for every $r \in \mathbb{N}$ there exist n, $k \in \mathbb{N}$ with $k > r$ and $A \in \Sigma$ with $A \subset H_k \setminus W$, $|m_n(A)| \not\leq \sigma_p$. By (3.116), we get $|m_n(A \setminus H_q)| \not\leq \sigma_p$ for q sufficiently large.

We find a set $A_1 \in \Sigma$ and three integers $k_1 > 1$, $n_1 \in \mathbb{N}$ and $q_1 > \max\{k_1, n_1\}$, with $A_1 \subset H_{k_1} \setminus W$, $|m_{n_1}(A_1)| \not\leq \sigma_p$ and $|m_{n_1}(A_1 \setminus H_{q_1})| \not\leq \sigma_p$. By (3.114) we find a $h_1 > q_1$ with

$$|m_n(A)| \leq \sigma_p \tag{3.117}$$

whenever $k \geq h_1$ and $A \subset H_k \setminus W$.

Furthermore, there are $A_2 \in \Sigma$, $k_2 > h_1$, $n_2 \in \mathbb{N}$ and $q_2 > \max\{k_2, n_2\}$, with $A_2 \subset H_{k_2} \setminus W$ and

$$|m_{n_2}(A_2)| \not\leq \sigma_p; \quad |m_{n_2}(A_2 \setminus H_{q_2})| \not\leq \sigma_p. \tag{3.118}$$

From (3.117) and (3.118) it follows that $k_2 > k_1$.

Proceeding by induction, it is possible to construct a sequence $(A_k)_k$ in Σ and three strictly increasing sequences in $\mathbb{N}$, $(k_r)_r$, $(n_r)_r$, $(q_r)_r$, with $q_r > k_r > q_{r-1}$, $q_r > n_r$, $A_r \subset H_{k_r} \setminus W$, $|m_{n_r}(A_r \setminus H_{q_r})| \not\leq \sigma_p$ for all $r \geq 2$. This is absurd, since the sets $A_r \setminus H_{q_r}$, $r \in \mathbb{N}$, are disjoint, and because the measures m_n, $n \in \mathbb{N}$, are globally uniformly (s)-bounded on Σ with respect to the order sequence $(\sigma_p)_p$. $\square$

We now deal with some properties of (ℓ)-group-valued regular set functions. Let $\Sigma \subset \mathcal{P}(G)$ be a σ-algebra, and $\mathcal{G}$, $\mathcal{H} \subset \Sigma$ are two lattices, satisfying the same properties as above.

Lemma 3.4 (see also Boccuto and Candeloro (2010), Lemma 4.2) *Let R, G, Σ, $\mathcal{G}$, $\mathcal{H}$ be as above, and $m:\Sigma \to R$ be a regular measure. Then for each $A \in \Sigma$ we get:*

$$(m^{(\pm)})_{\Sigma}(A) = (m^{(\pm)})_{\mathcal{H}}(A), \; v_{\Sigma}(m)(A) = v_{\mathcal{H}}(m)(A). \tag{3.119}$$

(positive, negative parts and semivariation with respect to Σ and $\mathcal{H}$)

Moreover, for every $V \in \mathcal{G}$ one has also

$$(m^{(\pm)})_{\Sigma}(V) = (m^{(\pm)})_{\mathcal{G}}(V) \text{ and } v_{\Sigma}(m)(V) = v_{\mathcal{G}}(m)(V) \tag{3.120}$$

(positive, negative parts and semivariation with respect to $\mathcal{G}$ and Σ)

Finally for all $K \in \mathcal{H}$ we get:

$$\bigwedge_{H \in \mathcal{G}, K \subset H} \| m \|(H \setminus K) = 0. \tag{3.121}$$

Proof: We begin with the first part. To this aim, it is enough to show that

$$(m^{(\pm)})_{\Sigma}(A) \le (m^{(\pm)})_{\mathcal{H}}(A), \; v_{\Sigma}(m)(A) \le v_{\mathcal{H}}(m)(A).$$

Fix arbitrarily $A \in \Sigma$, and pick $B \subset A$, $B \in \mathcal{A}$: then there exists a sequence $(F_n)_n$ in $\mathcal{H}$, such that $F_n \subset F_{n+1} \subset B$ for all $n \in \mathbb{N}$ and $\bigwedge_n [v(m)(B \setminus F_n)] = 0$. Then we get $\bigwedge_n \| m \|(B \setminus F_n) = 0$: this clearly implies that

$$\bigwedge_n ((| m(B) | - | m(F_n) |) \vee (| m(F_n) | - | m(B) |)) = 0,$$

and hence $| m(B) | \le \bigvee_n | m(F_n) | \le v_{\mathcal{H}}(m)(A)$. So we have proved that, for every $A \in \Sigma$:

$$m^{(+)}(A) = \bigvee_{F \subset A, F \in \mathcal{H}} m(F) \le \bigvee_{F \subset A, F \in \mathcal{H}} m^{(+)}(F) \le m^{(+)}(A), \tag{3.122}$$

and similarly

$$m^{(-)}(A) = \bigvee_{F \subset A, F \in \mathcal{H}} (-m(F)) \leq \bigvee_{F \subset A, F \in \mathcal{H}} m^{(-)}(F) \leq m^{(-)}(A), \qquad \textbf{(3.123)}$$

$$v_{\Sigma}(m)(A) = \bigvee_{F \subset A, F \in \mathcal{H}} | m(F) | \leq \bigvee_{F \subset A, F \in \mathcal{H}} v_{\Sigma}(m)(F) \leq v_{\Sigma}(m)(A).$$

So, all inequalities in (3.122) and (3.123) are equalities. Since $m^{(\pm)}$ are positive measures, then we deduce that

$$\bigwedge_{F \in \mathcal{H}, F \subset A} \| m \|(A \setminus F) = 0 \text{ for all } A \in \Sigma. \qquad \textbf{(3.124)}$$

Let us consider an arbitrary element $K \in \mathcal{H}$: since all elements F of $\mathcal{H}$ are complements of elements of $\mathcal{G}$, by (3.124) we get

$$0 \leq \bigwedge_{H \in \mathcal{G}, K \subset H} \| m \|(H \setminus K) \leq \bigwedge_{F \in \mathcal{H}, F \subset G \setminus K} \| m \|((G \setminus K) \setminus F) = 0. \qquad \textbf{(3.125)}$$

Thus, all terms in (3.125) are equal to zero, and (3.121) is proved.

We now turn to (3.120): we just prove the last equality, since the others are similar. To this aim, fix an arbitrary element $V \in \mathcal{G}$, and set $S := v_{\mathcal{G}}(m)(V)$, $T := v_{\Sigma}(m)(V)$. Clearly $S \leq T$, so we just prove the converse inequality. Thanks to the previous step, we have $T = \bigvee_{F \in \mathcal{H}, F \subset V} | m(F) |$,

and hence all we must show is that $| m(F) | \leq S$ for any element $F \subset V$ with $F \in \mathcal{H}$.

So, let F be such a set; then, for every element $H \in \mathcal{G}$ with $F \subset H$, we have

$$| m(F) | = | m(H \cap V) | + | m(F) | - | m(H \cap V) | \leq S + | m(F) | - | m(H \cap V) |.$$

Since H is arbitrary, taking into account (3.25), we have:

$$| m(F) | - S \leq \bigwedge_{H \in \mathcal{G}, F \subset H} |m(F)| - | m(H \cap V)| \leq \bigwedge_{H \in \mathcal{G}, F \subset H} \| m \| (H \setminus F) = 0,$$

namely $| m(F) | \leq S$. Since F was arbitrary, this ends the proof. □

Before proving our versions of the Dieudonné theorem, we state the following

Theorem 3.5 (see also Boccuto and Candeloro (2010), Theorem 4.4) *Under the same assumptions above, suppose that* $m_j : \Sigma \to R$, $j \in \mathbb{N}$, *is an equibounded sequence of regular measures,* (RO)*-convergent to* m_0 *on* $\mathcal{G}$, *and let* $A \in \Sigma$, $(F_n)_n$ *in* $\mathcal{H}$, $(G_n)_n$ *in* $\mathcal{G}$ *satisfy* $F_n \subset F_{n+1} \subset A \subset G_{n+1} \subset G_n$ *for any* $n \in \mathbb{N}$. *Moreover, assume that* $\bigwedge_n [v(m_j)(G_n \setminus F_n)] = 0$ *for every* $j \in \mathbb{N}$. *Then,* $\bigwedge_n [\bigvee_j v(m_j)(G_n \setminus F_n)] = 0$.

Proof: First of all we observe that, by virtue of Lemma 3.4, $v := v_\Sigma$ and $v_\mathcal{G}$ are equivalent, because, in the involved semivariations, we deal with elements of $\mathcal{G}$.

By Theorem 3.1 there exists a meager set $N \subset \Omega$ such that the real-valued measures $m_j(\cdot)(\omega)$ are uniformly (s)-bounded on $\mathcal{G}$ for all $\omega \notin N$.

Fix arbitrarily $A \in \Sigma$ and let $(F_n)_n$, $(G_n)_n$ be as in the hypotheses. By arguing analogously as in (3.103-3.106), we find a meager set $N^* \subset \Omega$ (depending on A), with

$$\lim_n [v_\mathcal{G}(m_j(\cdot)(\omega))(G_n \setminus F_n)] = \inf_n [v_\mathcal{G}(m_j(\cdot)(\omega))(G_n \setminus F_n)] = 0$$

for every $j \in \mathbb{N}$ and $\omega \notin N^*$. By Lemma 3.2, we get

$$\inf_n \{\sup_j [v_\mathcal{G}(m_j(\cdot)(\omega))(G_n \setminus F_n)]\} = \inf_n \{\sup_j [v_\mathcal{G}(m_j(\cdot)(\omega))(G_n' \setminus W)]\} = 0 \qquad \textbf{(3.126)}$$

for each $\omega \notin N \cup N^*$.

The assertion follows from (3.126), proceeding again analogously as in (3.103-3.106). □

We now prove that, if a sequence $(m_j)_j$ of equibounded regular finitely additive measures (RO)-converges in $\mathcal{G}$, then they are uniformly regular and have pointwise limit on the whole of Σ.

Theorem 3.6 (see also Boccuto and Candeloro (2010), Theorem 5.1) *With the same notations as before, fix* $A \in \Sigma$, *let* $(G_n)_n$, $(F_n)_n$ *satisfy the hypotheses of Theorem 3.5. and suppose that* $(m_j)_j$ *is* (RO)*-convergent to* m_0 *on* $\mathcal{G}$. *Then the following assertions hold.*

(j) *The measures* m_j, $j \in \mathbb{N}$, *are uniformly regular.*

(jj) *The sequence* $(m_j(A))_j$ *is* (O)*-Cauchy in* R *for each* $A \in \Sigma$.

(jjj) *The measure* m_0, *defined by setting*

$$m_0(A) := (O)\lim_j m_j(A), \tag{3.127}$$

is regular on Σ.

Proof: (j) Uniform regularity of the measures m_j follows easily from Theorem 3.5.

(jj) Fix arbitrarily $A \in \Sigma$. By uniform regularity of m_j, $j \in \mathbb{N}$, there is a sequence $(G_n)_n$ in $\mathcal{G}$ with $A \subset G_{n+1} \subset G_n$ for all $n \in \mathbb{N}$ and

$$\bigwedge_n [\bigvee_j v(m_j)(G_n \setminus A)] = (O)\lim_n [\bigvee_j v(m_j)(G_n \setminus A)] = 0.$$

Let $(v_n)_n$ be an (O)-sequence with $|m_j(G_n) - m_j(A)| \le v_n$ for all $j, n \in \mathbb{N}$, and let $(p_l)_l$ be an (O)-sequence, related with (RO)-convergence of $(m_j)_j$ to m_0 on $\mathcal{G}$.

For every l, $n \in \mathbb{N}$ there exists $j^* \in \mathbb{N}$ with $|m_p(G_n) - m_q(G_n)| \le 2p_l$ whenever $p, q \ge j^*$. In particular, to each $n \in \mathbb{N}$ we can associate a positive integer $j_n > n$ with

$$\begin{aligned} |m_p(A) - m_q(A)| &\le |m_p(A) - m_p(G_n)| + |m_p(G_n) - m_q(G_n)| + |m_q(G_n) - m_q(A)| \\ &\le 2p_n + 2v_n \end{aligned}$$

for all $p,q \geq j_n$. Set $j_0 := 0$, $p_0 := p_1$, $v_0 := v_1$. We can assume $j_{n-1} < j_n$ for any $n \in \mathbb{N}$. For every j there is a positive integer $n = n(j)$ with $j_n \leq j < j_{n+1}$. Set $w_j := 2p_{n(j)} + 2v_{n(j)}$, $j \in \mathbb{N}$. Then $(w_j)_j$ is an (O)-sequence and that $|m_j(A) - m_{j+r}(A)| \leq w_j$ for all $j, r \in \mathbb{N}$. Therefore $(m_j(A))_j$ is (O)-Cauchy.

(jjj) For each fixed $A \in \Sigma$, define $m_0(A) := (O) \lim_j m_j(A)$. This limit exists in R, since by (jj) the sequence $(m_j(A))_j$ is (O)-Cauchy and from the fact that Dedekind completeness implies (O)-completeness. Regularity of m_0 is an easy consequence of definition of m_0 and uniform regularity of the measures m_j, $j \in \mathbb{N}$. □

Now we give a Vitali-Hahn-Saks-type theorem.

Theorem 3.7 (see Boccuto, Dimitriou and Papanastassiou (2011a), Theorem 3.3) *Let $m_j : \Sigma \to R$, $j \in \mathbb{N}$, be an equibounded sequence of finitely additive measures, $\lambda : \Sigma \to [0,+\infty]$ be a finitely additive measure, with the property that $(RO) \lim_j m_j = m_0$ and every m_j, $j \in \mathbb{N}$, is λ-absolutely continuous on Σ.*

Then the set functions m_j, $j \in \mathbb{N}$, are equiabsolutely continuous with respect to λ and m_0 is λ-absolutely continuous on Σ.

Proof: By Theorem 3.1, we get the existence of a meager set $N \subset \Omega$ such that the measures $m_j(\cdot)(\omega)$, $j \in \mathbb{N}$, are uniformly (s)-bounded on Σ for any $\omega \in \Omega \setminus N$.

Fix arbitrarily a decreasing sequence $(H_n)_n$ with $\lim_n \lambda(H_n) = 0$, and set $W := \bigcap_{n=1}^{\infty} H_n$. By λ-absolute continuity of m_j, $j \in \mathbb{N}$, $(O) \lim_n v(m_j)(H_n) = 0$, and hence $v(m_j)(W) = 0$ for all $j \in \mathbb{N}$. Thanks to λ-absolute continuity of m_j, $j \in \mathbb{N}$ and the Maeda-Ogasawara-Vulikh theorem we find a meager set $N' \subset \Omega$, $N' \supset N$, with

$$\lim_n [\sup_{A\in\Sigma, A\subset H_n} |m_j(A)(\omega)|] = \inf_n [\sup_{A\in\Sigma, A\subset H_n} |m_j(A)(\omega)|] = 0 \quad \textbf{(3.128)}$$

for any $\omega \in \Omega \setminus N'$ and $j \in \mathbb{N}$. Since $v(m_j)(W) = 0$, we find a meager set $N'' \subset \Omega$, $N'' \supset N'$, with

$$\sup_{A\in\Sigma, A\subset W} |m_j(A)(\omega)| = 0 \quad \textbf{(3.129)}$$

for every $\omega \in \Omega \setminus N''$ and $j \in \mathbb{N}$. From (3.128) and (3.129) it follows that

$$\lim_n [\sup_{A\in\Sigma, A\subset H_n \setminus W} |m_j(A)(\omega)|] = \inf_n [\sup_{A\in\Sigma, A\subset H_n \setminus W} |m_j(A)(\omega)|] = 0$$

for all $\omega \in \Omega \setminus N''$ and $j \in \mathbb{N}$. By Lemma 3.2, for all $\omega \notin N''$ we get:

$$\lim_n (\sup_j [\sup_{A\in\Sigma, A\subset H_n \setminus W} |m_j(A)(\omega)|]) = \inf_n (\sup_j [\sup_{A\in\Sigma, A\subset H_n \setminus W} |m_j(A)(\omega)|]) = 0$$

and hence, taking into account (3.129), we obtain

$$\lim_n (\sup_j [\sup_{A\in\Sigma, A\subset H_n} |m_j(A)(\omega)|]) = \inf_n (\sup_j [\sup_{A\in\Sigma, A\subset H_n} |m_j(A)(\omega)|]) = 0 \quad \textbf{(3.130)}$$

for every $\omega \notin N''$. From (3.130) and the Maeda-Ogasawara-Vulikh theorem we get

$$\{\bigwedge_n (\bigvee_j [\bigvee_{A\in\Sigma, A\subset H_n} |m_j(A)|])\}(\omega) = 0$$

for each $\omega \notin N''$. By a density argument, it follows that

$$0 = \bigwedge_n (\bigvee_j [\bigvee_{A\in\Sigma, A\subset H_n} |m_j(A)|]) = \bigwedge_n (\bigvee_j v(m_j)(H_n)) = (O)\lim_n (\bigvee_j v(m_j)(H_n)),$$

getting uniform λ-absolute continuity of the m_j's. From this and (RO)-convergence of $(m_j)_j$ to m_0 we get λ-absolute continuity of m_0. □

Similarly as Theorem 3.7, it is possible to prove the following results.

Theorem 3.8 (Nikodým convergence theorem) (see Boccuto, Dimitriou and Papanastassiou (2010c), Theorem 3.6) *Assume that* $(m_j : \Sigma \to R)_j$ *is an equibounded sequence of* σ*-additive measures, and suppose that there exists a set function* $m_0 : \Sigma \to R$ *such that* $(RO) \lim_j m_j = m_0$. *Then* m_0 *is* σ*-additive and the* m_j*'s are uniformly* σ*-additive.*

Theorem 3.9 (see Boccuto and Dimitriou (2011c), Theorem 3.6) *Let* τ *be a Fréchet-Nikodým topology on* Σ *and* $m_n : \Sigma \to R$, $n \in \mathbb{N}$, *be an equibounded sequence of finitely additive and* τ*-continuous measures on* Σ, (RO)*-convergent to* $m_0 : \Sigma \to R$. *Then the measures* m_n, $n \in \mathbb{N}$, *are uniformly* τ*-continuous and* m_0 *is* τ*-continuous on* Σ.

3.4. Decomposition Theorems for (ℓ)-Group-Valued Measures

We now deal with different types of convergence and decomposition theorems (Lebesgue, Sobczyk-Hammer, Yosida-Hewitt). We give a uniform picture of the topic, by using the so-called σ*-ideal principle*: this tool is the basis of several decomposition theorems. For further aspects and developments of the decomposition theory in the setting of lattice groups or vector lattices and filter convergence, see also Candeloro and Sambucini (2014a).

3.4.1. Lebesgue-Type Decompositions

Let R be a weakly σ-distributive and super Dedekind complete (ℓ)-group. Given a σ-algebra $\Sigma \subset \mathcal{P}(G)$, a lattice $\mathcal{E} \subset \Sigma$, and any set function $m : \Sigma \to R$, we denote by $m_{|\mathcal{E}}$ the restriction of m to $\mathcal{E}$.

A σ*-ideal* is a family $\mathcal{J}$ of elements from Σ, closed under countable unions and arbitrary intersections. We recall the σ*-ideal principle*, which will be useful to prove the next results in the context of (ℓ)-groups.

Theorem 4.1 (see Boccuto and Candeloro (2005), Theorem 2.11) *Suppose that* $m_n : \Sigma \to R$, $n \in \mathbb{N}$, *is an increasing sequence of positive* σ*-additive measures,* $(\mathcal{J}_n)_n$ *is a decreasing sequence of* σ*-ideals in* Σ, *and* $\mathcal{J} := \bigcap_{n=1}^{\infty} \mathcal{J}_n$.

Then the following conditions are equivalent.

4.1.1) *There exists an element* $J \in \mathcal{J}$ *with*

$$m_n(J) = \bigvee \{m_n(F) : F \in \mathcal{J}_n\}$$

for every $n \in \mathbb{N}$.

4.1.2) *For each* $n \in \mathbb{N}$ *and* $A \in \mathcal{J}_n$ *with* $m_n(A) \neq 0$, *there is a subset* $B \subset A$, $B \in \mathcal{J}_{n+1}$, *with* $m_{n+1}(B) \neq 0$.

Proof: For every $n \in \mathbb{N}$, set $\alpha_n = \bigvee \{m_n(F) : F \in \mathcal{J}_n\}$. We first show that for any n there is $E_n \in \mathcal{J}_n$, with $m_n(E_n) = \alpha_n$. Choose $n \in \mathbb{N}$: since R is super Dedekind complete, we find a sequence $(F_j^n)_j$ in $\mathcal{J}_n$, with $\alpha_n = \bigvee \{m_n(F_j^n) : j \in \mathbb{N}\}$. Without restriction, let us assume that $(F_j^n)_j$ is increasing, and denote by E_n the union of F_j^n, as j varies in $\mathbb{N}$. Observe that E_n belongs to $\mathcal{J}_n$, and

$$m_n(E_n) = (O)\lim_j m_n(F_j^n) = \alpha_n.$$

Now we prove that 4.1.2) implies 4.1.1). For every n, put $J_n := \bigcup_{k=n}^{\infty} E_k$. Since $E_k \in \mathcal{J}_k \subset \mathcal{J}_n$ for any $k \geq n$, then $J_n \in \mathcal{J}_n$, and thus $m_n(J_n) = m_n(E_n)$. Therefore $J := \bigcap_{n=1}^{\infty} J_n$ belongs to $\mathcal{J}$. We now show that $m_n(J) = m_n(J_n)$ for any $n \in \mathbb{N}$.

Without restriction, we will prove this only for $n = 1$. To prove that $m_1(J) = m_1(J_1)$, it is enough to show that $m_1(J_1 \setminus J_h) = 0$ for every $h > 1$. If $m_1(J_1 \setminus J_2) \neq 0$, then $A := J_1 \setminus J_2 \in \mathcal{J}_1$. By 4.1.2) there is $B \in \mathcal{J}_2$, with $B \subset A$ and $m_2(B) \neq 0$: so $B \cup J_2 \in \mathcal{J}_2$ and $m_2(B \cup J_2) \neq m_2(J_2)$, obtaining a contradiction with maximality of J_2. Thus $m_1(A) = 0$ and $m_1(J_1) = m_1(J_2)$. A similar argument shows that $m_2(J_2) = m_2(J_3)$, hence $m_1(J_2 \setminus J_3) \leq m_2(J_2 \setminus J_3) = 0$. By induction we obtain $m_1(J_1) = m_1(J_h)$ for any $h > 1$.

We now prove that 4.1.1) implies 4.1.2). If 4.1.1) holds and 4.1.2) is not fulfilled, then there exist an integer n and a set $A \in \mathcal{J}_n$ with $m_n(A) \neq 0$ and $m_{n+1}(B) = 0$ for any set $B \in \mathcal{J}_{n+1}$, $B \subset A$.

Let J be as in 4.1.1). As $J \cap A$ belongs to $\mathcal{J}_{n+1}$, then $m_n(J \cap A) = m_{n+1}(J \cap A) = 0$. So, $J \cup A \in I_n$ and $m_n(J \cup A) = m_n(J) + m_n(A) \neq \alpha_n$, which contradicts maximality of J. □

Let $\mathcal{L} \subset \mathcal{P}(G)$ be an algebra, and $\sigma(\mathcal{L})$ be the σ-algebra generated by $\mathcal{L}$. Also in proving the decomposition theorems in (ℓ)-group setting, we will use the Stone Isomorphism technique. Let Q^* be the Stone space associated with $\mathcal{L}$, $\mathcal{Q}$ be the algebra of all open-closed subsets of Q^*, $\psi : \mathcal{L} \to \mathcal{Q}$ be the Stone isomorphism and $m : \mathcal{L} \to R$ any finitely additive measure. The measure $m \circ \psi^{-1} : \mathcal{Q} \to R$ admits a globally σ-additive measure $M : \sigma(\mathcal{Q}) \to R$, which is called the *Stone extension* of m.

Let $m, \lambda : \mathcal{L} \to R$ be two finitely additive bounded measures. Then m is said to be *singular* with respect to λ (shortly, $m \perp_{\|\cdot\|} \lambda$) iff there exists an (O)-sequence $(b_j)_j$ in R such that for every $j \in \mathbb{N}$ there is $F_j \in \mathcal{L}$, with $\|\lambda\|(F_j) \leq b_j$ and $\|m\|(G \setminus F_j) \leq b_j$. Moreover, m is said to be *orthogonal* to λ ($m \perp \lambda$) iff there exists $F \in \mathcal{L}$ with $\|m\|(F) = \|\lambda\|(G \setminus F) = 0$.

Observe that orthogonality implies singularity. Moreover, we have the following:

Proposition 4.2 (see Boccuto and Candeloro (2005), Proposition 4.4) *Let m, $\lambda : \Sigma \to R$ be two positive σ-additive measures, with $m \perp_{\|\cdot\|} \lambda$. Then $m \perp \lambda$.*

Proof: Without restriction, we suppose that m and λ are positive. Set $u = m(G) \vee \lambda(G)$, and let $(b_j)_j$ be an (O)-sequence in R associated with singularity. By Lemma 2.4 of Chapter 2, we find an $\omega \in \mathbb{N}^{\mathbb{N}}$ with the property that the sequence $(s_N)_N$ is an (O)-sequence (we put

$$s_N = u \wedge \sum_{n=N}^{\infty} b_{\omega(n)}, \quad N \in \mathbb{N}).$$

For any n there is $F_n \in \Sigma$, with $\lambda(F_n) \le b_{\omega(n)}$ and $m(G \setminus F_n) \le b_{\omega(n)}$. For $N \in \mathbb{N}$, set $B_N := \bigcup_{n=N}^{\infty} F_n$.

We obtain $\lambda(B_N) \le s_N$ and $m(G \setminus B_N) = 0$, since $m(G \setminus B_N) \le m(G \setminus F_n) \le s_n$ for all $n \ge N$. Thus the set $B = \bigcap_{N=1}^{\infty} B_N$ satisfies the required properties, since $\lambda(B) = m(G \setminus B) = 0$. □

We now present a Lebesgue decomposition result.

Theorem 4.3 (see also Boccuto and Candeloro (2004a), Theorem 6.3, Boccuto and Candeloro (2005), Theorem 4.5) *Let m, $\lambda : \Sigma \to R$ be two positive globally σ-additive measures. Then there exists a set $F \in \Sigma$ with $m|_F \ll \lambda$ and $m|_{G \setminus F} \perp \lambda$. (Such a decomposition is said to be* ***sectional****). Furthermore, if $\lambda \le m$ and $\lambda \ne 0$, then $m(F) \ne 0$.*

Proof: Let $\mathcal{J}$ be the σ-ideal of all λ-null sets. By Theorem 4.1, there is a set $E \in \mathcal{J}$ with $m(E) = \bigvee\{m(H) : H \in \mathcal{J}\}$. Then $F := G \setminus E$ is the requested set. The final assertion follows from the fact that λ is not orthogonal to m. □

The same argument proves the following result.

Theorem 4.4 (see Boccuto and Candeloro (2005), Theorem 4.6) *Assume that $m_n : \Sigma \to R$, $n \in \mathbb{N}$, is a sequence of globally σ-additive and positive measures.*

Then for every globally σ-additive and positive measure $\lambda : \Sigma \to R$ there is a set $V \in \Sigma$, with $m_n|_V \perp \lambda$, $m_n|_{G \setminus V} \ll \lambda$ for every $n \in \mathbb{N}$.

We now give the following:

Theorem 4.5 (see also Boccuto and Candeloro (2004a), Theorem 6.5, Boccuto and Candeloro (2005), Theorem 4.7) *Let $m_n : \Sigma \to R$, $n \in \mathbb{N}$, be an equibounded sequence of positive and globally σ-additive measures, and assume that $(RO)\lim_n m_n = m$ exists in R.*

Then m *is globally* σ*-additive, and for every positive and globally* σ*-additive measure* $\lambda:\Sigma\to R_0^+$ *we have:*

$$(RO)\lim_n m_n^a = m^a,\ (RO)\lim_n m_n^s = m^s,$$

where m_n^a *and* m_n^s *are the (globally) absolutely continuous and the singular part of* m_n*, and analogously for* m^a*,* m^s*.*

Proof: We apply Theorem 4.4 to the double sequence $(m_n^{(+)}, m_n^{(-)})_n$, and we find a set $U \in \Sigma$ independent of n, with

$$m_n^{(+)}|_U \perp \lambda,\ m_n^{(+)}|_{G\setminus U} \ll \lambda,$$

$$m_n^{(-)}|_U \perp \lambda,\ m_n^{(-)}|_{G\setminus U} \ll \lambda,\ n \in \mathbb{N}.$$

We get that $m_n|_U \perp \lambda$ and $m_n|_{G\setminus U} \ll \lambda$ for every n, and also $m|_U \perp \lambda$, $m|_{G\setminus U} \ll \lambda$.

Finally, from (RO)-convergence of $(m_n)_n$ to m it follows (RO)-convergence of $(m_n|_U)_n$ to $m|_U$ and $(m_n|_{G\setminus U})_n$ to $m|_{G\setminus U}$. □

We now prove the existence of a Lebesgue-type decomposition in the finitely additive setting.

Theorem 4.6 (see also Boccuto and Candeloro (2004a), Theorem 6.6, Boccuto and Candeloro (2005), Theorem 4.8) *Let* m*,* $\lambda:\mathcal{L}\to R$ *be two positive and globally* (s)*-bounded measures.*

Then $m = m^a + m^s$ *is a decomposition in two finitely additive positive measures, with* $m^a \ll \lambda$ *and* $m^s \perp_{\|\cdot\|} \lambda$*.*

Proof: Let Q^* be the Stone space in correspondence with $\mathcal{L}$, and ψ be the lattice isomorphism from $\mathcal{L}$ to the algebra $\mathcal{Q}$ of all open-closed subsets of Q^*. The measures $m_1 := m\circ\psi^{-1}$ and $\lambda_1 := \lambda\circ\psi^{-1}$ are globally σ-additive on $\mathcal{Q}$. Let us denote by M and P, respectively, the Carathéodory extensions of m_1 and of λ_1, defined on the σ-algebra $\sigma(\mathcal{Q})$. We apply Theorem 4.3, and let $M=M^a+M^s$ be the Lebesgue decomposition associated with P. Now, we restrict M^a and M^s to the algebra $\mathcal{Q}$, and let $m^a_{Q^*}$, $m^s_{Q^*}$ be the corresponding associated measures.

We claim that $m^a_{Q^*} \ll \lambda_1$ and $m^s_{Q^*} \perp \lambda_1$: from this it will follow that the measures

$$m^a := m^a_{Q^*} \circ \psi, \; m^s := m^s_{Q^*} \circ \psi,$$

give the requested decomposition of m.

Global absolute continuity of $m^a_{Q^*}$ follows from global absolute continuity (with respect to P) of the measure M^a. We now prove singularity of $m^s_{Q^*}$. We know that there exists a set $A \in \sigma(\mathcal{Q})$ with $P(Q^* \setminus A) = M(A) = 0$, however we need to find a similar set in $\mathcal{Q}$. By Theorem 3.29 of Chapter 2 there exists an (O)-sequence $(b_k)_k$ such that for any $k \in \mathbb{N}$ there exists a set $F \in \mathcal{Q}$ with

$$M(A \Delta F) \le b_k \text{ and}$$

$$P(A \Delta F) = P((Q^* \setminus A) \Delta (Q^* \setminus F)) \le b_k.$$

This implies that

$$P(Q^* \setminus F) = \lambda_1(Q^* \setminus F) \le b_k$$

and

$$M^s(F) = m^s_{Q^*}(F) \le b_k,$$

that is singularity of $m^a_{Q^*}$ with respect to λ_1. This ends the proof. □

We now prove the Lebesgue decomposition theorem in the finitely additive setting.

Theorem 4.7 (see also Boccuto and Candeloro (2004a), Theorem 6.9, Boccuto and Candeloro (2005), Theorem 4.9) *Let* $m_n : \Sigma \to R$, $n \in \mathbb{N}$, *be any equibounded sequence of globally* (s) *-bounded finitely additive measures, such that* $(RO)\lim_n m_n = m$ *exists in* R.

Then m *is globally* (s)*-bounded, and for every finitely additive positive measure* $\lambda : \Sigma \to R$ *we get* $(RO)\lim_n m_n^a = m^a$, $(RO)\lim_n m_n^s = m^s$, *where* m_n^a, m_n^s, m^a, m^s *have the same meaning as in Theorem* 4.5.

Proof: Let $\mathcal{Q}$ be the algebra of all open-closed sets of the Stone space associated with $\mathcal{L}$, and $\psi : \mathcal{L} \to \mathcal{Q}$ be the Stone Isomorphism. Let M_n, M, P be the globally σ-additive extensions of m_n, m, λ to $\sigma(\mathcal{Q})$, respectively (we observe that m is globally (s)-bounded, and hence its Stone extension M does exist).

We now prove that the measures M_n (RO)-converge to M in $\sigma(\mathcal{Q})$. From the assumptions, we know that there exists an (O)-sequence $(b_k)_k$ related to (RO)-convergence of the sequence $(m_n)_n$. From Theorem 3.29 of Chapter 2 we know that there exists an (O)-sequence $(c_k)_k$ such that, for every set $A \in \sigma(\mathcal{Q})$ and $k \in \mathbb{N}$, there is a set $F \in \mathcal{Q}$ with

$$v(M_n)(F\Delta A) \le c_k,$$

$$v(M)(F\Delta A) \le c_k \text{ for each } n.$$

Let us take $a_k = b_k + 2c_k$, $k \in \mathbb{N}$: then $(a_k)_k$ is an (O)-sequence. If we fix $A \in \sigma(\mathcal{Q})$ and $k \in \mathbb{N}$, then we find a set $F \in \mathcal{Q}$, satisfying the previous relations. Moreover there is $n_0 \in \mathbb{N}$, with

$$| M_n(F) - M(F) | \le b_k \text{ for all } n \ge n_0.$$

For such n's we get:

$$| M_n(A) - M(A) | \le v(M_n)(A\Delta F) + v(M)\,(A\Delta F) + | M_n(F) - M(F) | \le$$

$\leq 2c_k + | M_n(F) - M(F) | \leq 2c_k + b_k = a_k$.

Now we apply Theorem 4.5 to the sequence $(M_n)_n$, obtaining the existence of a set $U \in \sigma(\mathcal{Q})$ such that

a) $M_n|_U \perp P$, $M_n|_{Q^* \setminus U} \ll P$,

b) $(RO)\lim_n M_n|_U = M|_U$, $(RO)\lim_n M_n|_{Q^* \setminus U} = M|_{Q^* \setminus U}$.

As in Theorem 4.6, if we denote by m^a_{n,Q^*}, m^s_{n,Q^*}, $m^a{}_{Q^*}$, $m^s{}_{Q^*}$ respectively the restrictions to $\mathcal{Q}$ of $M_n|_{Q^* \setminus U}$, $M_n|_U$, $M|_{Q^* \setminus U}$, $M|_U$, then $m^a_n := m^a_{n,Q^*} \circ \psi$, $m^s_n := m^s_{n,Q^*} \circ \psi$ are the Lebesgue decompositions of the measures m_n with respect to λ, and $m^a := m^a{}_{Q^*} \circ \psi$, $m^s := m^s{}_{Q^*} \circ \psi$ is the Lebesgue decomposition of m. It is not difficult to see that (RO)-convergence of m^a_n to m^a and of m^s_n to m^s follow from (RO)-convergence of $M_n|_{Q^* \setminus U}$ and $M_n|_U$, respectively. □

3.4.2. Sobczyk-Hammer-Type Decompositions

We now consider Sobczyk-Hammer-type decompositions of measures into an atomic and a non-concentrated part.

Let $\mathcal{L}$ be an algebra of parts of a nonempty set *G*, *R* be a super Dedekind complete lattice group and $m: \mathcal{L} \to R$ be a finitely additive and positive measure, then m is *non-concentrated* iff

$$\bigwedge_{P \in \Pi} \left(\bigvee_{D \in P} m(D) \right) = 0, \tag{3.131}$$

where P is any partition of G formed by elements of $\mathcal{L}$, and the infimum is in the class Π of such partitions (see also Boccuto (1995c), Boccuto and Candeloro (1990 and 1993)). A positive finitely additive measure m is said to be *atomic* iff for any non-concentrated finitely additive and positive measure $\lambda: \mathcal{L} \to R$ with $\lambda \leq m$ we get $\lambda = 0$. A finitely additive bounded measure $m: \mathcal{L} \to R$ is *non-concentrated* (resp. *atomic*) iff $\| m \|$ is.

The positive finitely additive bounded measures $m_n : \mathcal{L} \to R$, $n \in \mathbb{N}$, are *uniformly non-concentrated* iff $\bigwedge_{P\in\Pi}[\bigvee_{D\in P}(\bigvee_n m_n(D))] = 0$, where P, Π are as in (3.131). We have the following characterization.

Proposition 4.8 (see Boccuto and Candeloro (2008), Proposition 3.2) *A finitely additive positive measure $m : \mathcal{L} \to R$ is non-concentrated if and only if there is an (O)-sequence $(b_j)_j$ in R with the property that for every $j \in \mathbb{N}$ there is a finite partition $Q_j = \{ D_1,\dots,D_{h_j} \}$ with*

$$m(D_i) \le b_j \quad \textit{for every } i = 1,\dots,h_j. \tag{3.132}$$

Moreover, the finitely additive positive measures $m_n : \mathcal{L} \to R$, $n \in \mathbb{N}$, are uniformly non-concentrated if and only if there is an (O)-sequence $(b_j)_j$ in R with the property that for any $j \in \mathbb{N}$ there is a finite partition $Q_j = \{ D_1,\dots,D_{h_j} \}$ with $\bigvee_n m_n(D_i) \le b_j$ for each $i = 1,\dots,h_j$.

Proof: We prove only (3.132), since the other part is analogous. Pick a positive non-concentrated finitely additive measure m. By super Dedekind completeness of R, a sequence $(Q_j)_j$ of finite partitions of G can be found, with $\bigwedge_j(\bigvee_{D\in Q_j} m(D)) = 0$. We can suppose that every element of Q_j is union of some suitable elements belonging to Q_{j+1}. For each $j \in \mathbb{N}$, set $b_j := \bigvee_{D\in Q_j} m(D)$. We have that $(b_j)_j$ is an (O)-sequence, and thus the sequence $(Q_j)_j$ is the requested one.

Let now $(Q_j)_j$ be a countable family of finite partitions of G, fulfilling (3.132). Then $\bigwedge_j(\bigvee_{D\in Q_j} m(D)) = 0$, and so $\bigwedge_{P\in\Pi}(\bigvee_{D\in P} m(D)) = 0$. Thus, m is non-concentrated.

□

Theorem 4.9 (see Boccuto and Candeloro (2008), Theorem 3.5) *Let m, $\lambda : \Sigma \to R$ be two positive and globally σ-additive measures, such that m is* *globally absolutely continuous with respect to λ and λ is non-concentrated. Then m is non-concentrated too.*

Proof: Let $u := m(G)+\lambda(G)$, and $(b_j)_j$ be an (O)-sequence related to global σ-additivity of m, global σ-additivity and non-concentratedness of λ. By Lemma 2.4 of Chapter 2 we find a subsequence $(b_{j_k})_k$ such that the sequence $\rho_N := u \wedge \sum_{k \geq N} b_{j_k}$, $N \in \mathbb{N}$, is an (O)-sequence. Let $b'_N := b_{j_N}$ and $\alpha_N := \bigvee \{m(A) : \lambda(A) \leq b'_N\}$, $N \in \mathbb{N}$: we claim that $(\alpha_N)_N$ is an (O)-sequence. First of all, note that $(\alpha_N)_N$ is decreasing. We now prove that $\alpha := \bigwedge_N \alpha_N$ is equal to 0. Otherwise we find a natural number N with $\alpha \not\leq b_N$, and so $\alpha_k \not\leq b_N$ for each k. Thus for each k there is an element $A_k \in \Sigma$, with $\lambda(A_k) \leq b'_k$ and $m(A_k) \not\leq b_N$. For every s, set $A^*_s := \bigcup_{k \geq s} A_k$: thus $\lambda(A^*_s) \leq u \wedge \sum_{k \geq s} b'_k \leq \rho_s$ and

$$m(A^*_s) \not\leq b_N \tag{3.133}$$

for every s. Setting $A := \bigcap_{s=1}^{\infty} A^*_s$, then $\lambda(A) = 0$ thanks to global σ-additivity, and thus $m(A) = 0$. So the sequence $(A^*_s \setminus A)_s$ is decreasing and its intersection is the empty set, and moreover $m(A^*_s \setminus A) = m(A^*_s)$ for every s. Thanks to global σ-additivity, we find an index τ_N with $m(A^*_{\tau_N}) \leq b_N$. This contradicts (3.133), and thus $\alpha = 0$.

We prove that m is non-concentrated. For each $N \in \mathbb{N}$ there is a partition P of G with $\lambda(D) \leq b'_N$ for each $D \in P$. So $m(D) \leq \alpha_N$ for every $D \in P$, and thus we get the assertion. □

Let now $m : \mathcal{L} \to R$ be a positive measure, $u \in R$, $u \geq 0$, and $A \in \mathcal{L}$. The set A is said to be u*-decomposable* (with respect to m) iff there exists a finite partition of $A := \{D_1, \ldots, D_k\}$ in $\mathcal{L}$, with $m(D_i) \leq u$ for every $i = 1, \ldots, k$. Note that a finitely additive positive measure m is non-concentrated if and only if there is an (O)-sequence $(r_j)_j$ in R such that G is r_j-decomposable for every $j \in \mathbb{N}$.

Let $m_n : \mathcal{L} \to R$, $n \in \mathbb{N}$, be an equibounded sequence of positive and finitely additive measures. For each $A \in \mathcal{L}$, put $M(A) := \bigvee_{n\in\mathbb{N}} m_n(A)$. A set A, u-decomposable with respect to m, is said to be *uniformly* u*-decomposable.* Note that the m_n's are uniformly non-concentrated if and only if there is an (O)-sequence $(r_j)_j$ with the property that G is uniformly r_j-decomposable for each $j \in \mathbb{N}$.

Lemma 4.10 (see Boccuto and Candeloro (2008), Lemma 3.7) *Let* $m_n : \mathcal{L} \to R$, $n \in \mathbb{N}$, *be an equibounded sequence of positive, non-concentrated and globally uniformly* (s)*-bounded measures, and let* $(r_j)_j$ *be an* (O)*-sequence associated with global uniform* (s)*-boundedness and non-concentratedness of* m_n*. Suppose that there are* $j \in \mathbb{N}$ *and* $A \in \mathcal{L}$*, such that* $M(A) \not\leq 2r_j$*. Then there is a set* $H \subset A$, $H \in \mathcal{L}$*, uniformly* r_j*-decomposable and with* $M(H) \not\leq 2r_j$*.*

Proof: Without loss of generality, we can assume $j = 1$. Let $r = r_1$ and let us deny the thesis. There is $A \in \mathcal{L}$, such that $M(A) \not\leq 2r$ and there are no uniformly r-decomposable set $H \in \mathcal{L}$, such that $M(H) \not\leq 2r$. We can and do assume that $m_1(A) \not\leq 2r$. Since m_1 is non-concentrated, there is a partition $D_A = \{D_1, \ldots, D_k\}$ of A, with $m_1(D_i) \leq r$ for every $i = 1, \ldots, k$. Thus there is $k_1 \leq k$ with $M(D_{k_1}) \not\leq 2r$, and so we find an $n_1 > 1$ with $m_{n_1}(D_{k_1}) \not\leq 2r$. Set $A_1 := D_{k_1}$, $B_1 := A \setminus A_1$. Then $m_1(B_1) \not\leq r$. As m_1, $m_2, \ldots, m_{n_1}$ are non-concentrated, then there exists a partition $D = \{ D_{1'}, D_{2'}, \ldots, D_{s'} \}$ of A_1, with $m_1(D_{i'}) \vee m_2(D_{i'}) \vee \ldots \vee m_{n_1}(D_{i'}) \leq r$ for every $i = 1, \ldots, s$. There is $k_2 \leq s$ with $M(D'_{k_2}) \not\leq 2r$, and so there is $n_2 > n_1$ with $m_{n_2}(D'_{k_2}) \not\leq 2r$. Set $A_2 := D'_{k_2}$, $B_2 := A_1 \setminus A_2$. Then $m_{n_1}(B_2) \not\leq r$. Since m_1, $m_2, \ldots, m_{n_2}$ are non-concentrated, we find a partition of A_2 and a set $A_3 \subset A_2$, $A_3 \in \mathcal{L}$, such that $M(A_3) \not\leq 2r$ and $m_{n_2}(A_3) \leq r$. If $B_3 := A_2 \setminus A_3$, then $m_{n_2}(B_3) \not\leq r$. By induction, we find an increasing $(n_h)_h$ in $\mathbb{N}$ and a decreasing sequence $(A_h)_h$ in $\mathcal{L}$, where $m_{n_h}(B_{h+1}) \not\leq r$ for every $h \in \mathbb{N}$, where $B_h := A_h \setminus A_{h+1}$. This is a contradiction with global uniform (s)-boundedness, and concludes the proof. □

Theorem 4.11 (see Boccuto and Candeloro (2008), Theorem 3.8) *If the m_n's are equibounded positive and finitely additive R-valued measures, non-concentrated and globally uniformly (s)-bounded, then the m_n's are uniformly non-concentrated.*

Proof: Thanks to equiboundedness and Lemma 2.5 of Chapter 2, there is an (O)-sequence $(r_j)_j$, associated with global uniform (s)-boundedness and non-concentratedness of each measure m_n. We claim that G is uniformly $2r_j$-decomposable for every $j \in \mathbb{N}$. For any natural number j, let $r = r_j$. By contradiction, assume that G is not uniformly $2r$-decomposable. So, $M(G) \not\leq 2r$. Thanks to Lemma 4.10, we find $A_1 \in \mathcal{L}$, such that $M(A_1) \not\leq 2r$ and uniformly r-decomposable. Thus $G \setminus A_1$ is not uniformly $2r$-decomposable, and so $M(G \setminus A_1) \not\leq 2r$. Thanks to Lemma 4.10, we find a uniformly r-decomposable set $A_2 \subset G \setminus A_1$ with $A_2 \in \mathcal{L}$ and $M(A_2) \not\leq 2r$. Note that A_1 and A_2 are uniformly r-decomposable and G is not uniformly r-decomposable, and hence $G \setminus (A_1 \cup A_2)$ is not uniformly $2r$-decomposable. There is a uniformly r-decomposable set A_3 in $\mathcal{L}$, such that $A_1 \cap A_3 = A_2 \cap A_3 = \emptyset$ and $M(A_3) \not\leq 2r$. So we find a disjoint sequence $(A_k)_k$ of uniformly r-decomposable sets in $\mathcal{L}$, with $M(A_k) \not\leq 2r$ for all $k \in \mathbb{N}$. So we obtain a contradiction with global uniform (s)-boundedness of the m_n's. Hence, G is uniformly $2r$-decomposable. This ends the proof. □

The next step is to prove some results involving Sobczyk-Hammer-type decompositions. We begin with the case of globally σ-additive measures.

Lemma 4.12 (see Boccuto and Candeloro (2008), Lemma 4.1) *Let $m : \Sigma \to R$ be a globally σ-additive, positive and non-atomic measure. Then there exists $F \in \Sigma$ such that $m_{|F}$ is non-concentrated and $m(F) \neq 0$.*

Proof: As m is not atomic, then there is a positive non-trivial non-concentrated measure $\mu : \Sigma \to R$, $\mu \leq m$, which turns to be globally σ-additive. Let H be the class of all μ-null sets, and put

$$\alpha = \bigvee \{m(V) : V \in H\}. \quad \textbf{(3.134)}$$

Since R is super Dedekind completeness, there is a sequence $(K_n)_n$ of elements of H, with $\alpha = \bigvee_n m(K_n)$. Without restriction, we suppose that $(K_n)_n$ is increasing. If $K = \bigcup_{n=1}^{\infty} K_n$, then we get $m(K) = \alpha$ and $\mu(K) = 0$.

We now claim that $F = G \setminus K$ is the requested set. Note that $m_{|F}$ is globally μ-absolutely continuous. Otherwise, if there is $W \in \Sigma$ with $\mu(W) = 0$ and $m_{|F}(W) \neq 0$, then we have

$$m((W \cap F) \cup K) = m(W \cap F) + m(K) > \alpha,$$

$$\mu((W \cap F) \cup K) = \mu(W \cap F) + \mu(K) = 0,$$

and so α is not the supremum in the relation (3.134), getting a contradiction. So, thanks to Theorem 4.9, $m_{|F}$ is non-concentrated. Furthermore we get $m(F) \neq 0$: otherwise $\mu \equiv 0$, and this is absurd. So the proof is finished. □

Proposition 4.13 (see Boccuto and Candeloro (2008), Proposition 4.2) *The sum of two positive R-valued non-concentrated (resp. atomic) globally σ-additive measures, defined on Σ, is non-concentrated (resp. atomic).*

Proof: It is not difficult to prove the assertion, when the measures involved are non-concentrated. We now turn to the atomic case. If m_1 and m_2 are atomic and $m_1 + m_2$ is not atomic, then there is a set F, satisfying the thesis of Lemma 4.12. We obtain that $m_{1|F}$ is non-concentrated, and so we deduce that $m_{1|F}$=0. In a similar way we can see that $m_{2|F}$=0. Thus $(m_1 + m_2)_{|F} = 0$, getting a contradiction.

□

We prove the Sobczyk-Hammer decomposition result.

Theorem 4.14 (see Boccuto and Candeloro (2005), Theorem 5.6, Boccuto and Candeloro (2008), Theorem 4.3) *If $m : \Sigma \to R$ is a positive globally σ-additive measure, then there is a set $E \in \Sigma$, such that $m_{|E}$ is non-concentrated and $m_{|G \setminus E}$ is atomic.*

Proof: In the case m is atomic, we take $E = \varnothing$. In the other case, let $\mathcal{J}$ be the ideal of the sets $A \in \Sigma$, such that $m_{|A}$ is non-concentrated. Note that $\mathcal{J}$ is a σ-ideal, thanks to Theorem 4.11. By using the σ-ideal principle (Theorem 4.1), we find a maximal set $E \in \mathcal{J}$. By Lemma 4.12, $m(E) \neq 0$, and so $m_{|G \setminus E}$ is atomic. □

Proposition 4.15 (see Boccuto and Candeloro (2005), Proposition 5.7, Boccuto and Candeloro (2008), Proposition 4.2) *Let $(m_n)_n$ be a sequence of positive and globally σ-additive measures, defined on a σ-algebra Σ and taking values in R. Then there exists a set $H \in \Sigma$ with the property that, for any $n \in \mathbb{N}$, $m_{n|H}$ is non-concentrated and $m_{n|G \setminus H}$ is atomic.*

Proof: Set $\mu_n = \sum_{i=1}^{n} m_i$ for any $n \in \mathbb{N}$. Note that the sequence $(\mu_n)_n$ is increasing. Let $\mathcal{J}_n$ be the σ-ideal of the sets $A \in \Sigma$, such that $\mu_{n|A}$ is non-concentrated. It is not difficult to see that $(\mathcal{J}_n)_n$ is decreasing. Now, fix n and $A \in \mathcal{J}_n$, with $\mu_n(A) \neq 0$: we take $\mu_{n|A}$ and $\mu_{n+1|A}$, and consider the Lebesgue decomposition of $\mu_{n+1|A}$ with respect to $\mu_{n|A}$. We find a measurable set $B \subset A$, with $\mu_{n+1|B} \ll \mu_{n|A}$, and $\mu_{n+1}(B) \neq 0$, thanks to Theorem 4.3. Then $B \in \mathcal{J}_{n+1}$ and 4.1.2) is fulfilled. So we find a single set H such that $\mu_{n|H}$ is non-concentrated and $\mu_{n|G \setminus H}$ is atomic for all n. The set H works also for the measures m_n. □

We turn to the following convergence theorem.

Theorem 4.16 (see Boccuto and Candeloro (2005), Theorem 5.8, Boccuto and Candeloro (2008), Theorem 4.5) *Let $(m_n)_n$ be an equibounded sequence of positive and globally σ-additive measures, defined on a σ-algebra Σ, with values in R and (RO)-convergent to m. Then m is a globally σ-additive measure. Moreover, if (m_n^1, m_n^2), (m^1, m^2) denote the sectional Sobczyk-Hammer decompositions of m_n and m, then $(m_n^1)_n$ and $(m_n^2)_n$ are (RO)-convergent to the measures m^1 and m^2, where are $n \in \mathbb{N}$, respectively.*

Proof: By virtue of Theorem 3.8, the measures m_n are globally uniformly (s)-bounded and so m is globally σ-additive. From Proposition 4.15 applied to

$(m_n)_n$, we find a set $H \in \Sigma$, in correspondence of which there is a sectional decomposition of the m_n, $n \in \mathbb{N}$, and of m. Taking into account (RO)-convergence, we obtain the requested result. □

We now consider finitely additive measures.

Theorem 4.17 (see Boccuto and Candeloro (2005), Theorem 5.9, Boccuto and Candeloro (2008), Theorem 4.6) *Let m be any positive globally (s)-bounded R-valued measure, defined on an algebra $\mathcal{L}$. Then there are two finitely additive and positive R-valued measures m^1 and m^2, such that m^1 is non-concentrated, m^2 is atomic and $m = m^1 + m^2$.*

Proof: Let Q^* be the Stone space related to $\mathcal{L}$, $\mathcal{Q}$ be the algebra of the open-closed subsets of Q^*, and ψ be the isomorphism from $\mathcal{L}$ into $\mathcal{Q}$. We know that there is a measure *M*, which is a globally σ-additive extension of the measure $m^0 := m \circ \psi^{-1}$, defined on $\sigma(\mathcal{Q})$. Thanks to Theorem 4.15, there is a Sobczyk-Hammer-type decomposition of *M*, $M = M_1 + M_2$. The measures M_1 and M_2 are non-concentrated and atomic, respectively. We denote the restrictions of M_1 and M_2 to $\mathcal{Q}$ by $m^1_{Q^*}$ and $m^2_{Q^*}$, respectively.

We claim that $m^1_{Q^*}$ is non-concentrated and and $m^2_{Q^*}$ is atomic, obtaining that $m^1 := m^1_{Q^*} \circ \psi$ and $m^2 := m^2_{Q^*} \circ \psi$ form the required decomposition. First, we prove that $m^1_{Q^*}$ is non-concentrated. Since M_1 is non-concentrated, we find an (O)-sequence $(a_j)_j$ with the property that for any $j \in \mathbb{N}$ there is a partition $\{D_1, \ldots, D_{h_j}\}$ with $M_1\,(D_k) \le a_j$ for each $k = 1, \ldots, h_j$. Thanks to Theorem 3.29 of Chapter 2, we find an (O)-sequence $(b_j)_j$ such that for every $D \in \sigma(\mathcal{Q})$ and $j \in \mathbb{N}$ there is $E \in \mathcal{Q}$ with $M_1(E \Delta D) \le b_j$. Moreover, thanks to Lemma 2.4 of Chapter 2, taking $u := m(G)$, we find a subsequence $(b_{j_l})_l$ and an (O)-sequence $(\rho_N)_N$, with $\rho_N := u \wedge \sum_{l=N}^{\infty} b_{j_l}$, $N \in \mathbb{N}$. We prove that $m^1_{Q^*}$ is non-concentrated with respect to the (O)-sequence $(a_N + \rho_N)_N$. Choose a natural number N: there is a partition $\{D_1,\ldots,D_{h_N}\}$ in $\sigma(\mathcal{Q})$, with $M_l(D_k) \le a_N$ for every k. For each $k \in [1, h_N]$ there is $E_k \in \mathcal{Q}$ with $M_l(E_k \Delta D_k) \le b_{j_{(N+k)}}$, and then

$$M_l\ (E_k) \le a_N + \rho_N \quad \text{for every } k.$$

Set

$$F_1 = E_1,\ F_2 = E_2 \setminus E_1,\ \ldots,\ F_{h_N} = E_{h_N} \setminus (E_1 \cup \ldots \cup E_{h_N - 1}),$$

and

$$F_{h_N+1} := Q^* \setminus \bigcup_{k=1}^{h_N} F_k = Q^* \setminus \bigcup_{k=1}^{h_N} E_k.$$

Then $F_k \in \mathcal{Q}$ for each $k \in \mathbb{N}$, and $m^1_{Q^*}(F_k) = M_l\ (F_k) \le b_N + \rho_N$ for all $k = 1,\ldots,h_N$. We get

$$m^1_{Q^*}(F_{h_N+1}) = M_l(Q^*) - \sum_{k=1}^{h_N} M_l\ (F_k) =$$

$$M_l\left((\bigcup_{k=1}^{h_N} D_k) \setminus (\bigcup_{k=1}^{h_N} E_k) \right) \le M_l(\bigcup_{k=1}^{h_N} (D_k \Delta E_k)) \le u \wedge \sum_{k=1}^{h_N} b_{j_{(N+k)}} \le \rho_N.$$

Thus the partition $\{F_1,\ldots,F_{h_N},F_{h_N+1}\}$ is such that

$$m^1_{Q^*}(F_k) \le \rho_N + a_N \text{ for every } k = 1,\ldots,h_N + 1,$$

and hence $m^1_{Q^*}$ is non-concentrated.

We now show that $m^2_{Q^*}$ is atomic. Let $\lambda : \mathcal{Q} \to R$ be a finitely additive, positive and non-concentrated measure, with $0 \le \lambda(A) \le m^2_{Q^*}(A)$ for all $A \in \mathcal{Q}$. The measure λ has a Carathéodory extension *P,* defined on $\sigma(\mathcal{Q})$. The non-concentratedness of P is a consequence of the fact that λ is non-concentrated. Moreover, $0 \le P(A) \le M_2(A)$ for all $A \in \sigma(\mathcal{Q})$. Thus, by atomicity of M_2, we have $P = 0$ on $\sigma(\mathcal{Q})$, and hence $\lambda = 0$ on $\mathcal{Q}$. This ends the proof. □

Let $m : \mathcal{L} \to R$ be a finitely additive and bounded measure, then m is said to be *non-concentrated* (resp. *atomic*) iff $m^{(+)} + m^{(-)}$ is non-concentrated (resp. atomic). We now recall the following Sobczyk-Hammer-type theorem in the finitely additive setting.

Theorem 4.18 (see Boccuto and Candeloro (2005), Theorem 5.10, Boccuto and Candeloro (2008), Theorem 4.8) *Let $(m_n)_n$ be an equibounded sequence of R-valued globally (s)-bounded measures, defined on a σ-algebra Σ, and $m := (RO)\lim_n m_n$. Then m is globally (s)-bounded, and $(m_n^1)_n$, $(m_n^2)_n$ are (RO)-convergent to m^1, m^2, where (m_n^1, m_n^2), (m^1, m^2) are respectively the non-concentrated and the atomic part of m_n and m.*

Proof: Let $\mathcal{Q}$ and ψ be as in the proof of Theorem 4.17, and M_n, $M_n^{(+)}$,..., be the σ-additive extensions of m_n, $m_n^{(+)}$,..., defined on $\sigma(\mathcal{Q})$.

By virtue of Theorem 1.12, $(M_n)_n$ (RO)-converges to M on the whole of $\sigma(\mathcal{Q})$. By Proposition 4.15 applied to $(M_n^{(+)})_n$ and $(M_n^{(-)})_n$, we find $H \in \sigma(\mathcal{Q})$ with the following properties:

4.18.1) $M_n^{(+)}|_H$ is non-concentrated, $M_n^{(+)}|_{Q^*\setminus H}$ is atomic;

4.18.2) $M_n^{(-)}|_H$ is non-concentrated, $M_n^{(-)}|_{Q^*\setminus H}$ is atomic;

4.18.3) $(RO)\lim_n M_n|_H = M|_H$, $(RO)\lim_n M_n|_{Q^*\setminus H} = M|_{Q^*\setminus H}$.

If m^1_{n,Q^*}, m^2_{n,Q^*}, $m^1{}_{Q^*}$, $m^2{}_{Q^*}$ are the restrictions to $\mathcal{Q}$ of $M_n|_H$, $M_n|_{Q^*\setminus H}$, $M|_H$, $M|_{Q^*\setminus H}$, then $m_n^1 := m^1_{n,Q^*} \circ \psi, m_n^2 := m^2_{n,Q^*} \circ \psi$ and $m^1 := m^1{}_{Q^*} \circ \psi$, $m^2 := m^2{}_{Q^*} \circ \psi$ form the Sobczyk-Hammer decompositions of m_n and m, respectively. Finally, since $M_n|_H$ and $M_n|_{Q^*\setminus H}$ (RO)-converge, then m_n^1 and m_n^2 (RO)-converge to m^1 and m^2, respectively. □

3.4.3. Yosida-Hewitt-Type Decompositions

We now prove some theorems relating the Stone extension of a measure m with its Yosida-Hewitt decomposition (see also Boccuto and Candeloro (2005), Brooks and Candeloro (2004)).

A finitely additive positive measure $m:\Sigma \to R$ is *(globally) purely finitely additive* iff for each (globally) σ-additive measure M with $0 \leq M \leq m$ we get M=0.

A finitely additive (s)-bounded measure $m:\Sigma \to R$ is *(globally) purely finitely additive* iff $\| m \|$ is.

The *Yosida-Hewitt decomposition* is formed by a globally σ-additive and a globally purely finitely additive part.

Let $m:\mathcal{L} \to R$ be a positive globally (s)-bounded measure, Q^* be the corresponding Stone space and $M: \sigma(\mathcal{Q}) \to R$ be the Stone extension of m. If $U \in \sigma(\mathcal{Q})$, then put

$$m_{:U}(A) = M(U \cap \psi(A)),\ A \in \Sigma.$$

Now we present the following

Proposition 4.19 (see Boccuto and Candeloro (2005), Proposition 6.2) *Let $m:\Sigma \to R$ be any positive and globally (s)-bounded measure. Then there exists a set $U \in \sigma(\mathcal{Q})$, such that $m_{:U}$ is globally σ-additive and $m_{:Q^*\setminus U} = m - m_{:U}$ is globally purely finitely additive.*

Proof: Let $\mathcal{J}$ be the ideal of all sets $V \in \sigma(\mathcal{Q})$ such that $m_{:V}$ is globally σ-additive. Since M is globally σ-additive, then $\mathcal{J}$ is a σ-ideal (see also Boccuto and Candeloro (2005)). Now, let U be any element in $\mathcal{J}$ maximizing M (see Theorem 4.1). We claim that $m_{:Q^*\setminus U}$ is globally purely finitely additive, so obtaining the assertion. If this is not true, then there is a positive non-trivial globally σ -additive measure $\beta:\Sigma \to R$, with $\beta \le m_{:Q^* \setminus U}$, which admits its corresponding Stone extension P. Note that $P \le M_{|Q^* \setminus U}$. Now, let us consider the Lebesgue decomposition $M_{|Q^* \setminus U}$ relatively to P: by Theorem 4.3, there is a subset $W \subset Q^* \setminus U$, such that $M_{|W} \ll P$. Thus we get $m_{:W} \ll \beta$, and so $m_{:W}$ is globally σ -additive. So $W \cup U$ contradicts maximality of U in $\mathcal{J}$, and this is impossible. □

Thus we obtain the Yosida-Hewitt decomposition of a globally (s)-bounded measure $m:\Sigma \to R$, not necessarily positive, by finding a set $U \in \sigma(\mathcal{Q})$ with the property that $\|m\|_{:U}$ is globally σ-additive and $\|m\|_{:Q^* \setminus U}$ is globally purely finitely additive. Then, $(m_{:U}, m_{:Q^* \setminus U})$ is the requested decomposition of m.

For every globally (s)-bounded measure $m:\Sigma \to R$, we denote by $R^{'}(m)$ its globally σ-additive part (see also Boccuto and Candeloro (2005), Brooks and Candeloro (2004)). Now we recall need also the following result.

Proposition 4.20 (see Boccuto and Candeloro (2005), Proposition 6.3) *Assume that $(m_n)_n$ is a sequence of globally (s)-bounded positive R-valued measures, defined on the σ-algebra Σ. Then we can find a set U in $\sigma(\mathcal{Q})$, with the property that $m_{n:U}$ is globally σ-additive and $m_{n:Q^* \setminus U}$ is globally purely finitely additive for any $n \in \mathbb{N}$.*

Proof: For each natural number n, let $\mu_n := \sum_{i=1}^{n} m_i$, pick $\mathcal{J}_n$, U_n as in Proposition 4.19, associated with μ_n, and let M_n be the Stone extension of μ_n. In order to apply Theorem 4.1, we claim that, for any $n \in \mathbb{N}$ and $A \in \mathcal{J}_n$ such that $M_n(A) \neq 0$, there exists $B \subset A$, $B \in \mathcal{J}_{n+1}$, with $M_{n+1}(B) \neq 0$. Indeed, taking the Lebesgue decomposition of $M_{n+1\,|A}$ with respect to $M_{n\,|A}$, thanks to Theorem 4.3 there is $B \subset A$, $B \in \sigma(\mathcal{Q})$, such that $M_{n+1}(B) \neq 0$ and $M_{n+1\,|B} \ll M_{n\,|A}$. Thus $\mu_{n+1:B} \ll \mu_{n:A}$, and hence $\mu_{n+1:B}$ is globally σ-additive. So $B \in \mathcal{J}_{n+1}$, and thus, by the σ-ideal principle, we find a set $U \in \sigma(\mathcal{Q})$ which gives the Yosida-Hewitt decomposition of the measures μ_n, $n \in \mathbb{N}$. The set U works even for the m_n's. This ends the proof. □

Theorem 4.21 (see Boccuto and Candeloro (2005), Theorem 6.4) *Let $m_n : \Sigma \to R$, $n \in \mathbb{N}$, be an equibounded sequence of positive and (s)-bounded measures, and $m := (RO)\lim_n m_n$. Then, the sequence $(R'(m_n))_n$ (RO)-converges to $R'(m)$.*

Proof: By Proposition 4.20, we find a set $U \in \sigma(\mathcal{Q})$ with $m_{n:U} = R'(m_n)$ for every $n \in \mathbb{N}$. We observe that the sequence $(M_n(B))_n$ is (RO)-convergent to $M(B)$ for every $B \in \sigma(\mathcal{Q})$: this follows from Theorem 1.12. Thus, we get

$$(RO)\lim_n M_n(U \cap \psi(A)) = M(U \cap \psi(A)) \text{ for all } A \in \Sigma,$$

which proves the assertion. □

We note that, using similar techniques as above, Theorem 4.21 holds even for any globally (s)-bounded measure m. Indeed it is possible to find a set U as in Proposition 4.20, in correspondence with $m_n^{(+)}$, $m_n^{(-)}$, $n \in \mathbb{N}$, and $m^{(+)}$, $m^{(-)}$.

Now we treat a slightly different Yosida-Hewitt-type decomposition, in which σ-additivity is meant in the classical sense (see also Brooks and Candeloro (2004)).

Let R be any Dedekind complete lattice group, $\Sigma \subset \mathcal{P}(G)$ be a σ-algebra, and Π be the class of the partitions π of G into a countable family $\{F_n : n \in \mathbb{N}\}$ of disjoint sets of Σ. We endow the set Π with the *refinement ordering*, namely if $\pi = \{F_n : n \in \mathbb{N}\}$ and $\pi' = \{E_k : k \in \mathbb{N}\}$ are two partitions, then we say that $\pi \prec \pi'$ iff every set F_n in π is the union of some sets E_k of π'.

For each partition $\pi \in \Pi$, $\pi = \{F_n : n \in \mathbb{N}\}$ and for every finitely additive positive measure $m : \Sigma \to R$, set $s(m,\pi) = \sum_{n=1}^{\infty} m(F_n)$. It is not difficult to see that this series converges to an element of R, and that $S \leq m(G)$.

Now we claim that the limit $\lim_{\pi \in \Pi} s(m,\pi)$ (the *Rickart integral* of m) exists in R. We will denote it by $\int_G dm$.

Proposition 4.22 (see also Brooks and Candeloro (2004), Proposition 2.7) *Let* $m:\Sigma \to R$ *be a finitely additive positive measure. Then the quantity* $\int_G dm$ *exists in* R.

Proof: The existence of the limit $\lim_{\pi \in \Pi} s(m,\pi)$ is a consequence of the fact that $\pi \prec \pi' \Rightarrow s(m,\pi') \le s(m,\pi)$. Moreover note that, if $F \in \Sigma$ is the union of a disjoint sequence $(E_k)_k$ from Σ, one has $m(F) \ge \sum_{k=1}^{\infty} m(E_k)$, getting monotonicity. □

It is not difficult to see the operator $m \mapsto \int_G dm$ defines is additive, positively homogeneous and monotone. If $F \in \Sigma$ and $m_{|F}(E) = m(E \cap F)$, $E \in \Sigma$, then $\int_F dm$ is the Rickart integral of $m_{|F}$.

It is not difficult to see that the set function $F \mapsto \int_F dm$ is finitely additive and that $\int_F dm \le m(F)$ for every $F \in \Sigma$. We denote this measure by $R(m)$, and call it the *Rickart indefinite integral* of m. We now prove that $R(m)$ determines a Yosida-Hewitt-type decomposition of m in the classical like sense (see also Brooks and Candeloro (2004), Congost Iglesias (1981), Schmidt (1986)).

Theorem 4.23 (see also Brooks and Candeloro (2004), Theorem 3.1) *The measure* $R(m)$ *and* $m - R(m)$ *are* σ*-additive and purely finitely additive, respectively.*

Proof: We prove that $R(m)$ is σ-additive. Let $\{H_k : k \in \mathbb{N}\}$ be a disjoint sequence in Σ, and let us denote by H the union of its elements. By positivity and finite additivity of $R(m)$ we have

$$\sum_{k=1}^{\infty} R(m)(H_k) \le R(m)(H).$$

Let now $\pi = \{F_n : n \in \mathbb{N}\}$ be any partition of G. We get: $R(m)(H) \leq \sum_{n=1}^{\infty} m\,(H \cap F_n)$. Let π be such that every H_k is the union of some element F_n. So, for any k we find a countable subset $J(k) \subset \mathbb{N}$ such that $H_k = \bigcup_{n \in J(k)} F_n$. So we have

$$R(m)(H) \leq \sum_{k=1}^{\infty} \sum_{n \in J(k)} m(F_n) = \sum_{n \in J(1)} m(F_n) + \sum_{k=2}^{\infty} \sum_{n \in J(k)} m(F_n),$$

and thus

$$R(m)(H) - \sum_{k=2}^{\infty} \sum_{n \in J(k)} m(F_n) \leq \sum_{n \in J(1)} m(F_n). \tag{3.135}$$

We now refine π letting all the sets F_n fixed, with the exception of those contained in H_1. Let Π_1 be the class of the partitions π, constructed in this way. By (3.135) we have

$$R(m)(H) - \sum_{k=2}^{\infty} \sum_{n \in J(k)} m(F_n) \leq \bigvee_{\pi \in \Pi_1} \sum_{n \in J(1)} m(F_n) = R(m)(H_1), \tag{3.136}$$

from which

$$R(m)(H) - R(m)(H_1) \leq \sum_{k=2}^{\infty} \sum_{n \in J(k)} m(F_n). \tag{3.137}$$

Let the elements F_n, $n \in J(2)$, vary and let us keep fixed the other sets F_n. From (3.137) we get

$$R(m)(H) - R(m)(H_1) - R(m)(H_2) \leq \sum_{k=3}^{\infty} \sum_{n \in J(k)} m(F_n).$$

Proceeding in this fashion, for every integer N we get

$$R(m)(H)-\sum_{k=1}^{N}R(m)(H_k)\leq \sum_{k=N+1}^{\infty}\sum_{n\in J(k)}m(F_n).$$

Taking the limits as $N\to+\infty$, we obtain

$$R(m)(H)-\sum_{k=1}^{\infty}R(m)(H_k)\leq 0,$$

and so we get σ-additivity of *m*.

We now prove that $m-R(m)$ is purely finitely additive. If $\nu:\Sigma\to R$ is any σ-additive positive measure, with $\nu\leq m-R(m)$, then $\nu+R(m)\leq m$, and so $R(\nu+R(m))\leq R(m)$.

By virtue of σ-additivity, $\nu+R(m)$ coincides with its Rickart indefinite integral. Thus $\nu+R(m)\leq R(m)$, that is $\nu=0$. □

CHAPTER 4

Filter/Ideal Limit Theorems

Abstract: We present recent versions of limit and boundedness theorems in the setting of filter convergence, for measures taking values in lattice or topological groups, in connection with suitable properties of filters. Some results are obtained by applying classical versions to a subsequence, indexed by a family of the involved filter: in this context, an essential role is played by filter exhaustiveness. We give also some basic matrix theorems for lattice group-valued double sequences, in the setting of filter convergence. We give some modes of continuity for measures with respect to filter convergence, some comparisons between filter exhaustiveness and filter (α)-convergence of measure sequences and some weak filter Cauchy-type conditions, in connection with integral operators.

Keywords: (Filter) continuous measure, Banach-Steinhaus theorem, basic matrix theorem, block-respecting filter, Brooks-Jewett theorem, Diagonal filter, Dieudonné theorem, Drewnowski theorem, equivalence, filter (α)-convergence, filter exhaustiveness, filter limit theorem, filter weak compactness, filter weak convergence, Nikodým boundedness theorem, Nikodým convergence theorem, *P*-filter, Schur theorem, topological group, Vitali-Hahn-Saks theorem.

In this chapter we consider some recent developments of filter convergence, especially concerning the importance which it plays in different kinds of limit theorems and related topics (see, for instance, Dimitriou (2011)). Note that, in general, when one treats filter convergence with respect to a given free filter of $\mathbb{N}$, in general it is impossible to obtain results analogous to the classical Brooks-Jewett, Vitali-Hahn-Saks, Nikodým convergence, Nikodým boundedness and Dieudonné-type theorems when $\mathcal{F} \neq \mathcal{F}_{\text{cofin}}$, where $\mathcal{F}_{\text{cofin}}$ is the filter of the cofinite subsets of $\mathbb{N}$ (see also Boccuto, Das, Dimitriou and Papanastassiou (2012), Boccuto, Dimitriou and Papanastassiou (2011b and 2012b)). However, under suitable hypotheses on the involved filter, it is possible to get still some results, concerning the behavior of a subsequence of the given sequence of measures, indexed by an element of the filter. In this spirit, different kinds of limit theorems have been proved with respect to filter convergence. First we deal with Schur-type theorems, and successively we investigate some classes of filters, for which some versions of Schur-type theorem hold, for measures with values in topological or lattice groups. Using some basic properties of diagonal and/or block-respecting filters and stationary subsets of $\mathbb{N}$, we use some sliding hump-type techniques and prove some Schur-type theorems. Further recent studies and developments of Schur theorems and related topics in the context of filter or ideal convergence can be found for instance in Filipów and Szuca (2010), Hernández, Galindo and Macario (1999). As consequences we give some Nikodým convergence, Nikodým boundedness, Vitali-Hahn-Saks and Dieudonné-type theorems for topological and lattice group-valued measures. These results were proved in Boccuto and Dimitriou (2013e and 2014a) and Boccuto, Dimitriou and Papanastassiou (2011c, 2012b and 2012e)). We give also some versions of these theorems, whose it is possible to give a direct proof without using the Schur theorem (see for instance Boccuto and Dimitriou (2013e and 2014c)).

We note that, in the particular case of positive measures, it is possible to prove several filter limit theorems by requiring that the involved filter is only diagonal, and not necessarily block-respecting (see also Boccuto and Dimitriou (2014a), Boccuto, Dimitriou and Papanastassiou (2011b)). In Boccuto, Das, Dimitriou and Papanastassiou (2012), Boccuto and Dimitriou (2011b-c, 2013c, 2014a and 2014c-d) some other versions of filter/ideal limit theorems for real-valued, topological and lattice group-valued measures were given. In this framework, we investigate the powerful concepts of (weak and uniform) filter exhaustiveness, which play a fundamental role. We give some conditions, which in general, when $\mathcal{F} \neq \mathcal{F}_{\text{cofin}}$, cannot be dropped, as we will show (see also Boccuto and Dimitriou (2011c and 2013c)). These theorems are formulated in the topological and lattice group context, when σ-additivity and related concepts are formulated in the classical like setting or with respect to a single (O)-sequence. We deal also with measures, continuous with respect to a general Fréchet-Nikodým topology. Similar equivalence results are given in Drewnowski (1972b) in the classical case for topological group-valued measures. In particular, when it is proved that the Nikodým convergence theorem implies the Brooks-Jewett theorem, we consider countably additive restrictions of finitely additive (s)-bounded topological group-valued measures, defined on suitable σ-algebras (see also Boccuto, Dimitriou and Papanastassiou (2010c, 2011a) for a lattice group version). However in the lattice group setting, in order to relate finitely and countably additive measures, it is not advisable to use an approach of this kind. Indeed, in topological groups, the involved convergences fulfil some suitable properties, which are not always satisfied by order convergence in lattice groups, because in general it does not have a topological nature. So, to prove our results, we use the Stone Isomorphism technique (see Chapter 2), with which it is possible to construct a countably additive extension of a finitely additive (s)-bounded measure, and to study the properties of the starting measures in connection with the corresponding ones of the considered extensions. In the topological group setting, it is possible to use both the technique of Drewnowski (1972b) and the Stone Isomorphism technique, to construct σ-additive measures by starting from finitely additive (s)-bounded measures (see also Candeloro (1985a) and Sion (1969 and 1973)), and for a sake of simplicity we prefer to deal with the Drewnowski-type technique about countably additive restrictions. For lattice group-valued measures, to prove that the Brooks-Jewett theorem implies the Nikodým theorem, when we treat uniform (s)-boundedness and σ-additivity formulated not necessarily with respect to a same order sequence, in general for technical reasons it is not advisable to consider a direct approach, and we use the Maeda-Ogasawara-Vulikh representation theorem for Dedekind complete lattice groups, by studying the properties of the corresponding real-valued measures. When we deal with a single (O)-sequence, it is possible to give direct proofs, and it is not always advisable to use the tool of the Maeda-Ogasawara-Vulikh representation theorem, because it yields informations in general only about convergence of suitable lattice group-valued sequences by means of convergence of suitable real-valued sequences.

We investigate also some basic matrix theorems, extending earlier results of Aizpuru and Nicasio-Llach (2008) and Aizpuru, Nicasio-Llach and Rambla-Barreno (2010) (see Boccuto, Dimitriou and Papanastassiou (2010b, 2012d)). Note that in general these kinds of theorems, in their ideal/filter formulation, do not give immediate results like in the classical case, since in lattice groups the nature of order convergence is in general not topological, and because filter convergence is not inherited by subsequences. Moreover, we deal with some modes of continuity for filter convergence associated with a pair of filters of $\mathbb{N}$, for lattice group-valued measures (see also Boccuto and Dimitriou (2013d)). We give also some comparison results on filter (α)-convergence (continuous convergence) and give some necessary and sufficient conditions for (absolute) continuity of the limit measure. We prove also some relations between filter exhaustiveness and filter continuous convergence (or filter (α)-convergence) for measures and some applications to integrals, extending some results of Diestel and Uhl (1977) and Abbott, Bator, Bilyeu and Lewis (1990). The concept of (α)-*convergence* or *continuous convergence* or *stetige Konvergenz* of real-valued function sequences has been known since the beginning of the last century (see for example Catathéodory (1929), Hahn (1921), Stoilov (1959)). This notion was formulated for ordered structures by Wolk (1975). For a recent literature see also Athanassiadou, Dimitriou, Papachristodoulos and Papanastassiou (2012), Beer and Levi (2009 and 2010), Boccuto and Dimitriou (2011b), Boccuto, Dimitriou and Papanastassiou and Wilczyński (2011, 2012 and 2014).

4.1. Filter Limit Theorems in Lattice Groups

We consider several versions and features of limit theorems in the setting of filter convergence. We begin with observing that in general it is not possible to formulate the versions of convergence theorems, for instance Brooks-Jewett, Vitali-Hahn-Saks or Nikodým-type theorems, analogously as the classical ones, by replacing the classical pointwise convergence of the involved measures with the weaker pointwise filter/ideal convergence, even in the real case (see also Boccuto, Das, Dimitriou and Papanastassiou (2012), Example 4.5, and Boccuto, Dimitriou and Papanastassiou (2011b), Remark 2.7).

Let $\mathcal{F} \neq \mathcal{F}_{\text{cofin}}$ be any free filter of $\mathbb{N}$. There exists an infinite set $H := \{h_1 < \ldots < h_s < h_{s+1} < \ldots\}$, such that $\mathbb{N} \setminus H$ is infinite and belongs to $\mathcal{F}$. If $n \notin H$ and $E \subset \mathbb{N}$, put $m_n(E) = 0$, and for $s \in \mathbb{N}$, $E \subset \mathbb{N}$, let $m_{h_s}(E) = 1$ if $s \in E$ and 0 elsewhere. We have $m_0(E) := (\mathcal{F}) \lim_n m_n(E) = 0$ for all $E \subset \mathbb{N}$. Note that the measures m_n are equibounded, positive and σ-additive, since for any $n \in \mathbb{N}$ and each disjoint sequence $(C_j)_j$ in $\mathcal{P}(\mathbb{N})$, $m_n(C_j)$ is not equal to zero at most for one j, because for each $s \in \mathbb{N}$, $m_n(\{s\}) \neq 0$ if and only if $n = h_s$. If $C_j := \{j\}$, $j \in \mathbb{N}$, then we get $1 \geq \sup_{n \in \mathbb{N}} m_n(C_j) \geq m_{h_j}(C_j) = 1$. So the equality $(\mathcal{F}) \lim_j [\sup_{n \in \mathbb{N}} m_n(C_j)] = 0$ does not hold. □

However, under suitable conditions, it is possible to prove some limit theorems for measures, related to pointwise filter/ideal convergence. In this chapter we will present several results in this direction.

4.1.1. Schur-Type Theorems and Consequences

We begin with a Schur-type theorem for real-valued measures. To this aim, we prove the following technical lemma.

Lemma 1.1 *Let* $(x_k)_k$ *be an element of* l^1. *Then*

$$\sum_{k=N}^{M} | x_k | \leq 2 \max_{S \subset [N,M]} | \sum_{k \in S} x_k | \tag{4.1}$$

for every N, $M \in \mathbb{N}$ *with* $N < M$.

Proof: Define $m : \mathcal{P}(\mathbb{N}) \to \mathbb{R}$ by setting $m(A) := \sum_{k \in A} x_k$, $A \subset \mathbb{N}$. By Lemma 3.4 of Chapter 2 we get $\sum_{k=N}^{M} |x_k| = \sum_{k=N}^{M} |\mu(\{k\})| \leq 2 \max_{S \subset [N,M]} |\mu(S)| = 2 \max_{S \subset [N,M]} |\sum_{k \in S} x_k|$. This ends the proof. □

To prove the following version of the Hahn-Schur theorem, a sliding hump technique is used.

Theorem 1.2 (see Boccuto, Das and Dimitriou (2012), Theorem 3.1) *Let* $(y_{n,k})_{n,k}$ *be a double sequence in* $\mathbb{R}$ *such that* $(y_{n,k})_k \in l^1$ *for each* $n \in \mathbb{N}$, *and* $\mathcal{F}$ *be a* P-*filter of* $\mathbb{N}$. *Suppose that there exist a sequence* $(y_k)_k \in l^1$, *with*

(a) $(\mathcal{F}) \lim_n y_{n,k} = y_k$ *for every* $k \in \mathbb{N}$

and an element $F \in \mathcal{F}$, *such that*

(b) $(\mathcal{F}) \lim_n \sum_{k \in E} y_{i_r,k} = \sum_{k \in E} y_k$ *for each strictly increasing sequence* $(i_r)_r$ *in* F *and for every* $E \subset \mathbb{N}$ *with the property that both* E *and* $\mathbb{N} \setminus E$ *are infinite.*

Then $(\mathcal{F}) \lim_n \sum_{k=1}^{\infty} |y_{n,k} - y_k| = 0$.

Proof: For each n, $k \in \mathbb{N}$ put $b_{n,k} := y_{n,k} - y_k$. First of all note that, by (a) and Proposition 2.19 of Chapter 2, there is a set $K \in \mathcal{F}$ with

$$\lim_{n \in K} b_{n,k} = 0 \tag{4.2}$$

for any $k \in \mathbb{N}$. So, in order to prove the result, it will be enough to show that

$$\lim_{n \in F \cap K} \sum_{k=1}^{\infty} |b_{n,k}| = 0, \tag{4.3}$$

since $F \cap K \in \mathcal{F}$ and by the first part of Proposition 2.17 of Chapter 2. If (4.3) is not true, then there is a positive real number C with the property that for every $q \in K$ there exists $l_q \in F \cap K$, $l_q > q$, with

$$\sum_{k=1}^{\infty} | b_{l_q,k} | > C. \tag{4.4}$$

From (4.2) and (4.4) it follows that there is a positive integer $i_1 \in F \cap K$, with

$$\sum_{k=1}^{\infty} | b_{i_1,k} | > C \text{ and } | b_{i_1,1} | \le \frac{C}{8}. \tag{4.5}$$

At the first step, let $k_1 := 1$. Since $(y_k)_k$ and $(y_{i_1,k})_k$ belong to l^1, then $\sum_{k=1}^{\infty} | b_{i_1,k} | < +\infty$, and so we can choose a natural number $k_2 > k_1$ with $\sum_{k=k_2}^{\infty} | b_{i_1,k} | \le \frac{C}{8}$. From this and (4.5) we obtain

$$\sum_{k=2}^{k_2-1} | b_{i_1,k} | > \frac{3C}{4}. \tag{4.6}$$

From (4.6) and Lemma 1.1 we have $\max_{A \subset [2,k_2-1]} | \sum_{k \in A} b_{i_1,k} | > \frac{3C}{8}$.

At the second step, taking $q = i_1$ in (4.4), we find an element $i_2 \in F \cap K$, $i_2 > i_1$, with

$$\sum_{k=1}^{\infty} | b_{i_2,k} | > C \text{ and } \sum_{k=1}^{k_2} | b_{i_2,k} | \le \frac{C}{8}: \tag{4.7}$$

such a choice is possible, by virtue of (4.4) and (4.2) respectively. By proceeding analogously as above, we can find an integer $k_3 > k_2$ with $\sum_{k=k_3}^{\infty} | b_{i_2,k} | \le \frac{C}{8}$. From this and (4.7) we get:

$$\sum_{k=k_2+1}^{k_3-1} | b_{i_2,k} | > \frac{3C}{4}. \quad \textbf{(4.8)}$$

From (4.8) and Lemma 1.1 it follows that $\max_{A \subset [k_2+1,k_3-1]} | \sum_{k \in A} b_{i_2,k} | > \frac{3C}{8}$.

By induction, we construct two strictly increasing sequences $(i_r)_r$ and $(k_r)_r$ in $F \cap K$ and $\mathbb{N}$, respectively, such that $k_1 = 1$ and

1.2.1) $\sum_{k=1}^{k_r} | b_{i_r,k} | \leq \frac{C}{8}$;

1.2.2) $\sum_{k=k_{r+1}}^{\infty} | b_{i_r,k} | \leq \frac{C}{8}$;

1.2.3) $\max_{A \subset [k_r+1,k_{r+1}-1]} | \sum_{k \in A} b_{i_r,k} | > \frac{3C}{8}$

for every $r \geq 2$. From 1.2.3) it follows that for such r 's there exists a set $E_r \in \mathcal{P}([k_r+1, k_{r+1}-1])$ with

1.2.4) $| \sum_{k \in E_r} b_{i_r,k} | > \frac{3C}{8}$.

Let now $E := \bigcup_{r=1}^{\infty} E_r$. Note that, by construction, E is infinite and $E \cap \{k_r : r \geq 2\} = \emptyset$, and hence $\mathbb{N} \setminus E$ is infinite too. By (b) and Proposition 2.18 of Chapter 2, in correspondence with the sequence $(i_r)_r$ and the set E there is a subsequence $(i_{r_s})_s$ of $(i_r)_r$, with $\lim_s (\sum_{k \in E} b_{i_{r_s},k}) = 0$, where the limit is intended in the usual sense. Thus, in correspondence with $\frac{C}{8}$, there is a positive integer $s_0 \in \mathbb{N}$ such that for every $s \geq s_0$ we get:

$$| \sum_{k\in E_{r_s}} b_{i_{r_s},k} |=| \sum_{k\in E} b_{i_{r_s},k} - \sum_{k\in E,k=1}^{k_{r_s}} b_{i_{r_s},k} - \sum_{k\in E,k=k_{r_s}+1}^{\infty} b_{i_{r_s},k} | \leq$$

$$\leq | \sum_{k\in E} b_{i_{r_s},k} | + \sum_{k=1}^{k_{r_s}} | b_{i_{r_s},k} | + \sum_{k=k_{r_s}+1}^{\infty} | b_{i_{r_s},k} | \leq \frac{3C}{8}.$$

This contradicts 1.2.4) and proves the theorem. □

In the next step, we present some Schur-type and limit theorems for lattice group-valued measures, for filters which satisfy suitable properties. The following version of the Schur Lemma for filter convergence extends Aviles Lopez, Cascales Salinas, Kadets and Leonov (2007), Theorem 3.5, to the setting of (ℓ)-groups. Observe that, if the involved filter is not block-respecting, then, even when $R=\mathbb{R}$, the Schur lemma in general does not hold (see Aviles Lopez, Cascales Salinas, Kadets and Leonov (2007), Remark 3.4). However, in the sequel, we will give some particular versions of limit theorems for positive measures, for filters which are not necessarily block-respecting.

Theorem 1.3 (see also Boccuto, Dimitriou and Papanastassiou (2012e), Lemma 3.1 and Theorem 3.1) *Let R be any Dedekind complete (ℓ)-group, $\mathcal{F}$ be a block-respecting filter of $\mathbb{N}$, $m_j : \mathcal{P}(\mathbb{N}) \to R$, $j\in\mathbb{N}$, be a sequence of σ-additive equibounded measures, and define $(\beta_{A,j})_{A\in\mathcal{P}(\mathbb{N}), j\in\mathbb{N}}$ by setting $\beta_{A,j} := m_j(A)$, $A\in\mathcal{P}(\mathbb{N}), j\in\mathbb{N}$. Suppose that:*

(i) $(D)\lim_j m_j(\{n\}) = 0$ *for each* $n\in\mathbb{N}$;

(ii) The family $(\beta_{A,j})_{A\in\mathcal{P}(\mathbb{N}), j\in\mathbb{N}}$ $(RD\mathcal{F})$*-converges to* 0.

Then we get:

(j) $(D\mathcal{F})\lim_j \sum_{n=1}^{\infty} | m_j(\{n\}) | = 0$.

(jj) If $\mathcal{F}$ *is also diagonal and* R *is super Dedekind complete and weakly* σ*-distributive, then* (ii) *implies* (j).

Proof: We begin with (j). Let $u := \bigvee_{j\in\mathbb{N}} v(m_j)(\mathbb{N})$: such an element exists in R, thanks to equiboundedness of the sequence $(m_j)_j$. For each $j \in \mathbb{N}$ let $(a_{t,l}^{(j)})_{t,l}$ be a (D)-sequence related with σ-additivity of m_j and the sequence $(H_n)_n$, where $H_n := [n,+\infty[$, $n \in \mathbb{N}$. For each $\varphi \in \mathbb{N}^{\mathbb{N}}$ and $j \in \mathbb{N}$ there is $\overline{n} \in \mathbb{N}$ (depending on φ and j) with

$$v(m_j)_j([\overline{n},+\infty[) \le \bigvee_{t=1}^{\infty} a_{t,\varphi(t+j)}^{(j)}. \tag{4.9}$$

By virtue of the Fremlin lemma there exists a (D)-sequence $(a_{t,l})_{t,l}$ with

$$u \wedge [\bigvee_q (\sum_{j=1}^{q} (\bigvee_{t=1}^{\infty} a_{t,\varphi(t+j)}^{(j)}))] \le \bigvee_{t=1}^{\infty} a_{t,\varphi(t)} \tag{4.10}$$

for every $\varphi \in \mathbb{N}^{\mathbb{N}}$. From (4.9) and (4.10) it follows that for each $\varphi \in \mathbb{N}^{\mathbb{N}}$ and $j \in \mathbb{N}$ there exists $\overline{n} \in \mathbb{N}$ with

$$v(m_j)_j([\overline{n},+\infty[) \le \bigvee_{t=1}^{\infty} a_{t,\varphi(t)}. \tag{4.11}$$

Let $(b_{t,l})_{t,l}$ be a regulator, satisfying the condition of $(RD\mathcal{F})$-convergence as in (ii).

Since $(D)\lim_j m_j(\{n\}) = 0$ for each $n \in \mathbb{N}$, then for any $n \in \mathbb{N}$ there exists a (D)-sequence $(c_{t,l}^{(n)})_{t,l}$ such that for every $\varphi \in \mathbb{N}^{\mathbb{N}}$ there is $\overline{j} \in \mathbb{N}$ with $|m_j(\{n\})| \le \bigvee_{t=1}^{\infty} c_{t,\varphi(t+n)}^{(n)}$ whenever $j \ge \overline{j}$. Thanks to equiboundedness of the sequence $(m_j)_j$ and the Fremlin lemma, there is a regulator $(c_{t,l})_{t,l}$ such that for every $\varphi \in \mathbb{N}^{\mathbb{N}}$ and $n \in \mathbb{N}$ there exists $\overline{j} \in \mathbb{N}$ with $|m_j(\{n\})| \le \bigvee_{t=1}^{\infty} c_{t,\varphi(t)}$ whenever $j \ge \overline{j}$. Set now

$$d_{t,l} = 6(a_{t,l} + b_{t,l} + c_{t,l}), \quad t,l \in \mathbb{N}. \tag{4.12}$$

Observe that, since $\sum_{n=1}^{\infty} | m_j(\{n\}) | = \| m_j \|(\mathbb{N})$ (see also Boccuto and Papanastassiou (2007), Proposition 3.3) and

$$v(m_j)(\mathbb{N}) \le \| m_j \|(\mathbb{N}) \le 2v(m_j)(\mathbb{N})$$

for every $j \in \mathbb{N}$ (see also (2.38)), then to prove (j) it is enough to demonstrate that

$$(D\mathcal{F}) \lim_j v(m_j)(\mathbb{N}) = 0. \quad \textbf{(4.13)}$$

We prove that the (D)-sequence $(d_{t,l})_{t,l}$ defined in (4.12) satisfies $(D\mathcal{F})$-convergence as in (4.13). Otherwise there exists $\varphi \in \mathbb{N}^{\mathbb{N}}$ with the property that the set

$$I^* := \{ j \in \mathbb{N} : v(m_j)(\mathbb{N}) \le \bigvee_{t=1}^{\infty} d_{t,\varphi(t)} \} \quad \textbf{(4.14)}$$

does not belong to $\mathcal{F}$. From this it follows that every element F of $\mathcal{F}$ is not contained in I^*, namely F has nonempty intersection with $\mathbb{N} \setminus I^*$. This means that $I := \mathbb{N} \setminus I^*$ is $\mathcal{F}$-stationary. Note that I is an infinite set, since $\mathcal{F}$ is a free filter.

Let $n_0 := 1$, $a := \bigvee_{t=1}^{\infty} a_{t,\varphi(t)}$, $b := \bigvee_{t=1}^{\infty} b_{t,\varphi(t)}$, $c := \bigvee_{t=1}^{\infty} c_{t,\varphi(t)}$, $d := \bigvee_{t=1}^{\infty} d_{t,\varphi(t)}$, where φ is as in (4.14). By σ-additivity of m_1, there is an integer $l(1) > 1$ with $v(m_1)(]l(1), +\infty[) \le a$. Furthermore, thanks to (i), there is an integer $n_1 > l(1)$ with $| m_s(L) | \le c$ for any $s \ge n_1$ and for each finite subset $L \subset [1, l(1)]$, and thus $v(m_s)([1, l(1)]) \le c$ for all $s \ge n_1$. At the next step, by σ-additivity of $m_1, \ldots, m_{n_1}$, we find a natural number $l(n_1) > n_1$, with $v(m_r)(]l(n_1), +\infty[) \le a$ for every $r \le n_1$, and an integer $n_2 > l(n_1)$ with $v(m_s)([1, l(n_1)]) \le c$ whenever $s \ge n_2$. By induction, we get two strictly increasing sequences $(n_h)_h$ and $(l(n_h))_h$, such that, for every $h \in \mathbb{N}$, $n_{h-1} < l(h) < n_h$, $v(m_r)(]l(n_h), +\infty[) \le a$ whenever $r \le n_h$, and $v(m_s)([1, l(n_h)]) \le d$ for all $s \ge n_{h+1}$. Without loss of generality, we can suppose that $I \cap [n_{h-1}, n_h[\ne \emptyset$ for every $h \in \mathbb{N}$. There is a set $J \in \mathcal{F}^*$, $J \subset I$, $J = \{ j_0, j_1,$

$j_2,\dots\}$, such that J intersects each interval $[n_h, n_{h+1}[$ in exactly one point. As $J \in \mathcal{F}^*$, then at least one of the two sets $J_1 := \{j_1, j_3, j_5, \dots\}$ and $J_2 := \{j_0, j_2, j_4, \dots\}$ is $\mathcal{F}$-stationary. Without loss of generality, we assume that $J_1 \in \mathcal{F}^*$. For every $h \in \mathbb{N}$, we have: $n_h \le j_h < n_{h+1}$,

$$v(m_{j_{2h-1}})(]l(n_{2h},+\infty[) \le a, \; v(m_{j_{2h-1}})([1, l(n_{2h-2}]) \le c. \tag{4.15}$$

From this, since $v(m_{j_{2h-1}})(\mathbb{N}) \not\le d$, for every h we get:

$$v(m_{j_{2h-1}})(]l(n_{2h-2}), l(n_{2h})]) \not\le a + b + c: \tag{4.16}$$

otherwise, from (4.15) and (4.16) we have $v(m_{j_{2h-1}})(\mathbb{N}) \le 2(a+b+c) = d$, a contradiction. By (4.16) there is a set $Q_h \subset]l(n_{2h-2}), l(n_{2h})]$ with

$$| m_{j_{2h-1}}(Q_h) | \not\le a + b + c. \tag{4.17}$$

Note that the sets Q_h, $h \in \mathbb{N}$, are pairwise disjoint. Set $H := \bigcup_{h=1}^{\infty} Q_h$. For every h we get

$$m_{j_{2h-1}}(H) = m_{j_{2h-1}}(H \cap [1, l(n_{2h-2})]) + m_{j_{2h-1}}(H \cap]l(n_{2h}), +\infty[) + m_{j_{2h-1}}(H \cap]l(n_{2h-2}), l(n_{2h})]) =$$

$$= m_{j_{2h-1}}(H \cap [1, l(n_{2h-2})]) + m_{j_{2h-1}}(H \cap]l(n_{2h}), +\infty[) + m_{j_{2h-1}}(Q_h),$$

and so we see that

$$| m_{j_{2h-1}}(H) - m_{j_{2h-1}}(Q_h) | \le a + c. \tag{4.18}$$

Thanks to (4.18), we obtain $| m_{j_{2h-1}}(H) | \not\le b$ for all h, otherwise $| m_{j_{2h-1}}(Q_h) | \le a + b + c$, which contradicts (4.17). By (ii), there exists a set $F \in \mathcal{F}$ with $| m_j(H) | \le b$ for all $j \in F$, and, since J_1 is $\mathcal{F}$-stationary, there is

$j_* \in F \cap J_1$. So we get $|m_{j_*}(H)| \not\leq b$ and $|m_{j_*}(H)| \leq b$, a contradiction. This proves (4.13), and hence (j).

(jj) Let $(b_{t,l})_{t,l}$ be a (D)-sequence, satisfying $(RD\mathcal{F})$-convergence in condition (ii). Observe that, for every $\mathcal{F}$-stationary set J, $(b_{t,l})_{t,l}$ satisfies also $(RD\mathcal{F}(J))$-convergence in (ii). Since R is super Dedekind complete and weakly σ-distributive, by Theorem 2.2 of Chapter 2 there is an (O)-sequence $(\sigma_p)_p$, satisfying condition (ii), when $(RD\mathcal{F})$-convergence is replaced with $(RO\mathcal{F})$- or $(RO\mathcal{F}(J))$-convergence. By Lemma 2.23 of Chapter 2 and the successive remark, for every $I \in \mathcal{F}^*$ there exists $J \in \mathcal{F}^*$, $J \subset I$, with $(RO)\lim_{j\in J} m_j(\{n\}) = 0$, $n \in \mathbb{N}$, with respect to $(\sigma_p)_p$. Again by Theorem 2.2 of Chapter 2, we find a regulator $(c_{t,l})_{t,l}$, such that for every $I \in \mathcal{F}^*$ there is $J \in \mathcal{F}^*$, $J \subset I$, with

$$(RD)\lim_{j\in J} m_j(\{n\}) = 0 \text{ for each positive integer } n \qquad \textbf{(4.19)}$$

with respect to $(c_{t,l})_{t,l}$. Let now $(a_{t,l}^{(j)})_{t,l}$, $n \in \mathbb{N}$, be regulators associated to σ-additivity of the measures m_j, and u, $(a_{t,l})_{t,l}$, be as in the proof of (j). Set $d_{t,l} := 2(a_{t,l} + b_{t,l} + c_{t,l})$, t, $l \in \mathbb{N}$. We prove that the regulator $(d_{t,l})_{t,l}$ satisfies the thesis of (jj). Otherwise, by proceeding analogously as in the proof of (j), we get the existence of a function $\varphi \in \mathbb{N}^{\mathbb{N}}$ and of an $\mathcal{F}$-stationary set $I \subset \mathbb{N}$, with $v(m_j)(\mathbb{N}) \not\leq \bigvee_{t=1}^{\infty} d_{t,\varphi(t)}$ for every $j \in I$. In correspondence with I, there is $J \in \mathcal{F}^*$, $J \subset I$, satisfying (4.19). Note that the sequence $v(m_j)(\mathbb{N})$, $j \in J$, does not $(\mathcal{F}(J))$-converge to 0 (see also Aviles Lopez, Cascales and Salinas, Kadets and Leonov (2007)). Since $J \in \mathcal{F}^*$ and $\mathcal{F}$ is block-respecting, then $\mathcal{F}(J)$ is block-respecting too. By (j) used with m_j, $j \in J$, and $\mathcal{F}(J)$, it follows that $(\mathcal{F}(J))\lim_{j\in J} v(m_j)(\mathbb{N}) = 0$, getting a contradiction. This proves (jj). □

Analogously as in Theorem 1.3, it is possible to prove the following version of the Brooks-Jewett theorem in the setting of filter convergence.

Theorem 1.4 (see also Boccuto, Dimitriou and Papanastassiou (2012e), Theorems 4.1 and 4.2) *Let $\mathcal{F}$ be a block-respecting filter of $\mathbb{N}$, and $\mu_j : \mathcal{P}(\mathbb{N}) \to R$, $j \in \mathbb{N}$, be a sequence of globally (s)-bounded measures. Set $\zeta_{A,j} := \sum_{k\in A} \mu_j(\{k\})$, $A \in \mathcal{P}(\mathbb{N})$, $j \in \mathbb{N}$, and assume that:*

(i) $(D)\lim_{j\in J}\mu_j(\{k\}) = 0$ *for each* $k \in \mathbb{N}$;

(ii) *The family* $(\zeta_{A,j})_{A\in\mathcal{P}(\mathbb{N}), j\in\mathbb{N}}$, $(RD\mathcal{F})$*-converges to zero.*

Then we get:

(j) $(D\mathcal{F})\lim_{j\in J}(\bigvee_{A\in\mathcal{P}(\mathbb{N})} | \sum_{k\in A}\mu_j(\{k\}) |) = 0.$

(jj) *If $\mathcal{F}$ is also diagonal and R is super Dedekind complete and weakly σ-distributive, then (ii) implies (j).*

As a consequence of Theorem 1.3, we will extend Brooks and Mikusiński (1970), Lemma 1, to the context of (ℓ)-groups.

Corollary 1.5 (Boccuto, Dimitriou and Papanastassiou (2012e), Corollary 3.1.1) *Let R be a super Dedekind complete and weakly σ-distributive (ℓ)-group and $(a_{j,n})_{j,n}$ be any bounded double sequence in R, suppose that $\sum_{n=1}^{\infty} | a_{j,n} | \in R$ for all $j \in \mathbb{N}$, and set $\alpha_{A,j} := \sum_{n\in A} a_{j,n}$, $A \in \mathcal{P}(\mathbb{N})$, $j \in \mathbb{N}$.*

If the family $(\alpha_{A,j})_{A\in\mathcal{P}(\mathbb{N}), j\in\mathbb{N}}$ $(RD\mathcal{F})$-converges to 0, then we get $(D\mathcal{F})\lim_j \sum_{n=1}^{\infty} | a_{j,n} | = 0.$

Proof: For every $A \subset \mathbb{N}$ and $j \in \mathbb{N}$, set $m_j(A) := \sum_{n\in A} a_{j,n}$. First of all observe that, by hypothesis, $(a_{j,n})_n \in l^1_R$ for all $j \in \mathbb{N}$. We now claim that

$$\| m_j \|(\mathbb{N}) = \sum_{n=1}^{\infty} | a_{j,n} | \text{ for every } j \in \mathbb{N}. \qquad \textbf{(4.20)}$$

Indeed, by Lemma 3.4 of Chapter 2 we get:

$$\sum_{k=1}^{n} | a_{j,k} | = \| m_j \|(\{1,\ldots,n\}) \leq \| m_j \|(\mathbb{N})$$

for every $j,n \in \mathbb{N}$, and hence $\sum_{n=1}^{\infty} | a_{j,n} | \leq \| m_j \|(\mathbb{N})$ for all $j \in \mathbb{N}$.

We now turn to the converse inequality. For every $j \in \mathbb{N}$ and $A \subset \mathbb{N}$ we get

$$\sum_{n \in A} a_{j,n} \leq \sum_{n \in A} a^{+}_{j,n} \leq \sum_{n=1}^{\infty} a^{+}_{j,n}$$

and

$$-\sum_{n \in A} a_{j,n} = \sum_{n \in A} (-a_{j,n}) \leq \sum_{n \in A} a^{-}_{j,n} \leq \sum_{n=1}^{\infty} a^{-}_{j,n}.$$

Note that all the involved quantities belong to R, since $(a_{j,n})_n \in l^1_R$ for all j. Taking the supremum as A varies in $\mathcal{P}(\mathbb{N})$ we obtain:

$$\bigvee_{A \subset \mathbb{N}} (\sum_{n \in A} a_{j,n}) \leq \sum_{n=1}^{\infty} a^{+}_{j,n},$$

$$\bigvee_{A \subset \mathbb{N}} (-\sum_{n \in A} a_{j,n}) \leq \sum_{n=1}^{\infty} a^{-}_{j,n};$$

$$\| m_j \|(\mathbb{N}) = \bigvee_{A \subset \mathbb{N}} (\sum_{n \in A} a_{j,n}) + \bigvee_{A \subset \mathbb{N}} (-\sum_{n \in A} a_{j,n}) \leq \sum_{n=1}^{\infty} a^{+}_{j,n} + \sum_{n=1}^{\infty} a^{-}_{j,n} = \sum_{n=1}^{\infty} | a_{j,n} |$$

for all $j \in \mathbb{N}$. Thus (4.20) is proved. We now prove σ-additivity of m_j (with respect to a single (O)-sequence). To this aim, it is enough to show that $(O) \lim_n \| m_j \|(H_n) = 0$ with respect to a common (O)-sequence, where $H_n = [n,+\infty[$, $n \in \mathbb{N}$. From Lemma 3.4 of Chapter 2 we get:

$$\| m_j \|(H_n) = \| m_j \|(\mathbb{N}) - \sum_{s=1}^{n-1} | a_{j,s} |, \; j, n \in \mathbb{N},$$

and hence

$$(O) \lim_n \| m_j \|(H_n) = \| m_j \|(\mathbb{N}) - (O) \lim_n \sum_{s=1}^{n-1} | a_{j,s} | = 0, \; j \in \mathbb{N},$$

obtaining σ-additivity (with respect to a single (O)-sequence). The assertion follows from this and Theorem 1.3. □

Corollary 1.6 (Boccuto, Dimitriou and Papanastassiou (2012e), Corollary 3.1.2) *Under the same notations and hypotheses as in Theorem 1.3, we get that the family*

$$(m_j(A))_{A \in \mathcal{P}(\mathbb{N}), j \in \mathbb{N}} \tag{4.21}$$

$(UD\mathcal{F})$*-converges to 0. Furthermore for every* $I \in \mathcal{F}^*$ *there exists an* $\mathcal{F}$*-stationary set* $J \subset I$ *with*

$$(D) \lim_k (\bigvee_{j \in J} v(m_j)([k, +\infty[)) = 0. \tag{4.22}$$

Proof: We first prove $(UD\mathcal{F})$-convergence to 0 of the family $(m_j(A))_{A \in \mathcal{P}(\mathbb{N}), j \in \mathbb{N}}$. By virtue of Theorem 1.3 and (4.20) it follows that

$$0 = (D\mathcal{F}) \lim_j \sum_{n=1}^{\infty} | m_j(\{n\}) | = (D\mathcal{F}) \lim_j \| m_j \|(\mathbb{N}) = 0,$$

and thus the family in (4.21) $(UD\mathcal{F})$-converges to 0. Similar conclusions hold even for the positive and negative parts and for the semivariations of the m_j's.

From this, by virtue of (4.20), Lemma 2.24 of Chapter 2 and by equivalence between $(O\mathcal{F})$- and $(D\mathcal{F})$-convergence (we also take into account super Dedekind completeness and weak σ-distributivity of R and use Theorem 2.2. of Chapter 2) it follows that there is an $\mathcal{F}$-stationary set $J \subset \mathbb{N}$ with

$$0 = (O)\lim_{j \in J} \sum_{n=1}^{\infty} | m_j(\{n\}) | = (O)\lim_{j \in J} \| m_j \|(\mathbb{N}) =$$

$$= (D)\lim_{j \in J} \sum_{n=1}^{\infty} | m_j(\{n\}) | = (D)\lim_{j \in J} \| m_j \|(\mathbb{N}). \tag{4.23}$$

Now, by virtue of (4.23), there exists a (D)-sequence $(h_{t,l})_{t,l}$ such that to every $\varphi \in \mathbb{N}^{\mathbb{N}}$ there is a positive integer $\overline{j}$, without loss of generality $\overline{j} \in J$, such that

$$| m_j(A) | \le \bigvee_{t=1}^{\infty} h_{t,\varphi(t)} \tag{4.24}$$

whenever $j \ge \overline{j}$, $j \in J$, and $A \subset \mathbb{N}$. Moreover, by virtue of σ-additivity of m_j, $j \in \mathbb{N}$, the equiboundedness and the Fremlin lemma, there is a (D)-sequence $(a_{t,l})_{t,l}$ in (4.11) is such that for all $\varphi \in \mathbb{N}^{\mathbb{N}}$ and $j \in \mathbb{N}$ there is $\overline{k} = \overline{k}(\varphi, j) \in \mathbb{N}$ with

$$| m_j(A) | \le \bigvee_{t=1}^{\infty} a_{t,\varphi(t)} \text{ for all } A \subset [\overline{k}, +\infty[. \tag{4.25}$$

Fix arbitrarily $\varphi \in \mathbb{N}^{\mathbb{N}}$, and let $\overline{j}$ as in (4.24). In correspondence with φ and $j = 1, \ldots, \overline{j} - 1$, there exist $\overline{k}_1, \ldots, \overline{k}_{\overline{j}-1}$ as in (4.25). If $k^* := \max\{\overline{k}_1, \ldots, \overline{k}_{\overline{j}-1}\}$, then we have

$$| m_j(A) | \le \bigvee_{t=1}^{\infty} a_{t,\varphi(t)} \text{ for all } A \subset [k^*, +\infty[\text{ and } j = 1, \ldots, \overline{j} - 1. \tag{4.26}$$

Moreover, for every $j \ge \overline{j}$, $j \in J$, and $A \subset [k^*, +\infty[$, we get

$$| m_j(A) | \le \bigvee_{t=1}^{\infty} h_{t,\varphi(t)} + \bigvee_{t=1}^{\infty} a_{t,\varphi(t)}. \tag{4.27}$$

From (4.26), (4.27) and Lemma 2.24 of Chapter 2 we obtain (4.22). □

Now, before proving some limit theorems, which are consequences of the Schur Lemma, we state some technical results. Here and in the sequel, Σ denotes a σ-algebra of parts of a nonempty set G.

Lemma 1.7 (see also Boccuto, Dimitriou and Papanastassiou (2011c), Lemma 2.4 and Boccuto, Dimitriou and Papanastassiou (2012c), Lemma 3.8) *Assume that R is any Dedekind complete lattice group, $m:\Sigma \to R$ is a finitely additive measure, $(H_k)_k$ is any decreasing sequence of Σ, and $m(F)=0$, where $F=\bigcap_{k=1}^{\infty} H_k$. Put $C_k := H_k \setminus H_{k+1}$, $k \in \mathbb{N}$, and let $\mathcal{K}$ and $\mathcal{L}$ be the σ-algebras generated by the sets C_k and H_k in H_1, respectively. Then we get*

$$v_{\mathcal{L}}(m)(H_k) = v_{\mathcal{K}}(m)(\bigcup_{l=k}^{\infty} C_l) \text{ for every } k \in \mathbb{N}.$$

Proof: Note that $\mathcal{L} = \mathcal{K} \cup \{X \cup F : X \in \mathcal{K}\}$. Set $\mathcal{K}_k := \{X \in \mathcal{K} : X \subset \bigcup_{l=k}^{\infty} C_k\}$, $k \in \mathbb{N}$. Taking into account finite additivity of m and since $m(F)=0$, we get:

$$v_{\mathcal{L}}(m)(H_k) = v_{\mathcal{L}}(m)(F \cup (\bigcup_{l=k}^{\infty} C_k)) = \bigvee_{X \in \mathcal{K}_k} [|m(X)| \vee |m(X \cup F)|]$$

$$= \bigvee_{X \in \mathcal{K}_k} |m(X)| \vee |m(X) + m(F))| = \bigvee_{X \in \mathcal{K}_k} |m(X)| = v_{\mathcal{K}}(m)(\bigcup_{l=k}^{\infty} C_l). \ \square$$

We now recall the following characterization of σ-additivity for measures defined on $\mathcal{P}(\mathbb{N})$ and taking values in lattice groups.

Proposition 1.8 (see also Boccuto, Dimitriou and Papanastassiou (2011c), Proposition 2.5, and Boccuto, Dimitriou and Papanastassiou (2012c), Proposition 2.10) *Let m be an R-valued measure m, defined on $\mathcal{P}(\mathbb{N})$. Then m is σ-additive if and only if*

$$(O)\lim_k v(m)([k,+\infty[) = \bigwedge_k v(m)([k,+\infty[) = 0.$$

Proof: Let $(H_k)_k$ be any decreasing sequence in $\mathcal{P}(\mathbb{N})$ with $\bigcap_{k=1}^{\infty} H_k = \emptyset$, and denote by $\mathcal{L}$ the σ-algebra generated by $(H_k)_k$ in H_1. Without loss of generality we can suppose that $H_k \subset [k,+\infty[$ for each $k \in \mathbb{N}$. We get $v_{\mathcal{L}}(m)(H_k) \le v(m)([k,+\infty[)$, and from this the “if” part follows. The “only if” part is easy. $\square$

Note that, proceeding analogously as in Proposition 1.8, it is possible to prove a similar result also concerning uniform σ-additivity of a sequence of R-valued measures, defined on $\mathcal{P}(\mathbb{N})$, $j \in \mathbb{N}$.

The following proposition links measures defined on $\mathcal{P}(\mathbb{N})$ with set functions defined on general σ-algebras.

Proposition 1.9 (see Boccuto, Dimitriou and Papanastassiou (2011c), Proposition 3.6) *Suppose that R is a Dedekind complete lattice group, $m : \Sigma \to R$ is a finitely additive measure, $(H_k)_k$ is a decreasing sequence in Σ, and put $C_k = H_k \setminus H_{k+1}$, $k \in \mathbb{N}$, and $\nu(A) = m(\bigcup_{k\in A} C_k)$, $A \subset \mathbb{N}$. Assume that*

$$\bigwedge_k v_{\mathcal{L}}(m)(H_k) = 0, \tag{4.28}$$

where $\mathcal{L}$ is the σ-algebra generated by $(H_k)_k$ in H_1. Then ν is a σ-additive measure on $\mathcal{P}(\mathbb{N})$.

Proof: Let $\mathcal{K}$ be the σ-algebra generated by $(C_k)_k$ in H_1, and set $F = \bigcap_{k=1}^{\infty} H_k$. The sets C_k are pairwise disjoint, and thanks to (4.28) we get $v_{\mathcal{L}}(m)(F) = 0$, and hence $m(F) = 0$. We prove that

$$v(\nu)([k,+\infty[) = \bigvee(|\nu(B)| : B \subset [k,+\infty[) = \bigvee(|m(C)| : C \in \mathcal{L}$$

$$\text{with } C \subset H_k) = v_{\mathcal{L}}(m)(H_k) \tag{4.29}$$

for every $k \in \mathbb{N}$. To check that $v(\nu)([k,+\infty[) \le v_{\mathcal{L}}(m)(H_k)$, it is sufficient to associate each set $B \subset [k,+\infty[$ with the corresponding set $C = \bigcup_{k\in B} C_k$. Note that $C \subset H_k$. On the other hand, observe that by Lemma 1.7 we have

$$v_{\mathcal{L}}(m)(H_k) = v_{\mathcal{K}}(m)(\bigcup_{l=k}^{\infty} B_l) \tag{4.30}$$

for every $k \in \mathbb{N}$.

If $C \in \mathcal{K}$, $C \subset \bigcup_{l=k}^{\infty} C_l$, then $C = \bigcup_{k \in B} C_k$ for a suitable set $B \subset [k,+\infty[$. The relation (4.29) is a consequence of this and (4.30). By virtue of (4.28) and (4.29) it follows that

$$\bigwedge_k v(\nu)([k,+\infty[) = 0. \ \Box \tag{4.31}$$

We now give a filter version of the Vitali-Hahn-Saks convergence theorem for finitely additive measures, which is a consequence of a filter version of the Schur theorem.

Theorem 1.10 (see also Boccuto, Dimitriou and Papanastassiou (2012e), Theorem 4.2)

Let R be a super Dedekind complete and weakly σ-distributive (ℓ)-group, $\mathcal{F}$ be a diagonal and block-respecting filter of $\mathbb{N}$, τ be a Fréchet-Nikodým topology on Σ, $m_j : \Sigma \to R$, $j \in \mathbb{N}$, be an equibounded sequence of τ-continuous finitely additive measures. Assume that the family $m_j(A)$, $A \in \Sigma$, $j \in \mathbb{N}$, $(RD\mathcal{F})$-converges to 0.

Then for each decreasing sequence $(H_k)_k$ from Σ such that $\tau\text{-}\lim_k H_k = \emptyset$ and for every $\mathcal{F}$-stationary set $I \subset \mathbb{N}$ there exists an $\mathcal{F}$-stationary set $J \subset I$, with

$$\bigwedge_k [\bigvee_{j \in J} v_{\mathcal{L}}(m_j)(H_k)] = 0, \tag{4.32}$$

where $\mathcal{L}$ is the σ-algebra generated by the sets $H_k, k \in \mathbb{N}$ in H_1.

Proof: Let I and $(H_k)_k$ be as in the hypotheses, and set $C_k = H_k \setminus H_{k+1}$ for all $k \in \mathbb{N}$. As m_j, $j = 0,1,\ldots$, is τ-continuous, we get

$$\bigwedge_k v_{\mathcal{L}}(m_j)(H_k) := \bigwedge_k \bigvee (| m_j(C) | : C \in \mathcal{L} \text{ with } C \subset H_k) = 0. \tag{4.33}$$

For each $A \in \mathcal{P}(\mathbb{N})$ and $j = 0,1,\ldots$, set

$$v_j(A) = m_j(\bigcup_{k \in A} C_k). \tag{4.34}$$

By Proposition 1.9, ν_j is a σ-additive measure. Uniform boundedness and $(RD\mathcal{F})$-convergence of the family $\nu_j(A)$, $A \in \mathcal{P}(\mathbb{N})$, $j \in \mathbb{N}$, to 0, follow from equiboundedness and $(RD\mathcal{F})$-convergence of the family $m_j(A)$, $A \in \Sigma$, $j \in \mathbb{N}$, to 0.

By applying Theorem 1.3 and taking into account (4.29), it follows that for every $\mathcal{F}$-stationary set $I \subset \mathbb{N}$ there exists an $\mathcal{F}$-stationary set $J \subset I$, satisfying (4.32). □

Analogously as in Theorem 1.10 it is possible to prove the following Nikodým convergence-type theorem in the filter setting.

Theorem 1.11 (see also Boccuto, Dimitriou and Papanastassiou (2012e), Theorem 3.2)

Let R, $\mathcal{F}$, Σ, $\mathcal{L}$ be as in Theorem 1.10, $m_j : \Sigma \to R$, $j \in \mathbb{N}$, be an equibounded sequence of σ-additive measures. Suppose that the family $m_j(A)$, $A \in \Sigma$, $j \in \mathbb{N}$, $(RD\mathcal{F})$-converges to 0.

Then for every decreasing sequence $(H_k)_k$ from Σ such that $\bigcap_{k=1}^{\infty} H_k = \emptyset$ and for every $\mathcal{F}$-stationary set $I \subset \mathbb{N}$ there exists an $\mathcal{F}$-stationary set $J \subset I$, with $\bigwedge_k [\bigvee_{j \in J} \nu_{\mathcal{L}}(m_j)(H_k)] = 0.$

The following result extends Rickart (1943), Lemma 2.2 to the setting of (ℓ)-groups.

Lemma 1.12 (see Boccuto, Dimitriou and Papanastassiou (2012e), Lemma 2.4) *Let $m : \Sigma \to R$ be a finitely additive globally (s)-bounded measure (with respect to a single regulator $(a_{t,l})_{t,l}$).*

Then for each disjoint sequence $(E_k)_k$ in Σ and for every $\varphi \in \mathbb{N}^{\mathbb{N}}$ there is a finite set of positive integers Δ_φ such that

$$v(m)(\bigcup_{k \in \Delta} E_k) \le \bigvee_{t=1}^{\infty} a_{t,\varphi(t)}$$

whenever $\Delta \subset \mathbb{N}$ is a finite set, such that $\Delta \cap \Delta_{\varphi} = \varnothing$ (here, the semivariation is intended with respect to Σ).

Proof: If the lemma is false, then there exists $\varphi \in \mathbb{N}^{\mathbb{N}}$ such that for every finite set $\Delta \subset \mathbb{N}$ there is a finite set $\Delta' \subset \mathbb{N}$ with $\Delta \cap \Delta' = \varnothing$ and

$$v(m)(\bigcup_{k \in \Delta'} E_k) \not\leq \bigvee_{t=1}^{\infty} a_{t,\varphi(t)}.$$

From this, proceeding by induction, it follows that there exists a disjoint sequence $(\Delta_n)_n$ of finite subsets of $\mathbb{N}$, with

$$v(m)(\bigcup_{k \in \Delta_n} E_k) \not\leq \bigvee_{t=1}^{\infty} a_{t,\varphi(t)}$$

for every $n \in \mathbb{N}$. Hence there exists a disjoint sequence $(T_n)_n$ in Σ with

$$T_n \subset \bigcup_{k \in \Delta_n} E_k \text{ and } | m(T_n) | \not\leq \bigvee_{t=1}^{\infty} a_{t,\varphi(t)} \text{ for all } n \in \mathbb{N}.$$

This contradicts globally (s)-boundedness of m, and concludes the proof. □

4.1.2. Other Nikodým and Brooks-Jewett-Type Theorems

We now prove some other versions of Nikodým convergence and Brooks-Jewett-type theorems, which use a direct approach, based on the sliding hump technique. For technical reasons, we prefer to deal with (D)-convergence, because it allows to apply the Fremlin lemma, replacing a *series* of regulators with a single (D)-sequence.

Theorem 1.13 (see Boccuto and Dimitriou (2014c), Theorem 3.2) *Let R be a Dedekind complete (ℓ)-group, $\mathcal{F}$ be a block-respecting filter of $\mathbb{N}$, $m_n : \Sigma \to R$, $n \in \mathbb{N}$, be a sequence of equibounded σ-additive measures, $(C_k)_k$ be a disjoint sequence in Σ, with*

(i) (D) $\lim_n m_n(C_k) = 0$ for any $k \in \mathbb{N}$, and

(ii) $(RD\mathcal{F})\lim_n m_n(\bigcup_{k\in P} C_k) = 0$ *for every* $P \in \mathcal{P}(\mathbb{N})$.

Then,

γ) *for every strictly increasing sequence* $(k_n)_n$ *in* $\mathbb{N}$ *we get*

$$(D\mathcal{F})\lim_n m_n(C_{k_n}) = 0; \qquad \textbf{(4.35)}$$

$\gamma\gamma$) *if* $\mathcal{F}$ *is also diagonal and* R *is super Dedekind complete and weakly* σ*-distributive, then* (ii) *implies* (4.35).

Proof: For each $n \in \mathbb{N}$ set $H_n := C_{k_n}$. Let $u := \bigvee_{A\in\Sigma, n\in\mathbb{N}} |m_n(A)|$: such an element exists in R, by equiboundedness. For any $n \in \mathbb{N}$, let $(a_{t,l}^{(n)})_{t,l}$ be a (D)-sequence associated with σ-additivity of m_n and the sequence $(H_n)_n$. For every $\varphi \in \mathbb{N}^{\mathbb{N}}$ and $n \in \mathbb{N}$ there is $\overline{l} \in \mathbb{N}$, with

$$|m_n(A)| \le \bigvee_{t=1}^{\infty} a_{t,\varphi(t+n)}^{(n)} \text{ for all } A \in \Sigma,\ A \subset \bigcup_{l=\overline{l}}^{\infty} H_l. \qquad \textbf{(4.36)}$$

By the Fremlin lemma there is a (D)-sequence $(a_{t,l})_{t,l}$ with

$$u \wedge \left(\sum_{n=1}^{q}\left(\bigvee_{t=1}^{\infty} a_{t,\varphi(t+n)}^{(n)}\right)\right) \le \bigvee_{t=1}^{\infty} a_{t,\varphi(t)} \qquad \textbf{(4.37)}$$

for each $q \in \mathbb{N}$ and $\varphi \in \mathbb{N}^{\mathbb{N}}$. Let $(b_{t,l})_{t,l}$ be a regulator, satisfying the condition of $(RD\mathcal{F})$-convergence as in (ii).

Since $(D)\lim_n m_n(C_k) = 0$ for every $k \in \mathbb{N}$, then for each $k \in \mathbb{N}$ there exists a regulator $(c_{t,l}^{(k)})_{t,l}$ such that for every $\varphi \in \mathbb{N}^{\mathbb{N}}$ there is $\overline{n} \in \mathbb{N}$ with $|m_n(C_k)| \le \bigvee_{t=1}^{\infty} c_{t,\varphi(t+k)}^{(k)}$ for all $n \ge \overline{n}$. Arguing analogously as above, by the

Fremlin lemma we find a regulator $(c_{t,l})_{t,l}$ such that for each $\varphi\in\mathbb{N}^{\mathbb{N}}$ and $k\in\mathbb{N}$ there is $\overline{n}\in\mathbb{N}$ with $|m_n(C_k)|\leq\bigvee_{t=1}^{\infty}c_{t,\varphi(t)}$ for all $n\geq\overline{n}$.

Again by the Fremlin lemma, there are two (D)-sequences $(d_{t,l})_{t,l}$, $(e_{t,l})_{t,l}$, with

$$u\wedge(\sum_{k=1}^{q}(\bigvee_{t=1}^{\infty}a_{t,\varphi(t+k+1)}))\leq\bigvee_{t=1}^{\infty}d_{t,\varphi(t)}, \tag{4.38}$$

$$u\wedge(\sum_{k=1}^{q}(\bigvee_{t=1}^{\infty}c_{t,\varphi(t+k)}))\leq\bigvee_{t=1}^{\infty}e_{t,\varphi(t)} \tag{4.39}$$

for every $q\in\mathbb{N}$ and $\varphi\in\mathbb{N}^{\mathbb{N}}$. For each t, $l\in\mathbb{N}$, set

$$z_{t,l}=6(b_{t,l}+d_{t,l}+e_{t,l}). \tag{4.40}$$

We prove that the (D)-sequence $(z_{t,l})_{t,l}$ satisfies the condition of $(D\mathcal{F})$-convergence in $\gamma)$. Otherwise there is $\varphi\in\mathbb{N}^{\mathbb{N}}$ with the property that $C:=\{n\in\mathbb{N}:|m_n(H_n)|\leq\bigvee_{t=1}^{\infty}z_{t,\varphi(t)}\}\notin\mathcal{F}$.

We get that $I:=\mathbb{N}\setminus C=\{n\in\mathbb{N}:|m_n(H_n)|\not\leq\bigvee_{t=1}^{\infty}z_{t,\varphi(t)}\}\in\mathcal{F}^*$: otherwise, there is $F\in\mathcal{F}$ with $I\cap F=\emptyset$, that is $F\subset C$. Thus we get $C\in\mathcal{F}$, a contradiction.

Let $N_0=1$. By σ-additivity of m_1, there exists a cofinite subset $P_1\subset\mathbb{N}$, with $N_0<p_1:=\min P_1$, and $v(m_1)(F_1)\leq\bigvee_{t=1}^{\infty}a_{t,\varphi(t+1)}$, where $F_1:=\bigcup_{l\in P_1}H_l$. By (i) there is an integer $N_1>p_1$ with $|m_n(H_l)|\leq\bigvee_{t=1}^{\infty}c_{t,\varphi(t+1)}$ whenever $n\geq N_1$ and $l=1,\ldots,p_1$. By σ-additivity of m_1, $m_2,\ldots,m_{N_1}$, there is a cofinite subset $P_2\subset P_1$, with $N_1<p_2:=\min P_2$, and $v(m_n)(F_2)\leq\bigvee_{t=1}^{\infty}a_{t,\varphi(t+2)}$ for every $n=1,2,\ldots,N_1$, where $F_2:=\bigcup_{l\in P_2}H_l$. Proceeding analogously as above, we find an integer $N_2>p_2$

with $|m_n(H_l)| \le \bigvee_{t=1}^{\infty} c_{t,\varphi(t+2)}$ whenever $n \ge N_2$ and $l = 1,\ldots,p_2$. By induction, we find a strictly decreasing sequence $(P_k)_k$ of cofinite subsets of $\mathbb{N}$, a strictly decreasing sequence $(F_k)_k$ in Σ and two strictly increasing sequences $(N_k)_k$, $(p_k)_k$ in $\mathbb{N}$ such that, for every $k \in \mathbb{N}$,

1.13.1) $N_k > p_k$, $p_{k+1} > N_k$, $p_k = \min P_k$; $F_k = \bigcup_{l \in P_k} H_l$;

1.13.2) $v(m_n)(F_{k+1}) \le \bigvee_{t=1}^{\infty} a_{t,\varphi(t+k+1)}$, and hence $v(m_n)(F_{k+1}) \le \bigvee_{t=1}^{\infty} d_{t,\varphi(t)}$, for every $n = 1,\ldots,N_k$;

1.13.3) $|m_n(H_l)| \le \bigvee_{t=1}^{\infty} c_{t,\varphi(t+k)}$, and so $|m_n(H_l)| \le \bigvee_{t=1}^{\infty} e_{t,\varphi(t)}$, whenever $n \ge N_k$ and $l = 1,\ldots,p_k$.

Since $\mathcal{F}$ is block-respecting, there is $J := \{j_1, j_2, \ldots\} \in \mathcal{F}^*$, $J \subset I$, with $N_k \le j_k < N_{k+1}$ for every $k \in \mathbb{N}$. As $J \in \mathcal{F}^*$, then either $J_1 := \{j_1, j_3, j_5, \ldots\} \in \mathcal{F}^*$ or $J_2 := \{j_2, j_4, j_6, \ldots\} \in \mathcal{F}^*$. Without loss of generality, let $J_1 \in \mathcal{F}^*$ (see also Aviles Lopez, Cascales Salinas, Kadets and Leonov (2007), and Boccuto, Dimitriou and Papanastassiou, 2012b and 2012e)). Put $A := \bigcup_{h=1}^{\infty} H_{j_{2h-1}}$. We have:

$$m_{j_1}(A) = m_{j_1}(H_{j_1}) + m_{j_1}(H_{j_3} \cup H_{j_5} \cup \ldots);$$

$$m_{j_{2h-1}}(A) = m_{j_{2h-1}}(H_{j_1} \cup H_{j_3} \cup \ldots \cup H_{j_{2h-3}}) + \tag{4.41}$$

$$+ m_{j_{2h-1}}(H_{j_{2h-1}}) + m_{j_{2h-1}}(H_{j_{2h+1}} \cup H_{j_{2h+3}} \cup \ldots), \quad h \ge 2.$$

Since $j_{2h-1} < N_{2h-1} < p_{2h}$ and

$$H_{j_{2h+1}} \cup H_{j_{2h+3}} \cup \ldots \subset \bigcup_{l=p_{2h+1}}^{\infty} H_l = F_{2h+1} \text{ for every } h \in \mathbb{N}, \tag{4.42}$$

from (4.42) and 1.13.2) we get

$$|m_{j_{2h-1}}(H_{j_{2h+1}} \cup H_{j_{2h+3}} \cup \ldots)| \leq \bigvee_{t=1}^{\infty} d_{t,\varphi(t)}. \qquad \textbf{(4.43)}$$

Moreover, since $j_{2h-3} < N_{2h-3} < p_{2h-2} < p_{2h-1}$ for every $h \geq 2$, from 1.13.3) we obtain

$$|m_{j_{2h-1}}(H_{j_1} \cup H_{j_3} \cup \ldots \cup H_{j_{2h-3}})| \leq \bigvee_{t=1}^{\infty} e_{t,\varphi(t)}. \qquad \textbf{(4.44)}$$

If $|m_{j_{2h-1}}(A)| \leq \bigvee_{t=1}^{\infty} b_{t,\varphi(t)}$, then from (4.41), (4.43) and (4.44) we have $|m_{j_{2h-1}}(H_{j_{2h-1}})| \leq \bigvee_{t=1}^{\infty} z_{t,\varphi(t)}$.

But we know that $|m_{j_{2h-1}}(H_{j_{2h-1}})| \not\leq \bigvee_{t=1}^{\infty} z_{t,\varphi(t)}$, and so we get a contradiction. Thus $|m_{j_{2h-1}}(A)| \not\leq \bigvee_{t=1}^{\infty} b_{t,\varphi(t)}$ for all $h \in \mathbb{N}$, and hence $L := \{l \in \mathbb{N} : |m_l(A)| \not\leq \bigvee_{t=1}^{\infty} b_{t,\varphi(t)}\} \in \mathcal{F}^*$.

Since, by (ii), $\mathbb{N} \setminus L \in \mathcal{F}$, we obtain that $L \cap (\mathbb{N} \setminus L) \neq \emptyset$, which is impossible. This proves $\gamma)$.

$\gamma\gamma)$ Set $H_n := C_{k_n}$, $n \in \mathbb{N}$, and let $(b_{t,l})_{t,l}$ be a regulator satisfying $(RD\mathcal{F})$-convergence in condition (ii). Of course, for every $\mathcal{F}$-stationary set J, $(b_{t,l})_{t,l}$ satisfies (ii) also with respect to $(RD\mathcal{F}(J))$-convergence. Since R is super Dedekind complete and weakly σ-distributive, by virtue of Theorem 2.2 of Chapter 2 we find an (O)-sequence $(\sigma_p)_p$ fulfilling condition (ii), when $(RD\mathcal{F})$-convergence is replaced by $(RO\mathcal{F})$- or $(RO\mathcal{F}(J))$-convergence. For every $I \in \mathcal{F}^*$ there is $J \in \mathcal{F}^*$, $J \subset I$, with $(RO)\lim_{n \in J} m_n(C_k) = 0$, $k \in \mathbb{N}$, with respect to $(\sigma_p)_p$ (see also Lemma 2.24 of Chapter 2). From this and Theorem 2.2 of Chapter 2 it follows that there is a regulator $(c_{t,l})_{t,l}$, such that for every $I \in \mathcal{F}^*$ there exists $J \in \mathcal{F}^*$, $J \subset I$, with

$$(RD)\lim_{n \in J} m_n(C_k) = 0, \quad k \in \mathbb{N}, \qquad \textbf{(4.45)}$$

with respect to $(c_{t,l})_{t,l}$. Let now $(a_{t,l}^{(n)})_{t,l}$, $n \in \mathbb{N}$, be a sequence of regulators in correspondence with σ-additivity of m_n, u be as in the proof of $\gamma)$, $(a_{t,l})_{t,l}$ be as in (4.37) and $(d_{t,l})_{t,l}$, $(e_{t,l})_{t,l}$, $(z_{t,l})_{t,l}$ be as in (4.38), (4.39), (4.40) respectively. We now prove that the regulator $(z_{t,l})_{t,l}$ satisfies $\gamma\gamma)$. Otherwise, by proceeding analogously as in the proof of $\gamma)$, we find $I \in \mathcal{F}^*$ and $\varphi \in \mathbb{N}^{\mathbb{N}}$ with $|m_n(H_n)| \not\leq \bigvee_{t=1}^{\infty} z_{t,\varphi(t)}$ for each $n \in I$. In correspondence with I, there is $J \in \mathcal{F}^*$, $J \subset I$, satisfying (4.45). Note that the sequence $m_n(H_n)$, $n \in J$, does not $(\mathcal{F}(J))$-converge to 0 (see also Aviles Lopez, Cascales Salinas, Kadets and Leonov (2007)). Since $J \in \mathcal{F}^*$ and $\mathcal{F}$ is block-respecting, then $\mathcal{F}(J)$ is block-respecting too (see also Aviles Lopez, Cascales Salinas, Kadets and Leonov (2007), Boccuto and Dimitriou (2014a)). By $\gamma)$ used with m_n, $n \in J$, and $\mathcal{F}(J)$, it follows that $(\mathcal{F}(J)) \lim_{n \in J} m_n(H_n) = 0$, getting a contradiction. This proves $\gamma\gamma)$.

□

A similar version of Theorem 1.13 holds for finitely additive measures.

Theorem 1.14 (see Boccuto and Dimitriou (2014c), Theorem 3.3) *Let R be a Dedekind complete (ℓ)-group, $(C_k)_k$ be as in Theorem 1.13, $\mathcal{F}$ be a block-respecting filter of $\mathbb{N}$, $m_n : \Sigma \to R$, $n \in \mathbb{N}$, be an equibounded sequence of finitely additive measures, and assume that*

(i) $(D) \lim_n m_n(C_k) = 0$ *for any* $k \in \mathbb{N}$;

(ii) $(RD\mathcal{F}) \lim_n \sum_{k \in P} m_n(C_k) = 0$ *for every* $P \subset \mathbb{N}$.

Then for every strictly increasing sequence $(k_n)_n$ *in* $\mathbb{N}$ *we get*

$$(D\mathcal{F}) \lim_n m_n(C_{k_n}) = 0. \qquad \textbf{(4.46)}$$

If $\mathcal{F}$ is also diagonal and R is super Dedekind complete and weakly σ-distributive, then (ii) implies (4.46).

Observe that, by means of similar techniques, in the case $\mathcal{F} = \mathcal{F}_{\text{cofin}}$ it is possible to prove the following Schur, Vitali-Hahn-Saks and Nikodým convergence-type theorems.

Theorem 1.15 (see also Boccuto, Dimitriou and Papanastassiou (2012c), Theorem 3.1)

Let $m_j : \mathcal{P}(\mathbb{N}) \to R$, $j \in \mathbb{N}$, *be an equibounded sequence of* σ*-additive measures, and assume that there exists* $m_0 : \mathcal{P}(\mathbb{N}) \to R$ *with* $(RD)\lim_j m_j = m_0$. *Then,*

$$(D)\lim_j \left(\sum_{n=1}^{\infty} | m_j(\{n\}) - m_0(\{n\}) |\right) = 0.$$

Moreover m_0 *is* σ*-additive,* $(UD)\lim_j m_j = m_0$ *and the measures* m_j *are uniformly* σ*-additive.*

Theorem 1.16 (see also Boccuto, Dimitriou and Papanastassiou (2012c), Theorem 3.2) *Let* τ *be a Fréchet-Nikodým topology on the* σ*-algebra* Σ, $m_j : \Sigma \to R$, $j \in \mathbb{N}$, *be an equibounded sequence of* τ*-continuous finitely additive measures. Assume that there exists* $m_0 : \Sigma \to R$ *with* $(RD)\lim_j m_j = m_0$. *Then* m_0 *is* τ*-continuous and the measures* m_j *are uniformly* τ*-continuous.*

Theorem 1.17 (see also Boccuto, Dimitriou and Papanastassiou (2012c), Theorem 3.3) *Let* G, Σ *be as in Theorem* 1.16, $m_j : \Sigma \to R$, $j \in \mathbb{N}$, *be an equibounded sequence of* σ*-additive measures, and let there exist* $m_0 : \Sigma \to R$ *such that* $(RD)\lim_j m_j = m_0$. *Then* m_0 *is* σ*-additive and the measures* m_j *are uniformly* σ*-additive.*

In the following limit theorems, which are formulated for positive (ℓ)-group-valued measures, the involved filter is required to be only diagonal, and not necessarily block-respecting. A meaningful example is the filter $\mathcal{F}_{\text{st}}$ of all subsets of $\mathbb{N}$ with asymptotic density 1 (see also Aviles Lopez, Cascales Salinas, Kadets and Leonov (2007), Kostyrko, Šalát and Wilczyński (2001)). We suppose that the involved measures are filter pointwise convergent not necessarily with respect to a single (O)-sequence or regulator.

We now give the following Brooks-Jewett-type theorems for (ℓ)-group-valued measures with respect to pointwise filter convergence. We use some properties of diagonal filters, positivity of the involved set functions and the Maeda-Ogasawara-Vulikh representation theorem in order to find a countable class of sets, in which it is possible to apply the classical version of the Brooks-Jewett theorem for real-valued measures defined on the σ-algebra generated by a disjoint sequence.

Theorem 1.18 (see also Boccuto, Dimitriou and Papanastassiou (2011b), Theorem 2.5)

Let $m_j : \Sigma \to R$, $j \in \mathbb{N}$, be a sequence of positive finitely additive measures and $\mathcal{F}$ be a diagonal filter of $\mathbb{N}$. Suppose that $m_0(E) := (O\mathcal{F})\lim_j m_j(E)$ exists in R for every $E \in \Sigma$, and that m_0 is σ-additive on Σ.

Then for every set $I \in \mathcal{F}^*$ and for each disjoint sequence $(C_k)_k$ in Σ there is a set $J \in \mathcal{F}^*$, $J \subset I$, such that

$$(O)\lim_k \Big(\bigvee_{j\in J} m_j(C_k)\Big) = 0.$$

Proof: Choose arbitrarily any disjoint sequence $(C_k)_k$ in Σ, set $C := \bigcup_{k=1}^{\infty} C_k$ and let $\mathcal{K}$ be the σ-algebra generated by the sequence $(C_k)_k$ in C. For any $B \in \mathcal{K}$ there is $P \subset \mathbb{N}$ with $B = \bigcup_{k\in P} C_k$. Let $\mathcal{A}$ be the countable family whose elements are C and all finite and cofinite unions (with respect to C) of elements of the sequence $(C_k)_k$. By hypothesis, we get

$$m_0(E) = (O\mathcal{F})\lim_j m_j(E) \text{ for all } E \in \mathcal{A}. \qquad \textbf{(4.47)}$$

By (4.47) and Lemma 2.24 of Chapter 2, there exists a set $J \in \mathcal{F}^*$, $J \subset I$, with $(O)\lim_{j\in J} m_j(E) = m_0(E)$ for each $E \in \mathcal{A}$. Hence the sequence $m_j(G)$, $j \in J \cup \{0\}$, is bounded in R, and so the family $m_j(E)$, $j \in J \cup \{0\}$, $E \in \Sigma$, is equibounded in R.

By the Maeda-Ogasawara-Vulikh representation theorem, we find a meager set $N' \subset \Omega$, such that

$$\sup_{j \in J \cup \{0\}} [\sup_{E \in \Sigma}(m_i(E)(\omega))] \in \mathbb{R} \text{ for each } \omega \notin N',$$

$$\lim_{j \in J \cup \{0\}} m_j(E)(\omega) = m_0(E)(\omega) \text{ for every } \omega \notin N' \text{ and } E \in \mathcal{A}.$$

Furthermore, by σ-additivity of m_0 and again by the Maeda-Ogasawara-Vulikh theorem, there exists a meager set $N \subset \Omega$, without loss of generality $N \supset N'$, with

$$\lim_k [m_0(\bigcup_{j>k} C_j)(\omega)] = 0 \text{ for all } \omega \notin N.$$

For each $\varepsilon > 0$ and $\omega \in \Omega \setminus N$ there exists $k_0(\varepsilon, \omega)$ such that $m_0(\bigcup_{k \ge k_0} C_k)(\omega) \le \varepsilon$, and so, by positivity of m_0, $m_0(\bigcup_{k \ge k_0, k \in P} C_k)(\omega) \le \varepsilon$. Moreover there is $j_0 \in J$, $j_0 = j_0(\varepsilon, \omega, k_0)$ with

$$| m_j(\bigcup_{k \le k_0, k \in P} C_k)(\omega) - m_0(\bigcup_{k \le k_0, k \in P} C_k)(\omega) | \le \varepsilon \text{ for all } j \in A,\ j \ge j_0,$$

$$| m_j(\bigcup_{k \le k_0} C_k)(\omega) - m_0(\bigcup_{k \le k_0} C_k)(\omega) | \le \varepsilon,\ | m_j(\bigcup_{k=1}^{\infty} C_k)(\omega) - m_0(\bigcup_{k=1}^{\infty} C_k)(\omega) | \le \varepsilon,$$

and therefore

$$| m_j(\bigcup_{k > k_0} C_k)(\omega) - m_0(\bigcup_{k > k_0} C_k)(\omega) | \le 2\varepsilon.$$

Let now $B \in \mathcal{K}$. For all $j \in J$, $j \ge j_0$ we get:

$$0 \le | m_j(B)(\omega) - m_0(B)(\omega) | = | m_j(\bigcup_{k \in P} C_k)(\omega) - m_0(\bigcup_{k \in P} C_k)(\omega) |$$

$$\le | m_j(\bigcup_{k \le k_0, k \in P} C_k)(\omega) - m_0(\bigcup_{k \le k_0, k \in P} C_k)(\omega) | + m_0(\bigcup_{k > k_0, k \in P} C_k)(\omega) + m_j(\bigcup_{k > k_0, k \in P} C_k)(\omega)$$

$$\leq | m_j(\bigcup_{k\leq k_0, k\in P} C_k)(\omega) - m_0(\bigcup_{k\leq k_0, k\in P} C_k)(\omega) | + m_0(\bigcup_{k>k_0} C_k)(\omega) + m_j(\bigcup_{k>k_0} C_k)(\omega)$$

$$\leq | m_j(\bigcup_{k\leq k_0, k\in P} C_k)(\omega) - m_0(\bigcup_{k\leq k_0, k\in P} C_k)(\omega) | + | m_j(\bigcup_{k>k_0} C_k)(\omega) - m_0(\bigcup_{k>k_0} C_k)(\omega) | +$$

$$+ 2\, m_0(\bigcup_{k>k_0} C_k)(\omega) \leq 5\varepsilon.$$

So, $\lim_j m_j(B)(\omega) = m_0(B)(\omega)$ for every $\omega \in \Omega \setminus N$ and $B \in \mathcal{K}$. Therefore the finitely additive real-valued measures $m_j(\cdot)(\omega)$, $j \in J$, $\omega \notin N$, satisfy the hypotheses of the classical Brooks-Jewett theorem on $\mathcal{K}$. Thus for each disjoint sequence $(C_k)_k$ in $\mathcal{K}$ we have $\lim_k [\sup_{j\in J} m_j(C_k)(\omega)] = 0$ for all $\omega \notin N$. Since N is meager, we obtain $(O)\lim_k (\bigvee_{j\in J} m_j(C_k)) = 0$. This ends the proof. □

With the same technique as in Theorem 1.18, assuming pointwise convergence in the classical like sense, we have the following

Theorem 1.19 (see also Boccuto (1996a), Theorem 2.5) *Let* $m_j : \Sigma \to R$, $j \in \mathbb{N}$, *be a sequence of positive finitely additive measures. Let* $m_0(E) := (O) \lim_j m_j(E)$ *exist in* R *for every* $E \in \Sigma$, *and* m_0 *be* σ*-additive on* Σ. *Then the measures* m_j, $j \in \mathbb{N}$, *are uniformly* (s)*-bounded on* Σ.

We now prove a Vitali-Hahn-Saks-type theorem.

Theorem 1.20 (see also Boccuto, Dimitriou and Papanastassiou (2011b), Theorem 2.6) *Let* $\mathcal{F}$ *be a diagonal filter of* $\mathbb{N}$, τ *be a Fréchet-Nikodým topology on* Σ, $m_j : \Sigma \to R$, $j \in \mathbb{N}$, *be a sequence of positive finitely additive* τ*-continuous measures. Let* $m_0(E) := (O\mathcal{F}) \lim_j m_j(E)$ *exist in* R *for every* $E \in \Sigma$, *and* m_0 *be* σ*-additive on* Σ. *Then for every* $I \in \mathcal{F}^*$ *and for each decreasing sequence* $(H_k)_k$ *in* Σ *with* τ-$\lim_k H_k = \emptyset$ *there is a set* $J \in \mathcal{F}^*$, $J \subset I$, *with*

$$(O)\lim_k (\bigvee_{j\in J} m_j(H_k)) = \bigwedge_k (\bigvee_{j\in J} m_j(H_k)) = 0.$$

Proof: Let τ, I, $(H_k)_k$ be as in the hypotheses, set $W := \bigcap_{k=1}^{\infty} H_k$, $C_k := H_k \setminus H_{k+1}$, $k \in \mathbb{N}$, and let $\mathcal{L}$ be the σ-algebra generated by the sequence $(C_k)_k$ and W in H_1. Proceeding analogously as in Theorem 1.18, we find a set $J \subset I$, $J \in \mathcal{F}^*$ and a meager set $N \subset \Omega$, such that the real-valued measures $m_j(\cdot)(\omega)$, $j \in J$, are uniformly (s)-bounded on $\mathcal{L}$ for every $\omega \in \Omega \setminus N$. Moreover, thanks to τ-continuity of m_j, we have $\bigwedge_k m_j(H_k) = 0$ for every $j \in \mathbb{N}$ and so there is a meager set $N_0 \subset \Omega$ with $\inf_k m_j(H_k)(\omega) = 0$ for any $j \in \mathbb{N}$ and $\omega \in \Omega \setminus N_0$. Fix $\omega \in \Omega \setminus (N \cup N_0)$. By Theorem 3.2 of Chapter 3 applied to $m_j(\cdot)(\omega)$, $j \in J$, we have

$$\lim_k (\sup_{j\in J} m_j(H_k)(\omega)) = \inf_k (\sup_{j\in J} m_j(H_k)(\omega)) = 0. \tag{4.48}$$

Since $N \cup N_0$ is meager, by virtue of the Maeda-Ogasawara-Vulikh representation theorem, from (4.48) we obtain $0 = (O)\lim_k (\bigvee_{j\in J} m_j(H_k)) = \bigwedge_k (\bigvee_{j\in J} m_j(H_k))$, that is the assertion. □

Analogously as in 1.20 it is possible to prove the following

Theorem 1.21 *Under the same above hypotheses and notations, let* $m_j : \Sigma \to R$, $j \in \mathbb{N}$, *be a sequence of positive* σ*-additive measures, such that* $m_0(E) := (O\mathcal{F})\lim_j m_j(E)$ *exists in* R *for each* $E \in \Sigma$, *and that* m_0 *is* σ*-additive on* Σ. *Then for every* $I \in \mathcal{F}^*$ *and for each decreasing sequence* $(H_k)_k$ *in* Σ *with* $\bigcap_{k=1}^{\infty} H_k = \emptyset$ *there is a set* $J \in \mathcal{F}^*$, $J \subset I$, *with*

$$(O)\lim_k (\bigvee_{j\in J} m_j(H_k)) = \bigwedge_k (\bigvee_{j\in J} m_j(H_k)) = 0.$$

4.1.3. Dieudonné-Type Theorems

We now prove the following

Theorem 1.22 (see also Boccuto, Dimitriou and Papanastassiou (2011c), Theorems 3.8, 3.10 and 3.15) *Let $\mathcal{F}$ be a diagonal filter of $\mathbb{N}$, $\mathcal{G}$, $\mathcal{H} \subset \Sigma$ be two lattices, such that the complement of any element of $\mathcal{H}$ belongs to $\mathcal{G}$, and $\mathcal{G}$ is closed under countable disjoint unions. Let $m_j : \Sigma \to R$, $j \in \mathbb{N}$, be a sequence of positive measures, such that $m_0(A) := (O\mathcal{F})\lim_j$ exists in R for every $A \in \mathcal{G}$. Suppose that for each $E \in \Sigma$ there are two sequences $(G_k)_k$, $(F_k)_k$ in $\mathcal{G}$, $\mathcal{H}$, respectively, with*

$$F_k \subset F_{k+1} \subset E \subset G_{k+1} \subset G_k \tag{4.49}$$

for all $k \in \mathbb{N}$, and

$$(O)\lim_k (m_j(G_k \setminus F_k)) = \bigwedge_k (m_j(G_k \setminus F_k)) = 0 \text{ for every } j \in \mathbb{N}. \tag{4.50}$$

Then,

(j) for each $I \in \mathcal{F}^*$ there is a set $J \in \mathcal{F}^*$, with

$$(O)\lim_k (\bigvee_{j\in J} m_j(G_k \setminus F_k)) = \bigwedge_k (\bigvee_{j \in J} m_j(G_k \setminus F_k)) = 0. \tag{4.51}$$

(jj) The limit measure $(O)\lim_{j\in J} m_j(E)$ exists in R for every $E \in \Sigma$.

Proof: (j) Let E, $(G_k)_k$, $(F_k)_k$ be as in the hypothesis, $\mathcal{L}$ be the σ-algebra generated by the sets $G_k \setminus F_k$, $k \in \mathbb{N}$, and $I \in \mathcal{F}^*$. Arguing analogously as in the proof of Theorem 1.18, we find a set $J \subset I$, $J \in \mathcal{F}^*$ and a meager set $N \subset \Omega$, such that the real-valued measures $m_j(\cdot)(\omega)$, $j \in J$, are uniformly (s)-bounded on $\mathcal{L}$ for every $\omega \in \Omega \setminus N$. Moreover, since $\bigwedge_k m_j(G_k \setminus F_k) = 0$ for every $j \in \mathbb{N}$, there exists a meager set $N_0 \subset \Omega$ with $\inf_k m_j(G_k \setminus F_k)(\omega) = 0$ for every $j \in \mathbb{N}$

and $\omega \in \Omega \setminus N_0$. Fix $\omega \in \Omega \setminus (N \cup N_0)$. By Theorem 3.2 of Chapter 3 applied to $m_j(\cdot)(\omega)$, $j \in J$, we get

$$\lim_k (\sup_{j\in J} m_j(G_k \setminus F_k)(\omega)) = \inf_k (\sup_{j\in J} m_j(G_k \setminus F_k)(\omega)) = 0. \tag{4.52}$$

Since $N \cup N_0$ is meager, from the Maeda-Ogasawara-Vulikh representation theorem and (4.52) we obtain $(O)\lim_k (\bigvee_{j\in J} m_j(G_k \setminus F_k)) = \bigwedge_k (\bigvee_{j\in J} m_j(G_k \setminus F_k)) = 0$, that is (4.50). This proves (j).

(jj) Choose $E \in \Sigma$, and let $(G_k)_k$, $(F_k)_k$ fulfil (4.49). By virtue of (4.49), for each $j \in \mathbb{N}$ there is an (O)-sequence $(y_k^{(j)})_k$ such that

$$| m_j(G_k) - m_j(F_k) | \leq y_k^{(j)} \tag{4.53}$$

for all $k \in \mathbb{N}$. Since $m_j(E) = (O)\lim_k m_i(G_k)$, $i \in \mathbb{N}$, from (4.53) we get

$$m_j(E) = (O)\lim_k m_j(F_k) \text{ for all } j \in \mathbb{N},$$

and hence for any $j \in \mathbb{N}$ there is an (O)-sequence $(w_k^{(j)})_k$ with

$$| m_j(F_k) - m_j(E) | \leq w_k^{(j)} \tag{4.54}$$

for all $k \in \mathbb{N}$. Thanks to (j), for each $\mathcal{F}$-stationary set $I \subset \mathbb{N}$ there are an $\mathcal{F}$-stationary set $D \subset I$ and an (O)-sequence $(z_k)_k$ with

$$\bigvee_{j\in D} m_j(G_k \setminus F_k) \leq z_k \tag{4.55}$$

for any $k \in \mathbb{N}$. From (4.55) we get

$$| m_j(G_k) - m_j(F_q) | \leq z_k \tag{4.56}$$

whenever $k,q \in \mathbb{N}$, $q \geq k$ and $j \in D$: this follows from the fact that $G_k \setminus F_q \subset G_k \setminus F_k$ for all $q \geq k$. So from (4.54) and (4.56), for every $j \in D$ and k, $q \in \mathbb{N}$ with $q \geq k$ we have

$$0 \leq | m_j(G_k) - m_j(E) | \leq | m_j(G_k) - m_j(F_q) | + | m_j(F_q) - m_j(E) | \leq z_k + w_q^{(j)}. \qquad \textbf{(4.57)}$$

Fix now $k \in \mathbb{N}$. Taking the (O)- limit as q tends to $+\infty$, from (4.57) we get

$$0 \leq | m_j(G_k) - m_j(E) | \leq z_k \text{ for all } k \in \mathbb{N},\ j \in D. \qquad \textbf{(4.58)}$$

Pick an (O)-sequence $(\sigma_l)_l$ associated with $(RO\mathcal{F})$-convergence of $(m_j(G_k))_{j,k\in\mathbb{N}}$ to $(m_0(G_k))_{k\in\mathbb{N}}$. There exists an $\mathcal{F}$-stationary set $J \subset D$ with $(RO)\lim_{j\in J} m_j(G_k) = m_0(G_k)$ with respect to k (see also Lemma 2.24 of Chapter 2). Thus for every l and $k \in \mathbb{N}$ there exists $j^* \in J$ such that $| m_p(G_k) - m_q(G_k) | \leq 2\sigma_l$ for each $p,q \in J$, $p,q \geq i^*$. From this and (4.58) it follows that to each $k \in \mathbb{N}$ we can associate a positive integer $i_k > k$ with

$$| m_p(E) - m_q(E) | \leq$$
$$| m_p(E) - m_p(G_k) | + | m_p(G_k) - m_q(G_k) | + | m_q(G_k) - m_q(E) | \leq 2\sigma_k + 2z_k$$

whenever $p,q \in J$, $p,q \geq i_k$. Put $j_0 = 0$, $\sigma_0 = \sigma_1$, $z_0 = z_1$. We can suppose that $j_{k-1} < j_k$ for every $k \in \mathbb{N}$. Set $J = \{r_i : i \in \mathbb{N}\}$: for each i there exists $k = k(i)$ such that $i_k \leq r_i < i_{k+1}$. Put $v_i = 2\sigma_{k(i)} + 2z_{k(i)}$, $i \in \mathbb{N}$. Note that $(v_i)_i$ is an (O)-sequence and that $| m_{r_i}(E) - m_{r_{i+s}}(E) | \leq v_i$ whenever $i,s \in \mathbb{N}$. Thus $(m_i(E))_{i\in J}$ is (O)-Cauchy and hence it has (O)-limit in R, since Dedekind completeness implies (O)-completeness. This ends the proof. □

4.1.4. The Uniform Boundedness Principle

We now present a version of the Nikodým boundedness theorem in the context of (ℓ)-groups and filter convergence, extending Boccuto and Candeloro (2004d), Theorem 4.6.

Theorem 1.23 (see also Boccuto, Dimitriou and Papanastassiou (2012b), Lemma 3.4, Boccuto and Dimitriou (2014c), Theorem 3.4) *Let R be any Dedekind complete (ℓ)-group, $u \in R$, $u > 0$, $U = [-u, u]$, $\mathcal{F}$ be a block-respecting filter of $\mathbb{N}$, $m_j : \Sigma \to R$, $j \in \mathbb{N}$, be a sequence of finitely additive measures, and assume that*

1.23.1) *for every disjoint sequence $(C_n)_n$ in Σ and $j \in \mathbb{N}$ there is $Q_j \subset \mathbb{N}$ with $\sum_{n \in Q} m_j(C_n) \in U$ for each $Q \subset Q_j$.*

Let $(C_n)_n$ be a disjoint sequence in Σ and $(w_n)_n$ be an increasing sequence of positive elements of R. For each $n \in \mathbb{N}$, set $W_n := [-w_n, w_n]$ and $V_n := nW_n + U$. Moreover suppose that:

(i) The set $\{m_n(C_p) : n \in \mathbb{N}\}$ is (PR)-eventually bounded by $(W_n)_n$ for each $p \in \mathbb{N}$;

(ii) The set $\{\sum_{p \in P} m_j(C_p) : n \in \mathbb{N}\}$ is (PR)-$\mathcal{F}$-bounded by $(W_n)_n$ for each $P \in \mathcal{P}(\mathbb{N})$.

Then we get:

(j) for every strictly increasing sequence $(l_n)_n$ in $\mathbb{N}$, the set $D := \{m_n(C_{l_n}) : n \in \mathbb{N}\}$ is (PR)-$\mathcal{F}$-bounded by $(V_n)_n$;

(jj) if $\mathcal{F}$ is also diagonal, then (ii) implies that D is (PR)-$\mathcal{F}$-bounded by $(V_n)_n$.

Proof: For every $n \in \mathbb{N}$, let $H_n := C_{l_n}$. If the thesis of the theorem is not true, then $I := \{n \in \mathbb{N} : m_n(H_n) \notin V_n\} \in \mathcal{F}^*$. Set $n_0 = 1$. By 1.23.1) there is a cofinite set $P_1 \subset \mathbb{N}$, with $1 < p_1 = \min P_1$ and $\sum_{r \in P} m_1(H_r) \in U$ for each $P \subset P_1$. By (i) there is $n_1 > p_1$ with $m_j(H_l) \in W_j$ for each $j \geq n_1$ and $l = 1, \ldots, n_1$. By induction, there are a strictly decreasing sequence $(P_k)_k$ of subsets of $\mathbb{N}$ and two strictly increasing sequences $(n_k)_k$, $(p_k)_k$ in $\mathbb{N}$ such that, for each $k \in \mathbb{N}$, $p_k = \min P_k$,

$n_k > p_k$, $p_{k+1} > n_k$, $\sum_{n\in P} m_r(H_n) \in U$ for all $r = 1,\ldots,n_k$ and $P \subset P_k$; $m_j(H_l) \in W_j$ for any $j \geq n_k$ and $l = 1,\ldots,p_k$.

As $\mathcal{F}$ is block-respecting, proceeding analogously as in the proof of Theorem 1.13, we find $J_1 := \{j_1, j_3, j_5, \ldots\} \in \mathcal{F}^*$, $J_1 \subset I$, with $n_k \leq j_k < n_{k+1}$ for every $k \in \mathbb{N}$. For any $h \in \mathbb{N}$ we have

$$\sum_{q=1}^{\infty} m_{j_{2h-1}}(H_{j_{2h+2q-1}}) \in U, \tag{4.59}$$

$m_{j_{2h-1}}(H_{j_l}) \in W_{2h-1}$, $h \geq 2$, $l = 1,3,\ldots,2h-3$, and

$$m_{j_{2h-1}}(H_{j_1} \cup H_{j_3} \cup \ldots \cup H_{j_{2h-3}}) \in (h-1)W_{2h-1}. \tag{4.60}$$

Put $P^* := \{j_{2q-1} : q \in \mathbb{N}\}$. If $\sum_{p\in P^*} m_{j_{2h-1}}(H_p) \in W_{j_{2h-1}}$, then from (4.59) and (4.60) we get $m_{j_{2h-1}}(H_{j_{2h-1}}) \in h\ W_{2h-1} + U \subset j_{2h-1}W_{j_{2h-1}} + U = V_{j_{2h-1}}$ and $m_{j_1}(H_{j_1}) \in W_{j_1} + U \subset j_1 W_{j_1} + U = V_{j_1}$. This contradicts $m_{j_{2h-1}}(H_{j_{2h-1}}) \notin V_{j_{2h-1}}$. Thus $\sum_{p\in P^*} m_{j_{2h-1}}(H_p) \notin W_{j_{2h-1}}$ for every $h \in \mathbb{N}$, and hence $\{l \in \mathbb{N} : m_l(A) \notin W_l\} \in \mathcal{F}^*$. From this, arguing as at the end of the proof of Theorem 1.13, (j), we obtain a contradiction, and this proves (j). From (j), proceeding analogously as in the proof of Theorem 1.13, (jj), we get (jj). □

Corollary 1.24 (see also Boccuto, Dimitriou and Papanastassiou (2012b), Corollary 3.6) *Let $\mathcal{F}$ be a diagonal and block-respecting filter of $\mathbb{N}$, $(a_{j,n})_{j,n}$ be any double sequence in R, and suppose that $\sum_{n\in A} a_{j,n} = (O)\lim_n \sum_{s\in A\cap[1,n]} a_{j,s} \in R$ for all $A \in \mathcal{P}(\mathbb{N})$ and $j \in \mathbb{N}$, with respect to a single sequence $(b_n)_n$. Let $(w_n)_n$ be an increasing sequence of positive elements of R, $W_n := [-w_n, w_n]$, $n \in \mathbb{N}$, and assume that the set $\{\sum_{n\in A} a_{j,n} : j \in \mathbb{N}\}$ is (PR)-$\mathcal{F}$-bounded by $(W_n)_n$ for any $A \subset \mathbb{N}$.*

If $U_1=[-b_1,b_1]$ *and* $V_n=n\,W_n+U_1$, $n\in\mathbb{N}$, $(l_n)_n$ *is a strictly sequence in* $\mathbb{N}$ *and* $(A_n)_n$ *is any disjoint sequence in* $\mathcal{P}(\mathbb{N})$, *then the set* $\{\sum_{l\in A_{l_n}} a_{n,l} : n\in\mathbb{N}\}$ *is* (PR)-$\mathcal{F}$*-bounded by* $(V_n)_n$.

The following result extends Boccuto and Candeloro (2004d), Proposition 4.5 and Theorem 2.178 of Chapter 1.

Theorem 1.25 (see also Boccuto, Dimitriou and Papanastassiou (2012b), Theorem 3.7)

Let $\mathcal{L}\subset\mathcal{P}(G)$ *be an algebra,* $m_i:\mathcal{L}\to R$, $i\in\mathbb{N}$, *be a sequence of finitely additive measures,* $\mathcal{F}$ *be any free filter of* $\mathbb{N}$, $(v_k)_k$ *be an increasing sequence of positive elements of* R, $V_k:=[-v_k,v_k]$, $k\in\mathbb{N}$, *such that for any disjoint sequence* $(A_j)_j$ *in* $\mathcal{L}$ *there is* $\overline{k}\in\mathbb{N}$ *with*

$$\{i\in\mathbb{N}:m_i(A_j)\in V_{\overline{k}}\}\in\mathcal{F} \text{ for all } j\in\mathbb{N}.$$

Then there is $k_0\in\mathbb{N}$ *with* $\{i\in\mathbb{N}:m_i(A)\in 2\,V_{k_0}\}\in\mathcal{F}$ *for every* $A\in\mathcal{L}$.

Proof: By hypothesis, there is $k_1\in\mathbb{N}$ such that $\{i\in\mathbb{N}:m_i(G)\in V_{k_1}\}\in\mathcal{F}$. Let us proceed by contradiction. If the thesis is not true, then there is $E_1\in\mathcal{L}$ such that $\{i\in\mathbb{N}:m_i(E_1)\in 2\,V_{k_1}\}\notin\mathcal{F}$. By finite additivity of m_i it follows that $\{i\in\mathbb{N}:m_i(G\setminus E_1)\in V_{k_1}\}\notin\mathcal{F}$. Note that either there is an element $A_1\in\mathcal{L}$ such that $\{i\in\mathbb{N}:m_i(A_1\cap E_1)\in V_{k_1}\}\notin\mathcal{F}$, or there exists $A_2\in\mathcal{L}$ such that $\{i\in\mathbb{N}:m_i(A_2\cap(G\setminus E_1))\in V_{k_1}\}\notin\mathcal{F}$. In the first case set $B_1:=E_1$; in the second one, put $B_1:=G\setminus E_1$. In both cases, set $A_1:=G\setminus B_1$. At the second step, there exists $k_2>k_1$ with $\{i\in\mathbb{N}:m_i(B_1)\in V_{k_2}\}\in\mathcal{F}$. So we find a set $E_2\in\mathcal{L}$ such that $\{i\in\mathbb{N}:m_i(B_1\cap E_2)\in 2\,V_{k_2}\}\notin\mathcal{F}$ and $\{i\in\mathbb{N}:m_i(B_1\setminus E_2)\in V_{k_2}\}\notin\mathcal{F}$. Now, we set $B_2:=B_1\setminus E_2$ if there exists $A_1\in\mathcal{L}$ such that $\{i\in\mathbb{N}:m_i(A_1\cap B_1\cap E_2)\in V_{k_2}\}\notin\mathcal{F}$, or $B_2:=B_1\cap E_2$ if there exists $A_2\in\mathcal{L}$ such that $\{i\in\mathbb{N}:m_i(A_2\cap(B_1\setminus E_2))\in V_{k_2}\}\notin\mathcal{F}$. In any case, let us define $A_2:=B_1\setminus B_2$. Proceeding by induction, we can construct a disjoint sequence $(A_j)_j$ in $\mathcal{L}$ and a strictly increasing sequence $(k_j)_j$ in $\mathbb{N}$, with $\{i \in \mathbb{N} : m_i(A_j) \in V_{k_j}\} \notin \mathcal{F}$ for every $j \in \mathbb{N}$. This contradicts the hypothesis and proves the theorem. $\square$

Let $R = \mathbb{R}$ and $(C_k)_k$ be a fixed disjoint sequence in a σ-algebra $\Sigma \subset \mathcal{P}(G)$. One can ask whether the $\mathcal{F}$-boundedness of the sets $\{m_i(C_k) : i \in \mathbb{N}\}$ for every $k \in \mathbb{N}$ is sufficient to get the order boundedness in the usual sense of the set $\{m_i(C_k) : i, k \in \mathbb{N}\}$. The answer is in general negative. Indeed, for example, let $G := \mathbb{N}$, $\Sigma := \mathcal{P}(\mathbb{N})$, $C_k := \{k\}$ for every $k \in \mathbb{N}$, $\mathcal{F} \neq \mathcal{F}_{\text{cofin}}$, $K := \{h_1 < \ldots < h_s < h_{s+1} < \ldots\}$ be an infinite set belonging to the dual ideal $\mathcal{I}$ of $\mathcal{F}$ and such that $\mathbb{N} \setminus K$ is infinite. For every $i \notin K$ and $E \subset \mathbb{N}$, set $m_i(E) = 0$. For any $s \in \mathbb{N}$ and $E \subset \mathbb{N}$, set $m_{h_s}(E) = s$ if $s \in E$ and 0 otherwise. It is easy to check that m_i is a σ-additive bounded positive real-valued measure for every $i \in \mathbb{N}$. Indeed, given $i \in \mathbb{N}$ and any disjoint sequence $(S_k)_k$ of subsets of $\mathbb{N}$, the quantity $m_i(S_k)$ can be different from zero at most for one index k, because for all $s \in \mathbb{N}$ we get that $m_i(\{s\}) \neq 0$ if and only if $i = h_s$. Moreover, since $m_i(E) = 0$ whenever $E \subset \mathbb{N}$ and $i \notin K$, then the set $\{m_i(C_k) : k \in \mathbb{N}\}$ is $\mathcal{F}$-bounded for all $k \in \mathbb{N}$. However, we get $\sup_{i,k \in \mathbb{N}} m_i(C_k) = +\infty$.

With a similar technique as in Theorem 1.23, in the classical case (that is, when $\mathcal{F} = \mathcal{F}_{\text{cofin}}$) it is possible to prove the following

Theorem 1.26 (see also Boccuto and Candeloro (2004d), Theorem 4.6) *Let R be any Dedekind complete lattice group, and let $m_j : \Sigma \to R$, $j \in \mathbb{N}$, be a sequence of finitely additive measures, globally (s)-bounded with respect to a single (O)-sequence. Let $(t_k)_k$ be a sequence in R, with $0 \le 2t_k \le t_{k+1}$ for every k and such that for each $A \in \Sigma$ there is a positive integer k_0 with*

$$|m_j(A)| \le t_{k_0} \quad \text{for every } j \in \mathbb{N}. \tag{4.61}$$

Then the set $\{m_j(A) : j \in \mathbb{N},\ A \in \Sigma\}$ is bounded in R.

Example 1.27 Observe that, if we require simply the (order) boundedness of the set $\{m_j(A): j\in \mathbb{N}\}$ for every $A\in \Sigma$ instead condition (4.61), in general Theorem 1.26 does not hold. For example, let Σ be the σ-algebra of all Borel subsets of $[0,1]$, $R=c_0$ be the Riesz space of all real sequences convergent to 0 with the usual order and e_i, $i\in \mathbb{N}$, be the sequence in c_0 with 1 in the i-th coordinate and 0 in the other coordinates. Define $m_j:\Sigma \to R$, $j\in \mathbb{N}$, by

$$m_j(A):=\sum_{i=1}^{j}\left(\int_A \sin(i\pi t)\, dt\right)e_i,\ \ A\in \Sigma .$$

The set $\{m_j(A): j\in \mathbb{N}, A\in \Sigma\}$ is not bounded in R (see also Swartz (1989a), Example 5).

The next step is to give some Banach-Steinhaus-type theorems for Riesz space-valued bounded operators with respect to diagonal and block-respecting filters.

An *operator* is a function $T: X\to R$, where X and R are a normed space and a Dedekind complete Riesz space respectively. A linear operator $T: X\to R$ is *bounded* (resp. (PR) *-bounded* by a bounding sequence $(t_k)_k$) iff $\{T(x): x\in X, \| x\| \le 1\}$ is order bounded (resp. (PR)-bounded by $(t_k)_k$). A linear operator $T: X\to R$ is called *continuous* iff there exists an (O)-sequence $(p_l)_l$ such that to every $l\in\mathbb{N}$ there corresponds a positive real number δ such that $|T(x)|\le p_l$ whenever $\| x\| < \delta$. By arguing similarly as in the classical case, it is easy to see that a linear operator $T: X\to R$ is continuous if and only if it is bounded.

We now state the following Banach-Steinhaus-type theorem.

Theorem 1.28 (see Boccuto, Dimitriou and Papanastassiou (2012b), Lemma 3.9 and Theorem 3.10) *Let $\mathcal{F}$ be a block-respecting filter of $\mathbb{N}$; $T_j: X\to R$, $j\in \mathbb{N}$, be a sequence of linear operators, continuous with respect to a same (O)-sequence; $(t_k)_k$ be a bounding sequence in R, such that the set $\{T_j(x_n): j,n\in\mathbb{N}\}$ is (PR)-eventually bounded by $([-t_k,t_k])_k$ with respect to $n\in\mathbb{N}$, and the set $\{T_j(x): j\in\mathbb{N}\}$ is (PR)-$\mathcal{F}$-bounded by $([-t_k,t_k])_k$ for all $x\in X$.*

Then there is a bounding sequence $(v_k)_k$ *such that for each sequence* $(x_n)_n$ *in* X *such that the series* $\sum_{n=1}^{\infty} x_n$ *norm converges unconditionally, the set* $\{T_n(x_n): n\in\mathbb{N}\}$ *is* (PR)-$\mathcal{F}$*-bounded by* $([-v_k, v_k])_k$. *If* $\mathcal{F}$ *is also diagonal, then (ii) implies the assertion.*

We now give the following version of the uniform boundedness principle, which does not involve the theory of filters (see also Swartz (1989a)).

Lemma 1.29 (see Boccuto, Dimitriou and Papanastassiou (2012b), Lemma 3.11) *Let* Λ *be any nonempty abstract set,* X *be a Banach space,* R *be any Dedekind complete Riesz space,* $T_i : X \to R$, $i\in\Lambda$, *be a family of linear bounded operators. Suppose that there exists a bounding sequence* $(t_k)_k$ *in* R *such that for every* $x\in X$ *the set* $\{T_i(x): i\in\Lambda\}$ *is* (PR)*-bounded by* $(t_k)_k$.

Then the set $\{T_i(x): x\in X, i\in\Lambda\}$ *is bounded in* R.

Proof: Let $(t_k)_k$ be a bounding sequence, satisfying the hypotheses of the lemma. For each $k\in\mathbb{N}$ set $A_k := \{x\in X : |T_i(x)|\le t_k \text{ for all } i\in\Lambda\}$. Since the set $\{T_i(x): i\in\Lambda\}$ is (PR)-bounded by $(t_k)_k$, then for every $x\in X$ there exists $k=k(x)\in\mathbb{N}$ such that $|T_i(x)|\le t_k$ for each $i\in\Lambda$. Hence, $X=\bigcup_{k\in\mathbb{N}} A_k$. Moreover, it is easy to check that A_k is closed for all fixed $k\in\mathbb{N}$. The assertion follows by proceeding similarly as in the classical case, using the Baire category theorem. □

Observe that, without assuming property (PR), in general our results do not hold. To this aim, it is enough to proceed analogously as in Example 1.27.

We now turn to a version of the Banach-Steinhaus theorem in the setting of filters and Riesz space-valued operators, which extends Kolk (2010), Theorem 3.

Theorem 1.30 (see Boccuto, Dimitriou and Papanastassiou (2012b), Theorem 3.13) *Let* X *be a separable Banach space,* $\mathcal{F}$ *be a* P*-filter,* $T_i : X\to R$, $i\in\mathbb{N}$, *be a sequence of linear bounded operators. Let* $\{\phi_j : j\in\mathbb{N}\}$ *be a dense subset of* X. *Then there are a bounding sequence* $(t_k)_k$ *and an* (O)*-sequence* $(\tau_p)_p$ *in* R

such that the sequence $(T_i(x))_i$ *is* (PR)*-bounded by* $(t_k)_k$ *and* $(O\mathcal{F})$*-convergent with respect to* $(\tau_p)_p$ *for every* $x \in X$ *if and only if there exists a positive element* $L \in R$ *with*

$$|T_i| \leq L \text{ for each } i \in \mathbb{N} \qquad \textbf{(4.62)}$$

and $(T_i(\phi_j))_i$ *is* $(O\mathcal{F})$*-convergent for any* $j \in \mathbb{N}$ *with respect to a single* (O)*-sequence.*

Furthermore, if $T(x) := (O\mathcal{F})\lim_i T_i(x)$, $x \in X$, *then* T *is a bounded linear operator.*

Proof: If there is a bounding sequence $(t_k)_k$ in R such that $(T_i(x))_i$ is (PR)-bounded by $(t_k)_k$ and $(O\mathcal{F})$-convergent for every $x \in X$, then by Lemma 1.29 the set $\{T_i(x) : x \in X, i \in \mathbb{N}\}$ is order bounded in R. So we get (4.62), and the "only if" part is proved.

We now turn to the "if" part. Since $\mathcal{F}$ is a P-filter, by Proposition 2.19 of Chapter 2 applied to the double sequence $(T_i(\phi_j))_{i,j}$ we get the existence of a set $B_0 \in \mathcal{F}$, a sequence $(z_j)_j$ in R and an (O)-sequence $(\sigma_p)_p$, such that for every p, $j \in \mathbb{N}$ there is $\bar{i} \in B_0$ with

$$|T_i(\phi_j) - z_j| \leq \sigma_p \qquad \textbf{(4.63)}$$

for all $i \geq \bar{i}$, $i \in B_0$.

Fix now $x \in X$: by hypothesis, $(T_i(x))_{i \in \mathbb{N}}$ is order bounded in R. Choose $\varepsilon > 0$: since the set $\{\phi_j : j \in \mathbb{N}\}$ is dense in X, there is $j \in \mathbb{N}$, depending on ε and x, with $\| x - \phi_j \| < \varepsilon$. Pick $p \in \mathbb{N}$: in correspondence with p and j, let $\bar{i} \in B_0$ satisfy (4.63). For any $i, k \geq \bar{i}$, $i, k \in B_0$, we get

$$0 \leq |T_i(x) - T_k(x)| \leq |T_i(x) - T_i(\phi_j)| + |T_i(\phi_j) - T_k(\phi_j)| + |T_k(x) - T_k(\phi_j)|$$

$$\leq |T_i| \, \| x - \phi_j \| + 2\sigma_p + |T_k| \, \| x - \phi_j \| \leq 2\, L\varepsilon + 2\,\sigma_p. \qquad \textbf{(4.64)}$$

This means that the sequence $(T_i(x))_{i\in B_0}$ is (O)-Cauchy, and thus (O)-convergent, by virtue of (O)-completeness of R. Since $(O\mathcal{F}')$-convergence implies $(O\mathcal{F})$-convergence (see Proposition 2.15 of Chapter 2), we get that $T(x) := (O\mathcal{F})\lim_i T_i(x)$, $x\in X$, exists in R (with respect to a single (O)-sequence). The (PR)-boundedness and $(O\mathcal{F})$-convergence follow from this, (4.62) and linearity of the T_i's. This concludes the proof of the "only if" part.

Let now $T(x) := (O\mathcal{F})\lim_i T_i(x)$, $x\in X$. Fix arbitrarily $x\in X$ with $\|x\|=1$. We have: $|T_i(x)|\le L$ for all $i\in\mathbb{N}$. Taking the $(O\mathcal{F})$-limit, we get: $|T(x)|\le L$. By arbitrariness of x, we obtain $|T|\le L$. Thus, the limit operator T is a bounded linear operator. This concludes the proof of the theorem. □

Observe that the sufficient part of Theorem 1.30 holds even when instead of (4.62) we require only that the set $\{|T_i|: i\in\mathbb{N}\}$ is $\mathcal{F}$-bounded. Indeed, in this case, there exist $L\in R$, $L>0$, and $A_0\in\mathcal{F}$ with $|T_i|\le L$ whenever $i\in A_0$. So it will be sufficient to argue analogously as in the proof of Theorem 1.30, by considering the set $A_0\cap B_0\in\mathcal{F}$ instead of B_0 (see also Boccuto, Dimitriou and Papanastassiou (2012b), Remark 3.14).

The next step is to prove a "dual version" of Theorem 1.30. Let R be a Dedekind complete Riesz space and X be a vector lattice endowed with a norm $\|\cdot\|$, such that $\|x\|\le\|y\|$ whenever x, $y\in X$ and $|x|\le|y|$. We say that the space R^* of all bounded operators $T: X\to R$ is *norm separable* iff there are a sequence $(f_j)_j$ in R^* and an (O)-sequence $(\tau_p)_p$ in R such that for every $x^*\in R^*$ there exists a subsequence $(h_j)_j$ of $(f_j)_j$ such that $(O)\lim_j |x^*-h_j|=0$ with respect to $(\tau_p)_p$, where $|x^*-h_j|$ denotes the following element of R:

$$|x^*-h_j| := \bigvee_{\|x\|=1} |x^*(x)-h_j(x)| = \bigvee_{x\neq 0} \frac{|x^*(x)-h_j(x)|}{\|x\|}.$$

In this case we say that the set $\{f_j : j\in\mathbb{N}\}$ is *norm dense* in R^*.

Theorem 1.31 (see Boccuto, Dimitriou and Papanastassiou (2012b), Theorem 3.15) *Under the hypotheses and notations above, let $\mathcal{F}$ be a P-filter. Assume that R^* is norm separable, and let $\{f_j : j \in \mathbb{N}\}$ be a norm dense subset of R^*. Furthermore, suppose that $(x_n)_n$ is an $\mathcal{F}$-bounded sequence in X, such that*

$$(O\mathcal{F})\lim_n f_j(x_n) \text{ exists in } R \tag{4.65}$$

for all $j \in \mathbb{N}$ with respect to a single (O)-sequence. Then $(O\mathcal{F})\lim_n x^*(x_n)$ exists in R for every $x^* \in R^*$ with respect to a common (O)-sequence.

Proof: By (4.65) and Proposition 2.19 of Chapter 2 applied to the double sequence $(f_j(x_n))_{j,n}$ there exist an element $B_0 \in \mathcal{F}$, a sequence $(z_j)_j$ in R and an (O)-sequence $(\sigma_p)_p$ such that for each p, $j \in \mathbb{N}$ there is $\overline{n} = \overline{n}(p, j) \in B_0$ with

$$| f_j(x_n) - z_j | \leq \sigma_p \tag{4.66}$$

whenever $n \geq \overline{n}$, $n \in B_0$. Moreover, since the sequence $(x_n)_n$ is $\mathcal{F}$-bounded in X, there exist a set $A_0 \in \mathcal{F}$ and a positive real number a, such that $\| x_n \| \leq a$ for each $n \in A_0$. Set $C_0 := A_0 \cap B_0$: note that $C_0 \in \mathcal{F}$ and, without loss of generality, we can suppose $\overline{n} \in C_0$.

Let $(\tau_p)_p$ be an (O)-sequence in R according to norm separability and fix arbitrarily $x^* \in R^*$: from norm separability, we find a subsequence $(h_j)_j$ of $(f_j)_j$ in R^* such that for each $p \in \mathbb{N}$ there is an integer $j = j(p)$ with $| x^* - h_r | \leq \tau_p$ for every $r \in \mathbb{N}$, $r \geq j$. Fix arbitrarily $p \in \mathbb{N}$, pick $j = j(p)$ as above and let $\overline{n} = \overline{n}(p, j(p)) \in C_0$ satisfy (4.66). For every $n \in A_0$ we have

$$| x^*(x_n) - h_j(x_n) | \leq | x^* - h_j | \, \| x_n \| \leq a\tau_p. \tag{4.67}$$

From (4.66) and (4.67) it follows that for any $n, m \geq \overline{n}$, $n, m \in C_0$, we get

$$|x^*(x_n)-x^*(x_m)| \;\leq\; |x^*(x_n)-h_j(x_n)|+|h_j(x_n)-h_j(x_m)|+|x^*(x_m)-h_j(x_m)|$$
$$\leq 2a\tau_p+2\sigma_p.$$

From this and (O)-completeness of R it follows that $(O)\lim_{n\in C_0} x^*(x_n)$ exists in R with respect to a single (O)-sequence. The assertion follows from Proposition 2.15 of Chapter 2. □

4.1.5. The Basic Matrix Theorem

We now prove some matrix theorems for lattice group-valued double sequences in the context of filter convergence. The classical basic matrix theorem of Antosik-Mikusiński-Swartz was extended in the topological group setting and even to Riesz space-valued double sequences (see Antosik and Swartz (1992), where the so-called "$(*)$-convergence" is used (see also Chapter 1). The basic matrix theorem was extended by A. Aizpuru and M. Nicasio-Llach (2008b) in the context of normed spaces and statistical convergence (see also Aizpuru and Gutiérrez-Dávila (2004)). In general, the nature of $(*)$-convergence is topological. However, there are Riesz spaces, which can be viewed as metrizable groups (with respect to a suitable topology), but such that order convergence is not generated by any topology: for example, the space $R=L^0([0,1],\mathcal{M},\nu)$ as in Example 2.181 of Chapter 1. Indeed, R can be metrized in order to obtain convergence in measure, though order convergence (which in this case coincides with (r)- and (D)-convergence) means almost everywhere convergence and is not topological, while $(*)$-convergence coincides with convergence in measure. For technical reasons, though in super Dedekind complete and weakly σ-distributive lattice groups the theories of order and (D)-convergence coincide, we deal with (D)-convergence because we consider *series* of regulators, which can be "replaced" with a single (D)-sequence by using the Fremlin lemma. We begin with a technical lemma on interchange of limits with respect to $(D\mathcal{F})$-convergence (for the classical version in the real setting, see also Dunford and Schwartz (1958), Lemma I.7.6).

Lemma 1.32 (see Boccuto, Dimitriou and Papanastassiou (2012d), Lemma 2.15) *Let R be a super Dedekind complete and weakly σ-distributive (ℓ)-group, $(x_{i,j})_{i,j}$ be a bounded double sequence of R, $\mathcal{F}$ be any free filter of $\mathbb{N}$, $\mathcal{I}$ be its* *dual ideal and* K *be any fixed element of* $\mathcal{F}$. *Set* $\mathcal{I}\times\mathcal{I} := \{D_1\times D_2 : D_1, D_2 \in \mathcal{I}\}$. *Suppose that*

(i) $(D\mathcal{F})\lim_i x_{i,j} = y_j$ *exists in* R *for all* $j\in\mathbb{N}$.

(ii) $(D\mathcal{F})\lim_j [\bigvee_{i\in K} | x_{i,j} - x_i |] = 0$.

Then the following results hold with respect to a single (D)*-sequence* $(b_{t,l})_{t,l}$.

(iii) In R, *there exists the limit* $a := (D\mathcal{F})\lim_j y_j$.

(iv) In R, *there exists* $b := (D\mathcal{F})\lim_i x_i$.

(v) There are an ideal $\mathcal{J}\subset\mathcal{P}(\mathbb{N}\times\mathbb{N})$ *and* $c\in R$ *with* $\mathcal{I}\times\mathcal{I}\subset\mathcal{J}$ *and* $(D\mathcal{S})\lim_{i,j} x_{i,j} = c$, *where* $\mathcal{S}$ *is the dual filter associated with* $\mathcal{J}$.

(vi) In R, *there exists* $d := (D\mathcal{F})\lim_i x_{i,i}$.

(vii) We get: $a = b = c = d$.

Proof: First of all note that by (i), thanks to boundedness of the given double sequence and the Fremlin lemma, there exists a regulator $(a_{t,l})_{t,l}$ such that for every $\varphi\in\mathbb{N}^{\mathbb{N}}$ and $j\in\mathbb{N}$ there is $D_j\in\mathcal{I}$ such that $| x_{i,j} - y_j | \leq \bigvee_{t=1}^{\infty} a_{t,\varphi(t)}$ whenever $i\notin D_j$. By (ii), without loss of generality, the regulator $(a_{t,l})_{t,l}$ can be chosen in such a way that for every $\varphi\in\mathbb{N}^{\mathbb{N}}$ there exists $D\in\mathcal{I}$ with

$$| x_{i,j} - x_i | \leq \bigvee_{t=1}^{\infty} a_{t,\varphi(t)} \qquad \textbf{(4.68)}$$

for all $j\notin D$ and $i\in K$. Now we prove (v). Let $j_0 := \min(\mathbb{N}\setminus D)$. Then by (4.68) we have:

$$|x_{i,j_0} - x_i| \leq \bigvee_{t=1}^{\infty} a_{t,\varphi(t)} \tag{4.69}$$

for all $i \in K$. By (4.68) and (4.69) we get

$$|x_{i,j} - x_{i,j_0}| \leq 2\bigvee_{t=1}^{\infty} a_{t,\varphi(t)}, \ j \notin D, i \in K. \tag{4.70}$$

By (i), there exists in R the limit $(D\mathcal{F})\lim_i x_{i,j_0} = y_{j_0}$ and so there is $D_{j_0} \in \mathcal{I}$ with

$$|x_{i,j_0} - y_{j_0}| \leq \bigvee_{t=1}^{\infty} a_{t,\varphi(t)} \quad i \notin D_{j_0}, i \in K. \tag{4.71}$$

Let $i_0 := \min(\mathbb{N} \setminus D_{j_0})$. By (4.71) we get:

$$|x_{i_0,j_0} - y_{j_0}| \leq \bigvee_{t=1}^{\infty} a_{t,\varphi(t)}. \tag{4.72}$$

By (4.71) and (4.72) we obtain:

$$|x_{i,j_0} - x_{i_0,j_0}| \leq 2\bigvee_{t=1}^{\infty} a_{t,\varphi(t)}, \ i \notin D_{j_0}, i \in K. \tag{4.73}$$

By (4.70) and (4.73) we get that

$$|x_{i,j} - x_{i_0,j_0}| \leq 4\bigvee_{t=1}^{\infty} a_{t,\varphi(t)}, \ i \notin D_{j_0}, \ i \in K, \ j \notin D. \tag{4.74}$$

Let now $i' \notin D_{j_0}$, $i' \in K$, $j' \notin D$. Then by (4.74) we have:

$$|x_{i_0,j_0} - x_{i',j'}| \leq 4\bigvee_{t=1}^{\infty} a_{t,\varphi(t)}. \tag{4.75}$$

Let $S := (D_{j_0} \cup (\mathbb{N} \setminus K)) \times D \in \mathcal{I} \times \mathcal{I}$ and $\mathcal{J} := \{\bigcup_{s=1}^{k} (A_s \times B_s) : A_s, B_s \in \mathcal{I}, s = 1, \ldots, k; \, k \in \mathbb{N}\}$.

Then $\mathcal{J}$ is an admissible ideal in $\mathbb{N}\times\mathbb{N}$ and $S\in\mathcal{J}$. By (4.75) we obtain that

$$|x_{i,j}-x_{i',j'}|\leq 8\bigvee_{t=1}^{\infty}a_{t,\varphi(t)} \tag{4.76}$$

for all (i,j), $(i',j')\notin S$, and by (4.76) the double sequence $(x_{i,j})_{i,j}$ is $(D\mathcal{S})$-Cauchy. By virtue of Proposition 2.10 of Chapter 2, the limit $c:=(D\mathcal{S})\lim_{i,j} x_{i,j}$ exists in R. Thus (v) is proved.

(vi) With the same notations as in the proof of (v), if i, $i'\notin D_{j_0}\cup D\cup(\mathbb{N}\setminus K)\in\mathcal{I}$, then from (4.74) and (4.75) it follows that

$$|x_{i,i}-x_{i',i'}|\leq 8\bigvee_{t=1}^{\infty}a_{t,\varphi(t)}.$$

Thus the sequence $(x_{i,i})_i$ is $(D\mathcal{F})$-Cauchy, and hence $(D\mathcal{F})\lim_{i} x_{i,i}$ exists in R and is equal to c.

We now prove (iii). By (v) there exists a (D)-sequence $(\alpha_{t,l})_{t,l}$ such that for any $\varphi\in\mathbb{N}^{\mathbb{N}}$ there is $S\in\mathcal{J}$ with

$$|x_{i,j}-c|\leq\bigvee_{t=1}^{\infty}\alpha_{t,\varphi(t)} \tag{4.77}$$

for all $(i,j)\notin S$. There is a positive integer k_0 with $S=\bigcup_{s=1}^{k_0}(A_s\times B_s)$, where A_s, $B_s\in\mathcal{I}$ for all $s=1,\ldots,k_0$. Moreover, by (i), for every $j\in\mathbb{N}$ there is $D_j\in\mathcal{I}$ with

$$|x_{i,j}-y_j|\leq\bigvee_{t=1}^{\infty}a_{t,\varphi(t)} \tag{4.78}$$

for all $i\notin D_j$. For each $i\notin(\bigcup_{s=1}^{k_0}A_s)\cup D_j\in\mathcal{I}$ and $j\notin(\bigcup_{s=1}^{k_0}B_s)\in\mathcal{I}$, by (4.77) and (4.78) we get:

$$|y_j - c| \leq |x_{i,j} - y_j| + |x_{i,j} - c| \leq \bigvee_{t=1}^{\infty} a_{t,\varphi(t)} + \bigvee_{t=1}^{\infty} \alpha_{t,\varphi(t)}. \quad \textbf{(4.79)}$$

By (4.79) and weak σ-distributivity of R, the element a as in (iii) exists in R, and $a = c$.

(iv) It is enough to proceed similarly as in (iii).

(vii) It is an obvious consequence of (iii), (iv), (v) and (vi). □

We now turn to our version of the basic matrix theorem in the lattice group context, in which the sliding hump technique is used.

Theorem 1.33 (see Boccuto, Dimitriou and Papanastassiou (2012d), Theorem 3.1) *Let $(x_{i,j})_{i,j}$ be a bounded double sequence in R, $\mathcal{F}$ be a P-filter of $\mathbb{N}$, $\mathcal{I}$ be its dual ideal, and let*

(i) $(D\mathcal{F})\lim_i x_{i,j} =: x_j$ exiss in R for all $j \in \mathbb{N}$;

(ii) $(D\mathcal{F})\lim_j x_{i,j} = 0$ for all $i \in \mathbb{N}$;

(iii) moreover, let there exist a regulator $(d_{t,l})_{t,l}$ such that for any infinite subset $B \subset \mathbb{N}$ there is an infinite subset $C \subset B$ such that the sequence $((\mathcal{F})\sum_{j \in C} x_{i,j})_i$ (D)-converges (with respect to $(d_{t,l})_{t,l}$). Then the following results hold:

(I) There exists $K \in \mathcal{F}$ with $(D\mathcal{F})\lim_i [\bigvee_{j \in K} |x_{i,j} - x_j|] = 0$;

(II) $(D\mathcal{F})\lim_j x_j = 0$;

(III) If $\mathcal{J} \subset \mathcal{P}(\mathbb{N}^2)$ is the ideal of $\mathbb{N}^2$ generated by the finite unions of the Cartesian products of the elements of $\mathcal{I}$ and $\mathcal{S}$ is its dual filter, then $(D\mathcal{S})\lim_{i,j} x_{i,j} = 0$;

(IV) $(D\mathcal{F})\lim_{i} = 0$;

(V) There is $A \in \mathcal{F}$ *with* $(D\mathcal{F})\lim_{j} [\bigvee_{i \in A} | x_{i,j} |] = 0$.

Proof: (I) First of all note that, by virtue of (ii) and Proposition 2.19 of Chapter 2 (see also Theorem 2.2 of Chapter 2), a set $K \in \mathcal{F}$ can be found, with

$$(D) \lim_{j \to +\infty, j \in K} x_{i,j} = 0 \qquad \textbf{(4.80)}$$

for all $i \in \mathbb{N}$, with respect to a single regulator $(\beta_{t,l})_{t,l}$. Let now $b_{t,l} = 2\beta_{t,l}$, $t, l \in \mathbb{N}$. From (4.80) it follows that for any $\varphi \in \mathbb{N}^{\mathbb{N}}$ and $i, k \in \mathbb{N}$ there is $s = s(i,k) \in K$ with

$$| x_{i,j} - x_{k,j} | \le \bigvee_{t=1}^{\infty} b_{t,\varphi(t)} \qquad \textbf{(4.81)}$$

for all $j \ge s$, $j \in K$. Let $u := \bigvee_{i,j} | x_{i,j} |$. By the Fremlin lemma there is a regulator $(b^*_{t,l})_{t,l}$ with

$$(2u) \wedge [\sum_{q=1}^{\infty} (\bigvee_{t=1}^{\infty} b_{t,\varphi(t+q)})] \le \bigvee_{t=1}^{\infty} b^*_{t,\varphi(t)} \qquad \textbf{(4.82)}$$

for all $\varphi \in \mathbb{N}^{\mathbb{N}}$. Moreover, by (i) and Proposition 2.19 of Chapter 2, we find a set $A \in \mathcal{F}$ such that $(D) \lim_{i \to +\infty, i \in A} x_{i,j} = x_j$ for all $j \in \mathbb{N}$ and with respect to a same (D)-sequence $(\alpha_{t,l})_{t,l}$. Let $A = \{q_1 < \ldots < q_i < \ldots\}$, and for the sake of simplicity, put $q_i = i$ for all i.

Again by the Fremlin lemma, proceeding analogously as above, taking into account boundedness of the double sequence $(x_{i,j})_{i,j}$, there is a regulator $(a^*_{t,l})_{t,l}$ such that for every $\varphi \in \mathbb{N}^{\mathbb{N}}$ and $s \in \mathbb{N}$ there is $p \in \mathbb{N}$ with
$\sum_{j \in K, j=1}^{s} | x_{i,j} - x_j | \le \bigvee_{t=1}^{\infty} a^*_{t,\varphi(t)}$ for all $i \ge p$.

Let now $a_{t,l} = 2a^*_{t,l}$, $t,l \in \mathbb{N}$. For any $\varphi \in \mathbb{N}^{\mathbb{N}}$ and $s \in \mathbb{N}$ there is $p = p(s) \in \mathbb{N}$ with

$$\sum_{j \in K, j=1}^{s} | x_{i,j} - x_{h,j} | \leq \bigvee_{t=1}^{\infty} a_{t,\varphi(t)} \tag{4.83}$$

for all $i,h \geq p$. Let $(d_{t,l})_{t,l}$ be as in (iii) and set $d^*_{t,l} := 2d_{t,l}$, $c_{t,l} := 6(a_{t,l} + b^*_{t,l} + d^*_{t,l})$, $t,l \in \mathbb{N}$. Let K be as in (4.80): we will prove that

$$(D)\lim_{i \in A}[\bigvee_{j \in K} | x_{i,j} - x_j |] = 0. \tag{4.84}$$

Note that, thanks to Proposition 2.16 of Chapter 2, (4.84) is enough to prove (I).

Before proving (4.84), we claim that for every $\varphi \in \mathbb{N}^{\mathbb{N}}$ there exists $i \in A$ such that the set

$$\{k \in A : \bigvee_{j \in K} | x_{i,j} - x_{k,j} | \not\leq \bigvee_{t=1}^{\infty} c_{t,\varphi(t)}\} \tag{4.85}$$

is finite. Otherwise, there is $\varphi \in \mathbb{N}^{\mathbb{N}}$ with the property that for every $i \in A$ there exist $k = k(i) \in A$, $k > i$ and $j \in K$ with

$$| x_{i,j} - x_{k,j} | \not\leq \bigvee_{t=1}^{\infty} c_{t,\varphi(t)}. \tag{4.86}$$

Choose arbitrarily $i_1 \in A$: there exist $k_1 = k(i_1) \in A$, $k_1 > i_1$ and $j_1 \in K$ with

$$| x_{i_1,j_1} - x_{k_1,j_1} | \not\leq \bigvee_{t=1}^{\infty} c_{t,\varphi(t)}. \tag{4.87}$$

Let $s_1 := s(i_1, k_1) \in K$ be as in (4.81): without loss of generality, we can suppose $s_1 > j_1$. We get

$|x_{i_1,j} - x_{k_1,j}| \le \bigvee_{t=1}^{\infty} b_{t,\varphi(t+1)}$ whenever $j \ge s_1$, $j \in K$. Let $p_1 := p(s_1)$ be as in (4.83). We obtain

$$\sum_{j\in K, j=1}^{s_1} |x_{p,j} - x_{q,j}| \le \bigvee_{t=1}^{\infty} a_{t,\varphi(t)} \text{ for any } p,q \ge p_1. \tag{4.88}$$

Let $i_2 \in A$ be with $i_2 > p_1$. Without loss of generality, we can choose $i_2 \in \{k(i) : i \in \mathbb{N}\}$. In correspondence with i_2 there are $k_2 = k(i_2) \in A$, $k_2 > i_2$, and $j_2 \in K$ such that

$$|x_{i_2,j_2} - x_{k_2,j_2}| \not\le \bigvee_{t=1}^{\infty} c_{t,\varphi(t)}. \tag{4.89}$$

Note that, by construction, $j_2 > s_1$. Let $s_2 := s(i_2,k_2) \in K$ be as in (4.81): without loss of generality, we can choose $s_2 > j_2$. We get $|x_{i_1,j} - x_{k_1,j}| \vee |x_{i_2,j} - x_{k_2,j}| \le \bigvee_{t=1}^{\infty} b_{t,\varphi(t+2)}$ whenever $j \ge s_2$, $j \in K$.

Proceeding by induction, we find four strictly increasing sequences, $(i_r)_r$ and $(k_r)_r$ in A, $(j_r)_r$ and $(s_r)_r$ in K, with the properties that: $i_r < k_r < i_{r+1}$, $j_r < s_r < j_{r+1}$ for all $r \in \mathbb{N}$; $i_r \in \{k(i) : i \in \mathbb{N}\}$ for any $r \ge 2$, and

(j) $\sum_{j\in K, j=1}^{s_{r-1}} |x_{i_r,j} - x_{k_r,j}| \le \bigvee_{t=1}^{\infty} a_{t,\varphi(t)}$,

(jj) $|x_{i_r,j_r} - x_{k_r,j_r}| \not\le \bigvee_{t=1}^{\infty} c_{t,\varphi(t)}$,

(jjj) $|x_{i_r,j_{r+h}} - x_{k_r,j_{r+h}}| \le \bigvee_{t=1}^{\infty} b_{t,\varphi(t+h)}$ for all $r \ge 2$ and $h \in \mathbb{N}$.

By virtue of (iii), in correspondence with $B := \{j_r : r \ge 2\}$ there exist an infinite set $C \subset B$ and a natural number n_0 such that

$$|(\mathcal{F})\sum_{j\in C}(x_{i_r,j} - x_{k_r,j})| \le \bigvee_{t=1}^{\infty} d^*_{t,\varphi(t)} \tag{4.90}$$

for all $r \geq n_0$. From (j), (jj), (jjj), if $s > r \geq n_0$ and $j_r \in C$, then

$$|x_{i_r,j_r} - x_{k_r,j_r}| \leq |\sum_{j\in C, j\in\{j_1,\ldots,j_s\}} (x_{i_r,j} - x_{k_r,j})| +$$

$$+ \sum_{j\in C, j\in\{j_1,\ldots,j_{r-1}\}} |x_{i_r,j} - x_{k_r,j}| + \sum_{j\in C, j\in\{j_{r+1},\ldots,j_s\}} |x_{i_r,j} - x_{k_r,j}| \tag{4.91}$$

$$\leq |\sum_{j\in C, j\in\{j_1,\ldots,j_s\}} (x_{i_r,j} - x_{k_r,j})| + \bigvee_{t=1}^{\infty} a_{t,\varphi(t)} + \sum_{h=1}^{s} (\bigvee_{t=1}^{\infty} b_{t,\varphi(t+h)}).$$

Taking the $(O\mathcal{F})$-limit as s tends to $+\infty$ in (4.91), and by Proposition 2.11 of Chapter 2, we get

$$|x_{i_r,j_r} - x_{k_r,j_r}| \leq |(\mathcal{F})\sum_{j\in C} (x_{i_r,j} - x_{k_r,j})| + \bigvee_{t=1}^{\infty} a_{t,\varphi(t)} + \bigvee_{s=1}^{\infty} (\sum_{h=1}^{s} (\bigvee_{t=1}^{\infty} b_{t,\varphi(t+h)}))$$

$$\leq \bigvee_{t=1}^{\infty} d^*_{t,\varphi(t)} + \bigvee_{t=1}^{\infty} a_{t,\varphi(t)} + \bigvee_{s=1}^{\infty} (\sum_{h=1}^{s} (\bigvee_{t=1}^{\infty} b_{t,\varphi(t+h)})),$$

that is

$$|x_{i_r,j_r} - x_{k_r,j_r}| - \bigvee_{t=1}^{\infty} d^*_{t,\varphi(t)} - \bigvee_{t=1}^{\infty} a_{t,\varphi(t)} \leq \bigvee_{s=1}^{\infty} (\sum_{h=1}^{s} (\bigvee_{t=1}^{\infty} b_{t,\varphi(t+h)})). \tag{4.92}$$

We have also

$$|x_{i_r,j_r} - x_{k_r,j_r}| - \bigvee_{t=1}^{\infty} d^*_{t,\varphi(t)} - \bigvee_{t=1}^{\infty} a_{t,\varphi(t)} \leq |x_{i_r,j_r} - x_{k_r,j_r}| \leq 2u. \tag{4.93}$$

From (4.82), (4.92) and (4.93) it follows that

$$|x_{i_r,j_r} - x_{k_r,j_r}| - \bigvee_{t=1}^{\infty} d^*_{t,\varphi(t)} - \bigvee_{t=1}^{\infty} a_{t,\varphi(t)} \leq \bigvee_{t=1}^{\infty} b^*_{t,\varphi(t)}, \tag{4.94}$$

and finally, if $r \geq n_0$ and $j_r \in C$, then we have

$$| x_{i_r, j_r} - x_{k_r, j_r} | \leq \bigvee_{t=1}^{\infty} d^*_{t,\varphi(t)} + \bigvee_{t=1}^{\infty} a_{t,\varphi(t)} + \bigvee_{t=1}^{\infty} b^*_{t,\varphi(t)} \leq \bigvee_{t=1}^{\infty} c_{t,\varphi(t)}. \qquad \textbf{(4.95)}$$

So (4.95) holds for infinitely many integers r. This contradicts (4.86) and proves (4.85).

We now prove (4.84). By (4.85), the family $\{(x_{i,j})_{i\in A} : j \in K\}$ is $(D\mathcal{F}_{\text{cofin}})$-Cauchy uniformly with respect to $j \in K$. Thus (4.84) follows from the last part of Proposition 2.14 of Chapter 2, since $(D\mathcal{F}_{\text{cofin}})$-convergence coincides with usual (D)-convergence. This ends the proof of (I).

(II) We have proved that $(D\mathcal{F})\lim_i [\bigvee_{j\in K} | x_{i,j} - x_j |] = 0$, and by (ii) we know that $(D\mathcal{F})\lim_j x_{i,j} = 0$ for every $i \in \mathbb{N}$. Thus by (iii), (iv) and (vii) of Lemma 1.32, by interchanging the role of the variables i and j, we get that $(D\mathcal{F})\lim_j x_j = 0$, that is (II).

(III) It is an immediate consequence of (I), (II) and Lemma 1.32.

(IV) It follows from (I) and (vi) of Lemma 1.32.

(V) In the proof of (I) we proved the existence of A, $K \in \mathcal{F}$ and of a regulator $(c_{t,l})_{t,l}$ such that $(D)\lim_{i\in A}[\bigvee_{j\in K} | x_{i,j} - x_j |] = 0$ with respect to the (D)-sequence $(c_{t,l})_{t,l}$. Moreover, by (II),

$0 = (D\mathcal{F})\lim_j x_j = (O\mathcal{F})\lim_j x_j$. Since $\mathcal{F}$ is a P-filter, by Proposition 2.17 of Chapter 2 we get $(O\mathcal{F}')\lim_j x_j = (D\mathcal{F}')\lim_j x_j = 0$. Thus we find a set $K_0 \in \mathcal{F}$, with $(D)\lim_{j\in K_0} x_j = 0$. Let $K' := K \cap K_0$: then $K' \in \mathcal{F}$. In order to prove the assertion, thanks to Proposition 2. 16 it is enough to show that

$$(D)\lim_{j\in K'}[\bigvee_{i\in A} | x_{i,j} |] = 0. \qquad \textbf{(4.96)}$$

To this aim observe that in correspondence with $\varphi \in \mathbb{N}^{\mathbb{N}}$ there is $\overline{i} \in A$ with

$$|x_{i,j} - x_j| \leq \bigvee_{t=1}^{\infty} c_{t,\varphi(t)}$$

whenever $i \in A$, $i \geq \overline{i}$ and $j \in K$ (and a fortiori $j \in K'$). Since $(D)\lim_{j \in K_0} x_j = 0$, there is a regulator $(\xi_{t,l})_{t,l}$ such that for every $\varphi \in \mathbb{N}^{\mathbb{N}}$ there exists $\overline{j} \in K_0$ with $|x_j| \leq \bigvee_{t=1}^{\infty} \xi_{t,\varphi(t)}$ for all $j \geq \overline{j}$, $j \in K_0$ (and a fortiori $j \in K'$). Note that, without loss of generality, the integer $\overline{j}$ can be taken in K'. Since (ii) holds, proceeding analogously as in the proof of (I) we get $(D)\lim_{j \in K'} x_{i,j} = 0$ for all $i \in \mathbb{N}$, with respect to a single regulator $(\beta_{t,l})_{t,l}$. Then, for every $\varphi \in \mathbb{N}^{\mathbb{N}}$ and $i = 1, \ldots, \overline{i} - 1$, $i \in A$, there is $j_i \in K'$ with $|x_{i,j}| \leq \bigvee_{t=1}^{\infty} \beta_{t,\varphi(t)}$ for any $j \geq j_i$, $j \in K'$. Let now $j^* := \max\{\overline{j}, \max_{i=1,\ldots,\overline{i}-1, i \in A} j_i\}$, and choose arbitrarily $i \in A$, $j \in K'$, $j \geq j^*$. If $i \geq \overline{i}$, then

$$|x_{i,j}| \leq |x_{i,j} - x_j| + |x_j| \leq \bigvee_{t=1}^{\infty} c_{t,\varphi(t)} + \bigvee_{t=1}^{\infty} \xi_{t,\varphi(t)} \leq \bigvee_{t=1}^{\infty} c_{t,\varphi(t)} + \bigvee_{t=1}^{\infty} \xi_{t,\varphi(t)} + \bigvee_{t=1}^{\infty} \beta_{t,\varphi(t)}.$$

If $i \leq \overline{i} - 1$, then $|x_{i,j}| \leq \bigvee_{t=1}^{\infty} \beta_{t,\varphi(t)} \leq \bigvee_{t=1}^{\infty} c_{t,\varphi(t)} + \bigvee_{t=1}^{\infty} \xi_{t,\varphi(t)} + \bigvee_{t=1}^{\infty} \beta_{t,\varphi(t)}$.

This proves the claim (4.96) and hence (V), and completes the proof of the theorem. □

Theorem 1.33 is an extension to the context of (ℓ)-groups and P-filters of Aizpuru and Nicasio-Llach (2008b), Theorem 4, which was formulated for normed spaces and $\mathcal{F} = \mathcal{F}_{\text{st}}$.

Furthermore observe that, if in the hypotheses of Theorem 1.33 we keep (i) and (iii), *fix* $K \in \mathcal{F}$ and replace (ii) with the condition

$$(D) \lim_{j \to +\infty, j \in K} x_{i,j} = 0 \qquad \textbf{(4.97)}$$

for all $i \in \mathbb{N}$ (without loss of generality with respect to a single regulator, see also Proposition 2.19 of Chapter 2), then the thesis of the theorem continues to hold, and the set K for which (I) is satisfied is just the element K of $\mathcal{F}$ fixed *a priori* in (4.97): indeed, it will be enough to repeat the same arguments of the proof of Theorem 1.33. In particular, if we take $K = \mathbb{N}$, instead of (ii) we get $(D)\lim_j x_{i,j} = 0$ for all $i \in \mathbb{N}$ (without loss of generality, with respect to a single regulator). Note that, by arguing analogously as in the proof of Theorem 1.33, it is possible to prove that (i), (ii') and (iii) imply that $(D\mathcal{F})\lim_i [\bigvee_{j\in\mathbb{N}} |x_{i,j} - x_j|] = 0$. Similarly, if in Theorem 1.33 we keep (ii) and (iii), fix $A \in \mathcal{F}$ and instead of (i) we require that $(D) \lim_{i\to+\infty, i\in A} x_{i,j} = x_j$ for all $j \in \mathbb{N}$ (again with respect to a single regulator, without restriction), then the set A for which (V) holds is just the mentioned element A of $\mathcal{F}$. In particular, if $A = \mathbb{N}$, then instead of (i) we will require that $(D)\lim_i x_{i,j} = x_j$ exists in R for all $j \in \mathbb{N}$. Proceeding analogously as in the proof of Theorem 1.33, we can prove that (i'), (ii) and (iii) imply that $(D\mathcal{F})\lim_j [\bigvee_{i\in\mathbb{N}} |x_{i,j}|] = 0$.

Observe that in general, even when $R = \mathbb{R}$, if we drop (iii) or (iii'), then (i') and (ii') do not imply (V'). Indeed, let $\mathcal{F} \neq \mathcal{F}_{\text{cofin}}$, $\mathcal{I}$ be its dual ideal, and $H := \{h_1 < \ldots < h_j < h_{j+1} < \ldots\}$ be an infinite set, belonging to $\mathcal{I}$. For every $j \in \mathbb{N}$, let us define $x_{h_j,j} := 1$; for the other choices of i and j, put $x_{i,j} := 0$. As $x_{i,j} = 0$ whenever $j \in \mathbb{N}$ and $i \in \mathbb{N} \setminus H$, and since $H \in \mathcal{I}$, then $\lim_i x_{i,j} = 0$ for all $j \in \mathbb{N}$ and $\lim_j x_{i,j} = 0$ for all $i \in \mathbb{N}$. So, (i') and (ii') are satisfied. However, $(x_{i,j})_{i,j}$ does not fulfil (V'), since $\sup_{i\in\mathbb{N}} |x_{i,j}| = 1$ for each $j \in \mathbb{N}$. Moreover, (iii) and (iii') do not hold, since for each strictly increasing sequence $(j_s)_s$ in $\mathbb{N}$ we get $\sum_{s=1}^{\infty} x_{h_{j_s}, j_s} = +\infty$ (see also Boccuto, Dimitriou and Papanastassiou (2012d), Example 3.4).

Proposition 1.34 *A series* $\sum_{j=1}^{\infty} a_j$ *is unconditionally convergent if and only if the limit*

$$(D)\lim_q \sum_{j\in A\cap[1,q]} a_j =: \sum_{j\in A} a_j$$

exists in R uniformly with respect to $A \subset \mathbb{N}$ and related to a single regulator independent of A.

Proof: We first prove the "only if" part. Let $(a_{t,l})_{t,l}$ be a regulator associated with unconditional convergence. By the Cauchy criterion it will be enough to prove that for every $\varphi \in \mathbb{N}^{\mathbb{N}}$ there exists $q \in \mathbb{N}$ with $|\sum_{j\in A\cap[r,s]} a_j| \le \bigvee_{t=1}^{\infty} a_{t,\varphi(t)}$ whenever $s > r > q$ and $A \subset \mathbb{N}$. Choose arbitrarily $\varphi \in \mathbb{N}^{\mathbb{N}}$, let $A_0 \in \mathcal{I}_{\text{fin}}$ be according to unconditional convergence, let $q = \max A_0$, pick $s > r > q$ and take $A_1 = A_0 \cup (A \cap [1,s])$, $A_2 = A_0 \cup (A \cap [1,r-1])$. We have:

$$|\sum_{j\in A_1} a_j - \sum_{j\in A_2} a_j| = |\sum_{j\in A\cap[r,s]} a_j| = |\sum_{j\in A\cap[1,s]} a_j - \sum_{j\in A\cap[1,r-1]} a_j| \le \bigvee_{t=1}^{\infty} a_{t,\varphi(t)},$$

We now turn to the "if" part. By hypothesis there is a regulator $(a_{t,l})_{t,l}$ such that for each $\varphi \in \mathbb{N}^{\mathbb{N}}$ there is $\overline{n} \in \mathbb{N}$ with $|\sum_{j\in A\cap[r,s]} a_j - \sum_{j\in B\cap[r,s]} a_j| \le |\sum_{j\in A\cap[r,s]} a_j| + |\sum_{j\in B\cap[r,s]} a_j| \le 2 \bigvee_{t=1}^{\infty} a_{t,\varphi(t)}$ for each $s > r > \overline{n}$ and A , $B \subset \mathbb{N}$. Let $A_0 = [1,\overline{n}]$ ad choose arbitrarily A_1 , $A_2 \in \mathcal{I}_{\text{fin}}$, with A_1 , $A_2 \supset A_0$. Let $s = \max\{\max A_1, \max A_2\}$. We get:

$$|\sum_{j\in A_1} a_j - \sum_{j\in A_2} a_j| = |\sum_{j\in A_1\cap[\overline{n}+1,s]} a_j - \sum_{j\in A_2\cap[\overline{n}+1,s]} a_j| \le 2 \bigvee_{t=1}^{\infty} a_{t,\varphi(t)}.$$

So the series $\sum_{j=1}^{\infty} a_j$ is unconditionally convergent. This ends the proof. □

Proposition 1.35 *Let $\sum_{j=1}^{\infty} a_j$ be an unconditionally convergent series, and set*

$m(A) := \sum_{j\in A} a_j$, $A \subset \mathbb{N}$. Then m is σ-additive on $\mathcal{P}(\mathbb{N})$.

Proof: Note that m is well-defined, thanks to Proposition 1.34. By virtue of Proposition 1.8 it will be enough to find a regulator $(\alpha_{t,l})_{t,l}$ such that for any $\varphi \in \mathbb{N}^{\mathbb{N}}$ there is $\overline{n} \in \mathbb{N}$ with

$$|\sum_{j\in A} a_j| \le \bigvee_{t=1}^{\infty} \alpha_{t,\varphi(t)} \text{ for every } n > \overline{n} \text{ and } A \subset [n,+\infty[. \qquad \textbf{(4.98)}$$

Let $(\alpha_{t,l})_{t,l}$ be a regulator associated with unconditional convergence, choose arbitrarily $\varphi \in \mathbb{N}^{\mathbb{N}}$ and let $A_0 \in \mathcal{I}_{\text{fin}}$ be according to unconditional convergence. Let $\overline{n} = \max A_0$, pick arbitrarily $q > n > \overline{n}$ and set $A_1 = A_0 \cup (A \cap [n,q])$. We get

$$|\sum_{j\in A\cap[1,q]} a_j| = |\sum_{j\in A\cap[n,q]} a_j| = |\sum_{j\in A_1} a_j - \sum_{j\in A_0} a_j| \le \bigvee_{t=1}^{\infty} \alpha_{t,\varphi(t)}. \qquad \textbf{(4.99)}$$

Taking in (4.99) the (D)-limit as q tends to $+\infty$, we obtain (4.98). This ends the proof. □

We say that a subset $\mathcal{W}$ of $\mathcal{P}(\mathbb{N})$ which contains the ideal $\mathcal{I}_{\text{fin}}$ of all finite subsets of $\mathbb{N}$ *satisfies property* (M) iff for every sequence $(F_k)_k$ of elements of $\mathcal{I}_{\text{fin}}$ with $\max F_k < \min F_{k+1}$ for each $k \in \mathbb{N}$, there exist a set $B \in \mathcal{W}$ and an infinite subset $M \subset \mathbb{N}$ with $\bigcup_{n\in M} F_n \subset B \subset \bigcup_{n\in\mathbb{N}} F_n$.

Some examples of families fulfilling property (M) can be found, for instance, in Aizpuru, Nicasio-Llach and Rambla-Barreno (2010), Example 2.1.

Lemma 1.36 *Let $\mathcal{F}$ be any free filter of $\mathbb{N}$, $\mathcal{W}$ be a family of subsets of $\mathbb{N}$, satisfying property (M), $(a_n)_n$ be a sequence of real numbers, and assume that the series $(\mathcal{F})\sum_{n\in B} a_n$*

$\mathcal{F}$-converges in $\mathbb{R}$ (that is, the sequence $(\sum_{i\in B\cap[1,n]} a_i)_n$ $\mathcal{F}$-converges in $\mathbb{R}$) for each $B \in \mathcal{W}$.

Then the series $\sum_{n\in A} a_n$ converges uniformly with respect to $A \subset \mathbb{N}$ in the usual sense.

Proof: By virtue of the Cauchy criterion, it is enough to prove that for every $\varepsilon > 0$ there is $n_0 \in \mathbb{N}$ with $|\sum_{n\in A\cap[q,r]} a_n| \le \varepsilon$ for each $r > q \ge n_0$ and $A \subset \mathbb{N}$. If we deny the thesis, then we find a positive real number ε and a sequence $(F_k)_k$ in $\mathcal{I}_{\text{fin}}$, with $\max F_k < \min F_{k+1}$ and $|\sum_{n\in F_k} a_n| > \varepsilon$ for each k. Now, proceeding analogously as in Aizpuru, Nicasio-Llach and Rambla-Barreno (2010), Lemma 2.2, set $G_k = \{n \in F_k : a_n \ge 0\}$ and $H_k = \{n \in F_k : a_n < 0\}$. Without loss of generality, we can suppose that the set $K := \{k \in \mathbb{N} : \sum_{n\in G_k} a_n > \varepsilon/2\}$ is infinite. Now, let us consider the sequence G_k, $k \in K$. Since $\mathcal{W}$ satisfies property (M), there are an element $B \in \mathcal{W}$ and an infinite set $P \subset K$ such that $\bigcup_{k\in P} G_k \subset B \subset \bigcup_{k\in K} G_k$. Since, by hypothesis, $(\mathcal{F})\sum_{n\in B} a_n$ exists in $\mathbb{R}$, then, by virtue of Proposition 1.6 of Chapter 2, there is an infinite subset $A \subset \mathbb{N}$ such that the sequence $(\sum_{i\in B\cap[1,n]} a_i)_{n\in A}$ is convergent in $\mathbb{R}$. Observe that for every $q \in A$ there is $k \in P$ with $q < \min F_k$ and if $r \in A$, $r > \max F_k$, then we get $\sum_{n\in[q,r]\cap B} a_n \ge \sum_{n\in G_k} a_n > \varepsilon/2$, and thus the series $(\sum_{i\in B\cap[1,n]} a_i)_{n\in A}$ is not Cauchy, obtaining a contradiction. This ends the proof. □

Theorem 1.37 *Let R be a super Dedekind complete and weakly σ-distributive lattice group, $\mathcal{F}$ be a free filter of $\mathbb{N}$, $\mathcal{W}$ satisfy property (M) and $(a_{i,j})_{i,j}$ be such that:*

1.37.1) *The limit $(D)\lim_i a_{i,j}$ exists in R for every $j \in \mathbb{N}$;*

1.37.2) *the series $(D\mathcal{F})\sum_{j\in B} a_{i,j}$ $(D\mathcal{F})$-converges (that is, the sequence $(\sum_{j\in B\cap[1,n]} a_{i,j})_n$ $(D\mathcal{F})$-converges) for each $B \in \mathcal{W}$ and $i \in \mathbb{N}$ with respect to a regulator independent of B and i;*

1.37.3) *the family $(\sum_{j\in B} a_{i,j})_{i\in\mathbb{N}, B\in\mathcal{I}_{\text{fin}}}$ is equibounded;*

1.37.4) *for every infinite subset $B \in \mathcal{W}$ the sequence $((D\mathcal{F})\sum_{j\in B} a_{i,j})_i$ $(D\mathcal{F})$-converges to 0 with respect to a regulator $(z_{t,l})_{t,l}$ independent of B.*

Then we get:

1.37.5) *the sequence* $(\sum_{j\in A} a_{i,j})_i$ $(D\mathcal{F})$*-converges for any subset* $A\subset\mathbb{N}$ *with respect to a* (D)*-sequence independent of* A;

1.37.6) *for any disjoint sequence* $(A_n)_n$ *in* $\mathcal{P}(\mathbb{N})$ *the sequence* $(\sum_{j\in A_n} a_{i,j})_i$ *(D)-converges uniformly with respect to* $n\in\mathbb{N}$ *and the sequence* $(\sum_{j\in A_n} a_{i,j})_n$ *(D)-converges to* 0 *uniformly with respect to* $i\in\mathbb{N}$.

Proof: By super Dedekind completeness and weak σ-distributivity of R, 1.37.2) and Proposition 2.9 of Chapter 2 (see also Boccuto (2003), Theorem 3.4), there is an (O)-sequence $(\sigma_p)_p$ such that the series $(O\mathcal{F})\sum_{j\in B} a_{i,j}$ $(RO\mathcal{F})$-converges with respect to $(\sigma_p)_p$ for each $B\in\mathcal{W}$ and $i\in\mathbb{N}$. Let us denote by $S^{(B,i)}$ its sum, and let Ω be as in the Maeda-Ogasawara-Vulikh theorem. There is a meager set $N\subset\Omega$ such that the sequence $(\sigma_p(\omega))_p$ is decreasing and $\inf_p \sigma_p(\omega)=\lim_p \sigma_p(\omega)=0$ for every $\omega\in\Omega\setminus N$. For each $B\in\mathcal{W}$, $i\in\mathbb{N}$ and $\omega\in\Omega\setminus N$ we get:

$$\mathcal{F}\ni\{n\in\mathbb{N}:|S^{(B,i)}-\sum_{j\in B\cap[1,n]} a_{i,j}|\leq\sigma_p\}=$$

$$=\{n\in\mathbb{N}:|S^{(B,i)}(\omega')-\sum_{j\in B\cap[1,n]} a_{i,j}(\omega')|\leq\sigma_p(\omega')\text{ for every }\omega'\in\Omega\setminus N\}\subset$$

$$\subset\{n\in\mathbb{N}:|S^{(B,i)}(\omega)-\sum_{j\in B\cap[1,n]} a_{i,j}(\omega)|\leq\sigma_p(\omega)\},$$

and hence $\{n\in\mathbb{N}:|\sum_{j\in B\cap[1,n]} a_{i,j}(\omega)-S^{(B,i)}(\omega)|\leq\sigma_p(\omega)\}\in\mathcal{F}$.

Thus the series $(\mathcal{F})\sum_{j\in B} a_{i,j}(\omega)$ $\mathcal{F}$-converges in $\mathbb{R}$ for each $B\in\mathcal{W}$, $i\in\mathbb{N}$ and $\omega\in\Omega\setminus N$. From this and Lemma 3.3 it follows that the series $\sum_{j\in A} a_{i,j}(\omega)$ converges, and hence it is Cauchy, in the usual sense uniformly with respect to $A\subset\mathbb{N}$ for every $i\in\mathbb{N}$ and $\omega\in\Omega\setminus N$. By the Maeda-Ogasawara-Vulikh theorem and taking into account 1.37.3), for every $i\in\mathbb{N}$ there is a meager set $N_i\subset\Omega$, without loss of generality $N_i\supset N$, such that, for every $i\in\mathbb{N}$ and $\omega\in\Omega\setminus N_i$,

$$0=\inf_{s\in\mathbb{N}}(\sup_{n\geq s}(\sup_{t\in\mathbb{N}}(\sup_{A\subset\mathbb{N}}|\sum_{j\in A\cap[n,n+t]}[a_{i,j}(\omega)]|)))=\sup_{s\in\mathbb{N}}(\inf_{n\geq s}(\sup_{t\in\mathbb{N}}(\sup_{A\subset\mathbb{N}}|\sum_{j\in A\cap[n,n+t]}[a_{i,j}(\omega)]|)))=$$

$$=\inf_{s\in\mathbb{N}}(\sup_{n\geq s}(\sup_{t\in\mathbb{N}}(\sup_{A\subset\mathbb{N}}|\sum_{j\in A\cap[n,n+t]}a_{i,j}|(\omega))))=\sup_{s\in\mathbb{N}}(\inf_{n\geq s}(\sup_{t\in\mathbb{N}}(\sup_{A\subset\mathbb{N}}|\sum_{j\in A\cap[n,n+t]}a_{i,j}|(\omega))))$$

$$=\bigwedge_{s\in\mathbb{N}}(\bigvee_{n\geq s}(\bigvee_{t\in\mathbb{N}}(\bigvee_{A\subset\mathbb{N}}|\sum_{j\in A\cap[n,n+t]}a_{i,j}|)))=\bigvee_{s\in\mathbb{N}}(\bigwedge_{n\geq s}(\bigvee_{t\in\mathbb{N}}(\bigvee_{A\subset\mathbb{N}}|\sum_{j\in A\cap[n,n+t]}a_{i,j}|))). \quad \textbf{(4.100)}$$

By (4.100), since Dedekind completeness of R implies (D)-completeness of R uniformly to respect to a parameter (see also Boccuto (1997)), for any $i\in\mathbb{N}$ the series $\sum_{j\in A}a_{i,j}$ (D)-converges uniformly with respect to $A\subset\mathbb{N}$, with a related regulator $(\alpha_{t,l}^{(i)})_{t,l}$. If $u=\bigvee_{i\in\mathbb{N},B\in\mathcal{I}_{\text{fin}}}(\sum_{j\in B}a_{i,j})$, as

$$|\sum_{j\in A}a_{i,j}|=(O)\lim_q|\sum_{j\in A\cap[1,q]}a_{i,j}|=(D)\lim_q|\sum_{j\in A\cap[1,q]}a_{i,j}|$$

for any $A\subset\mathbb{N}$, from 1.37.3) we get $u=\bigvee_{i\in\mathbb{N},A\subset\mathbb{N}}(\sum_{j\in A}a_{i,j})$. By the Fremlin lemma there is a regulator $(b_{t,l})_{t,l}$ such that, for any $\varphi\in\mathbb{N}^{\mathbb{N}}$ and $q\in\mathbb{N}$, $(2u)\wedge(\sum_{i=1}^{q}(\bigvee_{t=1}^{\infty}\alpha_{t,\varphi(t+i)}^{(i)}))\leq\bigvee_{t=1}^{\infty}b_{t,\varphi(t)}$, and hence for every $\varphi\in\mathbb{N}^{\mathbb{N}}$ and i, $r\in\mathbb{N}$ there is a positive integer $\overline{h}$ with

$$|\sum_{j\in C}(a_{i,j}-a_{r,j})|\leq(2u)\wedge(\bigvee_{t=1}^{\infty}\alpha_{t,\varphi(t+i)}^{(i)})\leq\bigvee_{q=1}^{\infty}((2u)\wedge(\sum_{i=1}^{q}(\bigvee_{t=1}^{\infty}\alpha_{t,\varphi(t+i)}^{(i)})))\leq\bigvee_{i=1}^{\infty}b_{t,\varphi(t)} \quad \textbf{(4.101)}$$

whenever $C\subset]\overline{h},+\infty[$. By virtue of 1.37.1), for every $j\in\mathbb{N}$ there exists a (D)-sequence $(\beta_{t,l}^{(j)})_{t,l}$, such that the sequence $(a_{i,j})_i$ (D)-converges with respect to $(\beta_{t,l}^{(j)})_{t,l}$. By the Fremlin lemma, taking into account 1.37.3) and proceeding analogously as in (4.101), we find a regulator $(c_{t,l})_{t,l}$ such that for each $\varphi\in\mathbb{N}^{\mathbb{N}}$ and $s\in\mathbb{N}$ there is $\overline{h}\in\mathbb{N}$ with

$$|\sum_{j\in D}(a_{p,j}-a_{q,j})|\leq\bigvee_{t=1}^{\infty}c_{t,\varphi(t)} \quad \textbf{(4.102)}$$

whenever $p,q\geq\overline{h}$ and $D\subset[1,s]$. For all $t,l\in\mathbb{N}$ put $e_{t,l}:=6(z_{t,l}+b_{t,l}+c_{t,l})$ and $d_{t,l}:=2e_{t,l}$.

Now we prove that for any subset $A\subset\mathbb{N}$ the sequence $(\sum_{j\in A}a_{i,j})_i$ is (D)-Cauchy with respect to the regulator $(d_{t,l})_{t,l}$. Without loss of generality, we can suppose that A is infinite. By contradiction, assume that there exists $\varphi\in\mathbb{N}^{\mathbb{N}}$ such that for every $i\in\mathbb{N}$ there is $k\in\mathbb{N}$, $k>i$, with

$$|\sum_{j\in A}(a_{i,j}-a_{k,j})|\not\leq\bigvee_{t=1}^{\infty}d_{t,\varphi(t)}. \quad \textbf{(4.103)}$$

At the first step, let $i_1=1$ and $k_1>1$ be such that $|\sum_{j\in A}(a_{i_1,j}-a_{k_1,j})|\not\leq\bigvee_{t=1}^{\infty}d_{t,\varphi(t)}$. There is $l_1>1$ with

$$|\sum_{j\in C}(a_{i_1,j}-a_{k_1,j})|\leq\bigvee_{t=1}^{\infty}b_{t,\varphi(t)} \quad \textbf{(4.104)}$$

whenever $C\subset[l_1+1,+\infty[$. Then we get

$$|\sum_{j\in A\cap[1,l_1]}(a_{i_1,j}-a_{k_1,j})|\not\leq\bigvee_{t=1}^{\infty}e_{t,\varphi(t)}, \quad \textbf{(4.105)}$$

otherwise

$$|\sum_{j\in A}(a_{i_1,j}-a_{k_1,j})| \leq |\sum_{j\in A\cap[1,l_1]}(a_{i_1,j}-a_{k_1,j})|+|\sum_{j\in A,j>l_1}(a_{i_1,j}-a_{k_1,j})|$$
$$\leq\bigvee_{t=1}^{\infty}e_{t,\varphi(t)}+\bigvee_{t=1}^{\infty}c_{t,\varphi(t)}\leq\bigvee_{t=1}^{\infty}d_{t,\varphi(t)},$$

which contradicts (4.103). Thus (4.105) is satisfied. Moreover, by (4.102), there is $r_1>k_1$ with

$$|\sum_{j\in D}(a_{p,j}-a_{q,j})|\leq\bigvee_{t=1}^{\infty}c_{t,\varphi(t)} \text{ whenever } p,q\geq r_1 \text{ and } D\subset[1,l_1]. \tag{4.106}$$

At the second step, let $i_2>r_1$ and $k_2>i_2$ be with

$$|\sum_{j\in A}(a_{i_2,j}-a_{k_2,j})|\not\leq\bigvee_{t=1}^{\infty}d_{t,\varphi(t)}. \tag{4.107}$$

There is an integer $l_2>l_1$ with $|\sum_{j\in C}(a_{i_2,j}-a_{k_2,j})|\leq\bigvee_{t=1}^{\infty}b_{t,\varphi(t)}$ whenever $C\subset[l_2+1,+\infty[$. Then

$$|\sum_{j\in A\cap[l_1+1,l_2]}(a_{i_2,j}-a_{k_2,j})|\not\leq\bigvee_{t=1}^{\infty}e_{t,\varphi(t)}: \tag{4.108}$$

otherwise we have

$$|\sum_{j\in A}(a_{i_2,j}-a_{k_2,j})|\leq|\sum_{j\in A\cap[l_1+1,l_2]}(a_{i_2,j}-a_{k_2,j})|+|\sum_{j\in A,j\leq l_1}(a_{i_2,j}-a_{k_2,j})|+$$

$+|\sum_{j\in A,j>l_2}(a_{i_2,j}-a_{k_2,j})|\leq\bigvee_{t=1}^{\infty}e_{t,\varphi(t)}+\bigvee_{t=1}^{\infty}b_{t,\varphi(t)}+\bigvee_{t=1}^{\infty}c_{t,\varphi(t)}\leq\bigvee_{t=1}^{\infty}d_{t,\varphi(t)}$, which contradicts (4.107). Thus (4.108) is fulfilled. By induction, we find three strictly increasing sequences in $\mathbb{N}$, $(i_r)_r$, $(k_r)_r$, $(l_r)_r$, such that $i_r<k_r<i_{r+1}$ for every $r\in\mathbb{N}$, and

(*j*) $|\sum_{j\in D}(a_{i_r,j}-a_{k_r,j})|\leq\bigvee_{t=1}^{\infty}c_{t,\varphi(t)}$ whenever $D\subset[1,l_{r-1}]$,

(*jj*) $|\sum_{j\in C}(a_{i_r,j}-a_{k_r,j})|\leq\bigvee_{t=1}^{\infty}b_{t,\varphi(t)}$ whenever $C\subset[l_r+1,+\infty[$,

(*jjj*) $|\sum_{j\in F_r}(a_{i_r,j}-a_{k_r,j})|\not\leq\bigvee_{t=1}^{\infty}e_{t,\varphi(t)}$ if $F_r=A\cap[l_{r-1}+1,\ldots,l_r]\in\mathcal{I}_{\text{fin}}$ for each $r\geq 2$.

If we consider the sequence $(F_r)_r$, then by property (M) there are a set $W\in\mathcal{W}$ and an infinite subset $P\subset\mathbb{N}$ with $\bigcup_{r\in P}F_r\subset W\subset\bigcup_{r\in\mathbb{N}}F_r$. The sequence

$((\mathcal{F})\sum_{j\in W} a_{i,j})_i$ is (D)-Cauchy with respect to the regulator $(z_{t,l})_{t,l}$. Note that the quantities $\sum_{j\in W} a_{i,j}$, $i\in\mathbb{N}$, exist in R, and so they coincide with $(\mathcal{F})\sum_{j\in W} a_{i,j}$. Thus the sequence $(\sum_{j\in W} a_{i,j})_i$ is (D)-Cauchy with respect to the regulator $(z_{t,l})_{t,l}$. From this and (j), (jj), for $r\in P$ with r large enough we get

$$|\sum_{j\in F_r}(a_{i_r,j}-a_{k_r,j})|\le|\sum_{j\in W}(a_{i_r,j}-a_{k_r,j})|+$$
$$|\sum_{j\in W, j\le l_{r-1}}(a_{i_r,j}-a_{k_r,j})|+|\sum_{j\in W, j>l_r}(a_{i_r,j}-a_{k_r,j})|$$

$$\le\bigvee_{t=1}^{\infty} z_{t,\varphi(t)}+\bigvee_{t=1}^{\infty} b_{t,\varphi(t)}+\bigvee_{t=1}^{\infty} c_{t,\varphi(t)}\le\bigvee_{t=1}^{\infty} e_{t,\varphi(t)},$$

which contradicts (jjj). Thus for any $A\subset\mathbb{N}$ the sequence $(\sum_{j\in A} a_{i,j})_i$ is (D)-Cauchy with respect to the regulator $(d_{t,l})_{t,l}$, which is independent of A. From this, since Dedekind completeness implies (D)-completeness uniformly with respect to a parameter (see also Boccuto (1997)), we get that the sequence $(\sum_{j\in A} a_{i,j})_i$ is (D)-convergent with respect to a single regulator independent of A, that is 1.37.5). Now for every $i\in\mathbb{N}$ let us define $m_i:\mathcal{P}(\mathbb{N})\to R$ by setting $m_i(A):=\sum_{j\in A} a_{i,j}$, $A\in\mathcal{P}(\mathbb{N})$. We have proved that the sequence $(m_i)_i$ is (RD)-Cauchy and so it is (RD)-convergent. By unconditional convergence of the series $m_i(A):=\sum_{j\in A} a_{i,j}$ and Propositions 1.34 and 1.35, the measure m_i is well-defined and σ-additive for each $i\in\mathbb{N}$. Since the family $(\sum_{j\in A} a_{i,j})_{i\in\mathbb{N}, A\subset\mathbb{N}}$ is equibounded, then the sequence $(m_i)_i$ is equibounded. Note that (RO)- and (RD)-convergence coincide, by Proposition 2.9 of Chapter 2 (see also Boccuto (2003), Theorem 3.4). Thus, by Theorem 1.8 of Chapter 3, the m_i's are uniformly σ-additive, the limit measure m_0 is σ-additive and the sequence $(m_i)_i$ (RD)-converges uniformly to m_0. From this we get 1.37.6). □

Corollary 1.38 *Under the same hypotheses and notations as in Theorem* 1.37, *instead of* (iii) *let us suppose that* $(a_{i,j})_i$ *is* $(O\mathcal{F})$*-convergent for every* $j\in\mathbb{N}$.

Then there exist a sequence $(a_j)_j$ *in* R *and a set* $K \in \mathcal{F}$ *such that*

$$(D)\lim_{i\in K}[\bigvee_{j\in\mathbb{N}} | a_{i,j} - a_j |] = 0 \text{ and } (D)\lim_{j\in\mathbb{N}}[\bigvee_{i\in K} | a_{i,j} |] = 0. \quad \textbf{(4.109)}$$

Moreover, if $\mathcal{J}$ *is the ideal of* $\mathbb{N}^2$ *generated by the finite unions of the Cartesian products of the elements of the dual ideal* $\mathcal{I}$ *of* $\mathcal{F}$, *and* $\mathcal{S}$ *is the dual filter of* $\mathcal{J}$, *then* $(D\mathcal{S})\lim_{i,j} a_{i,j} = 0$.

Proof: By Proposition 2.19 of Chapter 2 there is a set $K \in \mathcal{F}$, such that the double sequence $(a_{i,j})_{i\in K, j\in\mathbb{N}}$ satisfies (iii) with respect to a single regulator. By proceeding analogously as in Theorems 1.33 and 1.37, we get (4.109). The last part follows from this and Lemma 1.32. □

4.2. Filter Exhaustiveness and Convergence Theorems

In this section we deal with the tool of filter/ideal exhaustiveness for measures, investigated in the real case and in the lattice group context in Boccuto, Das, Dimitriou and Papanastassiou (2012), Boccuto and Dimitriou (2011b-c), Boccuto, Dimitriou, Papanastassiou and Wilczyński (2011-2012) and in the topological group setting in Boccuto and Dimitriou (2014a). This tool is very important to get results on the existence of the filter/ideal limit measure (note that, in several versions of convergence theorems, the existence of the limit measure with respect to filter/ideal convergence is assumed as *hypothesis*). We consider a sequence $(m_n)_n$ of measures, and give some results on the existence of the filter limit measures and uniform (s)-boundedness, uniform σ-additivity, uniform absolute continuity and uniform regularity of a subsequence of the type m_n, $n \in M_0$, where M_0 is a suitable element of the involved filter.

4.2.1. Filter Exhaustiveness

Let R be a Dedekind complete (ℓ)-group, $\mathcal{F}$ be a free filter of $\mathbb{N}$, $\Sigma \subset \mathcal{P}(G)$ be a σ-algebra, and $\lambda : \Sigma \to [0,+\infty)$ be a finitely additive measure, such that Σ is separable with respect to the Fréchet-Nikodým topology generated by λ (shortly, λ-*separable*). Let $\{F_j : j \in \mathbb{N}\}$ be a countable λ-dense subset of Σ. Assume that $m_n : \Sigma \to R$, $n \in \mathbb{N}$, is a sequence of finitely additive measures.

We say that $(m_n)_n$ is *λ-$\mathcal{F}$-exhaustive at* $E \in \Sigma$ iff there exists an (O)-sequence $(\sigma_p)_p$ (depending on E) such that for each $p \in \mathbb{N}$ there exist $\delta > 0$ and $V \in \mathcal{F}$ (depending on p and E) with $|m_n(E) - m_n(F)| \leq \sigma_p$ whenever $|\lambda(E) - \lambda(F)| \leq \delta$ and $n \in V$. The sequence $(m_n)_n$ is said to be *λ-$\mathcal{F}$-exhaustive on* Σ iff it is λ-$\mathcal{F}$-exhaustive at E for every $E \in \Sigma$.

The sequence $(m_n)_n$ is *λ-globally $\mathcal{F}$-exhaustive on* Σ iff there is an (O)-sequence $(\sigma_p)_p$ such that for any $p \in \mathbb{N}$ and $E \in \Sigma$ there are $\delta > 0$ and $V \in \mathcal{F}$ (depending on p and E) with $|m_n(E) - m_n(F)| \leq \sigma_p$ whenever $n \in V$ and $F \in \Sigma$ is such that $|\lambda(E) - \lambda(F)| \leq \delta$.

We say that $(m_n)_n$ is *λ-uniformly $\mathcal{F}$-exhaustive on* Σ iff there exists an (O)-sequence $(\sigma_p)_p$ such that for every $p \in \mathbb{N}$ there exist $\delta > 0$ and $V \in \mathcal{F}$ (depending only on p), with $|m_n(E) - m_n(F)| \leq \sigma_p$ whenever $E, F \in \Sigma$ with $|\lambda(E) - \lambda(F)| \leq \delta$ and for any $n \in V$.

We say that $(m_n)_n$ is *λ-weakly $\mathcal{F}$-exhaustive on* Σ iff there is an (O)-sequence $(\sigma_p)_p$ such that for any $p \in \mathbb{N}$ and $E \in \Sigma$ there is a positive real number δ (depending on p and E) with the property that for any $F \in \Sigma$ with $|\lambda(E) - \lambda(F)| \leq \delta$ there is a set $V \in \mathcal{F}$ with $|m_n(E) - m_n(F)| \leq \sigma_p$ whenever $n \in V$.

A sequence $(m_n)_n$ is said to be *λ-exhaustive at* E (resp. *λ-exhaustive, λ-globally exhaustive, λ-uniformly exhaustive on* Σ, *λ-weakly exhaustive on* Σ) iff it is λ-$\mathcal{F}_{\text{cofin}}$-exhaustive at E (resp. λ-$\mathcal{F}_{\text{cofin}}$-exhaustive, λ-globally $\mathcal{F}_{\text{cofin}}$-exhaustive, λ-uniformly $\mathcal{F}_{\text{cofin}}$-exhaustive on Σ, λ-weakly $\mathcal{F}_{\text{cofin}}$-exhaustive on Σ) respectively.

Observe that, when $R = \mathbb{R}$, the concepts of filter exhaustiveness and the related ones coincide with the respective (and related) notions of global filter exhaustiveness.

The following result about filter exhaustive measures extends Diestel (1984), Proposition VII.9 to the filter setting.

Proposition 2.1 (see also Boccuto and Dimitriou (2014d), Proposition 2.2) *If* $\lambda:\Sigma \to [0,+\infty)$ *and* $m_n:\Sigma \to R$, $n \in \mathbb{N}$, *are finitely additive measures, then the following are equivalent:*

(a) $(m_n)_n$ *is* λ-$\mathcal{F}$*-exhaustive at some* $E \in \Sigma$*;*

(b) $(m_n)_n$ *is* λ-$\mathcal{F}$*-exhaustive at* $\varnothing$*;*

(c) $(m_n)_n$ *is* λ*-uniformly* $\mathcal{F}$*-exhaustive on* Σ*;*

(d) $(m_n)_n$ *is* λ*-globally* $\mathcal{F}$*-exhaustive on* Σ.

Proof: (a) $\Rightarrow$ (b) Let $E \in \Sigma$ satisfy (a), $(\sigma_p)_p$ be an (O)-sequence and, in correspondence with $p \in \mathbb{N}$, δ and V be according to λ-$\mathcal{F}$-exhaustiveness of $(m_n)_n$ at E. Choose arbitrarily $A \in \Sigma$, with $\lambda(A) \le \delta$. Since λ is finitely additive and positive, then λ is monotone too, and hence we get

$$\lambda((E \cup A)\Delta E) = \lambda(A \setminus E) \le \lambda(A) \le \delta, \qquad \textbf{(4.110)}$$

$$\lambda((E \setminus A)\Delta E) = \lambda(E \setminus (E \setminus A)) = \lambda(E \cap A) \le \lambda(A) \le \delta.$$

By finite additivity of the m_n's and λ-$\mathcal{F}$-exhaustiveness of $(m_n)_n$ at E, from (4.110) we have

$$| m_n(A) | = | m_n(E \cup A) - m_n(E \setminus A) |$$
$$\le | m_n(E \cup A) - m_n(E) | + | m_n(E) - m_n(E \setminus A) | \le 2\sigma_p$$

for every $n \in V$.

(b) $\Rightarrow$ (c) Let $(\tau_p)_p$ be an (O)-sequence and, in correspondence with $p \in \mathbb{N}$, δ and V be according to λ-$\mathcal{F}$-exhaustiveness of $(m_n)_n$ at $\varnothing$. Pick arbitrarily C, $D \in \Sigma$ with $\lambda(C \Delta D) \le \delta$. Then we have

$$\lambda(C \setminus D) + \lambda(D \setminus C) = \lambda(C \Delta D) \le \delta,$$

and a fortiori $\lambda(C \setminus D) \le \delta$ and $\lambda(D \setminus C) \le \delta$. Thus for every $n \in V$ we get

$$|m_n(C)-m_n(D)|=|m_n(C\setminus D)+m_n(D\setminus C)|\leq|m_n(C\setminus D)|+|m_n(D\setminus C)|\leq 2\tau_p.$$

(c) $\Rightarrow$ (d) $\Rightarrow$ (a) Straightforward. □

We now prove the following result about extensions of filter limit measures, taking values in lattice groups.

Theorem 2.2 (see also Boccuto and Dimitriou (2011b), Theorem 3.3, Boccuto and Dimitriou (2011c), Theorem 3.8) *Under the same hypotheses and notations above, let $(m_n)_n$ be a sequence of finitely additive measures, λ-weakly $\mathcal{F}$-exhaustive on Σ, such that the limit measure $m(F_j):=(RO\mathcal{F})\lim\limits_n m_n(F_j)$, $j\in\mathbb{N}$, exists in R with respect to a single (O)-sequence. Then there is a finitely additive extension $m_0:\Sigma\to R$ of m, with $(RO\mathcal{F})\lim\limits_n m_n=m_0$ with respect to a single (O)-sequence $(b_p)_p$.*

Proof: Let $(\sigma_p)_p$ be an (O)-sequence associated with λ-weak $\mathcal{F}$-exhaustiveness, and choose arbitrarily $E\in\Sigma$. For each $p\in\mathbb{N}$ there exists a $\delta>0$ such that for every $F\in\Sigma$ with $|\lambda(E)-\lambda(F)|\leq\delta$ there is a set $V_F\in\mathcal{F}$ with $|m_n(E)-m_n(F)|\leq\sigma_p$ whenever $n\in V_F$. By virtue of λ-separability of Σ, there is $\overline{j}\in\mathbb{N}$ with $|\lambda(E)-\lambda(F_{\overline{j}})|\leq\delta$. By the Cauchy criterion 2.12 of Chapter 2 there is an (O)-sequence $(\zeta_p)_p$ such that for all j, $p\in\mathbb{N}$ there is a set $W_p^{(j)}\in\mathcal{F}$ with $|m_k(F_j)-m_n(F_j)|\leq\zeta_p$ whenever k, $n\in W_p^{(j)}$. In particular we get

$$\begin{aligned}&|m_k(E)-m_n(E)|\leq\\ &\leq|m_k(E)-m_k(F_{\overline{j}})|+|m_k(F_{\overline{j}})-m_n(F_{\overline{j}})|+|m_n(F_{\overline{j}})-m_n(E)|\leq\\ &2\sigma_p+\zeta_p\end{aligned}$$

for all k, $n\in V_{F_{\overline{j}}}\cap W_p^{(\overline{j})}$. By the Cauchy criterion, this proves the existence of a set function $m_0:\Sigma\to R$, extending m, with $(RO\mathcal{F})\lim\limits_n m_n=m_0$ with respect to a single (O)-sequence $(b_p)_p$. It is not difficult to check that m_0 is finitely additive on Σ. □

The next goal is to give some sufficient conditions on an $(RO\mathcal{F})$-convergent sequence m_n, $n\in\mathbb{N}$, of (ℓ)-group-valued measures for the existence of a set M_0 of $\mathcal{F}$ such that the subsequence m_n, $n\in M_0$, is uniformly (s)-bounded (resp. uniformly σ-additive, uniformly absolutely continuity, uniformly regular). These results yield also a sufficient condition for (s)-boundedness (resp. σ-additivity, absolute continuity, regularity) of the limit measure.

Lemma 2.3 (see also Boccuto and Dimitriou (2011b), Theorem 3.4, Boccuto and Dimitriou (2011c), Lemma 3.9) *Let $\mathcal{F}$ be a P-filter of $\mathbb{N}$, $m_n:\Sigma\to R$, $n\in\mathbb{N}$, be a λ-uniformly $\mathcal{F}$-exhaustive sequence of finitely additive measures, and suppose that*

$$(RO\mathcal{F})\lim_n m_n(F_j) =: m(F_j),\ \ j\in\mathbb{N},$$

with respect to a single (O)-sequence.

Then there are a set $M_0\in\mathcal{F}$ and a finitely additive extension m_0 of m, defined on Σ, with $(O)\lim_{n\in M_0} m_n(E)=m_0(E)$ for all $E\in\Sigma$ with respect to a same (O)-sequence.

Proof: By λ-uniform $\mathcal{F}$-exhaustiveness of m_n, $n\in\mathbb{N}$, there is an (O)-sequence $(\sigma_p)_p$ with the property that for each $p\in\mathbb{N}$ there exist $\delta>0$ and $M'_p\in\mathcal{F}$, with $|m_n(E)-m_n(F)|\le\sigma_p$ whenever $n\in M'_p$ and E, $F\in\Sigma$ with $|\lambda(E)-\lambda(F)|\le\delta$. As $\mathcal{F}$ is a P-filter, in correspondence with the sequence $(M'_p)_p$ there exists a sequence $(M_p)_p$ in $\mathcal{F}$ such that $\bigcap_{p=1}^{\infty} M_p\in\mathcal{F}$ and $M_p\Delta M'_p$ is finite for any $p\in\mathbb{N}$ and. Let $M:=\bigcap_{p=1}^{\infty} M_p$ and $Z_p:=M\setminus M'_p$, $p\in\mathbb{N}$. Then Z_p is finite for any $p\in\mathbb{N}$ and $|m_n(E)-m_n(F)|\le\sigma_p$ for each E, $F\in\Sigma$ that $|\lambda(E)-\lambda(F)|\le\delta$ and $n\in M\setminus Z_p$. Furthermore, thanks to Proposition 2.19 of Chapter 2, there are an (O)-sequence $(w_p)_p$ and $B_0\in\mathcal{F}$ such that for every j, $p\in\mathbb{N}$ there exists $\overline{n}\in B_0$ with $|m_n(F_j)-m(F_j)|\le w_p$ whenever $n\ge\overline{n}$, $n\in B_0$. Without loss of generality, we can take $\overline{n}\in B_0\cap M$. Set $M_0:=B_0\cap M$: we get $M_0\in\mathcal{F}$. The sequence m_n, $n \in M_0$, is λ-uniformly $\mathcal{F}_{\text{cofin}}$-exhaustive and also λ-weakly $\mathcal{F}_{\text{cofin}}$-exhaustive too, and $(O)\lim\limits_{n\in M_0} m_n(F_j) = m(F_j)$ with respect to a single (O)-sequence. From this and Theorem 2.2 applied to m_n, $n \in M_0$ and $\mathcal{F}_{\text{cofin}}$ we deduce the existence of a finitely additive extension m_0 of m, defined on Σ, with $(O)\lim\limits_{n\in M_0} m_n(E) = m_0(E)$ for each $E \in \Sigma$ and with respect to a single (O)-sequence (independent of E). So, M_0 is the requested set. □

Let R be any Dedekind complete (ℓ)-group and $\mathcal{F}$ be any free filter of $\mathbb{N}$. We say that a sequence of finitely additive measures $m_n : \Sigma \to R$, $n \geq 0$, together with λ, fulfils *property* (*) *with respect to R and* $\mathcal{F}$ iff it is λ-uniformly $\mathcal{F}$-exhaustive on Σ and $(O\mathcal{F})\lim\limits_{n} m_n = m_0$ with respect to a single (O)-sequence.

4.2.2. Stone Extensions and Equivalence Results Between Limit Theorems

The next step is to prove some equivalence results between different kinds of filter limit theorems for measures, taking values in (ℓ)-groups. Similar equivalence results were given by Drewnowski in (1972b), in the topological group setting. In particular, when it is proved that the Nikodým convergence theorem implies the Brooks-Jewett theorem, countably additive restrictions of finitely additive (s)-bounded topological group-valued measures are considered. However in lattice group setting, in order to relate finitely and countably additive measures, it is not advisable to use an approach of this kind. Indeed, in topological groups, the involved convergences fulfil some suitable properties (for example property $(\mathcal{U})$, which we will see in the sequel), which are not always satisfied by order convergence in (ℓ)-groups, because in general lattice groups does not have a topological nature. So, to prove our results, we use the Stone Isomorphism technique, by means of which it is possible to construct a (globally) σ-additive extension of a finitely additive (globally) (s)-bounded measure, and to study the properties of the starting measures in relation with the corresponding ones of the considered extensions. Moreover, to prove that the Brooks-Jewett theorem implies the Nikodým theorem, when we relate uniform (s)-boundedness to σ-additivity, when these concepts are intended not necessarily with respect to a same order sequence, in general for technical reasons it is not advisable to consider a direct approach, and we use the Maeda-Ogasawara-Vulikh representation theorem for

Dedekind complete (ℓ)-groups, studying the related properties of the corresponding real-valued measures. When we deal with a single (O)-sequence, it is possible to give direct proofs, and it is not always advisable to use the tool of the Maeda-Ogasawara-Vulikh theorem. We begin with the following

Lemma 2.4 (see also Boccuto and Dimitriou (2013c)) *Let $m_n : \Sigma \to R$, $n \in \mathbb{N}$, be a λ-uniformly $\mathcal{F}$-exhaustive sequence of finitely additive measures, and assume that*

$$(O\mathcal{F})\lim_n m_n(F_j) =: m(F_j) \quad \textit{for every } j \in \mathbb{N}$$

with respect to a single (O)-sequence. Then there exist a set $M_0 \in \mathcal{F}$ and a finitely additive measure $m_0 : \Sigma \to R$, which extends m and such that:

α) the measures m_n, $n \in M_0$, and m_0 satisfy together with λ property (*) with respect to R and $\mathcal{F}_{\text{cofin}}$;

$\alpha\alpha$) if Ω is as in the Maeda-Ogasawara-Vulikh representation theorem, then there is a meager set $N_0 \subset \Omega$ such that for each $\omega \in \Omega \setminus N_0$ the real-valued measures $m_n(\cdot)(\omega)$, $n \in M_0$, and $m_0(\cdot)(\omega)$, satisfy together with λ property (*) with respect to $\mathbb{R}$ and $\mathcal{F}_{\text{cofin}}$.

Proof: α) See Lemma 2.3.

$\alpha\alpha$) Let m_0, M_0 and $(\sigma_p)_p$ be associated with property (*) as in α), and $(b_p)_p$ be an (O)-sequence as in Theorem 2.2. By α) and the Maeda-Ogasawara-Vulikh representation theorem, there is a meager set $N_0 \subset \Omega$ with the property that the sequences $(b_p(\omega))_p$ and $(\sigma_p(\omega))_p$ are (O)-sequences in $\mathbb{R}$ for all $\omega \in \Omega \setminus N_0$, and in correspondence with $p \in \mathbb{N}$ and $E \in \Sigma$ there exists a natural number $\overline{n} \in M_0$ such that

$$|m_n(E)(\omega) - m_0(E)(\omega)| \le b_p(\omega) \text{ for each } n \ge \overline{n},\ n \in M_0 \text{ and } \omega \in \Omega \setminus N_0. \quad \textbf{(4.111)}$$

Moreover, since the sequence m_n, $n \in M_0$, is λ-uniformly $\mathcal{F}_{\text{cofin}}$-exhaustive and thanks to property α), for every $p \in \mathbb{N}$ there exist $Z_p \in \mathcal{I}_{\text{fin}}$ and $\delta > 0$ such that

$$| m_n(E)(\omega) - m_n(F)(\omega) | \leq \sigma_p(\omega) \tag{4.112}$$

for each E, $F \in \Sigma$, $| \lambda(E) - \lambda(F) | \leq \delta$, $n \in M \setminus Z_p$ and $\omega \in \Omega \setminus N_0$. Thanks to (4.111) and (4.112), we get $\lim_{n \in M_0} m_n(E)(\omega) = m(E)(\omega)$ for any $\omega \in \Omega \setminus N_0$, $E \in \Sigma$, and the set functions $m_n(\omega)$, $n \in M_0$, are uniformly $\mathcal{F}_{\text{cofin}}$-exhaustive. This ends the proof of $\alpha\alpha$). □

We use the tool of the Stone extension, in order to prove equivalence results between limit theorems.

Assume that R is a weakly σ-distributive and super Dedekind complete lattice group, $\lambda : \Sigma \to [0, +\infty]$, $m_n : \Sigma \to R$, $n \in \mathbb{N}$, are finitely additive measures, $\mathcal{Q}$ is the algebra of the open-closed sets of the Stone space Q^* associated with Σ, $\sigma(\mathcal{Q})$ is the σ-algebra generated by $\mathcal{Q}$, $\psi : \Sigma \to \mathcal{Q}$ is the Stone isomorphism. Let $(\lambda \circ \psi^{-1})(E) = \lambda(\psi^{-1}(E))$, $(m_n \circ \psi^{-1})(E) = m_n(\psi^{-1}(E))$, $n \in \mathbb{N}$, $E \in \mathcal{Q}$. The measures $\lambda \circ \psi^{-1}$ and $m_n \circ \psi^{-1}$ are globally σ-additive, and have unique globally σ-additive extensions P, P_n respectively, defined on $\sigma(\mathcal{Q})$, which are the *Stone extensions* of λ and m_n, respectively (see also Chapter 2). We now show that the Stone extensions satisfy property (*).

Theorem 2.5 (see Boccuto and Dimitriou (2013c), Theorem 3.2) *Let $\lambda : \Sigma \to \mathbb{R}$, $m_n : \Sigma \to R$, $n \geq 0$, be equibounded finitely additive measures, which fulfil property (*) with respect to R and $\mathcal{F}$ together with λ.*

Then $\sigma(\mathcal{Q})$ is P-separable, and the measures $P : \sigma(\mathcal{Q}) \to \mathbb{R}$, $P_n : \sigma(\mathcal{Q}) \to R$ fulfil property (*) with respect to R and $\mathcal{F}$ together with P.

Proof: We prove P-separability of $\sigma(\mathcal{Q})$. Choose arbitrarily $\varepsilon > 0$. By a density property (see also Theorem 3.27 of Chapter 2), for any $A \in \sigma(\mathcal{Q})$ there is $E \in \mathcal{Q}$ with $| P(A) - P(E) | \leq \frac{\varepsilon}{2}$. By λ-separability of Σ, there is a countable and λ-

dense set $\{F_j : j \in \mathbb{N}\}$ contained in Σ, and $\mathcal{P} := \{\psi(F_j) : j \in \mathbb{N}\}$ is a countable and P-dense set contained in $\mathcal{Q}$. Thus a positive integer j can be found, with the property that $| P(E) - P(\psi(F_j)) | \leq \frac{\varepsilon}{2}$, and hence $| P(A) - P(\psi(F_j)) | \leq \varepsilon$. Thus we get also that $\mathcal{P}$ is a countable and P-dense set, contained in $\sigma(\mathcal{Q})$.

We now prove P-uniform $\mathcal{F}$-exhaustiveness of the measures P_n, $n \in \mathbb{N}$. Let $(\sigma_p)_p$ be an (O)-sequence associated with λ-uniform $\mathcal{F}$-exhaustiveness of $(m_n)_n$, pick $p \in \mathbb{N}$ and let $\delta > 0$ and $D \in \mathcal{F}$ be associated with p.

Again thanks to Theorem 3.27 of Chapter 2, we get that for each $n \in \mathbb{N}$ there exists a regulator $(a_{t,l}^{(n)})_{t,l}$, such that for any $\varphi \in \mathbb{N}^{\mathbb{N}}$, A_1, $A_2 \in \sigma(\mathcal{Q})$ there exist $E_1^{(n)}$, $E_2^{(n)} \in \mathcal{Q}$ with

$$| P(E_s^{(n)}) - P(A_s) | \leq \frac{\delta}{3}, \quad | P_n (E_s^{(n)}) - P_n (A_s) | \leq \bigvee_{t=1}^{\infty} a_{t,\varphi(t+n)}^{(n)}, \; s = 1,2.$$

By the Fremlin Lemma applied to $u := \bigvee_{A \in \sigma(\mathcal{Q}), n \in \mathbb{N}} | P_n (A) |$ and the regulators $(a_{t,l}^{(n)})_{t,l}$, $n \in \mathbb{N}$, there is a regulator $(a_{t,l})_{t,l}$ in R such that

$$(2u) \wedge \left(\sum_{n=1}^{q} \left(\bigvee_{t=1}^{\infty} a_{t,\varphi(t+n)}^{(n)}\right)\right) \leq \bigvee_{t=1}^{\infty} a_{t,\varphi(t)} \text{ for each } q \in \mathbb{N}, \; \varphi \in \mathbb{N}^{\mathbb{N}},$$

and thus

$$| P_n (E_s^{(n)}) - P_n (A_s) | \leq \bigvee_{t=1}^{\infty} a_{t,\varphi(t)}, \; s = 1,2.$$

By weak σ-distributivity and super Dedekind completeness of R and Theorem 2.2 of Chapter 2 there is an (O)-sequence $(v_p)_p$, with the property that for each $p \in \mathbb{N}$ there is $\varphi_p \in \mathbb{N}^{\mathbb{N}}$ with

$$\bigvee_{t=1}^{\infty} a_{t,\varphi_p(t)} \leq v_p.$$

So for every p, $n \in \mathbb{N}$ and A_1, $A_2 \in \sigma(\mathcal{Q})$ there are $E_1^{(n)}$, $E_2^{(n)} \in \mathcal{Q}$ such that

$$|P_n(E_s^{(n)}) - P_n(A_s)| \le v_p,\ s = 1,2. \tag{4.113}$$

Furthermore we have

$$\begin{aligned}&|P(E_1^{(n)}) - P(E_2^{(n)})|\\ &\le |P(E_1^{(n)}) - P(A_1)| + |P(A_1) - P(A_2)| + |P(A_2) - P(E_2^{(n)})| \le \frac{3\delta}{3} = \delta,\end{aligned}$$

that is

$$|\lambda(\psi^{-1}(E_1^{(n)})) - \lambda(\psi^{-1}(E_2^{(n)}))| \le \delta.$$

Thus we get

$$\begin{aligned}&|P_n(A_1) - P_n(A_2)| \le\\ &|P_n(E_1^{(n)}) - P_n(A_1)| + |P_n(E_1^{(n)}) - P_n(E_2^{(n)})| + |P_n(A_2) - P_n(E_2^{(n)})| =\\ &= |P_n(E_1^{(n)}) - P_n(A_1)| + |m_n(\psi^{-1}(E_1^{(n)})) - m_n(\psi^{-1}(E_2^{(n)}))| +\\ &+ |P_n(A_2) - P_n(E_2^{(n)})| \le \sigma_p + 2v_p.\end{aligned}$$

Put $w_p := \sigma_p + 2v_p$, $p \in \mathbb{N}$. Observe that $(w_p)_p$ is an (O)-sequence, and for every $p \in \mathbb{N}$ there are $\delta > 0$, $D \in \mathcal{F}$ with the property that $|P_n(A_1) - P_n(A_2)| \le w_p$ for each $A_1, A_2 \in \Sigma(\mathcal{Q})$ such that $|P(A_1) - P(A_2)| \le \frac{\delta}{3}$ and $n \in D$.

So λ-uniform $\mathcal{F}$-exhaustiveness of $(m_n)_n$ implies P-uniform $\mathcal{F}$-exhaustiveness of $(P_n)_n$.

We now prove that

$$(O\mathcal{F})\lim_n P_n(A) = P_0(A) \text{ for every } A \in \sigma(\mathcal{Q}) \tag{4.114}$$

with respect to a single (O)-sequence. Since the set $\{\psi(F_j): j \in \mathbb{N}\}$ is countable and P-dense in $\sigma(\mathcal{Q})$, thanks (4.114) and Theorem 2.2 applied to $(P_n)_n$ we find an (O)-sequence $(\xi_p)_p$ and $M_0^* \in \mathcal{F}(\mathcal{I})$, with

$$(O) \lim_{n \in M_0^*} P_n(A) = P_0(A) \text{ for any } A \in \sigma(\mathcal{Q})$$

relatively to $(\xi_p)_p$. From this and Proposition 2.15 of Chapter 2 (see also Boccuto and Dimitriou (2013c), Proposition 2.4) we obtain (4.114). □

We now deal with equivalence between filter limit theorems in the (ℓ)-group setting. We begin with the following Brooks-Jewett-type theorem.

Theorem 2.6 (see also Boccuto and Candeloro (2004a), Theorem 6.8; Boccuto and Candeloro (2010), Theorem 3.1; Boccuto and Dimitriou (2011c), Theorem 3.4; Boccuto and Dimitriou (2013c), Theorem 3.3) *Assume that R is a Dedekind complete lattice group, and $m_n : \Sigma \to R$, $n \in \mathbb{N}$, is an equibounded sequence of finitely additive measures, (O)-convergent pointwise on Σ with respect to a single (O)-sequence to $m_0 : \Sigma \to R$.*

Then there is a meager set $N \subset \Omega$ such that the measures $m_n(\cdot)(\omega)$, $n \in \mathbb{N}$, are uniformly (s)-bounded on Σ for $\omega \in \Omega \setminus N$, and $m_0(\cdot)(\omega)$ is (s)-bounded. Furthermore the measures m_n are uniformly (s)-bounded and m_0 is (s)-bounded on Σ.

Finally, when R is weakly σ-distributive and super Dedekind complete and the measures m_n are globally (s)-bounded, then the measures m_n are globally uniformly (s)-bounded.

Now we prove the next limit theorems for lattice group-valued measures and their equivalence in the context of filter convergence (see for instance Boccuto and Dimitriou (2011c), Theorem 3.10).

Theorem 2.7 (Brooks-Jewett (BJ), see Boccuto and Dimitriou (2013c), Theorem 3.4) *Suppose that $\mathcal{F}$ is a P-filter of $\mathbb{N}$, R is a Dedekind complete lattice group, $\lambda : \Sigma \to [0,+\infty]$ is finitely additive, Σ is λ-separable, $m_0 : \Sigma \to R$, $m_n : \Sigma \to R$,*

$n \in \mathbb{N}$, is an equibounded sequence of finitely additive measures, fulfilling property (*) with respect to R and $\mathcal{F}$ together with λ.

Then we get the following results.

I) *There is $M_0 \in \mathcal{F}$ such that the measures m_n, $n \in M_0$, are uniformly (s)-bounded on Σ.*

II) *If R is weakly σ-distributive and super Dedekind complete and the measures m_n are globally (s)-bounded, then we can take M_0 such that the measures m_n, $n \in M_0$, are globally uniformly (s)-bounded on Σ.*

Theorem 2.8 (Vitali-Hahn-Saks (VHS), see Boccuto and Dimitriou (2013c), Theorem 3.5) *With the same notations and assumptions as in Theorem 2.7, assume that τ is a Fréchet-Nikodým topology on Σ. Then the following results hold.*

I) *If the measures m_n are τ-continuous, then a set $M_0 \in \mathcal{F}$ can be found, with the property that the measures m_n, $n \in M_0$, are uniformly τ-continuous.*

II) *If R is weakly σ-distributive and super Dedekind complete and the measures m_n are globally τ-continuous and globally (s)-bounded, then we can choose M_0 in such a way that the measures m_n, $n \in M_0$, are globally uniformly τ-continuous.*

Theorem 2.9 (Nikodým (N), see Boccuto and Dimitriou (2013c), Theorem 3.6)

I) *Under the same notations as above, if the measures m_n, $n \in \mathbb{N}$, are σ-additive, then there is $M_0 \in \mathcal{F}$, such that the measures m_n, $n \in M_0$, are uniformly σ-additive on Σ.*

II) *If R is weakly σ-distributive and super Dedekind complete and the measures m_n are globally σ-additive, then we can take M_0 in such a way that the measures m_n, $n \in M_0$, are globally uniformly σ-additive.*

Theorem 2.10 (Dieudonné (D), see Boccuto and Dimitriou (2013c), Theorem 3.7) *Under the same notations as in Theorem 2.7, we have the following results.*

I) *If the measures* m_n *are regular, then there exists* $M_0 \in \mathcal{F}$, *such that the measures* m_n, $n \in M_0$, *are uniformly regular and uniformly* (s)*-bounded.*

II) *If R is weakly* σ*-distributive and super Dedekind complete and the measures* m_n *are globally regular and globally* (s)*-bounded, then we can choose* M_0 *in order that the measures* m_n, $n \in M_0$, *are globally uniformly regular and globally uniformly* (s)*-bounded.*

Concerning Theorem 2.7 (BJ), note that there are $M_0 \in \mathcal{F}$ and $N_0 \subset \Omega$, which fulfil the thesis of Lemma 2.4. (BJ) follows from Theorem 1.18 applied to N_0 and to the sequence m_n, $n \in M_0$.

Now we show the equivalence between the theorems (BJ) II), (VHS) II), (N) II) and (D) II).

We prove (BJ) II) $\Rightarrow$ (VHS) II). Pick an equibounded sequence $m_n : \Sigma \to R$, $n \in \mathbb{N}$, of finitely additive, globally τ-continuous and globally (s)-bounded measures, which fulfil property (*) with respect to R and $\mathcal{F}$ together with λ. By (BJ) II), there is a set $M_0 \in \mathcal{F}$ such that the measures m_n, $n \in M_0$, are globally uniformly (s)-bounded, and hence there is an (O)-sequence $(\sigma_p)_p$, such that $(O) \lim_k [\bigvee_{n \in M_0} v(m_n)(C_k)] = 0$ for every disjoint sequence $(C_k)_k$ in Σ and with respect to $(\sigma_p)_p$. Choose a decreasing sequence $(H_k)_k$ from Σ, with τ-$\lim_k H_k = \emptyset$. From global τ-continuity of m_n, $n \in \mathbb{N}$, it follows that $(O) \lim_k v(m_n)(H_k) = 0$ for each $n \in \mathbb{N}$, with respect to a single (O)-sequence $(\zeta_p)_p$.

Thanks to Lemma 3.3 of Chapter 3, we get $(O)\lim_k (\bigvee_{n \in M_0} v(m_n)(H_k)) = 0$ with respect to $(\zeta_p)_p$, and hence uniform τ-continuity of the measures m_n, $n \in M_0$. Thus, (BJ) II) implies (VHS) II).

The implication (BJ) II) $\Rightarrow$ (D) II) can be proved similarly as (BJ) II) $\Rightarrow$ (VHS) II).

Now we turn to the implication (VHS) II) $\Rightarrow$ (N) II). Let $m_n : \Sigma \to R$, $n \in \mathbb{N}$, be a sequence of globally σ-additive measures, take the Fréchet-Nikodým topology τ generated by all order continuous submeasures, and pick a decreasing sequence $(H_k)_k$ in Σ such that $\tau\text{-}\lim_k H_k = \varnothing$ and $H = \bigcap_{k=1}^{\infty} H_k$. Then we get $\eta(H) = 0$ for each order continuous submeasure η, which implies that $H = \varnothing$. Thus the measures m_n are globally τ-continuous. As they are also globally (s)-bounded, then, thanks to (VHS) II), they are also globally uniformly τ-continuous, and hence globally uniformly σ-additive too.

(N) II) $\Rightarrow$ (BJ) II). Let $m_n : \Sigma \to R$, $n \in \mathbb{N}$, be an equibounded sequence of finitely additive globally (s)-bounded measures, which fulfil property (*) with respect to R and $\mathcal{F}$ together with λ. Let $P_n : \sigma(\mathcal{Q}) \to R$, $n \in \mathbb{N}$, $P: \sigma(\mathcal{Q}) \to R$, be the Stone extensions of m_n, λ, respectively. We get that $\sigma(\mathcal{Q})$ is P-separable, and for every $n \in \mathbb{N}$, P_n is a σ-additive measure, which fulfil property (*) with respect to R and $\mathcal{F}$ together with P. By virtue of (N) II), there exists a set $M_0 \in \mathcal{F}$, such that P_n, $n \in M_0$, is globally uniformly σ-additive, and therefore globally uniformly (s)-bounded. Thus the measures m_n, $n \in M_0$, are globally uniformly (s)-bounded on Σ.

(D) II) $\Rightarrow$ (BJ) II). Let Q be the Stone space associated with Σ and $\mathcal{G}^*$, $\mathcal{H}^* \subset \sigma(\mathcal{Q})$ be the sublattices of the open and of the closed subsets of Q, respectively. It is easy to check that the complement (with respect to Q) of every element of $\mathcal{H}^*$ belongs to $\mathcal{G}^*$, and that $\mathcal{G}^*$ is closed under countable disjoint unions.

Let $m_n : \Sigma \to R$, $n \ge 0$, be any equibounded sequence of globally (s)-bounded and finitely additive measures which fulfil property (*) with respect to R and $\mathcal{F}$ together with λ, and denote by $P_n : \sigma(\mathcal{Q}) \to R$ and $P: \sigma(\mathcal{Q}) \to [0,+\infty]$ be the Stone extensions of m_n and λ, respectively. Arguing analogously as in Theorem 3.1 of Chapter 2, it is possible to check that the measures P_n are globally regular on $\sigma(\mathcal{Q})$ (see also Boccuto and Dimitriou (2013c)). By proceeding similarly as in the implication (N) II) $\Rightarrow$ (BJ) II), thanks to (D) II), we find a set $M_0 \in \mathcal{F}$, such that P_n, $n \in M_0$, is globally uniformly (s)-bounded and globally uniformly regular on $\sigma(\mathcal{Q})$. This implies that m_n, $n \in M_0$, is globally uniformly (s)-bounded on Σ.

Now we claim that the theorems (BJ) I), (VHS) I), (N) I) and (D) I) are equivalent.

First, we prove the implication (BJ) I) $\Rightarrow$ (VHS) I). Let R be a Dedekind complete lattice group, Ω be as in the Maeda-Ogasawara-Vulikh representation theorem and $m_n : \Sigma \to R$, $n \geq 0$, be an equibounded sequence of finitely additive and τ-continuous measures, which fulfil property (*) with respect to R and $\mathcal{F}$ together with λ. Thanks to Lemma 2.4, $\alpha\alpha$), we find a meager set $N_0 \subset \Omega$ and a set $M_0 \in \mathcal{F}$ such that for every $\omega \in \Omega \setminus N_0$ the real-valued measures $m_n(\cdot)(\omega)$, $n \in M_0$, and $m_0(\cdot)(\omega)$ fulfil property (*) with respect to $\mathbb{R}$ and $\mathcal{F}_{\text{cofin}}$ together with λ. By applying (BJ) I) to $R = \mathbb{R}$ and $\mathcal{F} = \mathcal{F}_{\text{cofin}}$, we get that for every $\omega \in \Omega \setminus N_0$ there is a set $M_0^{(\omega)} \subset M_0$, such that $M_0 \setminus M_0^{(\omega)}$ is finite and $m_n(\omega)$, $n \in M_0^{(\omega)}$, are uniformly (s)-bounded, that is

$$\lim_k \Big(\sup_{n \in M_0^{(\omega)}} \Big[\sup_{A \in \Sigma, A \subset C_k} |m_n(A)(\omega)| \Big] \Big) = 0 \qquad \textbf{(4.115)}$$

for each disjoint sequence $(C_k)_k$ of elements of Σ. Choose $\varepsilon > 0$, $\omega \in \Omega \setminus N_0$ and a disjoint sequence $(C_k)_k$ from Σ. Thanks to (4.115) there is $k_0 \in \mathbb{N}$ such that

$$\sup_{A \in \Sigma, A \subset C_k} |m_n(A)(\omega)| \leq \varepsilon \qquad \textbf{(4.116)}$$

for each $k \geq k_0$ and $n \in M_0^{(\omega)}$. As the measures $m_n(\omega)$, $\omega \in \Omega \setminus N_0$, $n \in M_0 \setminus M_0^{(\omega)}$ are (s)-bounded and since $M_0 \setminus M_0^{(\omega)}$ is finite, then for every $n \in M_0 \setminus M_0^{(\omega)}$ there exists $k_n \in \mathbb{N}$ such that

$$\sup_{A \in \Sigma, A \subset C_k} |m_n(A)(\omega)| \leq \varepsilon \qquad \textbf{(4.117)}$$

whenever $k \geq k_n$. Let $k^* := \max\{k_0, k_n : n \in M_0 \setminus M_0^{(\omega)}\}$. By (4.116) and (4.117) we get

$$\sup_{A\in\Sigma, A\subset C_k} |m_n(A)(\omega)| \le \varepsilon \text{ for any } k \ge k^* \text{ and } n\in M_0.$$

So, $m_n(\cdot)(\omega)$, $\omega\in N\setminus N_0$, $n\in M_0$, is uniformly (s)-bounded.

Choose now a decreasing sequence $(H_k)_k$ in Σ, such that τ-$\lim_k H_k = \emptyset$. By τ-continuity of m_n, $n\in\mathbb{N}$, $(O)\lim_k v(m_n)(H_k) = 0$ for each $n\in\mathbb{N}$. By the Maeda-Ogasawara-Vulikh theorem there exists a meager set $N'\subset\Omega$, $N'\supset N_0$, depending on $(H_k)_k$, with

$$\lim_k [\sup_{A\in\Sigma, A\subset H_k} |m_n(A)(\omega)|] = \inf_k [\sup_{A\in\Sigma, A\subset H_k} |m_n(A)(\omega)|] = 0 \text{ for every } \omega\in\Omega\setminus N'. \quad \textbf{(4.118)}$$

Thanks to Lemma 3.2 of Chapter 3, we get

$$\lim_k (\sup_{n\in M_0} [\sup_{A\in\Sigma, A\subset H_k} |m_n(A)(\omega)|]) = \inf_k (\sup_{n\in M_0} [\sup_{A\in\Sigma, A\subset H_k} |m_n(A)(\omega)|]) = 0 \quad \textbf{(4.119)}$$

for each $\omega\in\Omega\setminus N'$. By (4.119) and the Maeda-Ogasawara-Vulikh theorem, we find a meager set $N''\subset\Omega$, $N''\supset N'$, such that

$$[\bigwedge_k (\bigvee_{n\in M_0} [\bigvee_{A\in\Sigma, A\subset H_k} |m_n(A)|])](\omega) = 0$$

for each $\omega\in\Omega\setminus N''$. By a density argument, we obtain

$$0 = \bigwedge_k (\bigvee_{n\in M_0} [\bigvee_{A\in\Sigma, A\subset H_k} |m_n(A)|]) = \bigwedge_k (\bigvee_{n\in M_0} v(m_n)(H_k)) = (O)\lim_k (\bigvee_{n\in M_0} v(m_n)(H_k)).$$

Thus the measures m_n's, $n\in M_0$, are uniformly τ-continuous.

The implication (BJ) I) $\Rightarrow$ (D) I) (resp. (VHS) I) $\Rightarrow$ (N) I)) can be proved similarly as (BJ) I) $\Rightarrow$ (VHS) I) (resp. (VHS) II) $\Rightarrow$ (N) II)).

(N) I) $\Rightarrow$ (BJ) I). Let $m_n : \Sigma \to R$, $n\in\mathbb{N}$, be an equibounded sequence of finitely additive measures, which fulfils property (*) with respect to R and $\mathcal{F}$ together with λ. Let Ω be as in the Maeda-Ogasawara-Vulikh representation theorem.

Thanks to Lemma 2.4 $\alpha\alpha$), a meager set $N_* \subset \Omega$ and a set $M_0 \in \mathcal{F}$ can be found, with the property that for each $\omega \in \Omega \setminus N_*$ the measures $m_n(\cdot)(\omega)$, $n \in M_0$, are (s)-bounded and fulfil property (*) with respect to $\mathbb{R}$ and $\mathcal{F}_{\text{cofin}}$ together with λ.

In correspondence with the σ-algebra Σ we associate the Stone space Q^* and the algebra $\mathcal{Q}$ of the open-closed subsets of Q^*. For every $n \geq 0$ and $\omega \in \Omega \setminus N_*$, let P, $P_{(n,\omega)}: \sigma(\mathcal{Q}) \to \mathbb{R}$ be the Stone extensions of $m_n(\cdot)(\omega)$, λ respectively.

Pick $\omega \in \Omega \setminus N_*$. Thanks to (N) used with $\mathcal{F}_{\text{cofin}}$, $\mathbb{R}$ and $\sigma(\mathcal{Q})$, there is a set $M_0^{(\omega)} \subset M_0$ such that $M_0 \setminus M_0^{(\omega)}$ is finite and $P_{(n,\omega)}, n \in M_0^{(\omega)}$, are uniformly σ-additive. Arguing analogously as in the implication (N) II) $\Rightarrow$ (BJ) II), we get that the measures $m_n(\cdot)(\omega)$, $n \in M_0^{(\omega)}$, defined on Σ, are uniformly (s)-bounded. So we have

$$\lim_k \Big(\sup_{n \in M_0^{(\omega)}} \Big[\sup_{A \in \Sigma, A \subset C_k} |m_n(A)(\omega)| \Big] = 0 \qquad \textbf{(4.120)}$$

for each disjoint sequence $(C_k)_k$ of elements of Σ. By (4.120), proceeding similarly as in (4.116) and (4.117), we get

$$\lim_k \Big(\sup_{n \in M_0} \Big[\sup_{A \in \Sigma, A \subset C_k} |m_n(A)(\omega)| \Big] = 0. \qquad \textbf{(4.121)}$$

By arbitrariness of $\omega \in \Omega \setminus N_*$, the Maeda-Ogasawara-Vulikh theorem and a density argument, proceeding similarly as in the implication (BJ) I) $\Rightarrow$ (VHS) I), from (4.121) we get

$$(O) \lim_k \Big[\bigvee_{n \in M_0} v(m_n)(C_k)\Big] = 0.$$

Since $(C_k)_k$ is arbitrary, the measures m_n, $n \in M_0$, are uniformly (s)-bounded. This proves that (N) I) $\Rightarrow$ (BJ) I).

(D) I) $\Rightarrow$ (BJ) I) With the same notations as above and by arguing analogously as in the implication (N) I) $\Rightarrow$ (BJ) I), we can see that the measures $P_{(n,\omega)}$, $n \in M_0^{(\omega)}$, are σ-additive, and also (s)-bounded and regular, since it is enough to proceed analogously as in the implication (D) II) $\Rightarrow$ (BJ) II), by taking $R=\mathbb{R}$. Thanks to (D) I), the set functions $P_{(n,\omega)}$, $n\in M_0^{(\omega)}$, are uniformly (s)-bounded and uniformly regular. This implies that the measures $m_n(\cdot)(\omega)$, $n\in M_0^{(\omega)}$, are uniformly (s)-bounded. Proceeding as in (4.116) and (4.117), we obtain uniform (s)-boundedness of $m_n(\cdot)(\omega)$, $n\in M_0$. By arguing similarly as in the implication (N) I) $\Rightarrow$ (BJ) I), we obtain uniform (s)-boundedness of m_n, $n\in M_0$. This ends the proof. □

In general, λ-uniform $\mathcal{F}$-exhaustiveness cannot be eliminated. Indeed we have the following

Example 2.11 (see also Boccuto and Dimitriou (2011c), Example 3.11, Boccuto and Dimitriou (2011c), Remark 3.8 (c)) Let $\Sigma=\mathcal{P}(\mathbb{N})$, $R=\mathbb{R}$, $\mathcal{F}$ be a free filter of $\mathbb{N}$ which is not an ultrafilter (for example, we can take the filter $\mathcal{F}_{\text{st}}$ of all subsets of $\mathbb{N}$ of asymptotic density one), and let us define $\lambda:\Sigma\to\mathbb{R}$ by setting $\lambda(A)=\sum_{n\in A}\frac{1}{2^n}$, $A\in\Sigma$. It is not difficult to check that λ is σ-additive, and Σ is λ-separable (note that $\mathcal{I}_{\text{fin}}$ is countable and dense in Σ, if we consider the Fréchet-Nikodým topology associated with λ). Let δ_n, $n\in\mathbb{N}$, be the Dirac measures: for each $A\subset\mathbb{N}$, $n\in\mathbb{N}$ let $\delta_n(A)=1$ when $n\in A$ and $\delta_n(A)=0$ when $n\in\mathbb{N}\setminus A$. We observe that δ_n is σ-additive for each $n\in\mathbb{N}$, and $\lim_n \delta_n(W)=0$ (and a fortiori $(\mathcal{F})\lim_n \delta_n(W)=0$ for every $W\in\mathcal{I}_{\text{fin}}$.

For each $\delta>0$ there is a cofinite set $Z_\delta\subset\mathbb{N}$ with $\lambda(Z_\delta)<\delta$ and

$$|\lambda(E)-\lambda(F)|=\lambda(E\Delta F)\le\lambda(Z_\delta)<\delta \text{ for each } E,\ F\in\Sigma \text{ such that } E\cup F\subset Z_\delta.$$

Note that for every infinite subset $M\subset N$, and a fortiori for each $M\in\mathcal{F}$, there are $\overline{n}\in M$ and E, $F\in\Sigma$ such that $E\cup F\subset Z_\delta$ and $\overline{n}\in E\setminus F$, and hence $|\delta_{\overline{n}}(E)-\delta_{\overline{n}}(F)|=\delta_{\overline{n}}(E)=1$. So the measure sequence δ_n, $n\in\mathbb{N}$, is not λ-uniformly $\mathcal{F}$-exhaustive. Furthermore note that, since $\mathcal{F}$ is not an ultrafilter, there exists a set $C \subset \mathbb{N}$ with $C \notin \mathcal{F}$ and $\mathbb{N} \setminus C \notin \mathcal{F}$. Thus $\delta_n(C) = 1$ when $n \in C$ and $\delta_n(C) = 0$ when $n \notin C$. Choose $l \neq 0$ and let $\varepsilon_0 := \frac{|l|}{2} > 0$. In this case we get

$|\delta_n(C) - l| = |l| > \varepsilon_0$ for every $n \in \mathbb{N} \setminus C$.

Thus

$B := \{n \in \mathbb{N} : |\delta_n(C) - l| \leq \varepsilon_0\} \notin \mathcal{F}$,

since B is contained in C and $C \notin \mathcal{F}$. When $l = 0$, set $\varepsilon_0 = \frac{1}{2}$. Then $|\delta_n(C)| = 1 > \varepsilon_0$ for each $n \in C$, and so we have

$Y := \{n \in \mathbb{N} : |\delta_n(C)| \leq \varepsilon_0\} \notin \mathcal{F}$,

since Y is contained in $\mathbb{N} \setminus C$ and $\mathbb{N} \setminus C \notin \mathcal{F}$. Thus, $(\mathcal{F}) \lim_n \delta_n(C)$ does not exist in $\mathbb{R}$. We get $\sup_{n \in M} \delta_n(\{k\}) = 1$ for every infinite subset $M \subset \mathbb{N}$ and $k \in M$. So the measures δ_n, $n \in M$, are not uniformly (s)-bounded.

Without requiring filter exhaustiveness, it is possible to prove the following theorem on the existence of the filter limit measure.

Theorem 2.12 (see Boccuto and Dimitriou (2011b), Theorem 3.5) *Let R be a super Dedekind complete and weakly σ-distributive (ℓ)-group, $\Sigma \subset \mathcal{P}(G)$ be a σ-algebra, $\mathcal{L}$ be an algebra of sets which generates Σ, and assume that $m_n : \Sigma \to R$, $n \in \mathbb{N}$, is an equibounded sequence of uniformly σ-additive measures, such that the limit $(O\mathcal{F}) \lim_n m_n(E)$ exists in R for every $E \in \mathcal{L}$. Then $(O\mathcal{F}) \lim_n m_n(E)$ exists in R for each $E \in \Sigma$.*

Proof: Let $\Pi := \{E \in \Sigma : (O\mathcal{F}) \lim_n m_n(E)$ exists in $R\}$. By hypothesis, $\mathcal{L} \subset \Pi$. If we prove that $\Pi = \Sigma$, then we get the assertion. To this aim, it will be enough to show that, if $(E_r)_r$ is any monotone sequence of elements of Π and $E := \lim_r E_r \in \Sigma$ in the set-theoretic sense, then $E \in \Pi$.

Since $E_r \in \Pi$ for all $r \in \mathbb{N}$, then the sequence $(m_n(E_r))_n$ is $(O\mathcal{F})$-Cauchy, and so for any $r \in \mathbb{N}$ there exists an (O)-sequence $(\sigma_l^{(r)})_l$ such that for all r, $l \in \mathbb{N}$ there is an element $W_{r,l} \in \mathcal{F}$ such that $| m_p(E_r) - m_q(E_r) | \le \sigma_l^{(r)}$ whenever p, $q \in W_{r,l}$. Since $(m_n)_n$ is equibounded, the family $\{\sigma_l^{(r)} : l,\ r \in \mathbb{N}\}$ satisfies the hypotheses of Lemma 2.5 of Chapter 2. So there is an (O)-sequence $(\tau_s)_s$ such that for every r, $s \in \mathbb{N}$ there exists $l = l_{r,s} \in \mathbb{N}$ with $\sigma_l^{(r)} \le \tau_s$. Then we get $| m_p(E_r) - m_q(E_r) | \le \tau_s$ for all $p, q \in W'_{r,s}$, where $W'_{r,s} := W_{r,l_{r,s}}$, r, $s \in \mathbb{N}$. Moreover, since $(m_n)_n$ is uniformly σ-additive, there exists an (O)-sequence $(\pi_s)_s$ such that for each $s \in \mathbb{N}$ there is an integer $\bar{r}$ with $| m_n(E_{\bar{r}}) - m_n(E) | \le \pi_s$ for all $n \in \mathbb{N}$. Thus for every p, $q \in W'_{\bar{r},s}$ we get

$$| m_p(E) - m_q(E) | \le | m_p(E) - m_p(E_{\bar{r}}) | + | m_p(E_{\bar{r}}) - m_q(E_{\bar{r}}) | + | m_q(E_{\bar{r}}) - m_q(E) |$$
$$\le 2\pi_s + \tau_s .$$

So the sequence $(m_n(E))_n$ is $(O\mathcal{F})$-Cauchy, and hence the limit $(O\mathcal{F})\lim_n m_n(E)$ exists in R. Thus we obtain the assertion. □

4.3. Modes of Continuity of Measures

4.3.1. Filter Continuity

We present some results about modes of continuity of lattice group-valued measures with respect to filter convergence (see also Boccuto and Dimitriou (2013d)). Let $\Sigma \subset \mathcal{P}(G)$ be a σ-algebra, $\lambda : \Sigma \to \mathbb{R}$ be a non-negative finitely additive measure, $d_\lambda(A,B) := \lambda(A \Delta B)$, A, $B \in \Sigma$ be the (pseudo)-λ-distance. A finitely additive measure $m : \Sigma \to R$ is *λ-continuous* at $A \in \Sigma$, iff

$$(O)\lim_p \Big(\bigvee_{B \in \Sigma, d_\lambda(A,B) \le 1/p} | m(B) - m(A) | \Big) = 0,$$

namely there is an (O)-sequence $(\sigma_p)_p$ in R such that for any $p \in \mathbb{N}$ there is $q = q(p) \in \mathbb{N}$ with

$|m(B)-m(A)| \le \sigma_p$ whenever $d_\lambda(A,B) \le 1/q$.

A measure $m:\Sigma \to R$ is *λ-continuous* on Σ iff it is λ-continuous at every $A \in \Sigma$.

Note that, arguing analogously as in Proposition 2.1, it is possible to see that a measure $m:\Sigma \to R$ is λ-continuous at some $A \in \Sigma$ if and only if m is λ-continuous at $\varnothing$, if and only if m is λ-continuous on Σ with respect to a single order sequence, if and only if m is globally λ-absolutely continuous on Σ.

Let $\mathcal{F}_1$, $\mathcal{F}_2$ be two free filters of $\mathbb{N}$, and $A \in \Sigma$. A finitely additive measure $m:\Sigma \to R$ is *$(\mathcal{F}_1,\mathcal{F}_2)$-$\lambda$-continuous at A* iff there is an (O)-sequence $(\sigma_p)_p$ in R such that for each sequence $(A_n)_n$ in Σ with $(\mathcal{F}_1)\lim_n d_\lambda(A_n,A)=0$ we have $(O\mathcal{F}_2)\lim_n m(A_n)=m(A)$ with respect to $(\sigma_p)_p$. We say that m is *$(\mathcal{F}_1,\mathcal{F}_2)$-$\lambda$-continuous on* Σ, iff it is $(\mathcal{F}_1,\mathcal{F}_2)$-$\lambda$-continuous at every $A \in \Sigma$.

Let $m, m_n:\Sigma \to R$, $n \in \mathbb{N}$, be finitely additive measures. We say that $(m_n)_n$ *λ-$(\mathcal{F}\alpha)$-converges to m at $E \in \Sigma$* iff $(\mathcal{F})\lim_n m_n(E_n)=m(E)$ for every sequence $(E_n)_n$ in Σ with $(\mathcal{F})\lim_n d_\lambda(E_n,E)=0$. The sequence $(m_n)_n$ *λ-$(\mathcal{F}\alpha)$-converges to m on Σ* iff it λ-$(\mathcal{F}\alpha)$-converges to m at every $E \in \Sigma$. If $\mathcal{F}=\mathcal{F}_{\text{cofin}}$, then we simply say that $(m_n)_n$ *λ-(α)-converges to m at $E \in \Sigma$* and $(m_n)_n$ *λ-(α)-converges to m* on Σ respectively.

The following result extends Baláž, Červeňanskij, Kostyrko and Šalát (2002), Theorem 3, and Sleziak (2003), Theorems 2.2 and 2.3 and Corollary 2.4, to lattice group-valued measures.

Theorem 3.1 (Boccuto and Dimitriou (2013d), Theorem 2.1) *Let $\lambda:\Sigma \to [0,+\infty[$, $m:\Sigma \to R$ be two finitely additive measures, and fix $A \in \Sigma$. Then the following properties hold:*

(a) *If $\mathcal{F}_1 \subset \mathcal{F}_2$, then m is $(\mathcal{F}_1,\mathcal{F}_2)$-$\lambda$-continuous at A if and only if m is λ-continuous at A.*

(b) *If* $\mathcal{F}_1 \setminus \mathcal{F}_2 \neq \emptyset$, *then* m *is* $(\mathcal{F}_1, \mathcal{F}_2)$ - λ *-continuous at* A *if and only if* $m(B) = 0$ *for every* $B \in \Sigma$.

Proof: We begin with (a). We first prove the "if" part. Let $\mathcal{F}_1 \subset \mathcal{F}_2$, $(\sigma_p)_p$ be an order sequence in R associated with λ-continuity of m and $(A_n)_n$ be a sequence in Σ, with $(\mathcal{F}_1) \lim\limits_n d_\lambda(A_n, A) = 0$. Fix arbitrarily $p \in \mathbb{N}$. By H1), we get

$$\{n \in \mathbb{N} : d_\lambda(A_n, A) \leq 1/p\} \subset \{n \in \mathbb{N} : | m(A_n) - m(A) | \leq \sigma_p\} \in \mathcal{F}_1 \subset \mathcal{F}_2,$$

that is $(O\mathcal{F}_2) \lim\limits_n m(A_n) = m(A)$.

We now turn to the "only if" part. Suppose that m is $(\mathcal{F}_1, \mathcal{F}_2)$-$\lambda$-continuous at A. We claim that m is λ-continuous at A. By $(\mathcal{F}_1, \mathcal{F}_2)$-$\lambda$-continuity of m, there is an (O)-sequence $(\sigma_p)_p$ in R with $(O\mathcal{F}_2) \lim\limits_n m(A_n) = m(A)$ with respect to $(\sigma_p)_p$ whenever $(\mathcal{F}_1) \lim\limits_n d_\lambda(A_n, A) = 0$. We prove that $(\sigma_p)_p$ satisfies the condition H1). If not, then there are $\overline{p} \in \mathbb{N}$ and a sequence $(A_q)_q$ in Σ, with $d_\lambda(A_q, A) \leq 1/q$ and $| m(A_q) - m(A) | \not\leq \sigma_{\overline{p}}$ for each $q \in \mathbb{N}$. Since $\lim\limits_q d_\lambda(A_q, A) = 0$, then $(\mathcal{F}_1) \lim\limits_q d_\lambda(A_q, A) = 0$ and so, by $(\mathcal{F}_1, \mathcal{F}_2)$ - λ - continuity of m, $(O\mathcal{F}_2) \lim\limits_q m(A_q) = m(A)$ with respect to $(\sigma_p)_p$. Hence, $\emptyset = \{q \in \mathbb{N} : | m(A_q) - m(A) | \leq \sigma_{\overline{p}}\} \in \mathcal{F}_2$, which is absurd. This concludes the proof of (a).

We now turn to (b), and prove only the "only if" part, since the "if" part is straightforward. Choose arbitrarily $B \in \Sigma$, and let us show that $m(B) = m(A)$. From this and arbitrariness of $B \in \Sigma$ it will follow that $m(B) = m(\emptyset) = 0$ for all $B \in \Sigma$. If $m(B) \neq m(A)$, then for every (O)-sequence $(\sigma_p)_p$ there is $\overline{p} \in \mathbb{N}$ with $| m(B) - m(A) | \not\leq \sigma_{\overline{p}}$. By hypothesis, there is a set $C \in \mathcal{F}_1 \setminus \mathcal{F}_2$. Note that C is infinite, since otherwise, as $\mathcal{F}_1$ is free, we get $\mathbb{N} \setminus C \in \mathcal{F}_1$ and also $\emptyset \in \mathcal{F}_1$, which is impossible. Set $C := \{n_k : k \in \mathbb{N}\}$. Let $(Z_k)_k$ be a sequence in Σ, with $\lim\limits_k d_\lambda(Z_k, A) = 0$. For each $n \in \mathbb{N}$, let

$$A_n := \begin{cases} Z_k, & \text{if } n = n_k, \\ B, & \text{if } n \in \mathbb{N} \setminus C. \end{cases}$$

For each $p \in \mathbb{N}$ there is $k_0 \in \mathbb{N}$ with $d_\lambda(A_{n_k}, A) = d_\lambda(Z_k, A) \leq 1/p$ whenever $k \geq k_0$, and so the set $\{n \in \mathbb{N} : d_\lambda(A_n, A) \leq 1/p\}$ contains $C \cap (\mathbb{N} \setminus \{n_1, \ldots, n_{k_0 - 1}\}) \in \mathcal{F}_1$. Thus $(\mathcal{F}_1) \lim_n d_\lambda(A_n, A) = 0$, and hence, thanks to $(\mathcal{F}_1, \mathcal{F}_2)$ - λ -continuity of m, there exists an (O) -sequence $(\sigma_p)_p$ with $(O\mathcal{F}_2) \lim_n m(A_n) = m(A)$ with respect to $(\sigma_p)_p$, namely

$$D_p := \{n \in \mathbb{N} : | m(A_n) - m(A) | \leq \sigma_p\} \in \mathcal{F}_2 \quad \text{for each } p \in \mathbb{N}.$$

As $m(A_n) = m(B)$ for every $n \in \mathbb{N} \setminus C$, it follows that $D_{\bar{p}} \subset C$, and thus $C \in \mathcal{F}_2$, obtaining a contradiction. This proves the "only if" part. □

We now turn to a limit theorem, which yields continuity of the limit measure.

Theorem 3.2 (see also Boccuto and Dimitriou (2014d), Theorem 2.5) *Let $\mathcal{F}$ be any free filter of $\mathbb{N}$, $\lambda : \Sigma \to [0, +\infty)$, $m_n : \Sigma \to R$, $n \in \mathbb{N}$, be finitely additive measures, such that $(m_n)_n$ ($RO\mathcal{F}$)-converges to a set function $m : \Sigma \to R$ on Σ with respect to an (O)-sequence $(\sigma^*_p)_p$. Then the following are equivalent:*

(a) *$(m_n)_n$ is λ-weakly $\mathcal{F}$-exhaustive on Σ;*

(b) *m is λ-continuous on Σ.*

Proof: (a) $\Rightarrow$ (b) Let $(\sigma_p)_p$ be an (O)-sequence related with λ-weak $\mathcal{F}$-exhaustiveness at $E \in \Sigma$, and pick arbitrarily $p \in \mathbb{N}$. By hypothesis there exists a positive real number δ, fulfilling the condition of λ-weak $\mathcal{F}$-exhaustiveness. Fix arbitrarily $A \in \Sigma$ with $| \lambda(A) - \lambda(E) | \leq \delta$: there exists a set $F_1 \in \mathcal{F}$ with $| m_n(A) - m_n(E) | \leq \sigma_p$ for each $n \in F_1$. Moreover, there is $F_2 \in \mathcal{F}$ with $| m_n(A) - m(A) | \vee | m_n(E) - m(E) | \leq \sigma^*_p$ for any $n \in F_2$. For each $n \in F_1 \cap F_2$ we get:

$$|m(A)-m(E)|\leq|m_n(A)-m(A)|+|m_n(A)-m_n(E)|+|m_n(E)-m(E)|\leq 2\sigma_p^*+\sigma_p.$$

Hence $|m(A)-m(E)|\leq 2\sigma_p^*+\sigma_p$ for each $A\in\Sigma$ with $|\lambda(A)-\lambda(E)|\leq\delta$, getting λ-continuity of m at E.

(b) $\Rightarrow$ (a) By λ-continuity of m on Σ there exists an (O)-sequence (τ_p) in R (without loss of generality independent of $E\in\Sigma$) such that for every $p\in\mathbb{N}$ and $E\in\Sigma$ there is $\delta>0$ with

$|m(A)-m(E)|\leq\tau_p$ for every $A\in\Sigma$ with $|\lambda(A)-\lambda(E)|\leq\delta$. By $(RO\mathcal{F})$-convergence of $(m_n)_n$ to m with respect to the (O)-sequence $(\sigma_p^*)_p$, in correspondence with p, A and E there exists a set $F^*\in\mathcal{F}$ with $|m_n(A)-m(A)|\vee|m_n(E)-m(E)|\leq\sigma_p^*$ for any $n\in F^*$. Thus for every $n\in F^*$ we have

$$|m_n(A)-m_n(E)|\leq|m_n(A)-m(A)|+|m(A)-m(E)|+|m_n(E)-m(E)|\leq 2\sigma_p^*+\tau_p,$$

namely λ-weak $\mathcal{F}$-exhaustiveness of $(m_n)_n$ at E. □

4.3.2. Filter (α)-Convergence

We now give some results related with filter exhaustiveness of measures, in connection with filter (α)-convergence.

Let Σ be a σ-algebra of parts of an abstract nonempty set G and E be a fixed element of Σ. A sequence of measures $m_n:\Sigma\to\mathbb{R}$, $n\in\mathbb{N}$, is said to *λ-$(\mathcal{F}\alpha)$-converge to a measure $m:\Sigma\to\mathbb{R}$ at E* iff $(\mathcal{F})\lim_n m_n(E_n)=m(E)$ whenever $(\mathcal{F})\lim_n d_\lambda(E_n,E)=0$. We say that the sequence $m_n:\Sigma\to\mathbb{R}$, $n\in\mathbb{N}$, *λ-$(\mathcal{F}\alpha)$-converges* to $m:\Sigma\to\mathbb{R}$ iff it converges to m at every $E\in\Sigma$. We denote by *λ-(α)-convergence* the λ-$(\mathcal{F}_{\text{cofin}}\,\alpha)$-convergence.

Theorem 3.3 (Boccuto, Das, Dimitriou and Papanastassiou (2012), Theorem 4.14) *Under the same notations as above, let $H_0\in\Sigma$, and $m,m_n:\Sigma\to\mathbb{R}$, $n\in\mathbb{N}$.*

If $(\mathcal{F})\lim_n m_n(H_0) = m(H_0)$ *and* $(m_n)_n$ *is* λ*-*$\mathcal{F}$*-exhaustive at* H_0*, then* $(m_n)_n$ λ*-*$(\mathcal{F}\alpha)$*-converges to* m *at* H_0.

Proof: Assume that $(\mathcal{F})\lim_n m_n(H_0) = m(H_0)$, and choose arbitrarily $\varepsilon > 0$. By virtue of λ-$\mathcal{F}$-exhaustiveness of $(m_n)_n$ at H_0, there are $\delta > 0$ and $F_1 \in \mathcal{F}$ with

$$| m_n(H) - m_n(H_0) | \leq \frac{\varepsilon}{2}$$

whenever $n \in F_1$ and $| \lambda(H) - \lambda(H_0) | < \delta$.

By $\mathcal{F}$-convergence of $(m_n(H_0))_n$ to $m(H_0)$ there exists an element $F_2 \in \mathcal{F}$ such that

$$| m_n(H_0) - m(H_0) | \leq \frac{\varepsilon}{2}$$

for any $n \in F_2$. As $(\mathcal{F})\lim_n d_\lambda(H_n, H_0) = 0$, there is an element $F_3 \in \mathcal{I}$ with $d_\lambda(H_n, H_0) < \delta$ whenever $n \in F_3$. Thus for each $n \in F_1 \cap F_2 \cap F_3$ we get

$$| m_n(H_n) - m(H_0) | \leq | m_n(H_n) - m_n(H_0) | + | m_n(H_0) - m(H_0) | \leq \varepsilon \ \square$$

One can ask whether the converse of Theorem 4.3 holds. To this aim, suppose that there exists a partition of the type $\mathbb{N} = \bigcup_{k=1}^{\infty} \Delta_k$, with the property that

$$\mathcal{I} = \{A \subset \mathbb{N}: A \text{ intersects at most a finite number of } \Delta_k\text{'s}\}, \qquad \textbf{(4.122)}$$

where $\mathcal{I}$ denotes the dual ideal of $\mathcal{F}$.

Examples 3.4

(see also Boccuto, Das, Dimitriou and Papanastassiou (2012), Remarks 4.15) (a) The filter $\mathcal{F}_{\text{cofin}}$ satisfies condition (4.122): indeed it is enough to take $\Delta_k = \{k\}$ for all $k \in \mathbb{N}$. Moreover, if for every $k \in \mathbb{N}$ the set Δ_k in (4.122) is infinite, then the ideal $\mathcal{I}$ associated to the Δ_k's is not a P-ideal (see also Kostyrko, Šalát and Wilczyński (2001), Example 3.1 (g)).

(b) Let $\mathcal{F}$ and $\mathcal{I}$ be as in (4.122), and $(A_j)_j$ be any sequence of subsets of $\mathbb{N}$, with $A_j \notin \mathcal{I}$ for all $j \in \mathbb{N}$. We claim that there exists a disjoint sequence $(B_j)_j$ in $\mathcal{I}$, with $B_j \subset A_j$ for every $j \in \mathbb{N}$ and $\bigcup_{j=1}^{\infty} B_j \notin \mathcal{I}$. Indeed, first of all observe that there exists an infinite subset $P_1 \subset \mathbb{N}$ with $A_1 = \bigcup_{k \in P_1} (A_1 \cap \Delta_k)$. There is a nonempty finite set $Q_1 \subset P_1$, with the property that the set $B_1 = \bigcup_{k \in Q_1} (A_1 \cap \Delta_k)$ belongs to $\mathcal{I}$. Let $P_2 \subset \mathbb{N}$ be such that $A_2 = \bigcup_{k \in P_2} (A_2 \cap \Delta_k)$: since $A_2 \notin \mathcal{I}$, we get that P_2 is infinite. There exists a nonempty finite set $Q_2 \subset P_2 \setminus Q_1$, such that the set $B_2 = \bigcup_{k \in Q_2} (A_2 \cap \Delta_k)$ belongs to $\mathcal{I}$. At the j-th step, we take $P_j \subset \mathbb{N}$ with $A_j = \bigcup_{k \in P_j} (A_j \cap \Delta_k)$. There exists a nonempty finite set $Q_j \subset P_j \setminus (\bigcup_{s=1}^{j-1} Q_s)$, with $B_j = \bigcup_{k \in Q_j} (A_j \cap \Delta_k) \in \mathcal{I}$. Since the sets Q_j, $j \in \mathbb{N}$, are by construction nonempty and pairwise disjoint and $\bigcup_{j=1}^{\infty} Q_j$ is infinite, we get that the sets B_j, $j \in \mathbb{N}$, are pairwise disjoint and $\bigcup_{j=1}^{\infty} B_j \notin \mathcal{I}$.

We now prove the converse of Theorem 4.3, when (4.122) is satisfied.

Theorem 3.5 (Boccuto, Das, Dimitriou and Papanastassiou (2012), Theorem 4.16) *Let* $\mathcal{F}$ *be as in* (4.122)*,* $H_0 \in \Sigma$*, and* m*,* $m_n : \Sigma \to \mathbb{R}$*,* $n \in \mathbb{N}$*. If* $(m_n)_n$ λ*-*$(\mathcal{F}\alpha)$*-converges to* m *at* H_0*, then* $(\mathcal{F})\lim_n m_n(H_0) = m(H_0)$ *and* $(m_n)_n$ *is* λ*-*$\mathcal{F}$*-exhaustive at* H_0.

Proof: First of all, note that λ-$(\mathcal{F}\alpha)$-convergence of $(m_n)_n$ to m at H_0 implies that $(\mathcal{F})\lim_n m_n(H_0) = m(H_0)$: indeed, in the definition of λ - $(\mathcal{F}\alpha)$ - convergence, it is enough to consider the sequence $(H_n)_n$, defined by setting $H_n := H_0$ for each $n \in \mathbb{N}$.

Now we prove that $(m_n)_n$ is λ-$\mathcal{F}$-exhaustive at H_0. For each $k \in \mathbb{N}$, set

$$V_k := \{H \in \Sigma : |\lambda(H) - \lambda(H_0)| < 1/k\} = \{H \in \Sigma : d_\lambda(H, H_0) < 1/k\}.$$

Let $\mathcal{I}$ be the dual ideal of $\mathcal{F}$. If $(m_n)_n$ is not λ-$\mathcal{F}$-exhaustive at H_0, then there is $\varepsilon > 0$ such that for every $k \in \mathbb{N}$ and $A \in \mathcal{I}$ there are $H \in V_k$ and $n \in \mathbb{N} \setminus A$ with $|m_n(H) - m_n(H_0)| \geq \varepsilon$. In correspondence with $A_0^k = \varnothing$, there exist $n_0^k \in \mathbb{N}$ and $H_0^k \in V_k$ with $|m_{n_0^k}(H_0^k) - m_{n_0^k}(H_0)| \geq \varepsilon$.

We now proceed by transfinite induction. Fix $k \in \mathbb{N}$. Suppose that n_β^k and H_β^k are defined for every $\beta \leq \alpha$, where α is an ordinal of first kind, and that $A_\alpha^k = \{n_\beta^k : \beta \leq \alpha\} \in \mathcal{I}$. Then we find an integer $n_{\alpha+1}^k \notin A_\alpha^k$ and a set $H^k_{n^k_{\alpha+1}} \in V_k$, with

$$|m_{n^k_{\alpha+1}}(H^k_{n^k_{\alpha+1}}) - m_{n^k_{\alpha+1}}(H_0)| \geq \varepsilon .$$

In this case, set $A_{\alpha+1}^k = A_\alpha^k \cup \{n_{\alpha+1}^k\}$. If α is a limit ordinal, n_β^k and H_β^k are defined for $\beta < \alpha$, put $A_\alpha^k = \bigcup_{\beta<\alpha} A_\beta^k$. This procedure ends at some countable limit ordinal α_k since otherwise we obtain a strictly increasing sequence $\{A_\alpha^k : \alpha < \omega_1\}$ of subsets of $\mathbb{N}$. Thus we get that $A_{\alpha_k}^k \notin \mathcal{I}$. Set $P_k := A_{\alpha_k}^k$ for each $k \in \mathbb{N}$. By virtue of Example 3.4 (b) there is a disjoint sequence $(B_k)_k$ in $\mathcal{I}$ with $B_k \subset P_k$ for all $k \in \mathbb{N}$ and $\bigcup_{k=1}^{\infty} B_k \notin \mathcal{I}$.

For every $n \in \mathbb{N}$ set $H_n = H^k_{n^k_{\alpha_k}}$ if $n = n^k_{\alpha_k}$ and $n \in B_k$, and $H_n = H_0$ if $n \notin \bigcup_{k=1}^{\infty} B_k$ (the integer k is unique, since the sets B_k are pairwise disjoint, and that $H^k_{n^k_{\alpha_k}} \in V_k$ for every $k \in \mathbb{N}$).

We get that $(\mathcal{F})\lim_n |\lambda(H_n) - \lambda(H_0)| = (\mathcal{F})\lim_n d_\lambda(H_n, H_0) = 0$. Indeed, for every $\delta > 0$, set $U := \{H \in \Sigma : |\lambda(H) - \lambda(H_0)| \leq \delta\}$. There exists $k_0 \in \mathbb{N}$ with $\{n : H_n \notin U\} \subset \bigcup_{k=1}^{k_0} B_k$ and $\bigcup_{k=1}^{k_0} B_k \in \mathcal{I}$. Since $(\mathcal{F})\lim_n m_n(H_0) = m(H_0)$, then

$$\{n\in\mathbb{N}: |m_n(H_0)-m(H_0)|\geq \frac{\varepsilon}{2}\}\in\mathcal{I}.$$

Since

$$L:=\{n\in\mathbb{N}: |m_n(H_n)-m_n(H_0)|\geq\varepsilon\}\supset\bigcup_{k=1}^{\infty}B_k,$$

then $L\notin\mathcal{I}$. So $\{n\in\mathbb{N}: |m_n(H_n)-m(H_0)|\geq\varepsilon/2\}\notin\mathcal{I}$, which contradicts λ-$(\mathcal{F}\alpha)$-convergence to m at H_0. Thus the theorem is completely proved. □

Example 3.6 When $\mathcal{F}\neq\mathcal{F}_{\text{cofin}}$, in general λ-$(\mathcal{F}\alpha)$-convergence does not imply λ-(α)-convergence. Indeed let $\Sigma=\mathcal{P}(\mathbb{N})$, $\lambda(A)=\sum_{n\in A}\frac{1}{2^n}$, $A\in\Sigma$, $\mathcal{I}$ be the dual ideal of $\mathcal{F}$, and $H:=\{q_1,\ldots,q_n,\ldots\}\in\mathcal{I}$ be an infinite set. Since $\mathcal{F}\neq\mathcal{F}_{\text{cofin}}$, such a H does exist. Set $m(A)=0$ for all $A\in\Sigma$, and $m_i(A)=0$ whenever $A\in\Sigma$ and $i\in\mathbb{N}\setminus H$. For each $n\in\mathbb{N}$, let δ_n be the Dirac measure defined by setting $\delta_n(A)=1$ if $A\in\Sigma$ and $n\in A$, and $\delta_n(A)=0$ if $A\in\Sigma$ and $n\notin A$. Let $E=\varnothing$ and $E_n=\{n\}$, $n\in\mathbb{N}$. We get $\lim_n\lambda(E_n)=0$, but $\delta_n(E_n)=1$ for all $n\in\mathbb{N}$. Set now $m_{q_n}(A)=\delta_n(A)$ for any $n\in\mathbb{N}$ and $A\in\Sigma$. It is easy to see that $(m_n)_n$ is λ-$(\mathcal{F}\alpha)$-convergent to 0. However, $\lim_n m_n(E_n)$ does not exist in the ordinary sense. □

So, we have the following relations between (filter) α-convergence and (filter) exhaustiveness.

Proposition 3.7 (see also Athanassiadou, Boccuto, Dimitriou and Papanastassiou (2012), Proposition 4.3) *Let $\mathcal{F}$ be a free filter, $\mathcal{F}\neq\mathcal{F}_{\text{cofin}}$, $\Sigma\subset\mathcal{P}(G)$ be a σ-algebra, $\lambda:\Sigma\to[0,+\infty)$, $m,m_n:\Sigma\to R$, $n\in\mathbb{N}$, be finitely additive measures, and let us consider the following statements:*

(i) $(m_n)_n$ is λ-$(\mathcal{F}\alpha)$-convergent to m;

(ii) $(m_n)_n$ is λ-(α)-convergent to m;

(iii) $(m_n)_n$ $\mathcal{F}$*-converges pointwise to* m *and* $(m_n)_n$ *is* λ*-*$\mathcal{F}$*-exhaustive on* Σ*;*

(iv) $(m_n)_n$ *converges pointwise to* m *and* $(m_n)_n$ *is* λ*-exhaustive on* Σ*.*

Then $(ii) \Leftrightarrow (iv)$, $(iv) \Rightarrow (iii) \Rightarrow (i)$, $(ii) \Rightarrow (i)$ *and* $(i) \not\Rightarrow (ii)$.

Example 3.8 (see also Athanassiadou, Boccuto, Dimitriou and Papanastassiou (2012), Remark 4.4) In Theorem 3.5 we proved implication $(i) \Rightarrow (iii)$ for filters, whose dual ideal fulfils (4.122). Note that, when the involved filter $\mathcal{F} \neq \mathcal{F}_{\text{cofin}}$ satisfies this property, the sequence m_n, $n \in \mathbb{N}$, defined in Example 3.6, is a λ-$\mathcal{F}$-exhaustive measure sequence, by virtue of Theorem 3.5. However, this sequence is not λ-exhaustive at $\varnothing$. Indeed, observe that for every $\delta > 0$ there is a cofinite set $E \subset \mathbb{N}$, with the property that $\lambda(E \Delta \varnothing) = \lambda(E) < \delta$. Note that for every cofinite subset $M \subset \mathbb{N}$ it is possible to find an integer $\overline{n}$ large enough with $q_{\overline{n}} \in M \cap E \cap H$, and so we get $1 = \delta_{\overline{n}}(E) = m_{q_{\overline{n}}}(E)$. □

With similar techniques as above, in Athanassiadou, Boccuto, Dimitriou and Papanastassiou (2012) some Ascoli and Helly-type theorems related with filter convergence and filter compactness for function sequences were proved, extending some results of Gregoriades and Papanastassiou (2008) (for other related results, see also Billingsley (1995), Theorems 8.2 and 8.3, Boccuto and Dimitriou (2014e), Duchoň and Maličký (2009), Collins and Zimmer (2007), Dugundji (1966), Kelley (1955), Theorems 7.17 and 7.21, Munkres (1975), Mynard (2013)).

4.3.3. Filter Weak Compactness and Weak Convergence of Measures

We now extend to the context of filter compactness some results proved in Diestel (1984), Chapter VII. We begin with a weak compactness and boundedness property for subsets of countably additive measures in the context of filter convergence, when we deal with a P-filter (see also Çakalli (2008)). Let $\mathrm{ca}(\Sigma)$ be the class of all σ-additive real-valued measures on a σ-algebra Σ.

Theorem 3.9 (see also Boccuto, Das, Dimitriou and Papanastassiou (2012), Theorem 4.4) *Let* $\mathcal{K}$ *be a subset of* $ca(\Sigma)$ *and* $\mathcal{F}$ *be a P-filter of* $\mathbb{N}$*. Then the following are equivalent:*

(a) $\mathcal{K}$ *is relatively weakly compact.*

(b) $\mathcal{K}$ *is relatively weakly* $\mathcal{F}$*-sequentially compact.*

(c) $\mathcal{K}$ *is bounded and the elements of* $\mathcal{K}$ *are uniformly* σ*-additive.*

(d) $\mathcal{K}$ *is bounded and there is a non-negative measure* $\lambda \in ca(\Sigma)$ *such that* $\mathcal{K}$ *is equicontinuous with respect to* λ *(that is for each* $E \in \Sigma$ *and* $\varepsilon > 0$ *there is* $\delta > 0$ *with* $|m(F) - m(E)| \le \varepsilon$ *whenever* $m \in \mathcal{K}$ *and* $|\lambda(F) - \lambda(E)| \le \delta$*).*

The following result is a characterization of weak convergence of countably additive measures and weak convergence in L^1 in terms of pointwise convergence under the hypothesis of filter exhaustiveness, and extends Diestel (1984), Theorem VII.11.

Theorem 3.10 (see also Boccuto, Das, Dimitriou and Papanastassiou (2012), Theorem 4.9) *Let* d_λ *be the pseudo-distance induced by a positive finitely additive measure* $\lambda : \Sigma \to \mathbb{R}$*, and let* (Σ, d_λ) *be* λ*- separable. If a sequence* $(m_n)_n$ *in* $ca(\Sigma)$ *weakly* $\mathcal{F}$*-converges to* $m \in ca(\Sigma)$*, then for each* $E \in \Sigma$*,* $m(E) = (\mathcal{F})\lim_n m_n(E)$*. Conversely, if* $(m_n)_n$ *is a sequence in* $ca(\Sigma)$ *such that there is a set function* $m : \Sigma \to \mathbb{R}$ *with* $m(E) = (\mathcal{F})\lim_n m_n(E)$ *for all* $E \in \Sigma$*,* $\mathcal{F}$ *is a* P*-filter and* $(m_n)_n$ *is* λ*-uniformly* $\mathcal{F}$*-exhaustive, then* $m \in ca(\Sigma)$ *and* $(m_n)_n$ *weakly* $\mathcal{F}$*-converges to* m *in* $ca(\Sigma)$*. Moreover, if* $\eta \in ca(\Sigma)$ *and* $f_n : G \to \mathbb{R}$*,* $n \in \mathbb{N}$*, is a sequence of functions belonging to* $L^1(\eta)$ *and such that the sequence* $m_n(E) := \int_E f_n \, d\eta$*,* $n \in \mathbb{N}$*,* $E \in \Sigma$*, is* η*-uniformly* $\mathcal{F}$*-exhaustive, then* $(f_n)_n$ *weakly* $\mathcal{F}$*-converges to* f *in* $L^1(\eta)$ *if and only if* $\int_E f \, d\eta = (\mathcal{F})\lim_n m_n(E)$ *for every* $E \in \Sigma$*.*

Proof: Since the functional $\nu \mapsto \nu(E)$ belongs to $ca(\Sigma)^*$ (the dual of $ca(\Sigma)$) for every $E \in \Sigma$, then the first part is straightforward.

Suppose that $(m_n)_n$ is a sequence in $ca(\Sigma)$ such that $m(E) = (\mathcal{F})\lim_n m_n(E)$ exists in $\mathbb{R}$ for each $E \in \Sigma$. Then by Lemma 2.3 used with $R = \mathbb{R}$, since the sequence $(m_n)_n$ is λ-uniformly $\mathcal{F}$-exhaustive, we get that $m \in ca(\Sigma)$ and there is an element $M \in \mathcal{F}$ with $\lim_{n \in M} m_n(E) = m(E)$ for all $E \in \Sigma$. Let us write the set $(m_n)_{n \in M}$ as $(\nu_k)_{k \in \mathbb{N}}$. Then we have $\lim_k \nu_k(E) = m(E)$ for all $E \in \Sigma$. Now, by the classical Nikodým boundedness theorem, the measures ν_k, $k \in \mathbb{N}$, are equibounded, and so $\sup_k \|\nu_k\|(G) < +\infty$, where the symbol $\|\cdot\|$ denotes the total variation. Hence the series $\sum_{k=1}^{\infty} \frac{\|\nu_k\|(\cdot)}{2^k}$ is absolutely convergent in the Banach space $\text{ca}(\Sigma)$, and η is its sum. Then, by the Radon-Nikodým theorem, for each $k \in \mathbb{N}$ there is a function $f_k \in L^1(\eta)$ with $\nu_k(E) = \int_E f_k \, d\eta$ (see also Diestel (1984)). Similarly, we get the existence of $f \in L^1(\eta)$ such that $m(E) = \int_E f \, d\eta$.

Now, proceeding analogously as in Diestel (1984), Theorem VII.11, it is possible to show that $(f_k)_k$ weakly converges to f in $L^1(\eta)$, which implies that $(\nu_k)_k$ weakly converges to m in $\text{ca}(\Sigma)$. But this implies that the sequence $(m_n)_n$ weakly $\mathcal{F}'$-converges (see also Chapter 2) and therefore it weakly $\mathcal{F}$-converges to m in $\text{ca}(\Sigma)$. Thus the first part is proved.

To prove the "if" part of the last statement of the theorem, proceeding similarly as above, we obtain the existence of an element $M \in \mathcal{F}$ such that $\lim_{n \in M} m_n(E) = m(E)$ in the usual sense for each $E \in \Sigma$, and hence the subsequence $(f_n)_{n \in M}$ weakly converges to f in $L^1(\eta)$ in the ordinary sense. Thus the sequence $(f_n)_{n \in \mathbb{N}}$ weakly $\mathcal{F}$-converges to f in $L^1(\eta)$. The "only if" part of the last statement is straightforward. □

Theorem 3.10 is a sufficient condition for σ-additivity of the $\mathcal{F}$-limit measure of a sequence of σ-additive measures, under the hypothesis of filter pointwise convergence and the condition of filter exhaustiveness. As a consequence of Theorem 3.10, we prove the next result, which extends Diestel (1984), Theorem VII.12.

Corollary 3.11 (see Boccuto, Das, Dimitriou and Papanastassiou (2012), Corollary 4.11) *Under the same notations as in Theorem* 3.10, *let* Σ *be* λ*-separable,* $(m_n)_n$ *be a* λ*-uniformly* $\mathcal{F}$*-exhaustive family and* $\mathcal{F}$ *be a* P*-filter.*

Then a sequence $(m_n)_n$ is weakly $\mathcal{F}$-Cauchy in $ca(\Sigma)$ if and only if it is weakly $\mathcal{F}$-convergent in $ca(\Sigma)$.

Moreover, if $\eta \in ca(\Sigma)$ is positive, $f_n : G \to \mathbb{R}$, $n \in \mathbb{N}$, is a function sequence in $L^1(\eta)$ and the sequence $m_n(E) := \int_E f_n \, d\eta$, $n \in \mathbb{N}$, $E \in \Sigma$, is η-uniformly $\mathcal{F}$-exhaustive, then $(f_n)_n$ is weakly $\mathcal{F}$-Cauchy in $L^1(\eta)$ if and only if it is weakly $\mathcal{F}$-convergent in $L^1(\eta)$.

Proof: Let $(m_n)_n$ be a weakly $\mathcal{F}$-Cauchy sequence in $\mathrm{ca}(\Sigma)$. Since each element $E \in \Sigma$ determines the member $\nu \mapsto \nu(E)$ of $\mathrm{ca}(\Sigma)^*$, then the sequence $(m_n(E))_n$ is $\mathcal{F}$-Cauchy in $\mathbb{R}$ for each $E \in \Sigma$. Then (see also Das and Ghosal (2010a-b and 2011), Dems (2005)) for every $E \in \Sigma$, $(m_n(E))_n$ is $\mathcal{F}$-convergent to a real number, which we call $m(E)$. So, a set function $m : \Sigma \to \mathbb{R}$ is defined. By Lemma 2.3, $m \in \mathrm{ca}(\Sigma)$. Thus, the sequence $(m_n)_n$ is weakly $\mathcal{F}$-convergent in $\mathrm{ca}(\Sigma)$. The converse implication of the first part is obvious.

The last assertion is a consequence of the first, by setting $m_n(E) := \int_E f_n \, d\eta$, $n \in \mathbb{N}$, $E \in \Sigma$, and taking into account the Radon-Nikodým theorem, because $L^1(\eta)$ is a closed subspace of $\mathrm{ca}(\Sigma)$ for every positive measure $\eta \in \mathrm{ca}(\Sigma)$ (see also Diestel (1984)). □

The next step is a result involving integral-type operators, which extends Abbott, Bator, Bilyeu and Lewis (1990), Lemma 2.3, to the setting of filters.

Let $G = (G, d)$ be a metric space. A sequence $f_n : G \to \mathbb{R}$, $n \in \mathbb{R}$, is $\mathcal{F}$-*exhaustive on* G iff for every $x \in G$ and $\varepsilon > 0$ there are $\delta > 0$ and $F \in \mathcal{F}$ with $|f_n(z) - f_n(x)| \le \varepsilon$ whenever $n \in F$ and $d(z, x) < \delta$.

The next result extends Balcerzak, Dems and Komisarski (2007), Proposition 5.

Proposition 3.12 (see also Athanassiadou, Boccuto, Dimitriou and Papanastassiou (2012), Proposition 2.6) *Let G be a metric space, $\mathcal{F}$ be any free filter of $\mathbb{N}$, $f_n : G \to \mathbb{R}$, $n \in \mathbb{N}$, be a function sequence, $\mathcal{F}$-convergent to $f \in \mathbb{R}^G$ pointwise and $\mathcal{F}$-exhaustive on G.*

Then f is continuous on G and $(f_n)_n$ $\mathcal{F}$-converges uniformly to f on every compact subset of G.

Theorem 3.13 (see also Boccuto, Das, Dimitriou and Papanastassiou (2012), Proposition 4.17) *Let (G,Σ,λ) be a measure space, assume that λ is regular and (G,d) is a complete metric space. Let $(m_l)_l$ be a uniformly $\mathcal{F}$-exhaustive sequence of σ-additive λ-continuous measures. Suppose that there exist a set $C_0 \in \mathcal{F}$ and $M' > 0$ with $\| m_l \|(G) \le M'$ for all $l \in C_0$, and that the sequence $(m_l(A))_l$ is weakly $\mathcal{F}$-Cauchy for all $A \in \Sigma$. Let $g \in L^\infty(\lambda)$, and assume that $(s_n)_n$ is a sequence of simple functions, $\mathcal{F}$-exhaustive on G and $\mathcal{F}$-convergent pointwise to g on G.*

Then the sequence $\left(\int_G g\, dm_l\right)_l$ is $\mathcal{F}$-Cauchy.

Proof: Fix arbitrarily $\varepsilon > 0$. Since the sequence $(m_l)_l$ is λ-uniformly $\mathcal{F}$-exhaustive, there exist a positive real number δ and a set $F \in \mathcal{F}$ such that, if $\lambda(A) < \delta$ and $l \in F$, then $\| m_l \|(A) < \varepsilon$. Without loss of generality we can suppose that $F \subset C_0$.

By regularity of λ, there is a set $A \in \Sigma$ such that $G \setminus A$ is compact and $\lambda(A) < \delta$. By Proposition 3.12, the sequence $(s_n)_n$ $\mathcal{F}$-converges uniformly to g on $G \setminus A$, and so there exists a set $C \in \mathcal{F}$ with $| s_n(t) - g(t) | \le \varepsilon$ whenever $t \in G \setminus A$ and $n \in C$. Let $n_0 = \min C$. Since $(m_l(A))_l$ is weakly $\mathcal{F}$-Cauchy for all $A \in \Sigma$, there is $C^* \in \mathcal{F}$, without loss of generality $C^* \subset C$, $C^* \subset F$, such that for every i, $j \in C^*$ we get

$$\left| \int_G s_{n_0}\, d(m_i - m_j) \right| \le \varepsilon,$$

and hence

$$0 \le \left| \int_G g\, dm_i - \int_G g\, dm_j \right| \le \left| \int_G (s_{n_0} - g)\, dm_i \right| + \left| \int_G s_{n_0}\, d(m_i - m_j) \right| + \left| \int_G (s_{n_0} - g)\, dm_j \right| \le$$

$$\le \left| \int_{G \setminus A} (s_{n_0} - g)\, dm_i \right| + \left| \int_A (s_{n_0} - g)\, dm_i \right| + \left| \int_{G \setminus A} (s_{n_0} - g)\, dm_j \right| + \left| \int_A (s_{n_0} - g)\, dm_j \right| + \varepsilon \le$$

$\leq \varepsilon \| m_i \| (G) + \varepsilon \| m_j \| (G) + 6\, \varepsilon \| g \|_\infty + \varepsilon.$

From this the assertion follows. □

Observe that the thesis of Proposition 4.13 can be interpreted as a weak $\mathcal{F}$-Cauchy-type condition in the space $L^\infty(\lambda)$: indeed, by virtue of the Riesz representation theorem, the dual of $L^\infty(\lambda)$ is isomorphic to the space of all σ-additive λ-continuous measures, and the integral is a functional which realizes such isomorphism. The topics and the tools about Proposition 3.13 are associated also with some properties of precompactness of sets of measures and uniform integrability (see for instance Diestel (1984), Diestel, Ruess and Schachermayer (1993), Randrianantoanina and Saab (1994), Rivera (2000)).

4.4. Topological Group-Valued Measures

4.4.1. Basic Properties

Limit theorems for filter convergence have been the object of several recent researches, not only for lattice group-valued measures, but also in the topological group context. In this setting, we consider some main properties of topological group-valued measures and different kinds of Schur, Brooks-Jewett, Vitali-Hahn-Saks, Dieudonné, Nikodým convergence and Nikodým boundedness theorem. As a particular case, we consider "positive" measures. We will prove also some equivalence results between filter limit theorems, using also in this setting the powerful tool of filter exhaustiveness, which allows to find a subsequence of the original sequence of measures, indexed by a suitable element of the involved filter, in which it is possible to apply some classical theorems (see for instance Candeloro (1985b), Candeloro and Letta (1985)), obtaining some results on the existence of limit measures and several convergence theorems. Note that, in the context of topological groups, it is sufficient to deal with a suitable basis of neighborhoods of zero, which allows us to give a direct approach to our theorems. Observe that, in this setting, in general it is advisable to use some techniques which in general are not similar to those used in the lattice group context, since we deal with a different kind of structure. For example, we will prove some equivalence results between filter limit theorems, in which we will apply some Drewnowski-type results about the existence of σ-additive restrictions of (s)-bounded measures (see also Drewnowski (1972a-b), without considering the Stone Isomorphism technique, though it is possible to get Stone-type extensions of (s)-bounded topological group-valued measures (see also Sion (1969 and 1973)). However, in topological groups, the Drewnowski-type technique is in general easier to handle than the Stone Isomorphism technique.

Let $\mathcal{F}$ be a free filter of $\mathbb{N}$, $R=(R,+)$ be a Hausdorff complete abelian topological group satisfying the first axiom of countability, with neutral element 0, and $\mathcal{J}(0)$ be a basis of closed and symmetric neighborhoods of 0 (see also Candeloro (1985a-c)).

A sequence $(x_n)_n$ in R $\mathcal{F}$*-converges to* $x_0 \in R$ iff for every $U \in \mathcal{J}(0)$, $\{n\in \mathbb{N}: x_n - x_0 \in U\} \in \mathcal{F}$, and we write $(\mathcal{F})\lim_n x_n = x_0$. Moreover, we say that a sequence $(B_n)_n$ of subsets of R $\mathcal{F}$*-converges to* 0 iff for each $U \in \mathcal{J}(0)$ the set $\{n\in \mathbb{N}: B_n \subset U\}$ belongs to $\mathcal{F}$, and we write $(\mathcal{F})\lim_n B_n = 0$. We say that $\lim_n x_n = x_0$ (resp. $\lim_n B_n = 0$) iff $(\mathcal{F}_{\text{cofin}})\lim_n x_n = x_0$ (resp. $(\mathcal{F}_{\text{cofin}})\lim_n B_n = 0$). Furthermore, we denote by $\sum_{n=1}^{\infty} B_n$ the set of all elements $b\in R$ of the type $b = \sum_{k=1}^{\infty} b_k := \lim_n \sum_{k=1}^{n} b_k$ as b_k varies in B_k, $k \in \mathbb{N}$. Note that the $\mathcal{F}$-limit is unique, since R is Hausdorff (see also Lahiri and Das (2005)). Observe that filter convergence satisfies the following property:

$(\mathcal{U})$ If each subsequence of a given sequence $(x_n)_n$ admits a sub-subsequence which

$\mathcal{F}$-converges to x_0, then $(\mathcal{F})\lim_n x_n = x_0$.

Otherwise, there exist $U \in \mathcal{J}(0)$ with $\mathcal{Z}(U) := \{n\in \mathbb{N}: x_n - x_0 \in U\} \notin \mathcal{F}$. Since $\mathcal{F}$ is free, the set $\mathcal{Y}(U) := \mathbb{N} \setminus \mathcal{Z}(U)$ is infinite. Let $\mathcal{Y}(U) := \{n_1 < n_2 < \ldots < n_k < \ldots\}$. The subsequence $(x_{n_k})_k$ does not have any sub-subsequence, $\mathcal{F}$-convergent to x_0.

Analogously as above, it is possible to prove that

($\mathcal{U}'$) If every subsequence of a given sequence $(B_n)_n$ of subsets of R admits a sub-subsequence $(B_{n_k})_k$ which $\mathcal{F}$-converges to x_0, then $(\mathcal{F})\lim_n B_n = 0$.

Note that, in general, property ($\mathcal{U}$) is not true in the lattice context: for instance, this is the case of the space $L^0([0,1],\mathcal{M},\nu)$ as in Example 2.181 of Chapter 1, endowed with the almost everywhere convergence (see also Luxemburg and Zaanen (1971) and Vulikh (1967)).

We now prove a Cauchy criterion for filter convergence. Similar results in the context of topological spaces can be found, for instance, in Das and Ghosal (2011).

Theorem 4.1 (see Boccuto and Dimitriou (2014), Theorem 2.2) *Let R and $\mathcal{F}$ be as above, $x \in R$ and $(x_n)_n$ be a sequence in R. Then the following are equivalent:*

(j) $(\mathcal{F})\lim_n x_n = x$;

(jj) *For every $U \in \mathcal{J}(0)$ there is $r \in \mathbb{N}$ with $\{n \in \mathbb{N} : x_n - x_r \in U\} \in \mathcal{F}$;*

(jjj) *For every $U \in \mathcal{J}(0)$ there is $F \in \mathcal{F}$ with $x_n - x_r \in U$ whenever n, $r \in F$.*

Proof: (jj) $\Rightarrow$ (j) Choose arbitrarily $U \in \mathcal{J}(0)$, let $U_0 \in \mathcal{J}(0)$ be with $3U_0 \subset U$, and $(U_p)_p$ be a decreasing countable basis of closed symmetric neighborhoods of 0. For each p, $q \in \mathbb{N}$ there are r_p, $r_q \in \mathbb{N}$ with

$$\{n \in \mathbb{N} : x_n - x_{r_p} \in U_p\} \cap \{n \in \mathbb{N} : x_n - x_{r_q} \in U_q\} \in \mathcal{F}.$$

So there exists $n_{p,q} \in \mathbb{N}$ with $x_{n_{p,q}} - x_{r_p} \in U_p$, $x_{n_{p,q}} - x_{r_q} \in U_q$, so that $x_{r_p} - x_{r_q} \in U_p + U_q$. Thus the sequence $(x_{r_p})_p$ is Cauchy in R in the classical sense and so it converges to an element $x \in R$, since R is complete. If $q \geq p$, we get $x_{r_p} - x_{r_q} \in 2U_p$. Taking the limit as q tends to $+\infty$, we obtain $x_{r_p} - x \in 2U_p$, since U_p is closed. Pick arbitrarily $p \in \mathbb{N}$. If $x_n - x_{r_p} \in U_p$, then

$x_n - x = x_n - x_{r_p} + x_{r_p} - x \in 3U_p,$

and thus

$\{n \in \mathbb{N} : x_n - x \in 3U_p\} \supset \{n \in \mathbb{N} : x_n - x_{r_p} \in U_p\}.$

Now, choose arbitrarily $U \in \mathcal{J}(0)$. There is $\overline{p} \in \mathbb{N}$ with $3U_{\overline{p}} \subset 3U_0 \subset U$, and so

$\{n \in \mathbb{N} : x_n - x \in U\} \supset \{n \in \mathbb{N} : x_n - x \in 3U_{\overline{p}}\} \supset \{n \in \mathbb{N} : x_n - x_{r_{\overline{p}}} \in U_{\overline{p}}\}.$

Since $\{n \in \mathbb{N} : x_n - x_{r_{\overline{p}}} \in U_{\overline{p}}\} \in \mathcal{F}$, then $\{n \in \mathbb{N} : x_n - x \in U\} \in \mathcal{F}$, and so $(\mathcal{F})\lim_n x_n = x$, that is (j).

$(j) \Rightarrow (jjj)$ Suppose that $(\mathcal{F})\lim_n x_n = x$, choose arbitrarily $U \in \mathcal{J}(0)$ and let $U^* \in \mathcal{J}(0)$ be such that $2U^* \subset U$. Then in correspondence with U^* there is $F \in \mathcal{F}$ with $x_n - x \in U^*$ for each $n \in F$, and so for every n, $r \in F$ we get $x_n - x_r \in U$.

$(jjj) \Rightarrow (jj)$ Pick arbitrarily $U \in \mathcal{J}(0)$. Then there exists $F \in \mathcal{F}$ with $x_n - x_r \in U$ for all n, $r \in F$. If $r_0 = \min F$. then $\{n \in \mathbb{N} : x_n - x_{r_0} \in U\} \supset F$, and so $\{n \in \mathbb{N} : x_n - x_{r_0} \in U\} \in \mathcal{F}$, since $F \in \mathcal{F}$. □

We now prove the next technical lemma.

Lemma 4.2 (see Boccuto and Dimitriou (2013e), Lemma 2.2, see also Aviles Lopez, Cascales and Salinas, Kadets and Leonov (2007)) *Let* $(x_{i,n})_{i,n}$ *be a double sequence in* R, *and* $\mathcal{F}$ *be a diagonal filter of* $\mathbb{N}$.

α) *If* $(\mathcal{F})\lim_{i \in \mathbb{N}} x_{i,n} = 0$ *for each* $n \in \mathbb{N}$, *then for every* $I \in \mathcal{F}^*$ *there exists* $J \in \mathcal{F}^*$, $J \subset I$ *such that* $\lim_{i \in J} x_{i,n} = 0$ *for each* $n \in \mathbb{N}$.

$\alpha\alpha$) *If* $(V_i)_i$ *is an increasing sequence in* $\mathcal{J}(0)$ *and* $(x_{i,n})_i$ *is* $\mathcal{F}$*-bounded by* $(V_i)_i$ *for every* $n\in\mathbb{N}$, *then for each* $I\in\mathcal{F}^*$ *there exists* $J\in\mathcal{F}^*$, $J\subset I$ *such that* $(x_{i,n})_i$ *is* $\mathcal{F}_{\text{cofin}}$*-bounded by* $(V_i)_i$.

Proof: α) Let $(U_p)_p$ be a countable basis of neighborhoods of 0. By hypothesis, for every n, $p\in\mathbb{N}$ we have $A_{n,p}:=\{i\in\mathbb{N}: x_{i,n}\in U_p\}\in\mathcal{F}$. Since $\mathcal{F}$ is diagonal, for each $I\in\mathcal{F}^*$ there is $J\in\mathcal{F}^*$, $J\subset I$, such that for every n, $p\in\mathbb{N}$ the set $J\setminus A_{n,p}$ is finite. Thus, for every n, $p\in\mathbb{N}$ there is $\bar{i}\in\mathbb{N}$ (without loss of generality $\bar{i}\in J$) with $x_{i,n}\in U_p$ for all $i\geq\bar{i}$, $i\in J$. This proves α).

The proof of $\alpha\alpha$) is analogous, taking the sets $A_n^*:=\{i\in\mathbb{N}: x_{i,n}\in V_i\}$, $n\in\mathbb{N}$. □

Also the following technical results hold (see Boccuto and Dimitriou (2014a), Proposition 2.6, Lahiri and Das (2005), Theorem 8 (i)).

Proposition 4.3 *Let* $(x_n)_n$ *be a sequence in* R, $(\mathcal{F})$*-convergent to* $x\in R$. *If* $\mathcal{F}$ *is a* P*-filter, then there exists an element* $E\in\mathcal{F}$, *with* $\lim_{n\in E} x_n = x$.

We now prove the following

Proposition 4.4 (see Boccuto and Dimitriou (2014a), Proposition 2.7) *Let* $(x_{i,n})_{i,n}$ *be a double sequence in* R, $\mathcal{F}$ *be any* P*-filter of* $\mathbb{N}$, *and suppose that* $(\mathcal{F})\lim_i x_{i,n} = x_n$ *for every* $n\in\mathbb{N}$. *Then there exists* $B_0\in\mathcal{F}$ *such that* $\lim_{i\in B_0} x_{i,n} = x_n$ *for all* $n\in\mathbb{N}$.

Proof: By hypothesis and Proposition 4.3 we get the existence of a sequence $(A_n)_n$ in $\mathcal{F}$, with $\lim_{i\in A_n} x_{i,n} = x_n$ for all $n\in\mathbb{N}$. As $\mathcal{F}$ is a P-filter, there is a sequence of sets $(B_n)_n$ in $\mathcal{F}$, such that $A_n\Delta B_n$ is finite for all $n\in\mathbb{N}$ and

$B_0 := \bigcap_{n=1}^{\infty} B_n \in \mathcal{F}$. Thus, since $\lim_{i \in A_n} x_{i,n} = x_n$ for all $n \in \mathbb{N}$, we get also $\lim_{i \in B_n} x_{i,n} = x_n$, and a fortiori $\lim_{i \in B_0} x_{i,n} = x_n$, for all n. □

We now consider some main properties of topological group-valued measures. A finitely additive measure $m : \Sigma \to R$ is (s)-*bounded* on the σ-algebra Σ, iff

$$\lim_k m(C_k) = 0 \text{ for each disjoint sequence } (C_k)_k \text{ in } \Sigma; \tag{4.123}$$

σ *-additive* on Σ, iff $m(\bigcup_{k=1}^{\infty} C_k) = \sum_{k=1}^{\infty} m(C_k) := \lim_i (\sum_{k=1}^{i} m(C_k))$ for any disjoint sequence $(C_k)_k$ in Σ.

For each σ-algebra $\mathcal{L} \subset \Sigma$, let us define $m^{\mathcal{L}}(A) := \{m(B) : B \in \mathcal{L}, B \subset A\}, A \in \mathcal{L}$. Moreover, put $m^+(A) := m^{\Sigma}(A) = \{m(B) : B \in \Sigma, B \subset A\}$, $A \in \Sigma$. Given two finitely additive measures, $m : \Sigma \to R$, $\lambda : \Sigma \to [0, +\infty]$, we say that m is λ-*absolutely continuous* on Σ, iff $\lim_k m^+(H_k) = 0$ for every decreasing sequence $(H_k)_k$ in Σ such that $\lim_k \lambda(H_k) = 0$.

We will see that m is λ-absolutely continuous if and only if $\lim_n m^+(A_n) = 0$ for any arbitrary sequence $(A_n)_n$ in Σ with $\lim_n \lambda(A_n) = 0$. We have seen that, in the lattice group setting, this property is in general not true, if it is intended in the classical like sense and not with respect to a single order sequence. We first extend Drewnowski (1972a), Lemma 4.6 to the topological group context.

Lemma 4.5 (see Boccuto and Dimitriou (2014a), Lemma 2.8) *Let $m : \Sigma \to R$ be an (s)-bounded measure and $(E_k)_k$ be any arbitrary sequence of elements of Σ. Then for every $U \in \mathcal{J}(0)$ there is $q \in \mathbb{N}$ with*

$$m^+(E_k \setminus \bigcup_{l=1}^{q} E_l) \subset U \text{ for each } k \geq q.$$

Proof: If we deny the thesis, then it is possible to find a neighborhood $U \in \mathcal{J}(0)$ and to construct a strictly increasing sequence $(r_h)_h$ in $\mathbb{N}$, with $m^+(B_h) \not\subset U$ for every $h \in \mathbb{N}$, where $B_h := E_{r_{h+1}} \setminus \bigcup_{l=1}^{r_h} E_l$. It is not difficult to see that the B_h's are pairwise disjoint, getting a contradiction with (s)-boundedness of m. □

We are in position to prove the following characterization of absolute continuity, using a technique similar to that of Drewnowski (1972a), Theorem 6.1 (a).

Theorem 4.6 (see Boccuto and Dimitriou (2014a), Theorem 2.9) *Let $\lambda : \Sigma \to [0,+\infty]$ be a finitely additive measure. An (s)-bounded measure $m : \Sigma \to R$ is λ-absolutely continuous if and only if $\lim\limits_n m^+(A_n) = 0$ for any sequence $(A_n)_n$ in Σ, such that $\lim\limits_n \lambda(A_n) = 0$.*

Proof: The "if" part is straightforward.

We now turn to the "only if" part. If we deny the thesis, then there exist: a neighborhood $U \in \mathcal{J}(0)$, a decreasing sequence $(U_h)_h$ in $\mathcal{J}(0)$, a sequence $(A_n)_n$ in Σ, with $2U_h \subset U_{h-1}$ for every $h \in \mathbb{N}$, $2U_0 \subset U$, $\lim\limits_n \lambda(A_n) = 0$ and $m^+(A_n) \not\subset U$ for each $n \in \mathbb{N}$. So, we can extract a subsequence $(A_{n_k})_k$ of $(A_n)_n$, with $\lambda(A_{n_k}) \le 2^{-k}$ for all $k \in \mathbb{N}$.

Let $E_k := A_{n_k}$. At the first step, by Lemma 4.5 applied to the sequence E_k, $k \in \mathbb{N}$, in correspondence with U_1 there exists $k_1 \in \mathbb{N}$, with

$$m^+(E_k) \subset m^+(E_k \setminus \bigcup_{l=1}^{k_1} E_l) + m^+(E_k \cap (\bigcup_{l=1}^{k_1} E_l)) \subset U_1 + m^+(E_k \cap (\bigcup_{l=1}^{k_1} E_l)) \text{ for every } k \ge k_1. \tag{4.124}$$

Put $B_1 := \bigcup_{l=1}^{k_1} E_l$. From (4.124) we deduce

$$m^+(E_k \cap B_1) \not\subset U_0 + U_1, \tag{4.125}$$

otherwise we should get $m^+(E_k) \subset U_0 + 2U_1 \subset 2U_0 \subset U$, a contradiction. Hence, from (4.125) we obtain $m^+(B_1) \not\subset U_0$.

Proceeding by induction, at the $h+1$-th step suppose that we have determined $k_1 < k_2 < \ldots < k_h \in \mathbb{N}$ and $B_1, \ldots B_h \in \Sigma$, with

$$B_0 := G, B_h = B_{h-1} \cap (\bigcup_{l=k_{h-1}+1}^{k_h} E_l), \; m^+(E_k \cap B_h) \not\subset U_0 + U_h \qquad \textbf{(4.126)}$$

for all $h \in \mathbb{N}$ and $k \geq k_h$. By Lemma 4.5 applied to the sequence $E_k \cap B_h$, $k = k_h + 1, \; k_h + 2, \ldots$, in correspondence with U_{h+1} we find an integer $k_{h+1} > k_h$, with

$$m^+(E_k \cap B_h) \subset m^+((E_k \cap B_h) \setminus \bigcup_{l=k_h+1}^{k_{h+1}} E_l) +$$

$$+ m^+((E_k \cap B_h) \cap (\bigcup_{l=k_h+1}^{k_{h+1}} E_l)) \subset U_{h+1} + m^+(E_k \cap B_{h+1}) \qquad \textbf{(4.127)}$$

whenever $k \geq k_{h+1}$, where $B_{h+1} = B_h \cap (\bigcup_{l=k_h+1}^{k_{h+1}} E_l)$. From (4.127) we obtain that $m^+(E_k \cap B_{h+1}) \not\subset U_0 + U_{h+1}$, otherwise we should have $m^+(E_k \cap B_h) \subset U_0 + 2U_{h+1} \subset U_0 + U_h$, which contradicts (4.126). Hence, $m^+(B_{h+1}) \not\subset U_0$.

By construction, $(B_h)_h$ is a decreasing sequence in Σ, $\lim_h \lambda(B_h) = 0$ and $m^+(B_h) \not\subset U_0$ for every $h \in \mathbb{N}$, which contradicts λ-absolute continuity of m.
□

Let τ be a Fréchet-Nikodým topology on Σ. A finitely additive measure $m : \Sigma \to R$ is said to be τ*-continuous* on Σ, iff $\lim_k m^+(H_k) = 0$ for every decreasing sequence $(H_k)_k$ of elements of Σ, with τ-$\lim_k H_k = \varnothing$.

Note that, when λ is a finitely additive non-negative real-valued measure defined on Σ and τ is the topology generated by the pseudo-λ-distance defined by $d_\lambda(A,B) := \lambda(A \Delta B)$, A, $B \in \Sigma$, then τ-continuity is equivalent to λ-absolute continuity (see also Boccuto, Das, Dimitriou and Papanastassiou (2012), Diestel and Uhl(1977)).

A finitely additive measure $m:\Sigma \to R$ is said to be *positive* iff every neighborhood W of 0 contains a neighborhood U_0 of 0 such that

for any $A \in \Sigma$ with $m(A) \in U_0$ and for each $B \in \Sigma$ with $B \subset A$, we get $m(B) \in U_0$. **(4.128)**

It is readily seen that, when $R = \mathbb{R}$, every measure, positive in the usual sense, is positive also in the sense of the definition given in (4.128).

It is not difficult to see that a finitely additive measure $m:\Sigma \to R$ is (s)-bounded on Σ if and only if $\lim_k m^+(C_k) = 0$ for all disjoint sequences $(C_k)_k$ in Σ. Otherwise, there exist a disjoint sequence $(C_k)_k$, a neighborhood $U \in \mathcal{J}(0)$ and two sequences $(n_k)_k$, $(B_k)_k$ in $\mathbb{N}$ and Σ, respectively, with $\lim_k n_k = +\infty$, $B_k \subset C_k$ and $m(B_k) \notin U$ for each $k \in \mathbb{N}$, getting a contradiction with (4.123), since the B_k's are pairwise disjoint.

We now give the following property of (s)-bounded topological group-valued measures (see also Candeloro (1985a), Sion (1969 and 1973)).

Proposition 4.7 (see Boccuto and Dimitriou (2014a), Proposition 2.10) *Let $m:\Sigma \to R$ be a σ-additive measure. Then*

$$\lim_k m^+(H_k) = 0 \qquad \textbf{(4.129)}$$

for each decreasing sequence $(H_k)_k$ in Σ, satisfying

$$m(B \cap (\bigcap_{k=1}^{\infty} H_k)) = 0 \quad \textit{for every } B \in \Sigma. \qquad \textbf{(4.130)}$$

Proof: Let m and $(H_k)_k$ be as in the hypothesis. First of all we prove that

$$\lim_k \Big(\bigcup_{p \ge q \ge k} m^+(H_q \setminus H_p) \Big) = 0,$$

that is for every $U \in \mathcal{J}(0)$ there is $\overline{k} \in \mathbb{N}$ with

$$m(E) \in U \text{ for any } p \ge q \ge \overline{k} \text{ and for each } E \in \Sigma \text{ such that } E \subset H_q \setminus H_p. \quad \textbf{(4.131)}$$

If (4.131) is not true, then there are a neighborhood $U \in \mathcal{J}(0)$, two sequences $(k_h)_h$, $(p_h)_h$ in $\mathbb{N}$, with $\lim_h k_h = +\infty$, a sequence $(B_h)_h$ in Σ, with $B_h \subset H_{k_h} \setminus H_{k_h + p_h}$ and $m(B_h) \notin U$ for every $h \in \mathbb{N}$. Without loss of generality, we can suppse that $k_{h+1} > k_h + p_h$ for every h. Thus the sequence $(B_h)_h$ is disjoint, and hence we obtain a contradiction with (s)-boundedness of m.

We now prove (4.129). To this aim, we claim that for each $U \in \mathcal{J}(0)$ there exists $\overline{k} \in \mathbb{N}$ with $m(E) \in U$ for all $E \subset H_k$ and $k \ge \overline{k}$. Set $H_\infty := \bigcap_{k=1}^\infty H_k$, $E' := E \setminus H_\infty$ and $E_p := E \setminus H_p$, $p \in \mathbb{N}$. Note that $(E_p)_p$ is an increasing sequence in Σ, and that $\bigcup_{p=1}^\infty E_p = E \setminus H_\infty = E'$. Let $\overline{k}$ be as in (4.131), and $p \ge k \ge \overline{k}$. Since $E_p \subset H_k \setminus H_p$, from (4.131) it follows that $m(E_p) \in U$. Since U is closed, by (4.130) we get that $m(E) = m(E') = \lim_p m(E_p) \in U$. This proves the claim, and so we obtain (4.129). □

The following result holds.

Theorem 4.8 (see Boccuto and Dimitriou (2014a), Theorem 2.11) *A finitely additive measure* $m : \Sigma \to R$ *is* σ*-additive on* Σ *if and only if* $\lim_k m^+(H_k) = 0$ *for each decreasing sequence* $(H_k)_k$ *in* Σ, *with* $\bigcap_{k=1}^\infty H_k = \varnothing$.

We now prove the following property of σ-additive topological group-valued measures.

Theorem 4.9 (see Boccuto and Dimitriou (2014a), Theorem 2.12) *Let $m:\Sigma \to R$ be a σ-additive measure, and $(E_k)_k$ be any sequence in Σ. Then we get*

$$m^+(\bigcup_{k=1}^{\infty} E_k) \subset \sum_{k=1}^{\infty} m^+(E_k). \quad \textbf{(4.132)}$$

Proof: Set $C_1 := E_1$, $C_k := E_k \setminus (\bigcup_{i=1}^{k-1} E_i)$, $k \geq 2$. Note that the sets C_k are disjoint and $\bigcup_{k=1}^{\infty} C_k = \bigcup_{k=1}^{\infty} E_k$. Choose arbitrarily $B \in \Sigma$, $B \subset \bigcup_{k=1}^{\infty} E_k$, and set $B_k := B \cap C_k$, $k \in \mathbb{N}$. Taking into account σ-additivity of m, we get

$$m(B) = \sum_{k=1}^{\infty} m(B_k) \in \sum_{k=1}^{\infty} m^+(C_k) \subset \sum_{k=1}^{\infty} m^+(E_k). \quad \textbf{(4.133)}$$

By (4.133) and arbitrariness of B we obtain (4.132). This ends the proof. □

We now turn to a Drewnowski-type theorem about the existence of σ-additive restrictions of (s)-bounded topological group-valued measures. This will be useful to prove some equivalence results between the filter limit theorems involved. We first recall the following

Theorem 4.10 (Candeloro and Letta (1985), Lemma 2.3) *Let $m:\Sigma \to R$ be an (s)-bounded measure. Then for each disjoint sequence $(C_k)_k$ in Σ there exists an infinite subset $P_0 \subset \mathbb{N}$, with*

$$\lim_r (\{m(\bigcup_{k \in Y, k \geq r} C_k) : Y \subset P_0\}) = 0, \quad \textbf{(4.134)}$$

and m is σ-additive on the σ-algebra generated by the sets C_k, $k \in P_0$.

Proof: Let $U \in \mathcal{J}(0)$, and $(U_h)_h$ be a sequence in $\mathcal{J}(0)$, with $U_{h+1} \subset U_h \subset U$ for every $h \in \mathbb{N}$, $(P_r^1)_r$ be a disjoint sequence of infinite subsets of $\mathbb{N}$, and for each r set $H_r^1 := \{C_n : n \in P_r^1\}$. The sequence $(H_r^1)_r$ in Σ is disjoint and hence, by (s)-boundedness of m, there is $r_1 \in \mathbb{N}$ with

$$\{m(\bigcup_{k\in Y,k\geq r} C_k):Y\subset P_r^1\}\subset U_1 \text{ for all } r\geq r_1.$$

Choose any infinite partition of $P_{r_1}^1$ into a sequence of disjoint infinite subsets $(P_r^2)_r$. For each r set $H_r^2:=\bigcup\{C_n:n\in P_r^1\}$. The sets H_r^2, $r\in\mathbb{N}$, are pairwise disjoint and contained in $H_{r_1}^1$ by construction. Again by (s)-boundedness of m, there exists an integer $r_2>r_1$ with

$$\{m(\bigcup_{k\in Y,k\geq r} C_k):Y\subset P_r^2\}\subset U_2 \text{ for every } r\geq r_2.$$

Proceeding by induction, we find a strictly increasing sequence $(r_l)_l$ in $\mathbb{N}$, a decreasing sequence $(P_{r_l}^l)_l$ of infinite subsets of $\mathbb{N}$, and a sequence $H_{r_l}^l:=\{C_n:n\in P_{r_l}^l\}$, $l\in\mathbb{N}$, satisfying

$$\{m(\bigcup_{k\in Y,k\geq r} C_k):Y\subset P_{r_l}^l\}\subset U_l \text{ for all } l\in\mathbb{N} \text{ and } r\geq r_l. \qquad \textbf{(4.135)}$$

Let n_1 be the first element of $P_{r_1}^1$, and by induction suppose that we have defined n_l and let n_{l+1} be the first element of $P_{r_l}^l$ strictly greater than n_l. Set $P_0:=\{n_l: l\in\mathbb{N}\}$. We claim that P_0 is the requested subset of $\mathbb{N}$. Indeed, first of all we observe that for each $l\in\mathbb{N}$ there is a sufficiently large $r'\in\mathbb{N}$ (depending on l), such that for each $r\geq r'$ we get $\{k\in P_0:k\geq r\}\subset\{k\in P_{r_l}^l:k\geq r\}$, and hence

$$\{m(\bigcup_{k\in Y,k\geq r} C_k):Y\subset P_0\}\subset\{m(\bigcup_{k\in Y,k\geq r} C_k):Y\subset P_{r_l}^l\}\subset U_l\subset U. \qquad \textbf{(4.136)}$$

From (4.135) and (4.136) it follows that

$$\lim_r(\{m(\bigcup_{k\in Y,k\geq r} C_k):Y\subset P_0\})=0, \text{ that is (4.134).}$$

We now turn to the last assertion. Let $C:=\bigcup_{l=1}^{\infty} C_{n_l}$, and choose any decreasing sequence $(H_s)_s$ in the σ-algebra $\mathcal{K}$ generated in C by the sets C_{n_l}, with

$\bigcap_{s=1}^{\infty} H_s = \emptyset$. For every $s \in \mathbb{N}$, let $r(s)$ be the greatest integer such that $H_s \subset \bigcup_{l \geq r(s)} C_{n_l}$. Note that $\lim_s r(s) = +\infty$. From this and (4.134) it follows that $\lim_s m^{\mathcal{K}}(H_s) = 0$, that is σ-additivity of m on $\mathcal{K}$. This completes the proof. □

A consequence of Theorem 3.10 is the following

Theorem 4.11 (see Boccuto and Dimitriou (2014a), Theorem 2.14) *Let $m_j : \Sigma \to R$, $j \in \mathbb{N}$, be a sequence of (s)-bounded measures. Then for any disjoint sequence $(C_k)_k$ in Σ there exists an infinite subset $P \subset \mathbb{N}$, such that $\lim_h (m_j(\bigcup_{k \in Y, k \geq h} C_k) : Y \subset P\}) = 0$ for every $j \in \mathbb{N}$, and each m_j is σ-additive on the σ-algebra generated by the sets C_k, $k \in P$.*

Proof: By Theorem 4.10 there is an infinite subset $P_1 \subset \mathbb{N}$ with

$$\lim_h (\{m_1(\bigcup_{k \in Y, k \geq h} C_k) : Y \subset P_1\}) = 0.$$

At the second step, there is an infinite subset $P_2 \subset P_1$ such that $\lim_h (\{m_2(\bigcup_{k \in Y, k \geq h} C_k) : Y \subset P_2\}) = 0.$

Proceeding by induction, we find a strictly increasing $(p_j)_j$ in $\mathbb{N}$ and a decreasing sequence of infinite subsets $P_j \subset \mathbb{N}$, with $p_j = \min P_j$ and

$$\lim_h (\{m_j(\bigcup_{k \in Y, k \geq h} C_k) : Y \subset P_j\}) = 0 \text{ for every } j \in \mathbb{N}. \quad \textbf{(4.137)}$$

Let $P := \{p_j : j \in \mathbb{N}\}$. For every $j \in \mathbb{N}$ there is $h' \in \mathbb{N}$ large enough (depending on j), such that for each $h \geq h'$ we get $\{k \in P : k \geq h\} \subset \{k \in P_j : k \geq h\}$, and hence

$$\{m_j(\bigcup_{k \in Y, k \geq h} C_k) : Y \subset P\} \subset \{m_j(\bigcup_{k \in Y, k \geq h} C_k) : Y \subset P_j\}. \quad \textbf{(4.138)}$$

From (4.137) and (4.138) it follows that

$$\lim_h(\{m_j(\bigcup_{k\in Y,k\geq h} C_k):Y\subset P\})=0 \text{ for all } j\in\mathbb{N}. \qquad \textbf{(4.139)}$$

We now turn to the last assertion. Let $C_* := \bigcup_{q\in P} C_q = \bigcup_{l=1}^{\infty} C_{p_l}$ and pick any decreasing sequence $(H_s)_s$ in the σ-algebra $\mathcal{L}$ generated in C_* by the sets C_{n_l}, with $\bigcap_{s=1}^{\infty} H_s = \emptyset$. For every $s\in\mathbb{N}$ let $h(s)$ be the greatest integer such that $H_s \subset \bigcup_{l\geq h(s)} H_{p_l}$. Observe that $\lim_s h(s) = +\infty$. From this and (4.139) it follows that $\lim_s m_j^{\mathcal{L}}(H_s)=0$ for every $j\in\mathbb{N}$, that is σ-additivity of every m_j on $\mathcal{L}$. This ends the proof. □

We say that the finitely additive measures $m_j:\Sigma\to R$, $j\in\mathbb{N}$, are *uniformly* (s)-*bounded* on Σ iff $\lim_k (\bigcup_{j=1}^{\infty} m_j^+(C_k))=0$ for each disjoint sequence $(C_k)_k$ in Σ. The m_j's are *uniformly* σ*-additive* on Σ iff $\lim_k (\bigcup_{j=1}^{\infty} m_j^+(H_k))=0$ for each decreasing sequence $(H_k)_k$ in Σ with $\bigcap_{k=1}^{\infty} H_k=\emptyset$. If λ is a finitely additive measure on Σ, then the measures m_j are said to be *uniformly* λ*-absolutely continuous* (or *equiabsolutely continuous with respect to* λ) on Σ iff $\lim_k (\bigcup_{j=1}^{\infty} m_j^+(H_k))=0$ for each decreasing sequence $(H_k)_k$ in Σ such that $\lim_k \lambda(H_k)=0$. If τ is a Fréchet-Nikodým topology on Σ, then the measures m_j are *uniformly* τ*-continuous* on Σ iff $\lim_k (\bigcup_{j=1}^{\infty} m_j^+(H_k))=0$ for every decreasing sequence $(H_k)_k$ in Σ such that $\tau\text{-}\lim_k H_k=\emptyset$.

We now recall the following property, which will be useful in the sequel in order to prove our limit theorems in the topological group setting.

Theorem 4.12 (Candeloro and Letta, Corollary 3.15) *Let* $m_j : \Sigma \to R$, $j \in \mathbb{N}$, *be a sequence of uniformly* (s)*-bounded measures,* $(H_k)_k$ *be a decreasing sequence in* Σ, *with* $\lim_k m_j^+(H_k) = 0$ *for every* $j \in \mathbb{N}$. *Then* $\lim_k (\bigcup_{j=1}^{\infty} m_j^+(H_k)) = 0$.

Note that, by arguing analogously as in Theorem 3.6, it is possible to prove the following

Theorem 4.13 (see Boccuto and Dimitriou (2014a), Theorem 2.16) *Let* $\lambda : \Sigma \to [0,+\infty]$ *be a finitely additive measure. A sequence* $m_j : \Sigma \to R$, $j \in \mathbb{N}$, *of uniformly* (s)*-bounded measures is equiabsolutely continuous with respect to* λ *if and only if* $\lim_n (\bigcup_j m_j^+(A_n)) = 0$ *for any sequence* $(A_n)_n$ *in* Σ, *with* $\lim_n \lambda(A_n) = 0$.

Indeed, it will be enough to consider the quantity $\bigcup_j m_j^+$ instead of m^+.

Let now $\mathcal{G}$, $\mathcal{H} \subset \Sigma$ be two lattices, such that $\mathcal{G}$ is closed with respect to countable disjoint unions, and the complement of every element of $\mathcal{H}$ belongs to $\mathcal{G}$. We say that $m : \Sigma \to R$ is *regular* on Σ iff for every $A \in \Sigma$ there exist two sequences $(G_k)_k$ in $\mathcal{G}$, $(F_k)_k$ in $\mathcal{H}$, with $F_k \subset F_{k+1} \subset A \subset G_{k+1} \subset G_k$ for every k and $\lim_k m^+(G_k \setminus F_k) = 0$. Observe that, if $m_j : \Sigma \to R$, $j \in \mathbb{N}$, are regular measures, then the sequences $(G_k)_k$, $(F_k)_k$ can be taken independently of j (see Candeloro and Letta (1985)). The measures $m_j : \Sigma \to R$, $j \in \mathbb{N}$, are *uniformly regular* on Σ iff for every $A \in \Sigma$ there are two sequences $(G_k)_k$ in $\mathcal{G}$, $(F_k)_k$ in $\mathcal{H}$, with $F_k \subset F_{k+1} \subset A \subset G_{k+1} \subset G_k$ for every k and $\lim_k (\bigcup_{j=1}^{\infty} m_j^+(G_k \setminus F_k)) = 0$.

We now prove the following relation between σ-additivity and regularity of measures.

Theorem 4.14 (see Boccuto and Dimitriou (2014a), Theorem 2.17) *Let* (G,d) *be a compact metric space,* Σ *be the* σ*-algebra of all Borel sets of* G, $\mathcal{G}$ *and* $\mathcal{H}$ *be the lattices of all open and all closed subsets of* G *respectively. Then a finitely additive measure* $m : \Sigma \to R$ *is regular if and only if it is* σ*-additive.*

Proof: We begin with the "if" part. Let $\mathcal{T} := \{A \in \Sigma$: for every $U \in \mathcal{J}(0)$ there are $D \in \mathcal{G}$, $F \in \mathcal{H}$ with $F \subset A \subset D$ and $m^+(D \setminus F) \subset U\}$. Observe that $\mathcal{H} \subset \mathcal{T}$. Indeed, pick arbitrarily $W \in \mathcal{H}$ and for each $k \in \mathbb{N}$ set $D_k := \{x \in G : d(x, W) < 1/k\}$, $W_k := D_k \setminus W$. Note that the sequence $(W_k)_k$ is decreasing, and $\bigcap_{k=1}^{\infty} W_k = \emptyset$. By σ-additivity of m, for every $U \in \mathcal{J}(0)$ there is $k_0 \in \mathbb{N}$, with $m^+(D_{k_0} \setminus W) \subset U$. Since $D_{k_0} \in \mathcal{G}$, $W \in \mathcal{H}$ and $W \subset D_{k_0}$, it follows that $W \in \mathcal{T}$.

We now prove that $\mathcal{T}$ is a σ-algebra. It is easy to see that, if $A \in \mathcal{T}$, then $G \setminus A \in \mathcal{T}$. Let now $(A_k)_k$ be a disjoint sequence of elements of $\mathcal{T}$, with $A := \bigcup_{k=1}^{\infty} A_k$. We claim that $A \in \mathcal{T}$.

Choose arbitrarily $U \in \mathcal{J}(0)$, let $(U_k)_k$ be a sequence in $\mathcal{J}(0)$, such that $2U_k \subset U_{k-1} \subset U_0 \subset U$ for every k and $2U_0 \subset U$. Note that $\sum_{k=1}^{n} U_k \subset U_0$ for all $n \in \mathbb{N}$, and hence, since U_0 is closed, we get also $\sum_{k=1}^{\infty} U_k \subset U_0$.

By hypothesis there are two sequences $(D_k)_k$ and $(F_k)_k$ in $\mathcal{G}$ and $\mathcal{H}$ respectively, with $F_k \subset A_k \subset D_k$ and $m^+(D_k \setminus F_k) \subset U_k$ for every k. Since $(F_k)_k$ is disjoint, by σ-additivity of m there exists $k_0 \in \mathbb{N}$ with

$$m^+\Big(\Big(\bigcup_{k=1}^{\infty} F_k\Big) \setminus \Big(\bigcup_{k=1}^{k_0} F_k\Big)\Big) = m^+\Big(\bigcup_{k=k_0+1}^{\infty} F_k\Big) \subset U_1.$$

Set $D := \bigcup_{k=1}^{\infty} D_k$, $F := \bigcup_{k=1}^{k_0} F_k$. Note that $F \subset A \subset D$, $D \in \mathcal{G}$, $F \in \mathcal{H}$, and by Theorem 3.9 we get

$$m^+(D \setminus F) \subset m^+\Big(D \setminus \Big(\bigcup_{k=1}^{\infty} F_k\Big)\Big) + m^+\Big(\Big(\bigcup_{k=1}^{\infty} F_k\Big) \setminus F\Big) \subset$$

$$\subset m^+\Big(\bigcup_{k=1}^{\infty} (D_k \setminus F_k)\Big) + U_0 \subset \sum_{k=1}^{\infty} m^+(D_k \setminus F_k) + U_0 \subset 2U_0 \subset U.$$

From this it follows that $A \in \mathcal{T}$, that is the claim. Therefore, $\mathcal{T}$ is a σ-algebra. Since $\mathcal{T} \supset \mathcal{H}$, then $\mathcal{T} = \Sigma$. Since R satisfies the first axiom of countability, there is a family $(U_k)_k$, which is a basis of neighborhoods of 0. In correspondence with U_k and every $A \in \Sigma$, there are $D_k^* \in \mathcal{G}$, $F_k^* \in \mathcal{H}$, with $F_k^* \subset A \subset D_k^*$ and $m^+(D_k^* \setminus F_k^*) \subset U_k$. For every $k \in \mathbb{N}$, let $D_k := \bigcap_{i=1}^k D_i^*$, $F_k := \bigcup_{i=1}^k F_i^*$. We get $F_k \subset F_{k+1} \subset A \subset G_{k+1} \subset G_k$, $D_k \in \mathcal{G}$, $F_k \in \mathcal{H}$, $m^+(D_k \setminus F_k) \subset m^+(D_k^* \setminus F_k^*) \subset U_k$, and so it follows that $\lim_k m^+(D_k \setminus F_k) = 0$. Thus, m is regular on Σ. This proves the "if" part.

We now turn to the "only if" part. Let $(C_k)_k$ be a disjoint sequence in Σ, and set $C := \bigcup_{k=1}^\infty C_k$. Fix arbitrarily $U \in \mathcal{J}(0)$, and let $(U_k)_k$ be a sequence in $\mathcal{J}(0)$, with $2U_k \subset U_{k-1}$ for every k and $2U_0 \subset U$. By hypothesis m is regular, and so in correspondence with C_k and U_k there exist $D_k \in \mathcal{G}$, $F_k \in \mathcal{H}$, with $F_k \subset C_k \subset D_k$ and $m^+(D_k \setminus C_k) \subset m^+(D_k \setminus F_k) \subset U_k$. Moreover there exists a set $K \in \mathcal{H}$, $K \subset C$, with $m^+(C \setminus K) \subset U_0$. Note that, since G is compact, K is also compact, and hence, since $K \subset \bigcup_{k=1}^\infty D_k \in \mathcal{G}$, there exists $N \in \mathbb{N}$ with $K \subset \bigcup_{k=1}^N D_k$. Choose arbitrarily $B \subset \bigcup_{k=N+1}^\infty C_k = C \setminus (\bigcup_{k=1}^N C_k)$. As $B \cap K \subset (\bigcup_{k=1}^N D_k) \setminus (\bigcup_{k=1}^N C_k)$, by Theorem 4.9 we get

$$m(B) = m(B \setminus K) + m(B \cap K) \in m^+(C \setminus K) + m^+((\bigcup_{k=1}^N D_k) \setminus (\bigcup_{k=1}^N C_k)) \subset$$

$$\subset m^+(C \setminus K) + m^+(\bigcup_{k=1}^N (D_k \setminus C_k)) \subset m^+(C \setminus K) + \sum_{k=1}^N m^+(D_k \setminus C_k) \subset$$

$$\subset m^+(C \setminus K) + \sum_{k=1}^N U_k \subset 2U_0 \subset U,$$

and hence $m^+(\bigcup_{k=N+1}^\infty C_k) \subset U$. This proves σ-additivity of m. □

Note that, arguing similarly as above, it is possible to prove that, under the same hypotheses as in Theorem 4.14, if $m_j : \Sigma \to R$, $j \in \mathbb{N}$, is a sequence of measures, then the set functions m_j are uniformly regular if and only if they are uniformly σ-additive.

Analogously as Proposition 2.20 of Chapter 2 it is possible to prove the following

Proposition 3.15 (see Boccuto and Dimitriou (2014a), Proposition 2.4) *If $\mathcal{F}$ is any free filter, x_n, $n \in \mathbb{N}$, is a sequence in R, $\mathcal{F}$-convergent to $x \in R$, and $J \in \mathcal{F}^*$, then the sequence x_n, $n \in J$, $(\mathcal{F}(J))$-converges to x.*

4.5. Filter Limit Theorems for Topological Group-Valued Measures

4.5.1. Schur-Type Theorems

We now prove some filter limit theorems for topological group-valued measures. We begin with a Schur theorem, which extends Aviles Lopez, Cascales Salinas, Kadets and Leonov (2007), Theorems 2.6 and 3.5. We use sliding hump techniques, diagonal arguments and the main properties of block-respecting and diagonal filters, and deduce Vitali-Hahn-Saks and Nikodým convergence-type theorems.

Theorem 5.1 (see Boccuto and Dimitriou (2014a), Theorem 3.1) *Let $\mathcal{F}$ be a block-respecting filter of $\mathbb{N}$, $m_j : \mathcal{P}(\mathbb{N}) \to R$, $j \in \mathbb{N}$, be a sequence of σ-additive measures, and let*

(i) $\lim\limits_j m_j(\{n\}) = 0$ *for any* $n \in \mathbb{N}$,

(ii) $(\mathcal{F}) \lim\limits_j m_j(A) = 0$ *for every* $A \subset \mathbb{N}$.

Then we have:

β) $(\mathcal{F}) \lim\limits_j m_j^+(\mathbb{N}) = 0$;

$\beta\beta$) *If $\mathcal{F}$ is also diagonal, then the only condition (ii) is sufficient to get β).*

Proof: We begin with proving $\beta)$. If $\beta)$ is not true, then there exists $U \in \mathcal{J}(0)$ such that

$$I^* := \{j \in \mathbb{N} : m_j^+(\mathbb{N}) \subset U\} \notin \mathcal{F}.$$

From this it follows that every element F of $\mathcal{F}$ is not contained in I^*, that is F has nonempty intersection with $\mathbb{N} \setminus I^*$: otherwise, if $F \in \mathcal{F}$ and $F \subset I^*$, then we should have $I^* \in \mathcal{F}$. Thus the set $I := \mathbb{N} \setminus I^*$ is $\mathcal{F}$-stationary. Note that I is an infinite set, since $\mathcal{F}$ is a free filter.

Let now $(U_k)_k$ be a decreasing sequence in $\mathcal{J}(0)$, with $2U_0 \subset U$ and $2U_k \subset U_{k-1}$ for every $k \in \mathbb{N}$, and put $n_0 := 1$. By σ-additivity of m_1, there exists an integer $l(1) > 1$ such that $m_1^+(]l(1), +\infty[) \subset U_1$. Moreover, by (i) there is $n_1 > l(1)$ with $m_s(L) \in U_1$ for all $s \geq n_1$ and for each finite subset $L \subset [1, l(1)]$, and thus $m_s^+([1, l(1)]) \subset U_1$ for any $s \geq n_1$. Furthermore, by σ-additivity of $m_1, \ldots, m_{n_1}$, we find a natural number $l(n_1) > n_1$, with $m_r^+(]l(n_1), +\infty[) \subset U_2$ for every $r \leq n_1$, and aan integer $n_2 > l(n_1)$ with $m_s^+([1, l(n_1)]) \subset U_2$ whenever $s \geq n_2$.

By induction, we construct two strictly increasing sequences $(n_h)_h$ and $(l(n_h))_h$, such that for every $h \in \mathbb{N}$, $n_{h-1} < l(h) < n_h$, $m_r^+(]l(n_h), +\infty[) \subset U_{h+1}$ for each $r \leq n_h$, and $m_s^+([1, l(n_h)]) \subset U_{h+1}$ whenever $s \geq n_{h+1}$. Without loss of generality, we can choose n_h with the property that the sets $I \cap [n_{h-1}, n_h[$, $h \in \mathbb{N}$, are nonempty and form a blocking of I. So there is an $\mathcal{F}$-stationary set $J \subset I$, such that J intersects each interval $[n_h, n_{h+1}[$ in exactly one point. Let $J = \{j_0, j_1, j_2, \ldots\}$. Since $J \in \mathcal{F}^*$, then at least one of the two sets $J_1 := \{j_1, j_3, j_5, \ldots\}$ and $J_2 := \{j_0, j_2, j_4, \ldots\}$ is $\mathcal{F}$-stationary. Without loss of generality, suppose that $J_1 \in \mathcal{F}^*$. Now, for each fixed natural number h, we have

$$m_{j_{2h-1}}^+(]l(n_{2h}, +\infty[) \subset U_{2h} \subset U_2, \; m_{j_h}^+([1, l(n_{2h-2}]) \subset U_{2h-1} \subset U_1. \qquad \textbf{(4.140)}$$

From this and since $m_{j_{2h-1}}^+(\mathbb{N}) \not\subset U$, for every positive integer h we get

$$m_{j_{2h-1}}^+(]l(n_{2h-2}), l(n_{2h})]) \not\subset U_0 : \qquad \textbf{(4.141)}$$

otherwise, from (4.140) and (4.141) we have

$$m^{+}_{j_{2h-1}}(\mathbb{N}) \subset U_0 + U_1 + U_2 \subset U_0 + 2U_1 \subset 2U_0 \subset U,$$

a contradiction. By (4.141) there is a set $Q_h \subset]l(n_{2h-2}), l(n_{2h})]$ with

$$m_{j_{2h-1}}(Q_h) \notin U_0. \qquad \textbf{(4.142)}$$

Note that the Q_h's are pairwise disjoint. Set now $H := \bigcup_{h=1}^{\infty} Q_h$. For every h we have

$$m_{j_{2h-1}}(H) = m_{j_{2h-1}}(H \cap [1, l(n_{2h-2})]) + m_{j_{2h-1}}(H \cap]l(n_{2h}), +\infty[) + m_{j_{2h-1}}(H \cap]l(n_{2h-2}), l(n_{2h})]) =$$

$$= m_{j_{2h-1}}(H \cap [1, l(n_{2h-2})]) + m_{j_{2h-1}}(H \cap]l(n_{2h}), +\infty[) + m_{j_{2h-1}}(Q_h),$$

and so we see that

$$m_{j_{2h-1}}(H) - m_{j_{2h-1}}(Q_h) \in U_1 + U_2. \qquad \textbf{(4.143)}$$

Thus we obtain $m_{j_{2h-1}}(H) \notin U_2$ for all h, otherwise

$$m_{j_{2h-1}}(Q_h) \in U_1 + U_2 + U_2 \subset U_1 + U_1 \subset U_0,$$

which contradicts (4.142). But by (ii), in correspondence with U_2 there exists an element $F \in \mathcal{F}$ with $m_j(H) \in U_2$ for all $j \in F$, and, since J_1 is $\mathcal{F}$-stationary, we get that F has at least an element j_* in common with J_1. So we have $m_{j_*}(H) \notin U_2$ and $m_{j_*}(H) \in U_2$, a contradiction. This proves β).

$\beta\beta$) Let $\mathcal{F}$ be a diagonal and block-respecting filter of $\mathbb{N}$. If the thesis is not true, then, proceeding analogously as in β), we get the existence of a neighborhood $U \in \mathcal{J}(0)$ and of an infinite $\mathcal{F}$-stationary set $I \subset \mathbb{N}$, with

$$m^{+}_{j}(\mathbb{N}) \not\subset U \text{ for every } j \in I. \qquad \textbf{(4.144)}$$

Since R satisfies the first axiom of countability, by (*ii*) and Lemma 4.2 α), in correspondence with I there is $J \in \mathcal{F}^*$, $J \subset I$, with $\lim_{j\in J} m_j(\{n\}) = 0$ for every $n \in \mathbb{N}$. Moreover, from (*ii*) and Proposition 4.15 it follows also that $(\mathcal{F}(J)) \lim_{j\in J} m_j(A) = 0$ for every $A \subset \mathbb{N}$. Furthermore observe that, since $J \in \mathcal{F}^*$ and $\mathcal{F}$ is block-respecting, then $\mathcal{F}(J)$ is block-respecting too. By β) applied to the sequence $m_j : \mathcal{P}(\mathbb{N}) \to R$, $j \in J$, and to the filter $\mathcal{F}(J)$ of J, we get $(\mathcal{F}(J)) \lim_{j\in J} m_j^+(\mathbb{N}) = 0$, contradicting (4.144). This ends the proof of $\beta\beta$). □

We now turn to the following Schur-type theorem, which will be useful in the sequel.

Theorem 5.2 (see Boccuto and Dimitriou (2014a), Theorem 3.2) *Let $\mathcal{F}$ be a diagonal filter of $\mathbb{N}$, $m_j : \mathcal{P}(\mathbb{N}) \to R$, $j \in \mathbb{N}$, be a sequence of σ-additive measures, and suppose that $(\mathcal{F}) \lim_{j} m_j^+(\mathbb{N}) = 0$.*

Then for every $I \in \mathcal{F}^*$ there is $J \subset I$, $J \in \mathcal{F}^*$, with $\lim_k (\bigcup_{j\in J} m_j^+([k,+\infty[)) = 0$.

Proof: For every j, $k \in \mathbb{N}$, let $x_{j,k} := m_j^+(\mathbb{N})$. Since R satisfies the first axiom of countability, from Lemma 4.2 α) it follows that for every $I \in \mathcal{F}^*$ there exists $J \subset I$, $J \in \mathcal{F}^*$, with $\lim_{j\in J} m_j^+(\mathbb{N}) = 0$. So, if $U \in \mathcal{J}(0)$ is chosen arbitrarily and $U_0 \in \mathcal{J}(0)$ is such that $2U_0 \subset U$, there is a natural number $\overline{j}$, without loss of generality $\overline{j} \in J$, with $m_j(A) \in U_0$ for every $j \geq \overline{j}$, $j \in J$, and $A \subset \mathbb{N}$. By σ-additivity of m_j, in correspondence with $j \in \mathbb{N}$ there exists $\overline{k}_j \in \mathbb{N}$ with $m_j(A) \in U_0$ for every $A \subset [\overline{k},+\infty[$. If $k^* := \max\{\overline{k}_1, \ldots, \overline{k}_{\overline{j}-1}\}$, then we get

$$m_j(A) \in U_0 \subset U \text{ for each } A \subset [k^*,+\infty[\text{ and } j \in [1, \overline{j}-1]. \qquad \textbf{(4.145)}$$

Moreover, we have

$$m_j(A) \in 2U_0 \subset U \text{ for every } A \subset [k^*,+\infty[\text{ and } j \geq \overline{j},\ j \in J. \qquad \textbf{(4.146)}$$

The assertion follows from (4.145) and (4.146). □

We now prove a Vitali-Hahn-Saks-type theorem, as a consequence of Theorems 5.1 and 5.2.

Theorem 5.3 (see Boccuto and Dimitriou (2014a), Theorem 3.3) *Let $\mathcal{F}$ be a diagonal and block-respecting filter of $\mathbb{N}$, τ be a Fréchet-Nikodým topology on Σ, $m_j : \Sigma \to R$, $j \in \mathbb{N}$, be a sequence of τ-continuous measures, with*

$$(\mathcal{F})\lim_j m_j(A) = 0 \quad \textit{for every} \quad A \in \Sigma. \qquad \textbf{(4.147)}$$

Then for each decreasing sequence $(H_k)_k$ in Σ with τ-$\lim\limits_k H_k = \emptyset$ and for every $\mathcal{F}$-stationary set $I \subset \mathbb{N}$ there is an $\mathcal{F}$-stationary set $J \subset I$ with $\lim\limits_k (\bigcup\limits_{j\in J} m_j^{\mathcal{L}}(H_k)) = 0$, where $\mathcal{L}$ is the σ-algebra generated by the H_k, $k \in \mathbb{N}$, in H_1.

Proof: Let I and $(H_k)_k$ be as in the hypotheses, set $C_k := H_k \setminus H_{k+1}$ for every $k \in \mathbb{N}$ and put $H_\infty := \bigcap_{k=1}^{\infty} H_k$. Since the m_j's are τ-continuous, we get

$$\lim_k m_j^+(H_k) = 0 \text{ for every } j \in \mathbb{N}. \qquad \textbf{(4.148)}$$

For $A \in \mathcal{P}(\mathbb{N})$ and $j \in \mathbb{N}$, set $\nu_j(A) = m_j(\bigcup\limits_{k\in A} C_k)$. We claim that the ν_j is σ-additive. We get

$$m_j^+(H_k) = \{m_j(B) : B \in \Sigma, B \subset H_k\} = \{m_j(B \setminus H_\infty) : B \in \Sigma, B \subset H_k\} =$$

$$= \{m_j(C) : C \in \Sigma, C \subset H_k \setminus H_\infty\} = m_j^+(H_k \setminus H_\infty) = m_j^+(\bigcup_{l=k}^{\infty} C_l) \qquad \textbf{(4.149)}$$

for every j, $k \in \mathbb{N}$. Arguing analogously as in (4.149), it is possible to prove also that

$$m_j^{\mathcal{L}}(H_k) = m_j^{\mathcal{K}}(\bigcup_{l=k}^{\infty} C_l), \tag{4.150}$$

where $\mathcal{K}$ is the σ-algebra generated by C_k, $k \in \mathbb{N}$, in H_1. From (4.148) and (4.149) it follows that $\lim_k m_j^+(\bigcup_{l=k}^{\infty} C_l) = 0$ for any decreasing sequence $(H_k)_k$ in Σ with τ-$\lim_k H_k = \varnothing$. From this, as

$$\nu_j^+([k,+\infty[) := \{\nu_j(D) : D \subset [k,+\infty[\} \subset m_j^+(\bigcup_{l=k}^{\infty} C_l) \text{ for every } j,\ k \in \mathbb{N},$$

we get

$$\lim_k \nu_j^+([k,+\infty[) = 0, \quad j \in \mathbb{N}. \tag{4.151}$$

We prove σ-additivity of ν_j. Let $(A_k)_k$ be a decreasing sequence in $\mathcal{P}(\mathbb{N})$ with $\bigcap_{k=1}^{\infty} A_k = \varnothing$. Without loss of generality, we assume that $A_k \supsetneq A_{k+1}$ for every k. We get $A_k \subset [k,+\infty[$ and $\nu_j^+(A_k) \subset \nu_j^+([k,+\infty[)$ for all j, $k \in \mathbb{N}$. From this and (4.151) we obtain $\lim_k \nu_j^+(A_k) = 0$, and thus σ-additivity of ν_j. Moreover, observe that, since m_j satisfies (4.147), then ν_j fulfils condition (*ii*) of Theorem 5.1. Since $\mathcal{F}$ is diagonal and block-respecting, by $\beta\beta$) of Theorem 5.1 we get $(\mathcal{F}) \lim_j m_j^+(\mathbb{N}) = 0$ for every $A \subset \mathbb{N}$. From this and Theorem 5.2, taking into account (4.150), it follows that for every $I \in \mathcal{F}^*$ there is $J \subset I$, $J \in \mathcal{F}^*$, with

$$0 = \lim_k (\bigcup_{j \in J} \nu_j^+([k,+\infty[)) = \lim_k (\bigcup_{j \in J} m_j^{\mathcal{K}}(\bigcup_{l=k}^{\infty} C_l)) = \lim_k (\bigcup_{j \in J} m_j^{\mathcal{L}}(H_k)).$$

This concludes the proof. □

Similarly as Theorem 5.3, it is possible to prove the following Nikodým convergence-type theorem. Note that in this case σ-additivity of ν_j is a direct consequence of σ-additivity of m_j.

Theorem 5.4 (see Boccuto and Dimitriou (2014a), Theorem 3.4) *Let $\mathcal{F}$ be as in Theorem 5.3, $m_j : \Sigma \to R$, $j \in \mathbb{N}$, be a sequence of σ-additive measures, satisfying* (4.147). *Then for each decreasing sequence $(H_k)_k$ in Σ with $\bigcap_{k=1}^{\infty} H_k = \emptyset$ and for every $I \in \mathcal{F}^*$ there exists $J \in \mathcal{F}^*$, $J \subset I$, with $\lim_k (\bigcup_{j \in J} m_j^{\mathcal{L}}(H_k)) = 0$.*

4.5.2. Other Types of Limit Theorems

We now give some other limit theorems for measures defined on an abstract σ-algebra Σ, which extends Candeloro (1985b), Theorem 2.6, and whose proof is direct and is given still by means of the sliding hump technique. We begin with the following Nikodým convergence-type theorem.

Theorem 5.5 (see Boccuto and Dimitriou (2013e), Theorem 3.1) *Let $\mathcal{F}$ be a block-respecting filter of $\mathbb{N}$, $m_j : \Sigma \to R$, $j \in \mathbb{N}$, be a sequence of σ-additive measures, $(A_n)_n$ be a disjoint sequence in Σ, with*

(*i*) $\lim_j m_j(A_n) = 0$ *for any* $n \in \mathbb{N}$, *and*

(*ii*) $(\mathcal{F}) \lim_j m_j(\bigcup_{p \in P} A_p) = 0$ *for every* $P \subset \mathbb{N}$. *Then,*

β) *For every strictly increasing sequence $(l_n)_n$ in $\mathbb{N}$ we get*

$$(\mathcal{F}) \lim_n m_n(A_{l_n}) = 0; \tag{4.152}$$

$\beta\beta$) *If $\mathcal{F}$ is also diagonal, then* (*ii*) *implies* (4.152).

Proof: Put $H_n := A_{l_n}$, $n \in \mathbb{N}$. If we deny the thesis, then there is $U \in \mathcal{J}(0)$ with $C := \{n \in \mathbb{N} : m_n(H_n) \in U\} \notin \mathcal{F}$. Note that $I := \mathbb{N} \setminus C = \{n \in \mathbb{N} : m_n(H_n) \notin U\} \in \mathcal{F}^*$: otherwise there is $F \in \mathcal{F}$ with $I \cap F = \emptyset$, namely $F \subset C$, and we obtain $C \in \mathcal{F}$, that is a contradiction.

Let $(U_k)_k$ be a decreasing sequence in $\mathcal{J}(0)$ with $U_0 = U$ and $2U_k \subset U_{k-1}$ for all $k \in \mathbb{N}$ (see also Candeloro (1985b)). It is not difficult to see that $lU_k \subset U_{k-l+1}$ for all k, $l \in \mathbb{N}$ with $l \leq k+1$.

Let $N_0 = 1$. By σ-additivity of m_1, there exists a cofinite subset $P_1 \subset \mathbb{N}$, with $N_0 < p_1 := \min P_1$, and $m_1^+(F_1) \subset U_1$, where $F_1 := \bigcup_{t\in P_1} H_t$. By (*i*) there is an integer $N_1 > p_1$ with $m_i(H_t) \in U_1$ whenever $i \geq N_1$ and $t = 1,\ldots,p_1$. By σ-additivity of m_1, $m_2,\ldots,m_{N_1}$, we find a cofinite subset $P_2 \subset P_1$, with $N_1 < p_2 := \min P_2$, and $m_r^+(F_2) \subset U_2$ for every $r = 1,\ldots,N_1$, where $F_2 := \bigcup_{t\in P_2} H_t$. Arguing as above, there exists $N_2 > p_2$ with $m_i(H_t) \in U_2$ whenever $i \geq N_2$ and $t = 1,\ldots,p_2$. Proceeding by induction, we find a strictly decreasing sequence $(P_k)_k$ of cofinite subsets of $\mathbb{N}$, a strictly decreasing sequence $(F_k)_k$ in Σ and two strictly increasing sequences $(N_k)_k$, $(p_k)_k$ in $\mathbb{N}$ such that, for every $k \in \mathbb{N}$,

5.5.1) $N_k > p_k$, $p_{k+1} > N_k$, $p_k = \min P_k$, $F_k = \bigcup_{t\in P_k} H_t$;

5.5.2) $m_r^+(F_{k+1}) \subset U_{k+1}$ for all $r = 1,\ldots,N_k$;

5.5.3) $m_i(H_t) \in U_k$ whenever $i \geq N_k$ and $t = 1,\ldots,p_k$.

Since $\mathcal{F}$ is block-respecting, there is $J := \{j_1, j_2,\ldots\} \in \mathcal{F}^*$, $J \subset I$, with $N_k \leq j_k < N_{k+1}$ for every $k \in \mathbb{N}$. As $J \in \mathcal{F}^*$, then either $J_1 := \{j_1, j_3, j_5,\ldots\} \in \mathcal{F}^*$ or $J_2 := \{j_2, j_4, j_6,\ldots\} \in \mathcal{F}^*$. Without loss of generality, we assume that $J_1 \in \mathcal{F}^*$. Put $A := \bigcup_{h=1}^{\infty} H_{j_{2h-1}}$. We get

$$m_{j_1}(A) = m_{j_1}(H_{j_1}) + m_{j_1}(H_{j_3} \cup H_{j_5} \cup \ldots);$$

$$m_{j_{2h-1}}(A) = m_{j_{2h-1}}(H_{j_1} \cup H_{j_3} \cup \ldots \cup H_{j_{2h-3}}) + \qquad \textbf{(4.153)}$$

$$+ m_{j_{2h-1}}(H_{j_{2h-1}}) + m_{j_{2h-1}}(H_{j_{2h+1}} \cup H_{j_{2h+3}} \cup \ldots),\ h \geq 2.$$

Since $j_{2h-1} < N_{2h-1} < p_{2h}$ and

$$H_{j_{2h+1}} \cup H_{j_{2h+3}} \cup \ldots \subset \bigcup_{l=p_{2h+1}}^{\infty} H_l = F_{2h+1} \text{ for every } h \in \mathbb{N}, \quad \textbf{(4.154)}$$

from (4.154) and 5.5.2) used with $k = 2h$ we obtain

$$m_{j_{2h-1}}(H_{j_{2h+1}} \cup H_{j_{2h+3}} \cup \ldots) \in U_{2h+1} \subset U_3. \quad \textbf{(4.155)}$$

Moreover, since $j_{2h-3} < N_{2h-3} < p_{2h-2} < p_{2h-1}$ for every $h \geq 2$, from 5.5.3) used with $k = 2h-1$ we get $m_{j_{2h-1}}(H_{j_l}) \in U_{2h-1}$, $h \geq 2$, $l = 1,3,\ldots,2h-3$, and hence

$$m_{j_{2h-1}}(H_{j_1} \cup H_{j_3} \cup \ldots \cup H_{j_{2h-3}}) \in (h-1)U_{2h-1} \subset U_h \subset U_3. \quad \textbf{(4.156)}$$

If $m_{j_{2h-1}}(A) \in U_1$, then from (4.41), (4.155) and (4.156) we have $m_{j_1}(H_{j_1}) \in U_1 + U_2 \subset U$ and $m_{j_{2h-1}}(H_{j_{2h-1}}) \in U_1 + U_2 + U_3 \subset 2U_1 \subset U$ for all $h \geq 2$. But we know that $m_{j_{2h-1}}(H_{j_{2h-1}}) \notin U$, and so we have a contradiction. Thus, we get that $m_{j_{2h-1}}(A) \notin U_1$ for all $h \in \mathbb{N}$, and so $L := \{l \in \mathbb{N} : m_l(A) \notin U\} \in \mathcal{F}^*$. Since, by (*ii*), $\mathbb{N} \setminus L \in \mathcal{F}$, we obtain $L \cap (\mathbb{N} \setminus L) \neq \emptyset$, which is absurd. This proves β).

We now prove $\beta\beta$). If we deny the thesis, then, proceeding analogously as in the proof of β), we find $I \in \mathcal{F}^*$ and $U \in \mathcal{J}(0)$ with $m_n(A_{l_n}) \notin U$ for each $n \in I$. By Lemma 2.23 of Chapter 2, there is $J \in \mathcal{F}^*$, $J \subset I$, with $\lim_{j \in J} m_j(A_{l_n}) = 0$ for any $n \in \mathbb{N}$. Note that the sequence $m_n(A_{l_n})$, $n \in \mathbb{N}$, does not $(\mathcal{F}(J))$-converge to 0 (see also Aviles Lopez, Cascales Salinas, Kadets and Leonov (2007)). Since $J \in \mathcal{F}^*$ and $\mathcal{F}$ is block-respecting, then, by Proposition 1.3 of Chapter 2, $\mathcal{F}(J)$ is block-respecting too. As $\mathcal{F}(J) \supset \mathcal{F}$, it is easy to see that $(A_{l_n})_n$ satisfies ii) with respect to $\mathcal{F}(J)$. By β) used with $\mathcal{F}(J)$ and $(A_{l_n})_n$, it follows that $(\mathcal{F}(J)) \lim_n m_n(A_{l_n}) = 0$, obtaining a contradiction. This proves $\beta\beta$). □

We have also the following Brooks-Jewett-type theorem.

Theorem 5.6 (see Boccuto and Dimitriou (2013e), Theorem 3.2) *Let $(A_n)_n$ be as in Theorem 5.5, $\mathcal{F}$ be a block-respecting filter of $\mathbb{N}$, $m_j : \Sigma \to R$, $j \in \mathbb{N}$, be a sequence of finitely additive (s)-bounded measures, and assume that*

(*i*) $\lim_j m_j(A_n) = 0$ *for any* $n \in \mathbb{N}$;

(*ii*) $(\mathcal{F}) \lim_j \sum_{p \in P} m_j(A_p) = 0$ *for every* $P \subset \mathbb{N}$.

Then for every strictly increasing sequence $(l_n)_n$ in $\mathbb{N}$ we get

$$(\mathcal{F}) \lim_n m_n(A_{l_n}) = 0. \tag{4.157}$$

If $\mathcal{F}$ is also diagonal, then (*ii*) *implies* (4.157).

With similar techniques, it is possible to prove the following Nikodým boundedness-type theorem for topological group-valued measures, extending Candeloro (1985b), Theorem 2.5 and Candeloro (1985c), Theorem 3.5.

Theorem 5.7 (see Boccuto and Dimitriou (2013e), Theorem 3.3) *Let $\mathcal{F}$ be a block respecting filter of $\mathbb{N}$, $m_j : \Sigma \to R$, $j \in \mathbb{N}$, be a sequence of σ-additive measures, and $(A_n)_n$ be a disjoint sequence in Σ. Let $U \in \mathcal{J}(0)$, $(W_n)_n$ be an increasing sequence in $\mathcal{J}(0)$, and set $V_n := n\, W_n + U$, $n \in \mathbb{N}$. Suppose that:*

(*j*) *The set $\{m_n(A_p) : n \in \mathbb{N}\}$ is eventually bounded by $(W_n)_n$ for each $p \in \mathbb{N}$;*

(*jj*) *The set $\{\sum_{p \in P} m_j(A_p) : n \in \mathbb{N}\}$ is $\mathcal{F}$-bounded by $(W_n)_n$ for any $P \subset \mathbb{N}$. Then,*

γ) For every strictly increasing sequence $(l_n)_n$ in $\mathbb{N}$, the set $D := \{m_n(A_{l_n}) : n \in \mathbb{N}\}$ is

$\mathcal{F}$-bounded by $(V_n)_n$.

$\gamma\gamma$) If $\mathcal{F}$ is also diagonal, then (jj) implies that D is $\mathcal{F}$-bounded by $(V_n)_n$.

Proof: For each $n \in \mathbb{N}$, let $H_n := A_{l_n}$. If γ) is not true, then $I := \{n \in \mathbb{N} : m_n(H_n) \notin V_n\} \in \mathcal{F}^*$. By σ-additivity of m_1, there is a cofinite set $P_1 \subset \mathbb{N}$, with $1 < p_1 = \min P_1$ and $m_1^+(F_1) \subset U$, where $F_1 := \bigcup_{t \in P_1} H_t$. By (*j*) there is $N_1 > p_1$ with $m_i(H_t) \in W_i$ for each $i \ge N_1$ and $t = 1, \ldots, N_1$. By induction, there are a strictly decreasing sequence $(F_k)_k$ in Σ and two strictly increasing sequence $(N_k)_k$, $(p_k)_k$ in $\mathbb{N}$ such that, for each $k \in \mathbb{N}$,

5.7.1) $N_k > p_k$, $p_{k+1} > N_k$, $m_r^+(F_{k+1}) \subset U_{k+1}$ for every $r = 1, \ldots, N_k$;

5.7.2) $m_i(H_t) \in W_i$ for any $i \ge N_k$ and $t = 1, \ldots, p_k$.

Since $\mathcal{F}$ is block-respecting, we find a set $J_1 := \{j_1, j_3, j_5, \ldots\} \in \mathcal{F}^*$, $J_1 \subset I$, such that $N_k \le j_k < N_{k+1}$ for every $k \in \mathbb{N}$. For any $h \in \mathbb{N}$ we have:

$$m_{j_{2h-1}}(H_{j_{2h+1}} \cup H_{j_{2h+3}} \cup \ldots) \in U; \qquad \textbf{(4.158)}$$

$m_{j_{2h-1}}(H_{j_l}) \in W_{2h-1}$, $h \ge 2$, $l = 1, 3, \ldots, 2h-3$, and

$$m_{j_{2h-1}}(H_{j_1} \cup H_{j_3} \cup \ldots \cup H_{j_{2h-3}}) \in (h-1) W_{2h-1}. \qquad \textbf{(4.159)}$$

Let now $A := \bigcup_{h=1}^{\infty} H_{j_{2h-1}}$. If $m_{j_{2h-1}}(A) \in W_{j_{2h-1}}$, then from (4.41), (4.158) and (4.159) we obtain $m_{j_{2h-1}}(A_{j_{2h-1}}) \in hW_{2h-1} + U \subset j_{2h-1} W_{j_{2h-1}} + U = V_{j_{2h-1}}$ and $m_{j_1}(A_{j_1}) \in W_{j_1} + U \subset j_1 W_{j_1} + U = V_{j_1}$. This contradicts the fact that $m_{j_{2h-1}}(H_{j_{2h-1}}) \notin V_{j_{2h-1}}$. Thus $m_{j_{2h-1}}(A) \notin W_{j_{2h-1}}$ for all $h \in \mathbb{N}$, and hence $\{l \in \mathbb{N} : m_l(A) \notin W_l\} \in \mathcal{F}^*$. From this, arguing as at the end of the proof of Theorem 3.20, β), we get a contradiction, and this proves γ). From γ), proceeding as in the proof of Theorem 3.20, $\beta\beta$), we get $\gamma\gamma$). □

4.5.3. Limit Theorems for Positive Measures

In the following theorems, which are formulated for positive topological group-valued measures, we suppose that the involved filter is only diagonal, and not necessarily block-respecting. We apply some countability properties of diagonal filters and some classical versions of limit theorems for topological group-valued measures defined on the σ-algebra generated by a disjoint sequence. This is possible thanks to positivity of measures. We begin with the following Brooks-Jewett-type theorem.

Theorem 5.8 (see Boccuto and Dimitriou (2014a), Theorem 3.5) *Let* $m_j : \Sigma \to R$, $j \in \mathbb{N}$, *be a sequence of positive* (s)*-bounded measures, and* $\mathcal{F}$ *be a diagonal filter of* $\mathbb{N}$. *Assume that* $m_0(E) := (\mathcal{F}) \lim_j m_j(E)$ *exists in* R *for any* $E \in \Sigma$, *and that* m_0 *is* σ*-additive and positive on* Σ.

Then for every disjoint sequence $(C_k)_k$ *in* Σ *and* $I \in \mathcal{F}^*$ *there exists* $J \in \mathcal{F}^*$, $J \subset I$, *with*

$$\lim_k (\bigcup_{j \in J} m_j^+(C_k)) = \lim_k \{m_j(C_k) : j \in J\} = 0.$$

Proof: Let $I \in \mathcal{F}^*$, $(C_k)_k$ be any disjoint sequence in Σ, and $\mathcal{K}$ be the σ-algebra generated by C_k, $k \in \mathbb{N}$, in $\bigcup_{k=1}^{\infty} C_k$. For every $B \in \mathcal{K}$ there exists $P \subset \mathbb{N}$ with $B = \bigcup_{k \in P} C_k$. Since $\mathcal{F}$ is diagonal, by Lemma 4.2 α) there is $J \in \mathcal{F}^*$, $J \subset I$, with

$$m_0(\bigcup_{k \in E} C_k) = \lim_j m_j(\bigcup_{k \in E} C_k) \quad \textbf{(4.160)}$$

for every $E \in \mathcal{I}_{\text{fin}} \cup \{\mathbb{N}\}$, where $\mathcal{I}_{\text{fin}}$ is the (countable) class of all finite subsets of $\mathbb{N}$. Moreover, by σ-additivity of m_0, we get

$$\lim_k (m_0^+(\bigcup_{l=k}^{\infty} C_l)) = 0. \quad \textbf{(4.161)}$$

Choose arbitrarily $U \in \mathcal{J}(0)$ and let $W \in \mathcal{J}(0)$ be such that $5W \subset U$. In correspondence with W let $U_0 \subset W$ be as in condition of positivity (4.128). In correspondence with U_0 there exists $k_0 \in \mathbb{N}$ with $m_0(\bigcup_{k>k_0} C_k) \in U_0$ and therefore, by positivity of m_0, $m_0(\bigcup_{k>k_0, k\in P} C_k) \in U_0$. Moreover there is $j_0 \in J$, $j_0 = j_0(U, k_0)$, such that for every $j \in J$ with $j \geq j_0$ we have

$$m_j(\bigcup_{k\le k_0, k\in P} C_k) - m_0(\bigcup_{k\le k_0, k\in P} C_k) \in U_0,\ m_j(\bigcup_{k\le k_0} C_k) - m_0(\bigcup_{k\le k_0} C_k) \in U_0,$$

$$m_j(\bigcup_{k=1}^{\infty} C_k) - m_0(\bigcup_{k=1}^{\infty} C_k) \in U_0,$$

$$m_j(\bigcup_{k>k_0} C_k) - m_0(\bigcup_{k>k_0} C_k) \in 2U_0,$$

and hence

$$m_j(\bigcup_{k>k_0} C_k) = m_j(\bigcup_{k>k_0} C_k) - m_0(\bigcup_{k>k_0} C_k) + m_0(\bigcup_{k>k_0} C_k) \in 3U_0.$$

By positivity of m_j, we have also $m_j(\bigcup_{k>k_0, k\in P} C_k) \in U_0$. So, taking into account symmetry of U_0, for every $B \in \mathcal{K}$, $B = \bigcup_{k\in P} C_k$, we get

$$m_j(B) - m_0(B) = m_j(\bigcup_{k\in P} C_k) - m_0(\bigcup_{k\in P} C_k) =$$

$$= m_j(\bigcup_{k\le k_0, k\in P} C_k) - m_0(\bigcup_{k\le k_0, k\in P} C_k) + m_j(\bigcup_{k>k_0, k\in P} C_k) - m_0(\bigcup_{k>k_0, k\in P} C_k) \in$$

$$\in 5U_0 \subset 5W \subset U.$$

Thus, $\lim_{j\in J} m_j(B) = m_0(B)$ for all $B \in \mathcal{K}$. Thus the finitely additive R-valued measures m_j, $j \in J$, satisfy the hypotheses of the classical version of the Brooks-Jewett theorem on $\mathcal{K}$ for topological group-valued measures (see also Candeloro

(1985b), Theorem 2.6, Candeloro and Letta (1985), Theorem 2.4). In particular, we get

$$\lim_k \Big(\bigcup_{j\in J} m_j^+(C_k)\Big) = \lim_k \{m_j(C_k) : j \in J\} = 0.$$

This ends the proof. □

With similar techniques as in Theorem 5.8 it is possible to prove the following Vitali-Hahn-Saks-type theorem.

Theorem 5.9 (see Boccuto and Dimitriou (2014a), Theorem 3.6) *Let Σ, $\mathcal{F}$ be as in Theorem 5.8, τ be a Fréchet-Nikodým topology on Σ, $m_j : \Sigma \to R$, $j \in \mathbb{N}$, be a sequence of positive finitely additive (s)-bounded and τ-continuous measures. Moreover, assume that $m_0(E) := (\mathcal{F}) \lim_j m_j(E)$ exists in R for each $E \in \Sigma$, and m_0 is σ-additive and positive on Σ.*

Then for any set $I \in \mathcal{F}^*$ and for any decreasing sequence $(H_k)_k$ in Σ with τ-$\lim_k H_k = \emptyset$ there exists a set $J \in \mathcal{F}^*$, $J \subset I$, with

$$\lim_k \Big(\bigcup_{j\in J} m_j^+(H_k)\Big) = \lim_k \{m_j(H_k) : j \in J\} = 0.$$

Proof: Let τ, I, $(H_k)_k$ be as in the hypotheses, put $H_\infty := \bigcap_{k=1}^\infty H_k$, $C_k := H_k \setminus H_{k+1}$, $k \in \mathbb{N}$, and let $\mathcal{L}$ be the σ-algebra generated by the sets C_k, $k \in \mathbb{N}$, and H_∞ in H_1. Proceeding analogously as in Theorem 5.8, by virtue of Candeloro, 1985b, Theorem 2.6 and Candeloro and Letta (1985), Theorem 2.4, there exists a set $J \subset I$, $J \in \mathcal{F}^*$, such that the measures m_j, $j \in J$, are uniformly (s)-bounded on $\mathcal{L}$. Moreover, by τ-continuity and positivity of the measures m_j, we get $\lim_k m_j^+(H_k) = 0$ for every $j \in \mathbb{N}$. By Theorem 4.12 applied to the sequence of measures $m_j : \Sigma \to R$, $j \in J$, we obtain that

$$\lim_k \Big(\bigcup_{j\in J} m_j^+(H_k)\Big) = \lim_k \{m_j(H_k) : j \in J\} = 0,$$

that is the assertion. □

Analogously as in Theorem 5.9 it is possible to obtain the following Nikodým convergence-type theorem.

Theorem 5.10 (see Boccuto and Dimitriou (2014a), Theorem 3.7) *Let* Σ, $\mathcal{F}$ *be as in Theorem* 5.9 *and* $m_j : \Sigma \to R$, $j \in \mathbb{N}$, *be a sequence of positive* σ*-additive measures. If* $m_0(E) := (\mathcal{F}) \lim_j m_j(E)$ *exists in* R *for each* $E \in \Sigma$, *and* m_0 *is* σ*-additive and positive on* Σ, *then for each* $I \in \mathcal{F}^*$ *and for every decreasing sequence* $(H_k)_k$ *in* Σ *with* $\bigcap_{k=1}^{\infty} H_k = \emptyset$ *there exists* $J \in \mathcal{F}^*$, $J \subset I$, *with*

$$\lim_k \left(\bigcup_{j \in J} m_j^+(H_k)\right) = \lim_k \{m_j(H_k) : j \in J\} = 0.$$

We now turn to a Dieudonné-type theorem in the context of topological groups.

Theorem 5.11 (see Boccuto and Dimitriou (2014a), Theorem 3.8) *Let* Σ, $\mathcal{F}$ *be as in Theorem* 5.9, $\mathcal{G}$, $\mathcal{H} \subset \Sigma$ *be as above,* $m_j : \Sigma \to R$, $j \in \mathbb{N}$, *be a sequence of positive regular measures, such that* $m_0(E) := (\mathcal{F}) \lim_j m_j(E)$ *exists in* R *for every* $E \in \Sigma$, *and* m_0 *is* σ*-additive and positive.*

Furthermore, let $A \in \Sigma$ *and* $(G_k)_k$, $(F_k)_k$ *be two sequences in* $\mathcal{G}$, $\mathcal{H}$ *respectively, with* $F_k \subset F_{k+1} \subset A \subset G_{k+1} \subset G_k$ *for each* $k \in \mathbb{N}$, *and*

$$\lim_k m_j(G_k \setminus F_k) = 0 \quad \textit{for every } j \in \mathbb{N}.$$

Then for each $I \in \mathcal{F}^*$ *there is* $J \in \mathcal{F}^*$, *with*

$$\lim_k \left(\bigcup_{j \in J} m_j^+(G_k \setminus F_k)\right) = 0. \qquad \textbf{(4.162)}$$

Proof: Let A, $(G_k)_k$, $(F_k)_k$ be as in the hypothesis, $\mathcal{L}$ be the σ-algebra generated by the sets $G_k \setminus F_k$, $k \in \mathbb{N}$, and $I \in \mathcal{F}^*$. Since the measures m_j, $j \in \mathbb{N}$, are (s)-bounded, then, arguing analogously as in the proof of Theorem 5.8, by Candeloro, 1985b, Theorem 2.6 and Candeloro and Letta (1985), Theorem 2.4, we find a set $J \subset I$, $J \in \mathcal{F}^*$, such that the measures m_j, $j \in J$, are uniformly (s)-bounded on $\mathcal{L}$. Moreover, by hypothesis and taking into account positivity of the measures m_j, we have $\lim_k m_j^+(G_k \setminus F_k) = 0$ for every $j \in \mathbb{N}$. From this and Theorem 4.12 applied to m_j, $j \in J$, we get (4.162). □

4.5.4. Filter Exhaustiveness and Equivalence Results

We deal with uniform filter exhaustiveness for sequences of measures in the topological group setting, and prove some results on existence of limit measures and some versions of convergence theorems. We consider a subsequence, indexed by a suitable element of the involved filter, in which we apply some classical limit theorems. We prove also some equivalence results between Brooks-Jewett, Vitali-Hahn-Saks, Nikodým and Dieudonné-type theorems, extending some theorems proved in Drewnowski (1972b).

Let $\mathcal{F}$ be a free filter of $\mathbb{N}$, $\Sigma \subset \mathcal{P}(G)$ be a σ-algebra and $\lambda : \Sigma \to [0,+\infty)$ be a finitely additive measure, such that Σ is separable with respect to the Fréchet-Nikodým topology generated by λ (shortly, λ*-separable*). Let $\{F_i : i \in \mathbb{N}\}$ be a countable λ-dense subset of Σ. Assume that $m_j : \Sigma \to R$, $j \in \mathbb{N}$, is a sequence of finitely additive measures.

Let $E \in \Sigma$ be fixed. We say that $(m_j)_j$ is λ-$\mathcal{F}$*-exhaustive at* E iff for every $U \in \mathcal{J}(0)$ there exist $\delta > 0$ and $V \in \mathcal{F}$ with $m_j(E) - m_j(F) \in U$ for each $F \in \Sigma$ with $|\lambda(E) - \lambda(F)| \le \delta$ and for any $j \in V$. The sequence $(m_j)_j$ is said to be λ-$\mathcal{F}$*-exhaustive on* Σ iff it is λ-$\mathcal{F}$-exhaustive at every $E \in \Sigma$.

We say that $(m_j)_j$ are λ*-uniformly* $\mathcal{F}$*-exhaustive on* Σ iff for every $U \in \mathcal{J}(0)$ there exist $\delta > 0$ and $V \in \mathcal{F}$ with $m_j(E) - m_j(F) \in U$ whenever $E, F \in \Sigma$ with $|\lambda(E) - \lambda(F)| \le \delta$ and for any $j \in V$.

We now prove the following result on extensions of filter limit measures in the topological group setting.

Theorem 5.12 (see Boccuto and Dimitriou (2014a), Theorem 4.1) *Let* $(m_j)_j$ *be a sequence of finitely additive measures,* λ*-uniformly* $\mathcal{F}$*-exhaustive on* Σ*, such that* $m(F_i) := (\mathcal{F})\lim_j m_j(F_i)$ *exists in* R *for every* $i \in \mathbb{N}$*. Then,*

(γ) *There is a finitely additive extension* $m_0 : \Sigma \to R$ *of* m*, with*

$(\mathcal{F})\lim_j m_j(E) = m_0(E)$ *for each* $E \in \Sigma$.

$(\gamma\gamma)$ *Moreover, if* $\mathcal{F}$ *is a* P*-filter, then there is a set* $M_0 \in \mathcal{F}$ *such that*

$\lim_{j \in M_0} m_j(E) = m_0(E)$ *for every* $E \in \Sigma$.

Proof: (γ) Choose arbitrarily $E \in \Sigma$ and $U \in \mathcal{J}(0)$, and let $U_0 \in \mathcal{J}(0)$ be with $3U_0 \subset U$. By hypothesis, there exist $\delta > 0$ and $V \in \mathcal{F}$ such that, if $|\lambda(E) - \lambda(F)| \le \delta$ and $j \in V$, then $m_j(E) - m_j(F) \in U_0$. By λ-separability of Σ, there is $\bar{i} \in \mathbb{N}$ with $|\lambda(E) - \lambda(F_{\bar{i}})| \le \delta$. By Theorem 4.1 there is a set $W^{(\bar{i})} \in \mathcal{F}$ with $m_j(F_{\bar{i}}) - m_l(F_{\bar{i}}) \in U$ whenever $j, l \in W^{(\bar{i})}$. In particular we get

$$m_j(E) - m_l(E) = m_j(E) - m_j(F_{\bar{i}}) + m_j(F_{\bar{i}}) - m_l(F_{\bar{i}}) + m_l(F_{\bar{i}}) - m_l(E) \in 3U_0 \subset U$$

for every $j, l \in V \cap W^{(\bar{i})}$. By Theorem 4.1, there is a set function $m_0 : \Sigma \to R$, extending m, with $(\mathcal{F})\lim_j m_j(E) = m_0(E)$. It is not difficult to see that m_0 is finitely additive on Σ. This proves (γ).

$(\gamma\gamma)$ Let $(U_p)_p$ be a base of neighborhoods of 0. By λ-uniform $\mathcal{F}$-exhaustiveness, for every $p \in \mathbb{N}$ there are a $\delta > 0$ and a set $M_p' \in \mathcal{F}$, with $m_j(E) - m_j(F) \in U_p$ whenever $E, F \in \Sigma$ with $|\lambda(E) - \lambda(F)| \le \delta$ and $j \in M_p'$. Since $\mathcal{F}$ is a P-filter, in correspondence with M_p' there exists $M_p \in \mathcal{F}$ such that $M_p \Delta M_p'$ is finite for each $p \in \mathbb{N}$ and $M := \bigcap_{p=1}^{\infty} M_p \in \mathcal{F}$. Let $Z_p := M \setminus M_p'$, $p \in \mathbb{N}$. Note that Z_p is finite for every $p \in \mathbb{N}$, and so we get

$m_j(E)-m_j(F)\in U_p$ whenever E, $F\in\Sigma$ with $|\lambda(E)-\lambda(F)|\le\delta$ and $j\in M\setminus Z_p$. Moreover, thanks to Proposition 4.4, there is a set $B_0\in\mathcal{F}$ such that for every j, $p\in\mathbb{N}$ there exists $\overline{j}\in B_0$ with $m_j(F_i)-m(F_i)\in U_p$ whenever $j\ge\overline{j}$, $j\in B_0$. Without loss of generality, we can take $\overline{j}\in B_0\cap M$. Set $M_0:=B_0\cap M$: we get $M_0\in\mathcal{F}$. The sequence m_j, $j\in M_0$, is λ-uniformly $\mathcal{F}_{\text{cofin}}$-exhaustive, and $\lim\limits_{j\in M_0} m_j(F_i)=m(F_i)$ for every $i\in\mathbb{N}$. From this and (γ) applied to m_j, $j\in M_0$ and $\mathcal{F}_{\text{cofin}}$, we find a finitely additive extension m_0 of m, defined on Σ, with

$$\lim_{j\in M_0} m_j(E)=m_0(E) \text{ for each } E\in\Sigma.$$

Thus M_0 is the requested set. □

The next step is to give some sufficient conditions on an $\mathcal{F}$-convergent sequence m_j, $j\in\mathbb{N}$, of topological group-valued measures, for the existence of a set $M_0\in\mathcal{F}$ such that the subsequence m_j, $j\in M_0$, is uniformly (s)-bounded (resp. uniformly σ-additive, uniformly τ-continuous, uniformly regular). These results yield also some sufficient conditions for (s)-boundedness (resp. σ-additivity, τ-continuity, regularity) of the limit measure. Observe that in this framework, as seen before, even when $R=\mathbb{R}$, the hypothesis of λ-uniform $\mathcal{F}$-exhaustiveness in general cannot be dropped. However, without requiring filter exhaustiveness, it is possible to prove the following theorem on the existence of the filter limit measure.

Theorem 5.13 (see Boccuto and Dimitriou (2014a), Theorem 4.2) *Let $\Sigma\subset\mathcal{P}(G)$ be a σ-algebra, $\mathcal{L}$ be an algebra generating Σ, $m_j:\Sigma\to R$, $j\in\mathbb{N}$, be a sequence of uniformly σ-additive measures, such that $(\mathcal{F})\lim\limits_j m_j(E)$ exists in R for each $E\in\mathcal{L}$. Then $(\mathcal{F})\lim\limits_j m_j(E)$ exists in R for any $E\in\Sigma$.*

Proof: Let $\Pi:=\{E\in\Sigma:(\mathcal{F})\lim\limits_j m_j(E) \text{ exists in } R\}$. By hypothesis, $\mathcal{L}\subset\Pi$. We now prove that, if $(E_r)_r$ is a monotone sequence of elements of Π with

$\lim_r E_r = E \in \Sigma$ in the set-theoretic sense, then $E \in \Pi$. From this it follows that $\Pi = \Sigma$, and hence we get the assertion.

Choose arbitrarily $U \in \mathcal{J}(0)$, and let $(U_r)_r$ be a family of elements of $\mathcal{J}(0)$, with $2U_r \subset U_{r-1}$ for each r and $2U_0 \subset U$. For every $r \in \mathbb{N}$, since $E_r \in \Pi$, the sequence $(m_j(E_r))_j$ is $\mathcal{F}$-convergent, and so by Theorem 4.1 there exists $W_r \in \mathcal{F}$ with

$$m_p(E_r) - m_q(E_r) \in U_r \text{ whenever } p,\ q \in W_r.$$

Moreover, since the measures m_j are uniformly σ-additive, there is $\overline{r} \in \mathbb{N}$ with $m_j(E_{\overline{r}}) - m_j(E) \in U_1$ for all $j \in \mathbb{N}$. Thus for every p, $q \in W_{\overline{r}}$ we get

$$m_p(E) - m_q(E) = [m_p(E) - m_p(E_{\overline{r}})] + [m_p(E_{\overline{r}}) - m_q(E_{\overline{r}})] +$$

$$+ [m_q(E_{\overline{r}}) - m_q(E)] \in 2U_1 + U_{\overline{r}} \subset 2U_1 + U_0 \subset 2U_0 \subset U.$$

By Theorem 3.1, the limit $(\mathcal{F})\lim_j m_j(E)$ exists in R. The assertion follows from arbitrariness of $E \in \Sigma$. □

A sequence of finitely additive measures $m_j : \Sigma \to R$, $j \geq 0$, together with a non-negative finitely real-valued measure λ defined on Σ, is said to *satisfy property* (*) *with respect to* R *and* $\mathcal{F}$ iff it is λ-uniformly $\mathcal{F}$-exhaustive on Σ and $(\mathcal{F})\lim_j m_j(E) = m_0(E)$ for any $E \in \Sigma$.

The next result is an immediate consequence of Lemma 5.12 $(\gamma\gamma)$.

Lemma 5.14 *Let* $m_j : \Sigma \to R$, $j \in \mathbb{N}$, *satisfy together with* λ *property* (*) *with respect to* R *and* $\mathcal{F}$. *Then there exists a set* $M_0 \in \mathcal{F}$ *such that the measures* m_j, $j \in M_0$, *and* m_0 *satisfy together with* λ *property* (*) *with respect to* R *and* $\mathcal{F}_{\text{cofin}}$.

We now deal with equivalence between filter limit theorems in the (ℓ)-group setting. We begin with recalling the following Brooks-Jewett-type theorem in the topological group context.

Theorem 5.15 (see Candeloro (1985b), Theorem 2.6, Candeloro and Letta (1985), Theorem 2.4) *Let $m_j : \Sigma \to R$, $j \in \mathbb{N}$, be a sequence of (s)-bounded measures, convergent pointwise on Σ to a measure m_0.*

Then the measures m_j, $j \in \mathbb{N}$, are uniformly (s)-bounded and m_0 is (s)-bounded on Σ.

We now prove the following filter limit theorems for topological group-valued measures and their equivalence. Note that, since we deal with topological group-valued measures, we can use a Drewnowski-type approach, considering suitable σ-additive restrictions of (s)-bounded measures, though it is possible to construct Stone-type extensions even for topological group-valued measures (see for instance Sion (1969 and 1973)). In what follows, let us assume that

H) $\lambda : \Sigma \to [0,+\infty]$ is a finitely additive measure, Σ is a λ-separable σ-algebra, $\mathcal{F}$ is a P-filter of $\mathbb{N}$, m_0, $m_j : \Sigma \to R$, $j \in \mathbb{N}$, are finitely additive measures, satisfying together with λ property (*) with respect to R and $\mathcal{F}$ on Σ, and Σ_0 is a sub-σ-algebra on Σ.

Theorem 5.16 (Brooks-Jewett (BJ)) *If each m_j is (s)-bounded on Σ_0, then there exists a set $M_0 \in \mathcal{F}$, such that the measures m_j, $j \in M_0$, are uniformly (s)-bounded on Σ_0.*

Theorem 5.17 (Vitali-Hahn-Saks (VHS)) *If every m_j is (s)-bounded and τ-continuous on Σ_0, then there exists a set $M_0 \in \mathcal{F}$, with the property that the measures m_j, $j \in M_0$, are uniformly (s)-bounded and uniformly τ-continuous on Σ_0.*

Theorem 5.18 (Nikodým (N)) *If each m_j is σ-additive on Σ_0, then there is $M_0 \in \mathcal{F}$, such that the measures m_j, $j \in M_0$, are uniformly σ-additive on Σ_0.*

Theorem 5.19 (Dieudonné (D)) *If each m_j is (s)-bounded and regular on Σ_0, then there is $M_0 \in \mathcal{F}$ with the property that the measures m_j, $j \in M_0$, are uniformly (s)-bounded and uniformly regular on Σ_0.*

To prove Theorem 5.16 (BJ), observe that there exists $M_0 \in \mathcal{F}$, satisfying the thesis of Lemma 5.14. The assertion of (BJ) follows by applying Theorem 5.15 to the sequence m_j, $j \in M_0$.

We now prove equivalence between (BJ), (VHS), (N) and (D).

We begin with the implication (BJ) $\Rightarrow$ (VHS). Let $m_j : \Sigma \to R$, $j \in \mathbb{N}$, be a sequence of measures, which fulfil property (*) with respect to R and $\mathcal{F}$ together with λ, (s)-bounded and τ-continuous on Σ_0. By (BJ), there is $M_0 \in \mathcal{F}$ such that the measures m_j, $j \in M_0$, are uniformly (s)-bounded on Σ_0.

So, $\lim_k \left(\bigcup_{j \in M_0} m_j^+(C_k) \right) = 0$ for every disjoint sequence $(C_k)_k$ in Σ_0.

Fix arbitrarily any decreasing sequence $(H_k)_k$ in Σ_0, with $\tau\text{-}\lim_k H_k = \varnothing$. By virtue of τ-continuity of each m_j, $j \in \mathbb{N}$, on Σ_0, we get $\lim_k m_j^+(H_k) = 0$ for every $j \in \mathbb{N}$.

By Theorem 4.12 we obtain $\lim_k \left(\bigcup_{j \in M_0} m_j^+(H_k) \right) = 0$, so getting uniform τ-continuity of m_j, $j \in M_0$, on Σ_0. Thus, (BJ) implies (VHS).

The proof of (BJ) $\Rightarrow$ (D) is similar to that of (BJ) $\Rightarrow$ (VHS).

We now prove (VHS) $\Rightarrow$ (N). Let τ be the Fréchet-Nikodým topology generated by the class of all order continuous submeasures defined on Σ_0. If $(H_k)_k$ is any decreasing sequence in Σ_0 with $\tau\text{-}\lim_k H_k = \varnothing$ and $H_\infty = \bigcap_{k=1}^{\infty} H_k$, then we have $\eta(H_\infty) = 0$ for each order continuous submeasure η defined on Σ_0, and so it follows that $H_\infty = \emptyset$. Thus we obtain that, if $(m_j)_j$ is a sequence of measures, σ-additive on Σ_0, then they are τ-continuous on Σ_0. Since every m_j is also (s)-bounded on Σ_0, then, by (VHS), $(m_j)_j$ is uniformly τ-continuous on Σ_0, and so also uniformly σ-additive. Thus, (VHS) implies (N).

We prove (N) $\Rightarrow$ (BJ). Let $m_j : \Sigma \to R$, $j \in \mathbb{N}$, be a sequence of (s)-bounded measures, fulfilling property (*) with respect to R and $\mathcal{F}$ together with λ.

Pick arbitrarily a disjoint sequence $(C_k)_k$ in Σ_0 and any subsequence $(C_{k_r})_r$ of $(C_k)_k$. By Theorem 4.11 there is a sub-subsequence $(C_{k_{r_s}})_s$, such that every m_j is σ-additive on the σ-algebra $\mathcal{L}$ generated by $(C_{k_{r_s}})_s$.

By (N) used with $\mathcal{F}$ and the sub-σ-algebra $\mathcal{L}$, where $\mathcal{L} \subset \Sigma_0 \subset \Sigma$, there exists a set $M^* \in \mathcal{F}$, such that the measures $m_{j|\mathcal{L}}$, $j \in M^*$, are uniformly σ-additive, and hence also uniformly (s)-bounded on $\mathcal{L}$. So we get that

$$\lim_s \Big(\bigcup_{j \in M^*} m_j(C_{k_{r_s}}) \Big) = 0. \qquad \textbf{(4.163)}$$

By arbitrariness of the subsequence $(C_{k_r})_r$ and property $(\mathcal{U}')$ used with $\mathcal{F} = \mathcal{F}_{\text{cofin}}$, from (4.163) it follows that

$$\lim_k \Big(\bigcup_{j \in M^*} m_j(C_k) \Big) = 0, \qquad \textbf{(4.164)}$$

and hence (N) implies (BJ).

We now prove (D) $\Rightarrow$ (BJ). Let $m_j : \Sigma \to R$, $j \geq 0$ be a sequence of measures, (s)-bounded on Σ_0 and fulfilling property (*) with respect to R and $\mathcal{F}$ together with λ. Of course, if we take $\mathcal{G} = \mathcal{H} = \Sigma_0$, then the measures m_j, $j \geq 0$, are regular on Σ_0 (with respect to this choice of $\mathcal{G}$ and $\mathcal{H}$). By virtue of (D) there exists a set $M_0 \in \mathcal{F}$, such that the measures m_j, $j \in M_0$, are uniformly (s)-bounded and uniformly regular on Σ_0. This proves that (D) implies (BJ). □

General Discussion

Abstract: We give a summary of the main concepts, ideas, tools and results of Chapters 2,3,4. In Chapter 2 we have presented the basic notions and results about filters/ideals, statistical and filter/ideal convergence, both in the real case and in abstract structures. In Chapter 3 we have given the classical limit theorems and the Nikodým boundedness theorem for lattice group-valued measures, different types of decompositions and the construction of optimal and Bochner-type integrals in the lattice group setting. In Chapter 4 we have proved different versions of Schur, Brooks-Jewett, Vitali-Hahn-Saks, Dieudonné, Nikodým convergence and boundedness theorems in the setting of filter convergence for lattice or topological group-valued measures, and also some different results on modes of continuity, filter continuous convergence, filter weak compactness and filter weak convergence of measures.

Keywords: (D)-convergence, Baire category theorem, Bochner integral, decomposition, Drewnowski technique, filter exhaustiveness, filter, filter/ideal convergence, Fremlin lemma, Ideal, lattice group, limit theorem, Maeda-Ogasawara-Vulikh theorem, optimal integral, order convergence, positive regular property, Stone Isomorphism technique, topological group, ultrafilter measures, uniform boundedness theorem.

Throughout this eBook, in Chapter 2, Section 1 we first have dealt with the statistical convergence and matrix methods with respect to positive regular matrices, which are particular cases of filter/ideal convergence, as well as ordinary and uniform asymptotic densities. Successively, we have treated the fundamental properties of ideals and filters, both of $\mathbb{N}$ and of an abstract directed set, considering several classes with some related properties. In particular, P-filters/ideals, diagonal and block-respecting filters are very important in limit theorems and uniform boundedness theorems with respect to filter convergence. We also dealt with the additive property of an ideal with respect to another ideal, extending the notion of P-ideal. We have considered some basic properties of filter/ideal convergence in the real context, giving some result on the existence of subsequences convergent in the classical sense, in connection with classical convergence along a suitable element of the filter involved. We have also presented a Cauchy criterion and some characterization of ultrafilters in terms of filter limits of bounded sequences and of limits of subsequences, dealing also with ultrafilter measures. We have considered even some other kinds of convergences which are not necessarily generated by filters/ideals, like almost and Single convergence, given some comparison results and presented some examples, showing the main differences between them. We have dealt also with the basic properties of filter compactness, and considered different kinds of closure.

Antonio Boccuto & Xenofon Dimitriou

In Chapter 2, Section 2, we have recalled some fundamental properties and basic results on lattice groups. In particular, we have dealt with order convergence, weakly σ-distributive lattice groups and (D)-convergence. Note that in lattice groups, as we showed, in general it is impossible to use the ε-technique, and so different kinds of convergences have been considered. Observe that order convergence of sequences implies always (D)-convergence, while the converse implication is true if and only if the involved lattice group is weakly σ-distributive. There are some contexts in which it is preferable to deal with (O)-convergence and some other situations in which it is more advisable to handle with (D)-convergence: this is the case, for example, when we deal with a sequence or a *series* of (D)-sequences and we need to dominate it with a single (D)-sequence. The tool of (D)-convergence, thanks to the very famous and powerful Fremlin Lemma, allows us to do this operation without requiring super Dedekind completeness of the involved lattice group, but assuming only Dedekind completeness and weak σ-distributivity. Observe that, in order to replace a sequence of (O)-sequences with a single order sequence, in general we need super Dedekind completeness and weak σ-distributivity of the involved lattice group. Note that, in super Dedekind complete and weakly σ-distributive (ℓ)-groups, the theories of order and (D)-convergence coincide. Another very fundamental and powerful tool, widely used in the lattice group theory, is the Maeda-Ogasawara-Vulikh representation theorem, which states that every Archimedean lattice group is algebraic and lattice isomorphic to a subgroup of continuous extended real-valued functions defined on a suitable compact extremely disconnected Hausdorff topological space Ω and which take the values $+\infty$ and $-\infty$ at most on a nowhere dense set, and that the lattice suprema/infima coincide with the pointwise suprema/infima in the complement of meager subsets of Ω. Thus it is possible to give several theorems for lattice group-valued measures by proving the corresponding ones for real-valued measures and taking into account the Maeda-Ogasawara-Vulikh theorem, since, by the Baire category theorem, the complement of every meager subset of Ω is dense in Ω. We have dealt also with the (PR) (positive regularity) property in lattice groups. We have considered filter convergence for sequences/nets in lattice groups, with respect to both order and (D)-convergence, as well as filter divergence. We have given some Cauchy criterion, and we have dealt with classical convergence/divergence in lattice group setting along suitable elements of the involved filter. We have also shown that, if we consider any fixed abstract directed set Λ and any (Λ)-free filter $\mathcal{F}$ of Λ, even in the setting of filter convergence, in super Dedekind complete and weakly σ-distributive lattice groups the theories of order and (D)-

convergence coincide. Moreover we have proved some other properties of filter convergence/boundedness in connection with diagonal filters, and we have extended to the lattice group context the notions of filter convergence/divergence for nets in the lattice group setting with respect to another filter of an abstract directed set Λ. We have extended to the lattice group setting the concept of convergence/divergence with respect to a pair of filters.

In Chapter 2, Section 3 we have treated the fundamental properties of lattice group-valued measures, in particular (uniform) (s)-boundedness, σ-additivity, regularity (with respect to both order and (D)-sequences), giving some comparison result between regularity and σ-additivity and relations between σ-additivity and (s)-boundedness. Note that, differently from Banach space-valued measures, every bounded lattice group-valued measure is (s)-bounded in the classical like sense, while the converse is in general not true. However, (s)-boundedness with respect to a single order or (D)-sequence implies boundedness, but the converse implication is in general not true. Furthermore, some properties of absolutely continuous lattice group-valued measures have been investigated. We have considered also some relations between finite and countable additivity for lattice group-valued measures. We have treated both the Drewnowski technique of finding countably additive restrictions of finitely additive measures, adapting it to the lattice group context, the Stone Isomorphism technique, and Carathéodory and Stone-type extensions for lattice group-valued measures, together with some related density properties. This argument is based on the fact that to every algebra $\mathcal{L}$ of subsets of an abstract nonempty set G it is possible to associate a compact totally disconnected topological space Q^* (that is, the *Stone space*) such that $\mathcal{L}$ is algebraically and lattice isomorphic to the algebra $\mathcal{Q}$ of all open-closed subsets of Q^*. Furthermore we have found some condition in order that every bounded function has a filter limit, both in the real setting and in the vector lattice context. Finally we have investigated the main properties of (ℓ)-group-valued measures, compatible with the operations of supremum and infimum, giving some extension results, some of which have been obtained by using the Stone Isomorphism technique.

In Chapter 3, Sections 1 and 3, using sliding hump arguments, density properties and the Stone Isomorphism techniques, we have proved different kinds of limit theorems for lattice group-valued measures, for example Brooks-Jewett, Vitali-Hahn-Saks, Nikodým convergence, Nikodým boundedness, Schur and Dieudonné-type theorems, both when the concepts of (s)-boundedness and the related ones are intended with respect to a single order or (D)-sequence, and when they are meant in the classical like sense. In this last case a crucial role is played by the Maeda-Ogasawara-Vulikh representation theorem.

In Chapter 3, Section 2, we have given also the construction of abstract Bochner-type integrals for Riesz space-valued functions with respect to σ-additive real-valued measures, and an optimal integral for real-valued functions with respect to vector lattice-valued measures, compatible with respect to the operation of supremum. We have extended the theory of Bochner and Chojnacki integration for functions and with respect to vector lattice-valued measures, in connection with abstract convergences satisfying suitable axioms, whose filter convergence is a particular case. Furthermore, we have dealt with ultrafilter measures, and studied the problem of defining limits by means of suitable integrals.

In Chapter 3, Section 3, using the previously proved limit theorems and the Stone Isomorphism technique, we have given some results about different types of convergence and decomposition theorems for lattice group-valued measures (Lebesgue, Sobczyk-Hammer, Yosida-Hewitt decompositions). We have dealt also with the Rickart integral in the lattice group context.

In Chapter 4, Section 1, we consider some recent developments of filter convergence, in particular different kinds of limit theorems and related topics, for lattice group-valued measures. Note that, in general, when one treats filter convergence with respect to a given free filter $\mathcal{F}$ of $\mathbb{N}$, with $\mathcal{F} \neq \mathcal{F}_{\text{cofin}}$ (recall that $\mathcal{F}_{\text{cofin}}$ is the filter of all cofinite subsets of $\mathbb{N}$), in general it is impossible to obtain results analogous to the classical limit theorems. However, under suitable hypotheses on the filters involved, it is possible to get some results of this kind, concerning the behavior of a subsequence of the given sequence of measures, indexed by an element of the filter. Thus we have proved different kinds of limit theorems with respect to filter convergence. First, we have dealt with Schur-type theorems and have investigated some classes of filters, for which some Schur-type theorems hold. We have used some properties of diagonal and/or block-respecting filters and some sliding hump arguments. As consequences, we have given some Nikodým convergence, Nikodým boundedness, Vitali-Hahn-Saks, Dieudonné-type theorems and uniform boundedness principle. We have proved even some versions of these kinds of theorems, whose it is possible to give a direct proof without using the Schur theorem. In the case of positive measures, we have given some limit theorems by requiring that the involved filter is only diagonal. Finally, we have proved some Moore-type interchanging limit theorem and some basic matrix theorems for lattice group-valued double sequences dealing with (D)-convergence with respect to P-filters, using sliding hump-type techniques.

In Chapter 4, Section 2 we have dealt with (uniform) filter exhaustiveness, which plays a fundamental role in proving different kinds of limit theorems and related results, and we have proved, using this tool, some Brooks-Jewett, Vitali-Hahn-Saks, Nikodým and Dieudonné-type theorems for lattice group-valued measures and their equivalence, by means of the Stone Isomorphism technique. We have given an example, by showing that in general the condition of filter exhaustiveness cannot be dropped.

In Chapter 4, Section 3 we have considered different types of modes of (filter) continuity for lattice group-valued measures and we have characterized continuity of the limit measure by means of filter weak exhaustiveness. Moreover we have investigated the filter continuous convergence (filter (α)-convergence), and have given some relations with filter exhaustiveness of pointwise filter convergence, using the tool of the transfinite induction. Furthermore we have proved some results on filter weak compactness and given some characterization of filter weak convergence of σ-additive measures and filter weak convergence in L^1 in terms of pointwise convergence and with filter exhaustiveness. Finally we have given a weak filter Cauchy-type condition in the space $L^\infty(G,\Sigma,\lambda)$, where (G,d) is a complete metric space, $\Sigma \subset \mathcal{P}(G)$ is a σ-algebra and $\lambda:\Sigma \to \mathbb{R}$ is a regular measure, using the fact that filter pointwise convergence and filter exhaustiveness of a function sequence imply continuity of the limit function and filter uniform convergence on compact sets.

In Chapter 4, Sections 4 and 5, we have extended to topological group-valued measures the main results of the first two sections of Chapter 4, investigating the main properties of such measures (Section 4) and proving the main theorems with respect to filter convergence (Section 5). We have used some Drewnowski-type techniques about the existence of suitable σ-additive restrictions of finitely additive measures in order to prove our results about equivalence of limit theorems for filter convergence in the topological group setting.

Appendix

Abstract: We present an abstract approach on probability measures, events and random variables, involving in particular lattice theory, distance functions, σ-additive extensions of finitely additive functions, some kinds of convergences in the lattice setting, which can be considered even in more abstract contexts. Furthermore we pose some open problems.

Keywords: Almost uniform convergence, attribute, Boolean algebra, Boolean σ-algebra, concept, distance function, duality principle, experiment, finitely additive function, lattice, normalized distance, object, order convergence, probability, random variable, regular lattice, subsemilattice, supersemilattice, σ-additive function, σ-regular lattice.

1.1. Random Variables

We deal with some topics on probability theory and random variables with an abstract approach, in which the involved probabilities are considered not necessarily as set functions, but as functions defined on Boolean algebras, and in which lattice theory is widely used (see also Chapter 1, § 1.1.2) to study fundamental properties of random variables in connection with various kinds of convergence, distance functions generated by probabilities and σ -additive extensions of finitely additive probability measures. These notions can be extended even in the setting of random variables, defined and/or taking values in more abstract structures. We treat also some relations between concepts, lattices and probabilities, together with possible developments (for a related primer on these subjects see also Wolff (1994)). These topics have several applications in various branches of Mathematics and different fields of sciences, for example in Statistics (see Kitsos (1989)), in Medicine and Biology (see Kitsos (2005 and 2012), Kitsos and Edler (2005), Kitsos and Sotiropoulos (2009)), in Geometry (see Müller and Kitsos (2004)). Finally, we pose some open problems.

Let L be a Boolean algebra with 0 and 1. A finitely additive measure $P: L \to \mathbb{R}$ is a *probability measure* iff $P(1)=1$, P is *strictly positive* (that is $P(a)=0$ if and only if $a=0$) and P is *finitely additive* (namely $P(a \vee b) = P(a) \oplus P(b)$ whenever $a \wedge b = 0$). A probability measure P is said to be *countably additive* or σ*-additive* on L, iff $P(a) = \sum_{n=1}^{\infty} P(a_n)$ whenever $(a_n)_n$ is a disjoint sequence in L such that $\bigvee_{n=1}^{\infty} a_n = a$ exists in R.

Antonio Boccuto & Xenofon Dimitriou

If L is a Boolean algebra and P is a probability measure, we can define a real-valued function by setting $d(a,b)=P(a\oplus b)$ for every a, $b\in L$. In Kappos (1969), Theorems II.2.1-II.2.3 it is proved that d is a distance and that in general the metric space (L,d) is not complete, and that by means of techniques similar to the classical ones it is possible to give a complete extension (L',d') of (L,d) and an extension P' of P, such that L' is a Boolean σ-algebra and P' is a σ-additive probability measure.

Let L be a Boolean σ-algebra with an associated probability measure P. An *experiment* a in L is a class of pairwise disjoint elements of L, different from 0, whose supremum is 1. Note that, since P is strictly positive, the cardinality of the elements of an experiment is finite or countable (see also Kappos (1969), IV.1.1). A *random variable* is a real valued function x on any experiment a in L defined by $a_j\in a\Rightarrow x(a_j)=\xi_j$, $j\in\mathbb{N}$.

A random variable is said to be *simple* iff it is defined on a finite experiment $a=\{a_1,\dots,a_k\}$. Note that, when $L=\Sigma$ is a σ-algebra of subsets of a nonempty set G, an experiment $\{a_1,a_2,\dots\}$ is a partition of G into a finitely or countably many elements of Σ and x can be viewed as a function $x:G\to\mathbb{R}$ defined by $x(t)=\xi_j$ whenever $t\in a_j$, $j\in\mathbb{N}$.

Given two sublattices $R_1\subset R_2$ of a lattice R, we say that R_1 is *regular* (resp. σ-*regular*) *in* R_2 iff for every decreasing net $(x_\lambda)_{\lambda\in\Lambda}$ in R_1, where $(\Lambda,\geq)$ is a direct set (resp. for each sequence $(x_n)_n$ in R_1) and $\bigwedge_\lambda x_\lambda=0$ (resp. $\bigwedge_n x_n=0$) in R_1, we have also $\bigwedge_\lambda x_\lambda=0$ (resp. $\bigwedge_n x_n=0$) in R_2. For example, if R is any Archimedean lattice group and $C_\infty(\Omega)$ is as in the Maeda-Ogasawara-Vulikh representation theorem, then R is regular in $C_\infty(\Omega)$.

Let Θ be a vector sublattice of the space Υ of all random variables (examples of such sublattices are the sets of all constant functions and that of all simple functions). A net x_λ, $\lambda\in\Lambda$, of elements of Υ is said to be (ΘO)-*convergent* to $x\in\Upsilon$ (shortly, $(\Theta O)\lim_\lambda x_\lambda=x$) iff there is a decreasing net $(\sigma_\lambda)_\lambda$ of elements of Θ, with $|x_\lambda-x|\leq\sigma_\lambda$ for every $\lambda\in\Lambda$ and $\bigwedge_\lambda\sigma_\lambda=0$ (here, the infimum is intended in Θ). It is not difficult to see that, when Θ is the space of all constant functions, (ΘO)-convergence coincides with uniform convergence.

The following result holds.

Theorem 1.1 (Kappos (1969), Theorems IV.4.1 and IV.4.3) *The (ΘO)-limit is unique if and only if Θ is regular in Υ. In this case, the (ΘO)-limit is compatible with respect to the operations of sum, difference, product, supremum and infimum.*

We say that a net x_λ, $\lambda \in \Lambda$, is (ΘO)-*Cauchy* iff $(\Theta O)\lim\limits_{\lambda_1,\lambda_2} \lim_{\lambda_1,\lambda_2} x_{\lambda_1,\lambda_2} = x$, that is iff there is a decreasing double net $(\sigma_{\lambda_1,\lambda_2})_{\lambda_1,\lambda_2}$ such that $| x_{\lambda_1} - x_{\lambda_2} | \leq \sigma_{\lambda_1,\lambda_2}$ for every $\lambda_1, \lambda_2 \in \Lambda$ and $\bigwedge_{\lambda_1,\lambda_2} \sigma_{\lambda_1,\lambda_2} = 0$, where the infimum is intended in Θ and with respect to the "componentwise directed" set $\Lambda \times \Lambda$, namely $(\lambda_1, \lambda_2) \geq (\zeta_1, \zeta_2)$ iff $\lambda_1 \geq \zeta_1$ and $\lambda_2 \geq \lambda_2$.

Theorem 1.2 (Kappos (1969), Theorems IV.4.6.1 - IV.4.6.5)

(a) *A sequence $(x_n)_n$ in Υ is (ΘO)-Cauchy if and only if there is a decreasing sequence $(\sigma_n)_n$ in Θ with $\bigwedge_n \sigma_n = 0$, where the infimum involved is in Θ, and $| x_n - x_{n+k} | \leq \sigma_n$ for every n, $k \in \mathbb{N}$.*

(b) *If $(x_n)_n \subset \Upsilon$ and $(\Theta O)\lim\limits_n x_n = x \in \Upsilon$, then $(\Upsilon O)\lim\limits_n x_n = x$.*

(c) *If $(x_n)_n \subset \Upsilon$ and $(x_n)_n$ is (ΘO)-Cauchy, then $(x_n)_n$ is (ΥO)-Cauchy.*

(d) *If $(x_n)_n \subset \Upsilon$, $(\Upsilon O)\lim\limits_n x_n = x \in \Upsilon$ and $(x_n)_n$ is (ΘO)-Cauchy, then $(\Theta O)\lim\limits_n x_n = x$.*

(e) *If $(x_n)_n \subset \Upsilon$ and $(\Theta O)\lim\limits_n x_n = x \in \Upsilon$, then $(x_n)_n$ is (ΘO)-Cauchy.*

Theorem 1.3 (Kappos (1969), Lemma IV.4.1) *A sequence $(x_n)_n$ in Υ is (ΘO)-Cauchy iff there exist an increasing sequence $(y_n)_n$ and a decreasing sequence $(z_n)_n$ in Υ with $y_n \leq x_n \leq z_n$ for every $n \in \mathbb{N}$ and $(\Theta O)\lim\limits_n (z_n - y_n) = 0$. In this* *case* $(y_n)_n$ *and* $(z_n)_n$ *are also* (ΘO)*-Cauchy, and, if* $(x_n)_n$ *is* (ΘO)*-convergent, then* $(y_n)_n$ *and* $(z_n)_n$ *are* (ΘO)*-convergent too.*

Corollary 1.4 (Kappos (1969), Theorem IV.4.2) *Every* (ΘO) *-fundamental sequence is bounded in* Υ.

We say that two sequences $(x_n)_n$ and $(y_n)_n$ in Υ are (ΘO)*-equivalent in* Υ iff $(\Theta O)\lim_n (x_n - y_n) = 0$. The following result holds.

Theorem 1.5 (see Kappos (1969), Theorems IV.4.10 and IV.4.11)

(a) *If* $(x_n)_n$ *is a* (ΥO)*-Cauchy sequence in* Υ*, then there is a* (ΥO)*-Cauchy* $(y_n)_n$ *of simple functions,* (ΥO)*-equivalent to* $(x_n)_n$ *in* Υ.

(b) *If* $(x_n)_n$ *is a* (ΥO) *-Cauchy sequence in* Υ *, and* Θ *is a* σ *-regular vector sublattice of* Υ *containing the constant functions, then there exists a* (ΘO)*-Cauchy sequence* $(y_n)_n$ *in* Υ*, which is* (ΥO)*-equivalent to* $(x_n)_n$ *in* Υ.

In Kappos (1969) the construction of the Dedekind completion $\overline{\Upsilon}$ of Υ is presented, and it is pointed out that Υ is Dedekind complete if and only if the associated Boolean σ-algebra is atomic. Moreover, it is shown that $\overline{\Upsilon}$ is both Dedekind complete (Kappos (1969), Theorem 5.3) and complete with respect to the $(\overline{\Upsilon}O)$-convergence in $\overline{\Upsilon}$ (Kappos (1969), Theorem 5.2). Furthermore, it is proved that $(\overline{\Upsilon}O)$-convergence of sequences in $\overline{\Upsilon}$ of sequences coincides with P-almost uniform convergence (a sequence $(x_n)_n$ in $\overline{\Upsilon}$ is said to be *almost uniformly convergent* to $x \in \overline{\Upsilon}$ iff for each $\varepsilon > 0$ there is an element a of the associated Boolean σ-algebra L, such that $P(a') \le \varepsilon$, where $a \oplus a' = 1$, and the sequence $(\chi_a x_n)_n$ converges uniformly to $\chi_a x$, where χ_a is the random variable which associates the values 1 and 0 to the elements a and a' respectively (Kappos (1969), Theorem 7.1)). In Kappos (1969) also some other kinds of convergences, like (r)- and (*)-convergence (see also Chapter 1), are considered. Moreover even random variables taking valued in abstract spaces, like for instance Banach spaces or lattice groups, are studied, together with properties on completion with respect to (O)-convergence and with respect to a norm, expectation and L^p spaces in the setting of Banach space-valued random variables.

1.2. Concept, Lattice and Probability, by X. Dimitriou and C. P. Kitsos

The sense of concept was recently applied in several fields. Kitsos and Sotiropoulos (2009) worked on the "concept" idea and defined an equivalence relation, the upper and lower concept of two given concepts. The distance between two sets can be an inspiration for thinking about order categorical data associated with this distance. As an example we mention the order and distance among cancer concepts, the case of leukemia.

Since concepts can be structured as a lattice (see also Ganter and Wille (1999)) the idea is to be connected with lattice theory and probability (see also Nation, unpublished course notes, available http://bigcheese.math.sc.edu/~mcnulty/algl atvar/lat.pdf).

a) Lattices

Lattice Theory is used in the foundation of Probability, for instance, in Kappos (1969).

A comprehensive survey related with concepts can be found in Ganter and Wille (1999). We refer to these two textbooks.

Definition 1 For an element u of a Dedekind complete lattice V we define $u_0 := \vee\{x \in V : x < u\}$ and $u^* := \wedge\{x \in V : u < x\}$. We say that u is $\vee$*-irreducible* (or supremum irreducible) iff u cannot be represented as supremum of strictly smaller elements. In this case, u_0 is the unique *lower neighbor* of u. Dually, u is called $\wedge$*-irreducible* (or infimum irreducible) iff cannot be represented as infimum of strictly greater elements. $J(V)$ denotes the set of all $\vee$-irreducible elements and $M(V)$ the set of all $\wedge$-irreducible elements of V.

A set $X \subseteq V$ is called *supremum-dense* (resp. *infimum-dense*) in *V*, iff every element from *V* can be represented as the supremum (resp. infimum) of a subset of X.

Proposition 2 *An element u of a finite lattice is* $\vee$*-irreducible if and only if it has exactly one lower neighbor, and* $\wedge$*-irreducible iff it has exactly one upper neighbor. Every supremum-dense subset contains all* $\vee$*-irreducible elements and every infimum-dense subset contains all* $\wedge$*-irreducible elements. Conversely, in a finite lattice V the set* $J(V)$ *is supremum-dense and* $M(V)$ *is infimum-dense.*

Definitions 3 A subset *U* of a Dedekind complete lattice *V* which is closed under suprema is called a $\vee$-*subsemilattice* of *V*.

Dually, a subset which is closed under infima is called a $\wedge$-*supersemilattice* of *V*.

A subset which is closed under both suprema and infima is called a *Dedekind complete sublattice*.

b) Concept Lattices of Contexts

Definition 4 A *concept* is a set function φ of a set *O* (the "objects") to another set *A* (the "attributes"). It can be expressed by the couple (O, A) or (O, A, φ) as a triple. That is, the set function φ is defined from the data set of objects or attributes under consideration. The data we examine (for example any quality attributes, yes or no to a disease, attributes of a given disease) acts as a generator φ of concepts.

When we deal with concepts and define intension, extension etc., we require that the simple laws of set theory are fulfilled.

Definition 5 The *union* of two concepts (O_1, A_1) and (O_2, A_2) is the concept defined by

$$(O_1 \cup O_2, A_1 \cap A_2),$$

where $\cup$ and $\cap$ are the usual set operations.

The *intersection* is defined as

$$(O_1 \cap O_2, A_1 \cup A_2).$$

Definition 6 Two concepts (O_1, A_1) and (O_2, A_2) are *equivalent* ($\sim$) iff $A_1 = A_2$, and we shall write

$$(O_1, A_1) \sim (O_2, A_2) \Leftrightarrow A_1 = A_2.$$

Proposition 7 It is possible to check that $\sim$ is an equivalence relation among concepts. So we create a partition of the concepts coming from the collected data. We call the set O the "extension" of a concept and the set A its "intension".

Now, from the definition of the "concept union" we realize that, by taking the union of two concepts, we find common attributes of another "greater" object (in other words, similarities). Similarly thinking about the operation we find that "less extension has greater intension".

We now treat the case of three concepts (O_1, A_1), (O_2, A_2), (O_3, A_3). In such a case we define four levels.

(*i*) The superordinated $(O_1, \varnothing)$ in the upper level.

(*ii*) The defined three (O_i, A_i), $i = 1,2,3$, in the second subordinated level.

(*iii*) The three concepts with two attributes in the third level.

(*iv*) The concept $(O_1, A_1 \cup A_2 \cup A_3)$ in the fourth level.

Fig. (**1**) describes the discussed structure above.

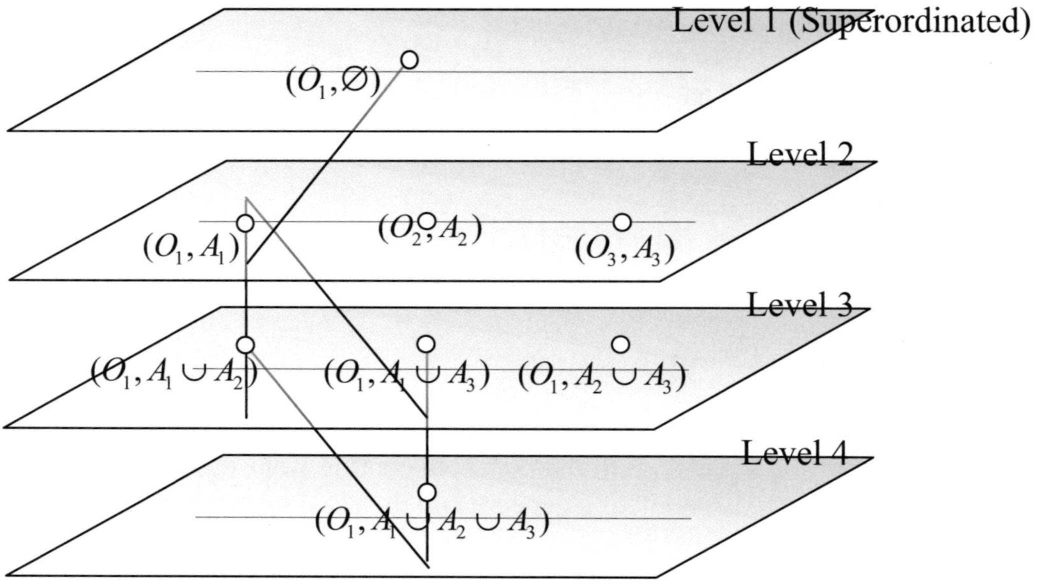

Fig. (1). From fourth level to superordinated.

Fig. (**2**) describes the dual problem: to build from the empty set, in the same number of levels, the subordinated.

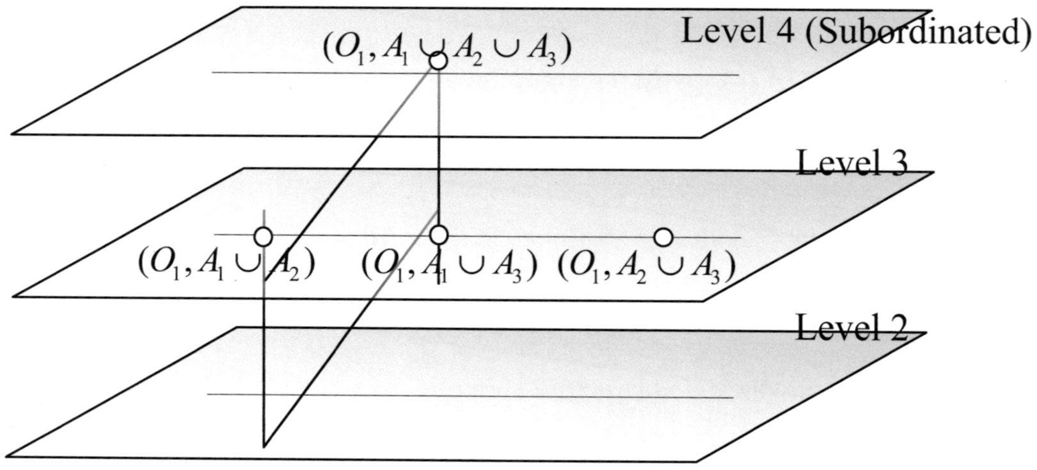

Fig. (2). From $(O_1,\varnothing)$ build up $(O_1, A_1 \cup A_2 \cup A_3)$.

The above discussion can be an inspiration for thinking about a definition of "distance" between the sets O_1 and O_2. Note that this distance satisfies all well-known axioms of the distance.

Definition 8 If (A_1, B_1) and (A_2, B_2) are concepts of a context, then (A_1, B_1) is called a *subconcept* of (A_2, B_2) provided that $A_1 \subseteq A_2$ (which is equivalent to $B_2 \subseteq B_1$). In this case (A_2, B_2) is called a *superconcept* of (A_1, B_1) and we write $(A_1, B_1) \leq (A_2, B_2)$. The relation $\leq$ is called the *hierarchical order* (or simply *order*) of the concepts involved. The set of all concepts of (G, M, I) ordered in this way is denoted by $B(G, M, I)$ and is called the concept lattice of the context (G, M, I).

Proposition 9 (*Duality Principle*) Let (G, M, I) be a context. Then (M, G, I^{-1}) is also a context, and $(B, A) \mapsto (A, B)$ is an isomorphism. In other words, if we exchange the roles of objects and attributes, we obtain the dual concept lattice.

Remark 10 The *normalized distance* D_N of two sets O_1 and O_2 is given by the formula

$$D_N(O_1, O_2) = \frac{\#(O_1 \Delta O_2)}{\#(O_1) + \#(O_2)},$$

where # and Δ denote the cardinality and the symmetric difference of the involved sets, respectively.

We try to connect the "concept" idea, which can be applied "elementarily" to categorical statistical data, with lattices, to formulate a probability definition for the data under consideration. Therefore a solid statistical background can be developed, in the investigation of concepts. The theory of fuzzy sets is also closely related.

Open Problems

1) Investigate the main properties of probability measures defined on more abstract structures and with values in different types of abstract spaces, and those of the corresponding expectations.

2) Study different modes of convergence for the elements of the space of random variables (also in abstract setting) with respect to a subspace.

3) Investigate some other kinds of "distance functions" in connection with more abstract set functions together with their main properties.

4) The limit theorems formulated in Boccuto, Dimitriou and Papanastassiou (2011b-c) for P-filters and the Schur, Dieudonné, Vitali-Hahn-Saks and Nikodým-type theorems, given for diagonal and block-respecting filters (see for instance Boccuto and Dimitriou (2013e, 2014a and 2014c), Boccuto, Dimitriou and Papanastassiou (2011c and 2012e)) were given by assuming as *hypotheses* some properties of the limit measures, like for example σ-additivity, absolute continuity, regularity, and so on. Find some condition to get these properties as *theses* rather than to require them as hypotheses.

5) In the filter limit theorems, find some different necessary or/and sufficient conditions on the involved filter to give further versions of such limits.

6) Prove some limit theorems for measures defined and/or with values in more abstract structures.

7) Find some limit theorems with respect to some other notions of (s)-boundedness, countable additivity, absolute continuity and regularity.

References

Aarnes, J. (1966). The Vitali-Hahn-Saks theorem for von Neumann algebras. *Math. Scand., 18*, 87-92.

Abbott, C., Bator, E. M., Bilyeu, R. G., & Lewis, P. L. (1990). Weak precompactness, strong boundedness, and weak complete continuity. *Math. Proc. Camb. Phil. Soc., 108,* 325-335.

Abraham, P. (1992). Saeki's improvement of the Vitali-Hahn-Saks-Nikodým theorem holds precisely for Banach spaces having cotype. *Proc. Amer. Math. Soc., 116*, 171-173.

Abraham, P. (1996). On the Vitali-Hahn-Saks-Nikodým theorem. *Quaestiones Math., 19*, 397-407.

Adamski, W., Gänssler, P., & Kaiser, S. (1976). On compactness and convergence in spaces of measures. *Math. Ann., 220,* 193-210.

Agbeko, N. K. (1994). On optimal averages. *Acta Math. Hungar., 63,* 133-147.

Agbeko, N. K. (1995). On the structure of optimal measures and some of its applications. *Publ. Math. Debrecen, 46,* 79-87.

Agbeko, N. K. (2000). How to characterize some properties of measurable functions. *Math. Notes Miskolc, 1,* 87-98.

Agbeko, N. K. (2009). *The role of the maximum operator in the theory of measurability and some applications.* Ph. D. Thesis, Department of Applied Mathematics, University of Miskolc.

Aizpuru, A. (1988). Relations between supremum and interpolation properties in Boolean algebras (Spanish). *Collect. Math., 39,* 115-125.

Aizpuru, A. (1992). On the Grothendieck and Nikodým properties of Boolean algebras. *Rocky Mt. J. Math., 22,* 1-10.

Aizpuru, A. (1996). The Nikodým property and local properties of Boolean algebras. *Colloq. Math., 71,* 79-85.

Aizpuru, A., Armario, R., García-Pacheco, F. J., & Pérez-Fernández, F. J. (2011). Banach limits and uniform almost summability. *J. Math. Anal. Appl., 379,* 82-90.

Aizpuru, A., Armario, R., & Pérez-Fernández, F. J. (2008). Almost summability and unconditionally Cauchy series. *Bull. Belg. Math. Soc. Simon Stevin, 15,* 635-644.

Aizpuru, A., García-Pacheco, F. J., & Pérez-Eslava, C. (2007). Matrix summability and uniform convergence of series. *Proc. Amer. Math. Soc., 135,* 3571-3579.

Aizpuru, A., & A. Gutiérrez-Dávila, A. (2004). On the interchange of series and some applications. *Bull. Belg. Math. Soc., 11,* 409-430.

Aizpuru, A., Gutiérrez-Dávila, A., & Pérez-Fernández, F. J. (2003). Boolean algebras and uniform convergence of series. *J. Math. Anal. Appl., 284,* 89-96.

Aizpuru, A., Gutiérrez-Dávila, A., & Sala, A. (2004). On the convergence and uconditionally convergence of series of operators on Banach spaces. *Extracta Math. 19,* 127-130.

Aizpuru, A., Gutiérrez-Dávila, A., & Sala, A. (2006). Unconditionally Cauchy series and Cesàro summability. *J. Math. Anal. Appl., 324,* 39-48.

Aizpuru, A., Moreno-Pulido, S., & Rambla-Barreno, F. (2013). Phillips lemma on effect algebras of sets. *Math. Slovaca, 63,* 639-646.

Aizpuru, A., & Nicasio-Llach, M. (2008a). Spaces of sequences defined by the statistical convergence. *Studia Sci. Math. Hungar., 45,* 519-529.

Aizpuru, A., & Nicasio-Llach, M. (2008b). About the statistical uniform convergence. *Bull. Braz. Math. Soc., 39,* 173-182.

Aizpuru, A., Nicasio-Llach, M., & Rambla-Barreno, F. (2010). A remark about the Orlicz-Pettis theorem and the statistical convergence. *Acta Math. Sinica, English Ser., 26,* 305-310.

Aizpuru, A., Nicasio-Llach, M., & Sala, A. (2010). A Remark about the statistical Cesàro summability and the Orlicz-Pettis theorem. *Acta Math. Hungar., 126*, 94-98. DOI: 10.1007/s10474-009-9021-1

Aizpuru, A., Nicasio-Llach, M., & Tamayo, M. (2007). The Hahn-Schur theorem on effect algebras. *Bull. Austral. Math. Soc., 75,* 161-168.

Aizpuru, A., Pérez-Eslava, C., & Seoane-Sepúlveda, J. B. (2009). Matrix summability methods and weakly unconditionally Cauchy series. *Rocky Mt. J. Math., 39,* 367-380.

Aizpuru, A., & Tamayo, M. (2006). Classical properties of measure theory on effect algebras. *Fuzzy Sets Systems, 157,* 2139-2143.

Antonio Boccuto & Xenofon Dimitriou

Albayrak, H., & Pehlivan, S. (2010). On the ideal convergence of subsequences and rearrangements of a real sequence. *Applied Math. Letters, 23,* 1203-1207.

Albuquerque, J. (1944). Ensembles de Borel. *Portug. Math., 4,* 161-198.

Aleksyuk, V. N. (1970). On the weak compactness of families of quasimeasures. On the interrelationship between metric and measure. *Sib. Math. J., 11,* 547-558.

Alexandroff, A. D. (1941). Additive set-functions in abstract spaces. *Math. Sbornik, 9 (51),* 563-628.

Alexiewicz, A. (1949). On sequences of operations. I. *Studia Math., 11,* 1-30.

Aliprantis, C. D., & Burkinshaw, O. (1985). *Positive operators.* New York, London: Acad. Press.

Aliprantis, C. D., & Burkinshaw, O. (1990). *Principles of Real Analysis.* New York, London: Acad. Press.

Aliprantis, C. D., & Burkinshaw, O. (2003). *Locally solid Riesz spaces with applications to Economics.* Math. Surveys Monographs, *105.* Rhode Island: Amer. Math. Soc., Providence.

Amir, D., & Lindenstrauss, J. (1968). The structure of weakly compact sets in Banach spaces. *Ann. Math., 88,* 35-46.

Anastassiou, G.A., & Duman, O. (2011). *Towards Intelligent Modeling: Statistical Approximation Theory.* Intelligent System Reference Library, *14.* Berlin, Heidelberg, New York: Springer-Verlag.

Andô, T. (1961). Convergent sequences of finitely additive measures. *Pacific J. Math., 11,* 395-404.

Angeloni, L., & Vinti, G. (2005). Rate of approximation for nonlinear integral operators with application to signal processing. *Different. Integral Equations, 18,* 855-890.

Antoni, J. (1986). On the A-continuity of real functions.II. *Math. Slovaca, 36,* 283-288.

Antoni, J., & Šalát, T. (1980). On the A-continuity of real functions. *Acta Math. Univ. Comenian., 39,* 159-164.

Antosik, P. (1971). On the Mikusiński diagonal theorem. *Bull. Acad. Polon. Sci. Sér. Sci. Math. Astronom. Phys., 19,* 305-310.

Antosik, P. (1972). A generalization of the diagonal theorem. *Bull. Acad. Polon. Sci. Sér. Sci. Math. Astronom. Phys., 20,* 373-377.

Antosik, P. (1973). Mapping from L-groups into topological groups. I, II. *Bull. Acad. Polon. Sci. Sér. Sci. Math. Astronom. Phys., 21,* 145-153, 155-160.

Antosik, P. (1976). A diagonal theorem for nonnegative matrices and equicontinuous sequences of mappings. *Bull. Acad. Polon. Sci. Sér. Sci. Math. Astronom. Phys., 24,* 855-859.

Antosik, P., & Swartz, C. (1983). The Nikodým boundedness theorem and the uniform boundedness principle. In: *Measure theory and its applications* (Sherbrooke, Que., 1982), (pp. 36-42). Lecture Notes Math., *1033.* Berlin, Heidelberg, New York: Springer-Verlag.

Antosik, P., & Swartz, C. (1984a). The Schur and Phillips lemmas for topological groups. *J. Math. Anal. Appl., 98,* 179-187.

Antosik, P., & Swartz, C. (1984b). A theorem on matrices and its applications in functional analysis. *Studia Math., 77,* 197-205.

Antosik, P., & Swartz, C. (1985a). *Matrix Methods in Analysis.* Lecture Notes Math., *1113.* Berlin, Heidelberg, New York: Springer-Verlag.

Antosik, P., & Swartz, C. (1985b). The Vitali-Hahn-Saks theorem for algebras. *J. Math. Anal. Appl., 106,* 116-119.

Antosik, P., & Swartz, C. (1991). Multiplier convergent series and the Schur theorem. *Southeast Asian Bull. Math., 15,* 173-182.

Antosik, P., & Swartz, C. (1992). The Nikodým convergence theorem for lattice-valued measures. *Rev. Roumaine Math. Pures Appl., 37,* 299-306.

Arvanitakis, A. D., & Stein J. D., Jr. (2008). A generalization of the uniform boundedness principle. *Int. J. Pure Appl. Math., 43,* 687-708.

Atanassov, K. (1999). *Intuitionistic Fuzzy Sets: Theory and Applications.* Heidelberg: Physica-Verlag.

Athanassiadou, E., Boccuto, A., Dimitriou, X., & Papanastassiou, N. (2012). Ascoli-type theorems and ideal (α)-convergence. *Filomat, 26,* 397-405. DOI 10.2298/FIL1202397A

Athanassiadou, E., Dimitriou, X., Papachristodoulos, Ch., & Papanastassiou, N. (2012). Strong (α)-convergence and ideal strong exhaustiveness of sequences of functions, *Int. J. Pure Appl. Math., 80,* 207-216.

Avallone, A. (2006). Cafiero and Nikodým boundedness theorems in effect algebras. *Ital. J. Pure Appl. Math., 20,* 203-214.

Avallone, A., Barbieri G., & Vitolo, P. (2003). Hahn decompositions of modular measures and applications. *Comment. Math., 43,* 149-168.

Avallone, A., Barbieri G., & Vitolo, P. (2008). On the Alexandroff decomposition theorem. *Math. Slovaca, 58,* 185-200.

Avallone, A., Barbieri G., & Vitolo, P. (2010). Central elements in pseudo *D*-lattices and Hahn decomposition theorem. *Boll. Un. Mat. Ital., 9,* 447-470.

Avallone, A., Barbieri G., Vitolo, P., & Weber, H. (2009). Decomposition of effect algebras and the Hammer-Sobczyk theorem. *Algebra Universalis, 60,* 1-18.

Avallone, A., & Lepellere, M. A. (1998). Modular functions: uniform boundedness and compactness. *Rend. Circ. Mat. Palermo Ser. II, 47,* 221-264.

Avallone, A., Rinauro, S., & Vitolo, P. (2007). Boundedness and convergence theorems in effect algebras. *Tatra Mt. Math. Publ., 35,* 159-174.

Avallone, A., & Vitolo, P. (2003). Decomposition and control theorems in effect algebras. *Sci. Math. Japon., 58,* 1-14.

Avallone, A., & Vitolo, P. (2005). Lattice uniformities on effect algebras. *Int. J. Theor. Phys., 44,* 793-806. DOI: 10.1007/s10773-005-7057-8

Avallone, A., & Vitolo, P. (2009). Lyapunov decomposition of measures on effect algebras. *Sci. Math. Japon., 69,* 633-641.

Avallone, A., & Vitolo, P. (2013). Lebesgue decomposition and Bartle-Dunford-Schwartz theorem in pseudo *D*-lattices. *Acta Math. Sci., 33 B,* 1-25.

Aviles Lopez, A., Cascales Salinas, B., Kadets, V., & Leonov, A. (2007). The Schur l_1 theorem for filters. *J. Math. Phys., Anal., Geom., 3,* 383-398.

Baayen, P. C., & Helmberg, G. (1965). On families of equi-uniformly distributed sequences in compact spaces. *Math. Ann., 161,* 255-278.

Baláž, V., Červeňanskij, J., Kostyrko, P., & Šalát, T. (2002). $\mathcal{I}$-convergence and $\mathcal{I}$-continuity of real functions. *Acta Math. (Nitra), 5,* 43-50.

Baláž, V., & Šalát, T. (2006). Uniform density *u* and corresponding $\mathcal{I}_u$-convergence. *Math. Commun., 11,* 1-7.

Balcerzak, M., & Dems, K. (2005). Some types of convergence and related Baire systems. *Real Anal. Exchange, 30,* 267-276.

Balcerzak, M., Dems, K., & Komisarski, A. (2007). Statistical convergence and ideal convergence for sequences of functions. *J. Math. Anal. Appl., 328,* 715-729.

Banach, S. (1922). Sur les opérations dans les ensemles abstraits et leur application aux équations intégrales. *Fund. Math., 3,* 133-181.

Banach, S. (1923). Sur le problème de la mesure. *Fund. Math., 4,* 7-33.

Banach, S. (1932). *Théorie des opérations linéaires.* Warsaw: Monografje Mat.

Banach, S., & Steinhaus, H. (1927). Sur le principle de la condensation de singularités. *Fund. Math., 9,* 50-61.

Barbieri, G. (2009a). On Dieudonné's boundedness theorem. *Boll. Un. Mat. Ital. II, 9,* 343-348.

Barbieri, G. (2009b). On the Dieudonné theorem. *Sci. Math. Japon., 44,* 403-408.

Barbieri, G. (2009c). Convergent regular measures on *MV*-algebras. *Kybernetika, 45,* 1052-1058.

Barbieri, G. (2009d). An extension theorem for modular measures on effect algebras. *Czech. Math. J., 59 (134),* 707-719.

Barbieri, G. (2014). A note on the Cafiero criterion in effect algebras. *Ital. J. Pure Appl. Math.,* in press.

Barbieri, G., Lepellere, M. A., & Weber, H. (2001). The Hahn decomposition theorem for fuzzy measures and applications. *Fuzzy Sets Systems, 118,* 519-528.

Barbieri, G., Valente, A., & Weber, H. (2012). Decomposition of (ℓ)-group-valued measures. *Czech. Math. J., 62,* 1085-1100.

Barbieri, G., & Weber, H. (2002). Measures on clans and on *MV*-algebras. In: Pap, E. (Ed.), *Handbook of measure theory, vol. II* (pp. 911-945). Amsterdam: Elsevier.

Bardaro, C., Boccuto, A., Dimitriou, X., & Mantellini, I. (2013). Abstract Korovkin-type theorems in modular spaces and applications. *Cent. Eur. J. Math., 11,* 1774-1784. DOI:10.2478/s11533-013-0288-7

Bardaro, C., Boccuto, A., Dimitriou, X., & Mantellini, I. (2013). Modular filter convergence theorems for abstract sampling-type operators. *Appl. Anal., 92,* 2404-2423. DOI:10.1080/00036811.2012.738480

Bardaro, C., & Mantellini, I. (2006a). Uniform modular integrability and convergence properties for a class of Urysohn integral operators in function spaces. *Math. Slovaca, 56,* 465-482.

Bardaro, C., & Mantellini, I. (2006b). Approximation properties in abstract modular spaces for a class of general sampling-type operators. *Appl. Anal., 85,* 383-413.

Bardaro, C., & Mantellini, I. (2009). A Korovkin theorem in multivariate modular function spaces. *J. Function Spaces Appl., 7,* 105-120.

Bardaro, C., & Mantellini, I. (2013). Asymptotic formulae for linear combinations of generalized sampling operators. *Zeitschrift Anal. Anw., 32,* 279-298.

Bardaro, C., Musielak, J., & Vinti, G. (2003). *Nonlinear Integral Operators and Applications.* Berlin: Walter de Gruyter & Co.

Bartle, C. (1966). *The elements of Integration and Lebesgue measure.* London, New York: John Wiley & Sons.

Bartle, R. G., Dunford, N., & Schwartz, J. T. (1955). Weak compactness and vector measures. *Canad. J. Math., 7,* 289-305.

Baumgartner, J. E., Taylor, A. D., & Wagon, S. (1982). Structural properties of ideals. *Dissertationes Math., 197.*

Beer, G. (2010). The Alexandroff property and the preservation of strong uniform continuity. *Appl. Gen. Topology, 11,* 117-133.

Beer, G., & Levi, S. (2009). Strong uniform continuity. *J. Math Anal. Appl., 350,* 568-589.

Beer, G., & Levi, S. (2010). Uniform continuity, uniform convergence, and shields. *Set-valued Anal., 18,* 251-275.

Behrends, E. (2000). Lectures on Rosenthal's l^1-theorem. *Rend. Ist. Mat. Univ. Trieste, 31, Suppl. 1,* 71-106.

Bell, H. T. (1973). Order summability and almost convergence. *Proc. Amer. Math. Soc., 38,* 548-552.

Bell, W. C., Bilyeu, R. G., & Lewis, P. W. (1986). Uniform exhaustivity and Banach lattices. *Ann. Mat. Pura Appl., 144,* 57-73.

Bennett, G., Boos, J., & Leiger, T. (2002). Sequences of 0's and 1's. *Studia Math., 149,* 75-99.

Bennett, G., & Kalton, N. J. (1974). Consistency theorems for almost convergence. *Trans. Amer. Math. Soc., 198,* 23-43.

Bennett, M. K., & Foulis, D. J. (1994). Effect algebras and unsharp quantum logics. *Found. Phys., 24,* 1331-1352.

Benvenuti, P., Mesiar, R., & Vivona, D. (2002). Monotone set functions-based integrals. In: Pap, E. (Ed), *Handbook of Measure Theory, vol. II* (pp. 1329-1379). Amsterdam: Elsevier.

Berberian, S. K. (1999). *Fundamental of Real Analysis.* Berlin, Heidelberg, New York: Springer-Verlag.

Bernau, S. J. (1965). Unique representation of Archimedean lattice groups and normal Archimedean lattice rings. *Proc. London Math. Soc., 15,* 599-631.

Bessaga, C., & Pełczyński, A. (1958). On bases and unconditional convergence of series in Banach spaces. *Studia Math., 17,* 151-164.

Bhardwaj, V. K., & Bala, I. (2007). On weak statistical convergence. *Int. J. Math. Math. Sci.*, Art. ID 38530, 9 pp.

Bhaskara Rao, K. P. S., & Bhaskara Rao, M. (1983). *Theory of Charges.* New York, London: Acad. Press.

Bhunia, S., & Das, P. (2011). Two-valued measures and summability of double sequences in asymmetric context. *Acta Math. Hungar., 130,* 167-187.

Billingsley, P. (1995). *Convergence and Probability Measures.* London, New York: John Wiley & Sons.

Bilyeu, R. G., & Lewis, P. W. (1980). Uniform differentiability, uniform absolute continuity and the Vitali-Hahn-Saks theorem. *Rocky Mt. J. Math., 10,* 533-557.

Birkhoff, G. (1967). *Lattice theory.* Rhode Island: Amer. Math. Soc., Providence.

Boccuto, A. (1993). Riesz Spaces, Integration and Sandwich Theorems. *Tatra Mt. Math. Publ., 3,* 213-230.

Boccuto, A. (1995a). Abstract integration in Riesz spaces. *Tatra Mt. Math. Publ., 5,* 107-124.

Boccuto, A. (1995b). On Stone-type extensions for group-valued measures. *Math. Slovaca, 45,* 309-315.

Boccuto, A. (1995c). Continuous invariant vector-valued means and applications. *Atti Semin. Mat. Fis. Univ. Modena, 43,* 229-244.

Boccuto, A. (1996a). Vitali-Hahn-Saks and Nikodým theorems for means with values in Riesz spaces. *Atti Semin. Mat. Fis. Univ. Modena, 44,* 157-173.

Boccuto, A. (1996b). Dieudonné-type theorems for means with values in Riesz spaces. *Tatra Mt. Math. Publ., 8,* 29-42.

Boccuto, A. (1997). Integration in Riesz spaces with respect to (*D*)-convergence. *Tatra Mt. Math. Publ., 10,* 33-54.

Boccuto, A. (2003). Egorov property and weak σ-distributivity in Riesz spaces. *Acta Math. (Nitra), 6,* 61-66.

Boccuto, A., & Candeloro, D. (1990). On the existence of continuous invariant means with respect to amenable semigroups. *Rend. Mat. (Roma), Ser. VII, 10,* 303-319.

Boccuto, A., & Candeloro, D. (1993). Invariant means with special properties. *Atti Semin. Mat. Fis. Univ. Modena, 41,* 167-180.

Boccuto, A., & Candeloro, D. (2002a). Dieudonné-type theorems for set functions with values in (ℓ)-groups. *Real Anal. Exchange, 27,* 473-484.

Boccuto, A., & Candeloro, D. (2002b). Uniform (s)-boundedness and convergence results for measures with values in complete (ℓ)-groups. *J. Math. Anal. Appl, 265,* 170-194.

Boccuto, A., & Candeloro, D. (2002c). Vitali and Schur-type theorems for Riesz space-valued set functions. *Atti Semin. Mat. Fis. Univ. Modena, 50,* 85-103.

Boccuto, A., & Candeloro, D. (2004a). Convergence and decompositions for (ℓ)-group-valued set functions. *Comment. Math., 44,* 11-37.

Boccuto, A., & Candeloro, D. (2004b). Itô formula in Riesz spaces. *Proceedings of the Tenth International Conference IPMU, Perugia, July 2004,* 1807-1814.

Boccuto, A., & Candeloro, D. (2004c). Kondurar Theorem in Riesz spaces. *Proceedings of the Tenth International Conference IPMU, Perugia, July 2004,* 1815-1822.

Boccuto, A., & Candeloro, D. (2004d). Uniform boundedness theorems in Riesz spaces. *Atti Semin. Mat. Fis. Univ. Modena Reggio Emilia, 52,* 369–382.

Boccuto, A., & Candeloro, D. (2005). A survey of decomposition and convergence theorems for (ℓ)-group-valued measures. *Atti Semin. Mat. Fis. Univ. Modena Reggio Emilia, 53,* 243-260.

Boccuto, A., & Candeloro, D. (2008). Sobczyk-Hammer decompositions and convergence theorems for measures with values in (ℓ)-groups. *Real Anal. Exchange, 33,* 91-106.

Boccuto, A., & Candeloro, D. (2009a). Integral and ideals in Riesz spaces. *Inform. Sci., 179,* 2891-2902.

Boccuto, A., & Candeloro, D. (2009b). Defining limits by means of integrals. *Operator Theory: Advances and Applications, 201,* 79-87.

Boccuto, A., & Candeloro, D. (2010). Some new results about Brooks-Jewett and Dieudonné-type theorems in (ℓ)-groups. *Kybernetika, 46,* 1049-1060.

Boccuto, A., & Candeloro, D. (2011). Uniform (s)-boundedness and regularity for (ℓ)-group-valued measures. *Cent. Eur. J. Math, 9,* 433-440. DOI:10.2478/s11533-010-0097-1

Boccuto, A., Candeloro, D., & Sambucini, A. R. (2007). Stieltjes-type integrals for metric semigroup-valued functions defined on unbounded intervals. *PanAmerican Math. J., 17,* 39-58.

Boccuto, A., Candeloro, D., & Sambucini, A. R. (2011). A Fubini theorem in Riesz spaces for the Kurzweil-Henstock integral. *J. Funct. Spaces Appl., 9,* 283-304. DOI: 10.1155/2011/158412

Boccuto, A., Candeloro, D., & Sambucini, A. R. (2014). Vitali-type theorems for filter convergence related to vector lattice-valued modulars and applications to stochastic processes. *J. Math. Anal. Appl.,* 419 (2014), 818-838. DOI 10.1016/j.jmaa.2014.05.014

Boccuto, A., Das, P., & Dimitriou, X. (2012). A Schur-type theorem for $\mathcal{I}$-convergence and maximal ideals. *Int. J. Pure Appl. Math., 81,* 517-529.

Boccuto, A., Das, P., Dimitriou, X., & Papanastassiou, N. (2012). Ideal exhaustiveness, weak convergence and weak compactness in Banach spaces. *Real Anal. Exchange, 37,* 389-410.

Boccuto, A., & Dimitriou, X. (2011a). Some properties of ideal α-convergence in (ℓ)-groups. *Int. J. Pure Appl. Math., 72,* 93-99.

Boccuto, A., & Dimitriou, X. (2011b). Ideal exhaustiveness and limit theorems for (ℓ)-groups. *Acta Math. (Nitra), 14,* 65-70.

Boccuto, A., & Dimitriou, X. (2011c). Some new results on limit theorems with respect to ideal convergence in (ℓ)-groups. *Atti Semin. Mat. Fis. Univ. Modena Reggio Emilia, 58,* 163-174.

Boccuto, A., & Dimitriou, X. (2013a). Modular convergence theorems for integral operators in the context of filter exhaustiveness and applications. *Mediterranean J. Math., 10,* 823-842.

Boccuto, A., & Dimitriou, X. (2013b). Modular filter convergence theorems for Urysohn integral operators and applications. *Acta Math. Sin. (Engl. Ser.), 29,* 1055-1066.

Boccuto, A., & Dimitriou, X. (2013c). Ideal limit theorems and their equivalence in (ℓ)-group setting. *J. Math. Research, 5,* 43-60.

Boccuto, A., & Dimitriou, X. (2013d). Modes of ideal continuity of (ℓ)-group-valued measures. *Intern. Math. Forum, 8,* 841-849.

Boccuto, A., & Dimitriou, X. (2013e). Limit theorems for topological group-valued maesures. *Acta Math. (Nitra), 16,* 37-43.
Boccuto, A., & Dimitriou, X. (2014a). Some new types of filter limit theorems for topological group-valued measures. *Real Anal. Exchange, 39,* 139-174.
Boccuto, A., & Dimitriou, X. (2014b). Rates of approximation for general sampling-type operators in the setting of filter convergence. *Appl. Math. Comput., 229,* 214-226.
Boccuto, A., & Dimitriou, X. (2014c). Nikodým-type theorems for lattice group-valued measures with respect to filter convergence. *Adv. Pure Math., 4,* 213-221.
Boccuto, A., & Dimitriou, X. (2014d). More on filter exhaustiveness of lattice group-valued measures. *Acta Math. (Nitra), 17,* 35-41.
Boccuto, A., & Dimitriou, X. (2014e). Ascoli-type theorems in the cone metric space setting, *J. Inequalities Appl.,* in press.
Boccuto, A., Dimitriou, X., & Papanastassiou, N. (2010a). Convergence theorems for the optimal integral in Riesz spaces. *Acta Math. (Nitra), 13,* 55-61.
Boccuto, A., Dimitriou, X., & Papanastassiou, N. (2010b). Unconditional convergence in lattice groups with respect to ideals. *Comment. Math., 50,* 161-174.
Boccuto, A., Dimitriou, X., & Papanastassiou, N. (2010c). Countably additive restrictions and limit theorems in (ℓ)-groups. *Atti Semin. Mat. Fis. Univ. Modena Reggio Emilia, 57,* 121-134.
Boccuto, A., Dimitriou, X., & Papanastassiou, N. (2011a). Addendum to: "Countably additive restrictions and limit theorems in (ℓ)-groups". *Atti Semin. Mat. Fis. Univ. Modena Reggio Emilia, 58,* 3-10.
Boccuto, A., Dimitriou, X., & Papanastassiou, N. (2011b). Brooks-Jewett-type theorems for the pointwise ideal convergence of measures with values in (ℓ)-groups. *Tatra Mt. Math. Publ., 49,* 17-26.
Boccuto, A., Dimitriou, X., & Papanastassiou, N. (2011c). Some versions of limit and Dieudonné-type theorems with respect to filter convergence for (ℓ)-group-valued measures. *Cent. Eur. J. Math., 9,* 1298-1311.
Boccuto, A., Dimitriou, X., & Papanastassiou, N. (2012a). Ideal convergence and divergence of nets in (ℓ)-groups. *Czech. Math. J., 62 (137),* 1073-1083.
Boccuto, A., Dimitriou, X., & Papanastassiou, N. (2012b). Uniform boundedness principle, Banach-Steinhaus and approximation theorems for filter convergence in Riesz spaces. In: Arhangel'skii, A. V., Khan, M. D., & Kočinac, Lj. D. R. (Eds.), *Proceedings of International Conference on Topology and its Applications ICTA 2011, Cambridge Sci. Publ.,* 45-58.
Boccuto, A., Dimitriou, X., & Papanastassiou, N. (2012c). Limit theorems in (ℓ)-groups with respect to (D)-convergence. *Real Anal. Exchange, 37,* 249-278.
Boccuto, A., Dimitriou, X., & Papanastassiou, N. (2012d). Basic matrix theorems for $\mathcal{I}$-convergence in (ℓ)-groups. *Math. Slovaca, 62,* 885-908.
Boccuto, A., Dimitriou, X., & Papanastassiou, N. (2012e). Schur lemma and limit theorems in lattice groups with respect to filters. *Math. Slovaca, 62,* 1145-1166.
Boccuto, A., Dimitriou, X., & Papanastassiou, N. (2012f). Modes of continuity involving almost and ideal convergence. *Tatra Mt. Math. Publ., 52,* 115-131. DOI: 10.2478/v10127-012-0032-x
Boccuto, A., Dimitriou, X., & Papanastassiou, N. (2014). Schur and matrix theorems with respect to $\mathcal{I}$-convergence. *Proceedings of the 13th PanHellenic Conference on Mathematical Analysis held in Ioannina, 28-29 May 2010,* in press.
Boccuto, A., Dimitriou, X., Papanastassiou, N., & Wilczyński, W. (2011). Ideal exhaustiveness, continuity and α-convergence for lattice group-valued functions. *Int. J. Pure Appl. Math., 70,* 211-227.
Boccuto, A., Dimitriou, X., Papanastassiou, N., & Wilczyński, W. (2012). Addendum to: "Ideal exhaustiveness, continuity and α-convergence for lattice group-valued functions". *Int. J. Pure Appl. Math., 75,* 383-384.
Boccuto, A., Dimitriou, X., Papanastassiou, N., & Wilczyński, W. (2014). Modes of ideal continuity and the additive property in the Riesz space setting. *J. Appl. Anal., 20,* 41-53.
Boccuto, A., Minotti, A. M., & Sambucini, A. R. (2013). Set-valued Kurzweil-Henstock integral in Riesz space setting. *PanAmer. Math. J., 23,* 57-74.
Boccuto, A., & Papanastassiou, N. (2007). Schur and Nikodým convergence-type theorems in Riesz spaces with respect to (r)-convergence. *Atti Sem. Mat. Fis. Univ. Modena e Reggio Emilia, 55,* 33-46.

Boccuto, A., & Riečan, B. (2006). Improper Kurzweil-Henstock integral for metric semigroup-valued functions. *Atti Sem. Mat. Fis. Univ. Modena e Reggio Emilia, 54,* 75-95.
Boccuto, A., & Riečan, B. (2008). The symmetric Choquet integral with respect to Riesz space-valued capacities. *Czech. Math. J. 58 (133),* 289-310.
Boccuto, A., & Riečan, B. (2009). The Concave Integral with respect to Riesz Space-valued capacities. *Math. Slovaca, 59,* 647-660.
Boccuto, A., & Riečan, B. (2010). On extension theorems for measures in (ℓ) -groups. *Math. Slovaca, 60,* 65-74.
Boccuto, A., Riečan, B., & Sambucini, A.R. (2010). On the product of *M*-measures in (ℓ) -groups. *Australian J. Math. Anal. Appl., 7,* Paper N. 9, 8 pp.
Boccuto, A., Riečan, B., & Vrábelová, M. (2009). *Kurzweil-Henstock Integral in Riesz Spaces.* Singapore: Bentham Science Publ.
Boccuto, A., & Sambucini, A.R. (1996a). On the De Giorgi-Letta integral with respect to means with values in Riesz spaces. *Real Anal. Exchange, 21,* 793-810.
Boccuto, A., & Sambucini, A.R. (1996b). The monotone integral with respect to Riesz space-valued capacities. *Rend. Mat. (Roma), 16,* 491-524.
Boccuto, A., & Sambucini, A.R. (1997). Comparison between different types of abstract integrals in Riesz spaces, *Rend. Circ. Mat. Palermo, Ser. II, 46,* 255-278.
Boccuto, A., & Skvortsov, V. A. (2006). Some applications of the Maeda-Ogasawara-Vulikh representation theorem to Differential Calculus in Riesz spaces. *Acta Math. (Nitra), 9,* 13-24.
Boccuto, A., & Skvortsov, V. A. (2009). Addendum to: "Some applications of the Maeda-Ogasawara-Vulikh representation theorem to Differential Calculus in Riesz spaces". *Acta Math. (Nitra), 12,* 39-46.
Bochner, S. (1939). Additive set functions on groups. *Ann. Math., 40,* 769-799.
Bochner, S. (1940). Finitely additive integral. *Ann. Math., 41,* 495-504.
Bochner, S., & Phillips, R. S. (1941). Additive set functions and vector lattices. *Ann. Math., 42,* 316-324.
Bogachev, V. I. (2007). *Measure Theory.* Berlin, Heidelberg, New York: Springer.
Bogdan, V. M., & Oberle, R. A. (1978). Topological rings of sets and the theory of vector measures. *Dissertationes Math., 154.*
Boos, J., & Cass, P. (2000). *Classical and modern methods in summability.* Oxford, New York: Oxford Univ. Press.
Boos, J., & Leiger, T. (2008). On some 'duality' of the Nikodým property and the Hahn property. *J. Math. Anal. Appl. 341,* 235-246.
Boos, J., & Seydel, D. (1999). Some remarks on invariant means and almost convergence. *J. Anal., 7,* 21-29.
Borsík, J., & Šalát, T. (1993). On *F*-continuity of real functions. *Tatra Mt. Math. Publ., 2,* 37-42.
Bourbaki, N. (1948). *Eléments de Mathématique, Livre III: Topologie générale. Act. Sci. Ind., 1045.* Paris: Hermann.
Bourbaki, N. (1950). Sur certains espaces vectoriels topologiques. *Ann. Inst. Fourier (Grenoble), 2,* 5-16.
Brooks, J. K. (1969a). Decomposition theorems for vector measures. *Proc. Amer. Math. Soc., 21,* 27-29.
Brooks, J. K. (1969b). On the Vitali-Hahn-Saks and Nikodým theorems. *Proc. Nat. Acad. Sci. U.S.A., 64,* 468-471.
Brooks, J. K. (1971). On the existence of a control measure for strongly bounded vector measures. *Bull. Amer. Math. Soc., 77,* 999-1001.
Brooks, J. K. (1972). Weak compactness in the space of vector measures. *Bull. Amer. Math. Soc., 78,* 284-287.
Brooks, J. K. (1973). Equicontinuous sets of measures and applications to Vitali's integral convergence theorem and control measures. *Adv. Math., 10,* 165-171.
Brooks, J. K. (1974). Interchange of limit theorems for finitely additive measures. *Rev. Roumaine Math. Pures Appl., 19,* 731-744.
Brooks, J. K. (1975). Sur les suites uniformément convergentes dans un espace de Banach. *C. R. Acad. Sci. Paris, 281,* 1037-1040.
Brooks, J. K. (1980). On a theorem of Dieudonné. *Adv. Math., 36,* 165-168.
Brooks, J. K. (2002). Stochastic processes and stochastic integration in Banach spaces. In: Pap., E. (Ed.), *Handbook on measure theory, vol. I* (pp. 449-502). Amsterdam: Elsevier.
Brooks, J. K. (2005). Equicontinuity in measure spaces and von Neumann algebras. *Positivity, 9,* 485-490.
Brooks, J. K., & Candeloro, D. (2004). The Rickart integral and Yosida-Hewitt decompositions. *Tatra Mt. Math. Publ., 28,* 227-240.

Brooks, J. K., & Chacon, R. V. (1980). Continuity and compactness of measures. *Adv. Math., 37,* 16-26.

Brooks, J. K., & Dinculeanu, N. (1974). Strong additivity, absolute continuity, and compactness in spaces of measures. *J. Math. Anal. Appl., 45,* 156-175.

Brooks, J. K., & Dinculeanu, N. (1991). Stochastic Integration in Banach Spaces. In: *Seminar on Stoch. Proc., 1990* (pp-27-115). Boston, Basel, Berlin: Birkhäuser.

Brooks, J. K., & Dinculeanu, N. (1995). Integration in Banach spaces. Application to stochastic integration. *Atti Sem. Mat. Fis. Univ. Modena, 43,* 317-361.

Brooks, J. K., & Jewett, R. S. (1970). On finitely additive vector measures. *Proc. Nat. Acad. Sci. U.S.A., 67,* 1294-1298.

Brooks, J. K., & Mikusiński, J. (1970). On some theorems in Functional Analysis. *Bull. Acad. Pol. Sciences Sér. Sci. Math. Astr. Phys., 18,* 151-155.

Brooks, J. K., Saitô, K., & Wright, J. D. M. (2003). Operator algebras and a theorem of Dieudonné. *Rend. Circ. Mat. Palermo Ser. II, 52,* 5-14.

Brooks, J. K., Saitô, K., & Wright, J. D. M. (2004). When absolute continuity on *C**-algebras is automatically uniform. *Quart. J. Math., 55,* 31-40.

Brooks, J. K., Saitô, K., & Wright, J. D. M. (2005). Operators on σ-complete *C**-algebras. *Quart. J. Math., 56,* 301-310.

Brooks, J. K., & Wright, J. D. M. (2001). Representing Yosida-Hewitt decompositions for classical and non-commutative vector measures. *Expo. Math., 19,* 373-383.

Brooks, J. K., & Wright, J. D. M. (2004). Convergence in the dual of a σ-complete *C**-algebra. *J. Math. Anal. Appl., 294,* 141-146.

Brown, T. C., & Freedman, A. R. (1990). The uniform density of sets of integers and Fermat's last theorem. *C. R. Math. Rep. Acad. Sci. Canada, 12,* 1–6.

Bruguera, M., Martín-Peinador, E., & Tarieladze, V. (2004). Eberlein-Šmulyan theorem for Abelian topological groups. *J. London Math. Soc., 70,* 341-355.

Buck, B. C. (1946). The measure theoretic approach to density. *Amer. J. Math., 68,* 560-580.

Buck, B. C. (1953). Generalized asymptotic density. *Amer. J. Math., 75,* 335-346.

Butzer, P. L. (1980). The Banach-Steinhaus theorem with rates, and applications to various branches of analysis. In v. Beckenbach, E. F. (Ed.), *General Inequalities II* (Proc. Second Intern. Conf., Oberwolfach, 1978), *Intern. Schriftenreihe Numer. Math., 47* (pp. 299-331). Boston, Basel, Berlin: Birkhäuser.

Butzer, P. L. (1983). A survey on the Whittaker-Shannon sampling theorem and some of its extension. *J. Math. Res. Exposition, 3,* 185-212.

Butzer, P. L., Higgins, J. R., & Stens, R. L. (2000). Sampling theory of Signal Analysis. In: Pier, J.-P. (Ed.), *Development of Mathematics 1950-2000* (pp.193-234). Boston, Basel, Berlin: Birkhäuser.

Butzer, P. L., & Nessel, R. J. (1971). *Fourier Analysis and Approximation. Vol. 1: One-dimensional theory.* Pure Appl. Math., *40.* New York, London: Acad. Press.

Caccioppoli, R. (1928). On the integration of discontinuous functions (Italian), *Rend. Circ. Mat. Palermo, 52,* 1-29.

Caccioppoli, R. (1929). Improper Stieltjes-type integrals. Extensions of the Vitali theorem (Italian), *Rend. Accad. Sci. Fis. Mat. Napoli, 35,* 34-40.

Cafiero, F. (1951). On the compact families of additive set functions (Italian). *Atti IV Congresso U.M.I.*, 30-40.

Cafiero, F. (1952). On the families of uniformly continuous additive set functions (Italian). *Rend. Acad. Lincei, 12,* 155-162.

Cafiero, F. (1953). On the exchange of limits under the sign of integral for sequences of Stieltjes-Lebesgue integrals in abstract spaces, with masses which vary with the integrands (Italian). *Rend. Sem. Mat. Univ. Padova, 22,* 223-245.

Cafiero, F. (1968). On uniform additivity (Italian). *Istituto Nazionale di Alta Matematica, Symposia Mathematica, 2,* 265-280.

Çakalli, H. (1996). On statistical convergence in topological groups. *Pure Appl. Math. Sci., 43*, 27-31.

Çakalli, H. (2008). Sequential definitions of compactness. *Appl. Math. Letters, 21,* 594-598.

Çakalli, H., & Hazarika, B. (2012). Ideal quasi-Cauchy sequences. *J. Inequal. Appl., 2012,* Art. 234, 11 pp. DOI:10.1186/1029-242X-2012-234

Candeloro, D. (1985a). Uniform (*s*)-boundedness and absolute continuity (Italian). *Boll. Un. Mat. Ital., 4-B,* 709-724.

Candeloro, D. (1985b). On the Vitali-Hahn-Saks, Dieudonné and Nikodým theorems (Italian). *Rend. Circ. Mat. Palermo Ser. II, Suppl. 8,* 439-445.

Candeloro, D. (1985c). Some theorems on uniform boundedness (Italian). *Rend. Accad. Naz. XL Mem. Mat., 9,* 249-260.

Candeloro, D. (2002). Convergence theorems for measures with values in Riesz spaces. *Kybernetika, 38,* 287-295.

Candeloro, D., & Letta, G. (1985). On Vitali-Hahn-Saks and Dieudonné theorems (Italian). *Rend. Accad. Naz. Sci. XL Mem. Mat., 9,* 203-213.

Candeloro, D., & Sambucini, A. R. (2014a). Filter convergence and decompositions for vector lattice-valued measures. *Mediterranean J. Math.,* in press. DOI:10.1007/s00009-014-0431-0

Candeloro, D., & Sambucini, A. R. (2014b). Order-type Henstock and McShane integrals in Banach lattice setting. In: Szakál, A. (Ed.), *Proceedings of the* SISY 2014 – IEEE *12th International Symposium on Intelligent Systems and Informatics.,* 55-59.

Candeloro, D., & Volčič, A. (2002). Radon-Nikodým theorems. In: Pap, E. (Ed.), *Handbook of Measure Theory, vol. I* (pp. 249-294). Amsterdam: Elsevier.

Carathéodory, C. (1929). Stetige Konvergenz und normale Familien von Funktionen. *Math. Ann., 101,* 515-533.

Carruth, J. H., Hildebrant, J. A., & Koch, R. J. (1983). *The theory of topological semigroups.* New York: Marcel Dekker.

Cartier, P., Illusie, L., Katz, N. M., Laumon, G., Manin, Yu., & Ribet, K. A. (Eds.) (1989 and 1990). *The Grothendieck Festschrift. Volls. I, II, III.* Progress in Mathematics, *86, 87, 88,* Boston, Basel, Berlin: Birkhäuser.

Cavaliere, P., & de Lucia, P. (2007). Regularity and exhaustivity for finitely additive functions. The Dieudonné's convergence theorem. *Sci. Math. Jpn., 66,* 313-323.

Cavaliere, P., & de Lucia, P. (2008). The Cafiero criterion on a Boolean ring. *Rend. Acc. Sci. Fis. Mat. Napoli, 75* 1-5.

Cavaliere, P., & de Lucia, P. (2009). On Brooks-Jewett, Vitali-Hahn-Saks and Nikodým convergence theorems for quasi-triangular functions. *Rend. Lincei Mat. Appl., 20,* 387-396.

Cavaliere, P., de Lucia, P., De Simone, A., & Ventriglia, F. (2013). On the Lebesgue decomposition for non-additive functions. *Positivity*, published online. DOI 10.1007/s11117-013-0257-1

Cavaliere, P., de Lucia, P., & Ventriglia, F. (2010). On Drewnowski lemma for non-additive functions and its consequences. *Positivity, 14,* 1-16.

Chang, C. C. (1958). Algebraic analysis of many valued logics. *Trans. Amer. Math. Soc., 88,* 467-490.

Chang, C. C. (1959). A new proof of the completeness of the Łukasiewicz axioms. *Trans. Amer. Math. Soc., 93,* 74-80.

Chetcuti, E., & Hamhalter, J. (2006). Vitali-Hahn-Saks theorem for vector measures on operator algebras. *Quart. J. Math., 57,* 479-493.

Chetcuti, E., & Hamhalter, J. (2009a). Non-commutative Vitali-Hahn-Saks theorem holds precisely for finite W^*-algebras. *Quart. J. Math., 60,* 45-51.

Chetcuti, E., & Hamhalter, J. (2009b). A non-commutative Brooks-Jewett theorem. *J. Math. Anal. Appl., 335,* 839-845.

Chojnacki, W. (1986). Sur un théorème de Day, un théorème de Mazur-Orlicz et une généralisation de quelques théorèmes de Silverman. *Colloq. Math., 50,* 257-262.

Choksi, J. R. (2001). Vitali's convergence theorem on term by term integration. *L'Enseignement Mathématique, 47,* 269-285.

Choquet, G. (1954). Theory of capacities. *Ann. Inst. Fourier,5,* 131-295.

Činčura, J., Šalát, T., Sleziak, M., & Toma, V. (2005). Sets of statistical cluster points and $\mathcal{I}$-cluster points. *Real Anal. Exchange, 30,* 565-580.

Cohn, D. L. (1993). *Measure Theory.* Boston, Basel, Berlin: Birkäuser.

Collins, J., & Zimmer, J. (2007). An asymmetric Arzelà-Ascoli theorem. *Topology Appl., 154,* 2312-2322.

Comfort, W. W. (1977). Ultrafilters: some old and some new results. *Bull. Amer. Math. Soc., 83,* 417-455.

Comfort, W. W. (1984). Topological groups. In: Kunen, K., & Vaughan, J. E. (Eds.), *Handbook of Set-theoretic Topology* (pp. 1143-1263). Amsterdam: North-Holland.

Comfort, W. W., & Negrepontis, S. (1974). *The theory of ultrafilters.* Grundlehren math. Wiss., *211.* Berlin, Heidelberg, New York: Springer-Verlag.

Congost Iglesias, M. (1981). Medidas y probabilidades en estructuras ordenadas. *Stochastica, 5,* 45-48.

Connor, J. (1988). The statistical and strong p-Cesàro convergence of sequences. *Analysis, 8,* 47-63.

Connor, J. (1990). Two valued measures and summability. *Analysis, 10,* 373-385.
Connor, J. (1992).*R*-type summability methods, Cauchy criteria, *P*-sets and statistical convergence. *Proc. Amer. Math. Soc., 115,* 319-327.
Connor, J., Fridy, J., & Kline, J. (1994). Statistically pre-Cauchy sequences. *Analysis, 14,* 311-317.
Connor, J., Ganichev, M., & Kadets, V. (2000). A characterization of Banach spaces with separable duals *via* weak statistical convergence. *J. Math. Anal. Appl., 244,* 251-261.
Connor, J., & Grosse-Erdmann, K. G. (2003). Sequential definitions of continuity for real functions. *Rocky Mt. J. Math., 33,* 93-121.
Connor, J., & Kline, J. (1996). On statistical limit points and the consistency of statistical convergence. *J. Math. Anal. Appl., 197,* 392-399.
Constantinescu, C. (1981a). On Nikodým boundedness theorem. *Libertas Math., 1,* 51-73.
Constantinescu, C. (1981b). Spaces of measures on topological spaces. *Hokkaido Math. J., 10, Special issue,* 89-156.
Constantinescu, C. (1989). Some properties of spaces of measures. *Atti Sem. Mat. Fis. Univ. Modena, Suppl. 35.*
d'Andrea, A. B., & de Lucia, P. (1991). The Brooks-Jewett theorem on an orthomodular lattice. *J. Math. Anal. Appl., 154,* 507-522.
d'Andrea, A. B., de Lucia, P., & Morales, P. (1991). The Lebesgue decomposition theorem and the Nikodým convergence theorem on an orthomodular poset. *Atti Sem. Mat. Fis. Univ. Modena, 39,* 73-94.
Darst, R. B. (1962a). A decomposition of finitely additive set functions. *J. Reine Angew. Math., 210,* 31-37.
Darst, R. B. (1962b). A decomposition for complete normed abelian groups with applications to spaces of additive set functions. *Trans. Amer. Math. Soc., 103,* 549-558.
Darst, R. B. (1966). A direct proof of Porcelli's condition for weak convergence. *Proc. Amer. Math. Soc., 17,* 1094-1096.
Darst, R. B. (1967). On a theorem of Nikodým with applications to weak convergence and von Neumann algebras. *Pacific J. Math., 23,* 473-477.
Darst, R. B. (1969). Properties of vector valued finitely additive set functions. *Proc. Amer. Math. Soc., 23,* 528-535.
Darst, R. B. (1970). The Vitali-Hahn-Saks and Nikodým theorems for additive set functions. *Bull. Amer. Math. Soc., 76,* 1297-1298.
Darst, R. B. (1973). The Vitali-Hahn-Saks and Nikodým theorems for additive set functions.II. *Bull. Amer. Math. Soc., 79,* 758-760.
Das, P. (2012). Some further results on ideal convergence in topological spaces. *Topology Appl., 159,* 2621-2626.
Das, P., & Ghosal, S. K. (2010a). On $\mathcal{I}$-Cauchy nets and completeness. *Topology Appl., 157,* 1152-1156.
Das, P., & Ghosal, S. K. (2010b). Some further results on $\mathcal{I}$-Cauchy sequences and condition (*AP*). *Comput. Math. Appl. 59,* 2597–2600.
Das, P., & Ghosal, S. K. (2011). When $\mathcal{I}$-Cauchy nets in complete uniform spaces are $\mathcal{I}$-convergent. *Topology Appl. 158,* 1529-1533.
Das, P., Kostyrko, P., Wilczyński, W., & Malík, P. (2008). $\mathcal{I}$-and $\mathcal{I}^*$-convergence of double sequences. *Math. Slovaca, 58,* 605-620.
Das, P., & Malík, P. (2008a). On the statistical and $\mathcal{I}$-variation of double sequences. *Real. Anal. Exchange, 33,* 351-363.
Das, P., & Malík, P. (2008b). On extremal $\mathcal{I}$-limit points of double sequences. *Tatra Mt. Math. Publ., 40,* 91-102.
Das, P., Pal, S. K., & Ghosal, S. K. (2011). Some further remarks on ideal summability in 2-normed spaces. *Appl. Math. Letters, 24,* 39-43.
Das, R., & Papanastassiou, N. (2004). Some types of convergence of sequences of real valued functions. *Real Anal. Exchange, 29,* 43-58.
Dashiell, F. K. (1981). Nonweakly compact operators for order-Cauchy complete *C*(*S*) lattices, with applications to Baire classes. *Trans. Amer. Math. Soc., 266,* 397-413.
Dawson, D. F. (1970). A generalization of a theorem of Hans Hahn concerning matrix summability. *Boll. Un. Mat. Ital., 3,* 349-356.
de Finetti, B. (1974). *Theory of probability: a critical introductory treatment, Vol. 1* (translated by Antonio Mach and Adrian Smith, with a foreword by D. V. Lindley). London, New York: John Wiley & Sons.
de Finetti, B. (1975). *Theory of probability: a critical introductory treatment, Vol. 2* (translated by Antonio Mach and Adrian Smith). London, New York: John Wiley & Sons.

Degla, G. (2000). A counterexample to "An extension of the Vitali-Hahn-Saks theorem" and a compactness result. *Proc. Amer. Math. Soc., 128,* 2553-2559.
de Lucia, P. (1985). *Finitely additive functions with values in a topological group (Italian).* Quaderni Un. Mat. Ital., *29.* Bologna: Pitagora. 1985.
de Lucia, P. (1998). Convergence theorems in orthomodular posets. *Atti Sem. Mat. Fis. Univ. Modena, Suppl. 46,* 171-179.
de Lucia, P., & Dvurečenskij, A. (1993a). Decompositions of Riesz space-valued measures on orthomodular posets. *Tatra Mt. Math. Publ., 2,* 229-239.
de Lucia, P., & Dvurečenskij, A. (1993b). Yosida-Hewitt decompositions of Riesz space-valued measures on orthoalgebras. *Tatra Mt. Math. Publ., 3,* 101-110.
de Lucia, P., & Morales, P. (1986). Equivalence of Brooks-Jewett, Vitali-Hahn-Saks and Nikodým convergence theorems for uniform semigroup-valued additive functions on a Boolean ring. *Ricerche Mat., 35,* 75-87.
de Lucia, P., & Morales, P. (1988). Some consequences of the Brooks-Jewett theorem for additive uniform semigroup-valued functions. *Conf. Semin. Mat. Univ. Bari, 227,* 23 pp.
de Lucia, P., & Morales, P. (1994). Noncommutative decomposition theorems in Riesz spaces. *Proc. Amer. Math. Soc., 120,* 193-202.
de Lucia, P., & Morales, P. (1998). Decomposition of group-valued measures on orthoalgebras. *Fund. Math., 158,* 109-124.
de Lucia, P., & Pap, E. (1995a). Nikodým convergence theorem for uniform space valued functions defined on *D*-posets. *Math. Slovaca, 45,* 367-376.
de Lucia, P., & Pap, E. (1995b). Noncommutative version of Nikodým boundedness theorem for uniform space-valued functions. *Int. J. Theor. Phys., 34,* 981-993.
de Lucia, P., & Pap, E. (1996). Lebesgue decomposition by σ-null-additive set functions using ideals. *Rend. Circ. Mat. Palermo Ser. II, 45,* 25-36.
de Lucia, P., & Pap, E. (2002). Convergence Theorems for Set Functions. In: Pap, E. (Ed.), *Handbook of Measure Theory, vol. I* (pp. 125-178). Amsterdam: Elsevier.
de Lucia, P., & Pap, E. (2003). Matrix method approach to Cafiero theorem. *Novi Sad J. Math., 33,* 117-127.
de Lucia, P., & Pap, E. (2008). Cafiero approach to the Dieudonné-type theorems. *J. Math. Anal. Appl., 337,* 1151-1157.
de Lucia, P., & Salvati, S. (1996). A Cafiero characterization of uniform (*s*)-boundedness. *Rend. Circ. Mat. Palermo, Suppl. 40,* 121-128.
Demirci, K. (2001). $\mathcal{I}$-limit superior and limit inferior. *Math. Commun., 6,* 165-172.
Dems, K. (2005). On *I*-Cauchy sequences. *Real Anal. Exchange, 30,* 123-128.
De Simone, A. (1998). Decomposition theorems in orthomodular posets: The problem of uniqueness. *Proc. Amer. Math. Soc., 126,* 2919-2926.
De Simone, A., & Navara, M. (2001). Yosida-Hewitt and Lebesgue decompositions of states on orthomodular posets. *J. Math. Anal. Appl., 255,* 74-104.
De Simone, A., & Navara, M. (2002). On the Yosida-Hewitt decomposition and Rüttimann decomposition of states. *Sci. Math. Jpn., 56,* 49-62.
De Simone, A., Navara, M., & Pták, P. (2002). The Vitali-Hahn-Saks theorem for the product of quantum logics. *Demonstratio Math., 35,* 717-725.
Dickmeis,W., & Nessel, R. J. (1981a). On uniform boundedness principles and Banach-Steinhaus theorems with rates. *Num. Funct. Anal. Optim., 3,* 19-52.
Dickmeis,W., & Nessel, R. J. (1981b). A unified approach to certain counterexamples in approximation theory in connection with a uniform boundedness principle with rates. *J. Approx. Theory, 31,* 161-174.
Dickmeis,W., & Nessel, R. J. (1985). A quantitative condensation of singularities on arbitrary sets. *J. Approx. Theory, 43,* 383-393.
Diestel, J. (1973a). Applications of weak compactness and bases to vector measures and vectorial integration. *Rev. Roumaine Math. Pures Appl., 18,* 211-224.
Diestel, J. (1973b). Grothendieck spaces and vector measures. In: Tucker, D. H., & Maynard, H. B. (Eds.), *Vector and operator valued measures and applications* (pp. 97-108). New York, London: Acad. Press.
Diestel, J. (1984). *Sequences and series in Banach spaces.* Berlin, Heidelberg, New York: Springer-Verlag.
Diestel, J. (1991). Uniform integrability: an introduction. *Rend. Ist. Mat. Univ. Trieste, 23,* 41-80.
Diestel, J., & Faires, B. (1974). On vector measures. *Trans. Amer. Math. Soc., 198,* 253-271.

Diestel, J., Ruess, W. M., & Schachermayer, W. (1993). Weak compactness in $L^1(\mu, X)$. *Proc. Amer. Math. Soc., 118,* 447-453.

Diestel, J., & Spalsbury, A. (2012). Finitely additive measures in action. In: Pérez-Fernández, F. J., & Rambla-Barreno, F. (Eds), *Advanced Courses of Mathematical Analysis IV. In memory of Professor Antonio Aizpuru Tomas. Proceedings of the Fourth International School held in Jerez de la Frontera, Spain, 8-12 September 2009* (pp. 58-115). Hackensack, NJ: World Scientific Publ. Co. Pte. Ltd.

Diestel J., & Swart, J. (2002). The Riesz theorem. In: Pap., E. (Ed.), *Handbook on measure theory, vol. I* (pp. 401-447). Amsterdam: Elsevier.

Diestel, J., & Uhl, J. J. (1977). *Vector Measures.* Rhode Island: Amer. Math. Soc. Providence.

Dieudonné, J. (1951a). Sur la convergence des suites de mesures de Radon. *An. Acad. Brasil. Ci., 23,* 21-38.

Dieudonné, J. (1951). Addition à mon article: " Sur la convergence des suites de mesures de Radon". *An. Acad. Brasil. Ci., 23,* 277-282.

Dimitriou, X. (2007). *From finite to countable additivity of set functions (Greek).* Master's thesis in Mathematics. Athens: University of Athens, Department of Mathematics.

Dimitriou, X. (2011). *Convergences of sequences of measures and of measurable functions.* Ph. D. thesis in Mathematics. Athens: University of Athens, Department of Mathematics.

Dimitriou, X., & Kitsos, C. P. (2014). Concept, Lattice and Probability, in preparation.

Dinculeanu, N. (1966). *Vector Measures.* Berlin: VEB Deutscher Verlag der Wissenshaften.

Dinculeanu, N., & Kluvánek, I. (1967). On vector measures. *Proc. London Math. Soc., 17,* 505-512.

Dixmier, J. (1969). *Les C*-algèbres et leurs représentations.* Paris: Gauthier-Villars.

Dobrakov, I. (1984). Uniform boundedness principle for exhaustive set functions. *Comment. Math., 24,* 201-205.

Doubrovsky, V. M. (1945). On some properties of completely additive set functions and passing to the limit under the integral sign (Russian). *Izv. Akad. Nauk SSSR, 9,* 311-320.

Doubrovsky, V. M. (1947a). On some properties of completely additive set functions and their application to generalization of a theorem of Lebesgue (Russian). *Math. Sbornik , N. S., 20,* 317-329.

Doubrovsky, V. M. (1947b). On the base of a family of completely additive set functions and the properties of uniform additivity and uniform continuity (Russian). *Dokl. Akad. Nauk SSSR, N. S., 58,* 737-740.

Drewnowski, L. (1972a). Topological rings of sets, Continuous set functions, Integration. I, II, III. *Bull. Acad. Polon. Sci. Sér. Sci. Math. Astronom. Phys., 20,* 269-276, 277-286, 439-445.

Drewnowski, L. (1972b). Equivalence of Brooks-Jewett, Vitali-Hahn-Saks and Nikodym theorems. *Bull. Acad. Polon. Sci. Sér. Sci. Math. Astronom. Phys., 20,* 725-731.

Drewnowski, L. (1973a). Uniform boundedness principle for finitely additive vector measures. *Bull. Acad. Polon. Sci. Sér. Sci. Math. Astronom. Phys., 21,* 115-118.

Drewnowski, L. (1973b). Decompositions of set functions. *Studia Math., 48,* 23-48.

Drewnowski, L. (1975). Un théorème sur les opérateurs de $l_\infty(\Gamma)$. *C. R. Acad. Sci. Paris, 281,* 967-969.

Drewnowski, L. (1981). Embeddings of topological semigroups in topological groups and semigroup-valued measures. *Demonstratio Math., 14,* 355-360.

Drewnowski, L., Florencio, M., & Paúl, P. J. (1994). Barrelled subspaces of spaces with subseries decompositions on Boolean rings of projections. *Glasgow Math. J., 36,* 57-69.

Drewnowski, L., Florencio, M., & Paúl, P. J. (1996). Some new classes of rings of sets with the Nikodým property. In: Dierolf, S., Dineen, S., & Domanski, P. (Eds.), *Functional Analysis, Proc. 1st Intern. Workshop on Funct. Anal., Trier Univ., Germany, Sept. 26-Oct. 1 (1994)* (pp. 143-152). Berlin: Walter de Gruyter & Co.

Duchoň, M., & Maličký, P. (2009). A Helly theorem for functions with values in metric spaces. *Tatra Mt. Math. Publ., 44,* 159-168.

Duchoň, M., & Riečan, B. (1996). On the Kurzweil-Stieltjes integral in ordered spaces. *Tatra Mt. Math. Publ., 8,* 133-142.

Dugundji, J. (1966). *Topology.* Boston: Allyn and Bacon, Inc.

Dunford, N., & Pettis, B. J. (1940). Linear operations on summable functions. *Trans. Amer. Math. Soc., 47,* 323-392.

Dunford, N., & Schwartz, J. T. (1958). *Linear Operators I. General Theory.* New York, London: Interscience Publ., Inc.

Dunford, N., & Schwartz, J. T. (1963). *Linear Operators II. Spectral Theory, Self Adjoint Operators in Hilbert Space.* New York, London: Interscience Publ., Inc.

Dunford, N., & Schwartz, J. T. (1971). *Linear Operators III. Spectral Operators.* New York, London: Interscience Publ., Inc.

Dvurečenskij, A. (2002). Measures on quantum structures. In: *E. Pap (Ed.), Handbook of Measure Theory, vol. II* (pp. 827-868). Amsterdam: Elsevier.

Dvurečenskij, A., de Lucia, P., & Pap, E. (1996). On a decomposition theorem and its applications. *Math. Japonica, 44,* 145-164.

Dvurečenskij, A., & Pulmannová, S. (2000). *New trends in Quantum Structures.* Dordrecht: Kluwer Acad. Publ.

Engelking, R. (1977). *General topology.* Warsaw: PWN-Polish Scientific Publishers.

Erné, M., & Gatzke, H. (1985). Convergence and continuity in partially ordered sets and semilattices. In: Hoffmann, R.-E., & Hofmann, K. H. (Eds.), *Continuous Lattices and Their Applications* (pp. 9-40). Lecture Notes Pure Appl. Math., *101.* New York: Marcel Dekker.

Fabian, M., Habala, P., Hájek, P., Montesinos-Santalucía, V., Pelant, J., & Zizler, V. (2001). *Functional Analysis and Infinite-Dimensional Geometry.* Berlin, Heidelberg, New York: Springer-Verlag.

Faires, B. (1974). On Vitali-Hahn-Saks-type theorems. *Bull. Amer. Math. Soc., 80,* 670-674.

Faires, B. (1976). On Vitali-Hahn-Saks-Nikodým-type theorems. *Ann. Inst. Fourier (Grenoble), 26,* 99-114.

Farah, I. (2000). Analytic quotients: Theory of liftings for quotients over analytic ideals on the integers. *Mem. Amer. Math. Soc., 148, n. 702.* www.math.yorku.ca/ ifarah/aq.html

Fast, H. (1951). Sur la convergence statistique. *Colloq. Math., 2,* 41-44.

Fefferman, C. (1967a). A Radon-Nikodým theorem for finitely additive set functions. *Pacific J. Math, 23,* 35-45.

Fefferman, C. (1967b). L^p spaces over finitely additive measures. *Pacific J. Math., 26,* 265-271.

Ferrando, J. C., & López-Pellicer, M. (1992). A note on a theorem of J. Diestel and B. Faires. *Proc. Amer. Math. Soc., 115,* 1077-1081.

Ferrando, J. C., & Sánchez Ruiz, L. M. (2004). A survey on recent advances on the Nikodým boundedness theorem and spaces of simple functions. *Rocky Mt. J. Math., 34,* 139-172.

Fichtenholtz, G., & Kantorovich, L. V. (1934). Sur les opérations linéaires. L'espace des fonctions bornées. *Studia Math., 5,* 65-98.

Filipów, R., Mrożek, N, Recław, I., & Szuca, P. (2007). Ideal convergence of bounded sequences. *J. Symbolic Logic, 72,* 501-512.

Filipów, R., & Szuca, P. (2010). Density versions of Schur's theorem for ideals generated by submeasures. *J. Combin. Theory, Ser. A, 117,* 943–956.

Filipów, R., & Szuca, P. (2012). Three kinds of convergence and the associated *I*-Baire classes. *J. Math. Anal. Appl., 391,* 1-9.

Filter, W. (1994). Representation of Archimedean Riesz spaces-a survey. *Rocky Mt. J. Math., 24,* 771-851.

Fleischer, I., & Traynor, T. (1989). Equicontinuity and uniform boundedness for non-commutative groups. *J. Math. Anal. Appl. 138,* 349-357.

Fletcher, P., & Lindgren, W. (1982). *Quasi uniform spaces.* Lecture Notes Pure Appl. Math., *77.* New York: Marcel Dekker.

Florencio, M., Paúl, P. J., & Virués, J. M. (1992). On the Mikusiński-Antosik diagonal theorem. *Bull. Acad. Polon. Sci. Math., 40,* 189-195.

Florescu, L. C., & Godet-Thobie, C. (2006). A version of biting lemma for unbounded sequences in L_E^1 with applications. *Math. Anal. Appl., AIP Conf. Proc., Amer. Inst. Phys., Melville NY,* 58-73.

Floyd, E. E. (1955). Boolean algebras with pathological order properties. *Pacific J. Math., 5,* 687-689.

Fox, G., & Morales, P. (1983). Théorèmes de Nikodym et de Vitali-Hahn-Saks pour les mesures à valeurs dans un sémigroupe uniforme. In: Belley, J.-M., Dubois, J., & Morales, P. (Eds), *Measure theory and its applications, Proc. Conf., Sherbrooke/Canada 1982* (pp. 199-208). Lecture Notes Math., *1033.* Berlin, Heidelberg, New York: Springer-Verlag.

Fréchet, M. (1921). Sur divers modes de convergence d'une suite de fonctions d'une variable. *Bull. Calcutta Math. Soc.,11,* 187-206.

Freedman, A. R., & Sember, J. J. (1981a). Densities and summability. *Pacific J. Math., 95,* 293-305.

Freedman, A. R., & Sember, J. J. (1981b). On summing sequences of 0's and 1's. *Rocky Mt. J. Math., 11,* 419-425.

Fremlin, D. H. (1974). *Topological Riesz spaces and Measure Theory.* Cambridge: Cambridge Univ. Press.

Fremlin, D. H. (1975). A direct proof of the Matthes-Wright integral extension theorem. *J. London Math. Soc., 11,* 276-284.

Fremlin, D. H. (2000). *Measure Theory, vol. 1. The irriducible minimum.* Colchester: Torres Fremlin.

Fremlin, D. H. (2001). *Measure Theory, vol. 2. Broad foundations.* Colchester: Torres Fremlin.

Fremlin, D. H. (2002). *Measure Theory, vol. 3. Measure algebras.* Colchester: Torres Fremlin.

Fremlin, D. H. (2003). *Measure Theory, vol. 4. Topological measure spaces.* Colchester: Torres Fremlin.

Fremlin, D. H. (2008). *Measure Theory, vol. 5. Set-theoretic Measure Theory.* Colchester: Torres Fremlin.

Fremlin, D. H. (2011). *Topological spaces after forcing.* http://www.essex.ac.uk/maths/people/fremlin/n05622.pdf

Freniche, F. J. (1984). The Vitali-Hahn-Saks theorem for Boolean algebras with the subsequential interpolation property. *Proc. Amer. Math. Soc., 92*, 362-366.

Fridy, J. (1985). On statistical convergence. *Analysis, 5,* 301-313.

Fridy, J. (1993). Statistical limit points. *Proc. Amer. Math. Soc., 118*, 1187-1192.

Fridy, J., & Miller, H. (1991). A matrix characterization of statistical convergence. *Analysis, 11*, 59-66.

Fridy, J., & Orhan, C. (1997). Statistical limit superior and limit inferior. *Proc. Amer. Math. Soc., 125*, 3625-3631.

Gadjiev, A. D., & Orhan, C. (2002). Some approximation theorems *via* statistical convergence. *Rocky Mt. J. Math., 32,* 129-138.

Gáliková, Z., László, B., & Šalát, T. (2002). Remarks on uniform density of sets of integers. *Acta Acad. Paedagog. Agriensis Sect. Mat., 29,* 3-13.

Gänssler, P. (1971a). Compactness and sequential compactness in spaces of measures. *Z. Wahrscheinlichkeitstheorie verw. Gebiete, 17*, 124-146.

Gänssler, P. (1971b). A convergence theorem for measures in regular Hausdorff spaces. *Math. Scand., 26*, 237-244.

Ganter, B., & Wille, R. (1999). *Formal Concept Analysis-Mathematical Foundations.* Berlin, Heidelberg, New York: Springer-Verlag.

García-Ferreira, S., & Kočinac, Lj. D. R.. (1996). Convergence with respect to ultrafilters: a survey. *Filomat, 10,* 1-32.

Gassó, M. T. (1992). Algebras with the local interpolation property. *Rocky Mt. J. Math., 22*, 181-196.

Gel'fand, I. M., & Naimark, M. A. (1943). On the imbedding of normed rings into the ring of operators on a Hilbert space. *Mat. Sbornik, 12*, 197-217.

Giannakoulias, E. (1987). Some properties of vector measures taking values in a topological vector space. *J. Austral. Math. Soc. (Ser. A), 43* (1987), 224-230.

Giannakoulias, E. (1994). Some results in vector measures. *J. Inst. Math. Comput. Sci., Math. Ser.,7,* 85-93.

Giannakoulias, E. (1996). Applications of bases to vector measures. *J. Inst. Math. Comput. Sci., Math. Ser., 7,* 207-219.

Giuliano-Antonini, R., & Grekos, G. (2007). Weighted uniform densities. *J. Théor. Nombres, Bordeaux, 19,* 191-204.

Giuntini, R., & Greuling, H. (1989). Toward a formal language for unsharp properties. *Found. Phys., 19,* 931-945.

Gordji, M. E., Sarabadan, S., & Arani, F. A. (2012). $\mathcal{I}$-convergence in 2-normed spaces. *Funct. Anal., Approx. Comput., 4,* 1-7.

Gould, G. G. (1965). Integration over vector-valued measures. *Proc. London Math. Soc., 15,* 193-225.

Gould, G. G. (1966). Extensions of vector-valued measures. *Proc. London Math. Soc., 16,* 685-704.

Graves, W. H. (1977). On the theory of vector measures. *Mem. Amer. Math. Soc., 12, n. 195.*

Graves, W. H. & Ruess, W. (1980). Compactness in spaces of vector-valued measures and a natural Mackey topology in spaces of bounded measurable functions. *Proc. Conf. on Integration, Topology and Geometry in Linear Spaces, Univ. North Carolina, Chapel Hill, N.C., 1979* (pp. 189–203). Contemporary Math., *2.* Providence, R.I.: Amer. Math. Soc.

Graves, W. H., & Wheeler, R. F. (1983). On the Grothendieck and Nikodým properties for algebras of Baire, Borel and universally measuable sets. *Rocky Mt. J. Math., 13,* 333-353.

Gregoriades, V., & Papanastassiou, N. (2008). The notion of exhaustiveness and Ascoli-type theorems. *Topology Appl., 155,* 1111-1128.

Grekos, G. (2005). On various definitions of density. *Tatra Mt. Math. Publ., 31,* 17-27.

Grekos, G., Toma, V., & Tomanová, J. (2010). A note on uniform or Banach density. *Ann. Math. Blaise Pascal, 17,* 153-163.

Grothendieck, A. (1953). Sur les applications linéaires faiblement compactes d'espaces of type *C*(*K*). *Canad. J. Math., 5,* 129-173.

Grothendieck, A. (1957). Un résultat sur le dual d'une *C**-algèbre. *J. Math. Pures Appl., 36,* 97-108.

Grothendieck, A. (1973). *Topological vector spaces.* New York, London, Paris: Gordon and Breach.

Guariglia, E. (1982). On a Nikodým theorem for topological group-valued functions (Italian). *Le Matematiche, 37,* 328-342.

Guariglia, E. (1990a). *k*-triangular functions on an orthomodular lattice and the Brooks-Jewett theorem. *Radovi Mat., 6,* 241-251.

Guariglia, E. (1990b). Uniform boundedness theorems for *k*-triangular set functions. *Acta Sci. Math., 54,* 391-407.

Guariglia, E. (1990c). On Dieudonné's boundedness theorem. *J. Math. Anal. Appl., 145,* 447-454.

Habil, E. D. (1995). D.Brooks-Jewett and Nikodým convergence theorems for orthoalgebras that have the weak subsequential interpolation property. *Int. J. Theor. Phys., 34,* 465-491.

Hadžić, O., & Pap, E. (1975). Some applications of the diagonal theorem in Functional Analysis (Serbo-Croatian). *Zb. Rad. Prirod.-Mat. Fak., Ser. Mat., 5,* 23-33.

Hahn, H. (1921). *Theorie der reellen Funktionen.* Berlin: Springer.

Hahn, H. (1922). Über Folgen linearer Operationen. *Monatshefte für Math. und Physik, 32,* 3-88.

Halmos, P. R. (1950). *Measure theory.* New York: D. Van Nostrand Company, Inc.

Haluška, J. (1993). On integration in complete vector lattices. *Tatra Mountains Math. Publ., 3,* 201-212.

Haluška, J., & Hutník, O. (2010). On convergences of measurable functions in Archimedean vector lattices. *Positivity, 14,* 515-527.

Hammer, P. C., & Sobczyk, A., (1944). The ranges of additive set functions. *Duke Math. J., 11,* 847-851.

Haydon, R. (1981). A non-reflexive Grothendieck space that does not contain l_∞. *Israel J. Math., 40,* 65-73.

Hejcman, J. (1959). Boundedness in uniform spaces and topological groups. *Czech. Math. J.,* 544-563.

Helly, E. (1912). Über lineare Funktionaloperationen. *Sitzsber. Akad. Wiss. Wien, 121,* 265-297.

Henriksen, M. (1959). Multiplicative summability methods and the Stone-Čech compactification. *Math. Zeitschr., 71,* 427-435.

Hernández, S., Galindo, J., & Macario, S. (1999). A characterization of the Schur property by means of the Bohr topology. *Topology Appl., 97,* 99-108.

Hernández-Lerma, O., & Lasserre, J. B. (1996). An extension of the Vitali-Hahn-Saks theorem. *Proc. Amer. Math. Soc., 124,* 3673-3676.

Hernández-Lerma, O., & Lasserre, J. B. (1998). Correction to "An extension of the Vitali-Hahn-Saks theorem". *Proc. Amer. Math. Soc., 126,* 949.

Hildebrandt, T. H. (1934). On bounded functional operations. *Trans. Amer. Math. Soc., 36,* 868-875.

Hoffmann-Jørgensen, J. (1971). Vector measures. *Math. Scand., 28,* 5-32.

Hoffmann-Jørgensen, J. (1972). Weak compactness and tightness of subsets of *M*(*X*). *Math. Scand., 31,* 127-150.

Horbaczewska, G., & Skalski, A. (2008). The Banach Principle for ideal convergence in the classical and noncommutative context. *J. Math. Anal. Appl., 342,* 1332-1341.

Huntington, E. V. (1904). Sets of independent postulates for the algebra of logic. *Trans. Amer. Math. Soc., 5,* 288-309.

Huntington, E. V. (1933a). New sets of independent postulates for the algebra of logic, with special reference to Whitehead and Russell's Principia mathematica. *Trans. Amer. Math. Soc., 35,* 274-304.

Huntington, E. V. (1933b). A correction. *Trans. Amer. Math. Soc., 35,* 557-558.

Huntington, E. V. (1933 c). A second correction. *Trans. Amer. Math. Soc., 35*, 971.

Iwiński, T. B. (1972). Some remarks on Toeplitz methods and continuity. *Comment. Math., 16*, 37-43.

Jacobs, K. (1978). *Measure and Integral.* New York, London: Acad. Press.

Jasiński, J., & Recław, I. (2008). On spaces with the ideal convergence property. *Colloq. Math., 111,* 43-50.

Jin, R. (2001). Nonstandard methods for upper Banach density problems. *J. Number Theory, 91,* 20-38.

Jordan, C. (1893 and 1896). *Cours d'analyse de l' École Polytechnique. Volls. 1-3* (2nd edition). Paris: Gauthier-Villars.

Joshi, K. D. (1983). *Introduction to general topology.* New Delhi: Wiley Eastern Ltd.

Kadec, M. I., & Pełczyński, A. (1962). Bases, lacunary sequences and complemented subspaces in the spaces L^p. *Studia Math., 21,* 161-176.

Kadets, V., & Leonov, A.(2007). Dominated convergence and Egorov theorems for filter convergence. *J. Math. Phys., Anal., Geom., 3,* 196-212.

Kadets, V., & Leonov, A. (2009). Weak and pointwise convergence in *C*(*K*) for filter convergence. *J. Math. Anal. Appl., 350,* 455-463.

Kadets, V., Leonov, A., & Orhan, C. (2010). Weak statistical convergence and weak filter convergence for unbounded sequences. *J. Math. Anal. Appl., 371,* 414-424.

Kadison, R. V. (1951). A representation theory for commutative topological algebra. *Mem. Amer. Math. Soc., 7.*
Kakutani, S. (1940). Weak topology, bicompact set and the principle of duality. *Proc. Imp. Acad. Tokyo, 16,* 63-67.
Kakutani, S. (1941). Concrete representation of abstract (M)-spaces (A characterization of the space of continuous functions). *Ann. Math., 42,* 994-1024.
Kalton, N. J. (1974). Topologies on Riesz groups and applications to Measure Theory. *Proc. London Math. Soc., 28,* 253-273.
Kalton, N. J. (1974/1975). Exhaustive operators and vector measures. *Proc. Edinburgh Math. Soc., 19,* 291-300.
Kappos, D. A. (1969). *Probability algebras and stochastic spaces.* New York, London: Acad. Press.
Katětov, M. (1968). Products of filters. *Comment. Math. Univ. Carolin., 9,* 173-189.
Kawabe, J. (2007). The Alexandroff theorem for Riesz space-valued non-additive measures. *Fuzzy Sets Systems, 158,* 2413-2421.
Kawabe, J. (2008). The Choquet integral in Riesz space. *Fuzzy Sets Systems, 159,* 629-645.
Kawabe, J. (2012). The bounded convergence theorem for Riesz space-valued Choquet integrals. *Bull. Malays. Math. Sci. Soc., 35,* 537-543.
Kelley, J. L. (1955). *General Topology.* Princeton, NJ: D. Van Nostrand Company, Inc.
Khurana, S. S. (1974). A vector form of Phillips' lemma. *J. Math. Anal. Appl., 48,* 666-668.
Khurana, S. S. (1976). Convergent sequences of regular measures. *Bull. Acad. Polon. Sci. Sér. Sci. Math. Astronom. Phys., 24,* 37-42.
Khurana, S. S. (1977). A note on a paper of J. D. Stein, Jr. *Proc. Amer. Math. Soc., 67,* 74-76.
Khurana, S. S. (1979). Some remarks on the sequences of exhaustive measures and uniform boundedness of measures. *Proc. Amer. Math. Soc., 73,* 207-208.
Khurana, S. S. (1980). Extension of group-valued regular Borel measures. *Math. Nachr., 97,* 159-165.
Khurana, S. S. (1981). Addendum to:"Extension of group-valued regular Borel measures". *Math. Nachr., 104,* 348.
Kitover, A. K., & Wickstead, A. W. (2005). Operator norm limits of order continuous operators. *Positivity, 9*, 341-355.
Kitsos, C. P. (1989). Fully sequential procedures in Nonlinear Design problems. *Computational Statistics and Data Analysis, 8,* 13-19.
Kitsos, C. P. (2005). Optimal design for Bioassays in carcinogenesis. In: Edler, L., & Kitsos, C. P. (Eds.), *Quantitative Methods for Cancer and Human Health Risk Assessment* (pp. 267-279). London, New York: Wiley.
Kitsos, C. P. (2012). *Cancer Bioassays : a statistical approach.* Saarbrücken: Lambert Acad. Publ.
Kitsos, C. P., & Edler, L. (2005). Cancer Risk assessment mixtures. In: Edler, L., & Kitsos, C. P. (Eds.), *Quantitative Methods for Cancer and Human Health Risk Assessment* (pp. 283-298). London, New York: Wiley. 283-298.
Kitsos, C. P., & Sotiropoulos, M. (2009). Statistical and Conceptual Distance Methods for Bioassays. In: Kitsos, C. P., & Caroni, C. (Eds.), *Proceedings ICCRA3 (3d International Conference on Cancer Risk Assessment), Porto Heli, Greece, 16-18 July 2009, e-Proccedings* (pp. 1-24).
Klimkin, V. M. (1975). Measures with values in topological groups. *Math. Notes, 17,* 471-475.
Kolk, E. (1991). The statistical convergence in Banach spaces. *Tartu Ül. Toimetised, 928,* 41-52.
Kolk, E. (1993). Matrix summability of statistically convergent sequences. *Analysis, 13,* 77-83.
Kolk, E. (1998). Inclusion relations between the statistical convergence and strong summability. *Acta Comment. Univ. Tartu. Math., 2,* 39-54.
Kolk, E. (2010). Banach-Steinhaus type theorems for statistical and $\mathcal{I}$-convergence with applications to matrix maps. *Rocky Mt. J. Math., 40,* 279-289.
Komisarski, A. (2008). Pointwise $\mathcal{I}$-convergence and $\mathcal{I}$-convergence in measure of sequences of functions. *J. Math. Anal. Appl., 340,* 770-779.
König, H. (2009). *Measure and Integration. An advanced course in basic procedures and applications* (2nd edition). Berlin, Heidelberg, New York: Springer-Verlag.
König, H. (2012). *Measure and Integration. Publications 1997-2011.* Boston, Basel, Berlin: Birkhäuser.
Kôpka, F., & Chovanec, F. (1994). *D*-posets. *Math. Slovaca, 44,* 21-34.
Kostyrko, P., Mačaj, M., Šalát, T., & Sleziak, M. (2005). $\mathcal{I}$-convergence and extremal $\mathcal{I}$-limit points. *Math. Slovaca, 55*, 443-464.

Kostyrko, P., Mačaj, M., Šalát, T., & Strauch, O. (2000). On statistical limit points. *Proc. Amer. Math. Soc, 129,* 2647-2654.
Kostyrko, P., Šalát, T., & Wilczyński, W. (2001). $\mathcal{I}$-convergence. *Real Anal. Exchange, 26,* 669-685.
Köthe, G. (1969). *Topological vector spaces. I.* Berlin, Heidelberg, New York: Springer-Verlag.
Krachounov, M. (2006). Intuitionistic probability and intuitionistic fuzzy sets. In: El-Darzi, E., Atanassov, R, & Chountas, P. (Eds), *First International Workshop on Intuitionistic Fuzzy Sets, Generalized Nets and Knowledge Engineering* (pp. 18-24). London: Univ. of Westminister.
Krein, M., & Krein, S. (1940). On an inner characteristic of the set of all continuous functions defined on a bicompact Hausdorff space. *C. R. (Doklady) Acad. Sci. URSS, 28,* 427-430.
Krein, M., & Krein, S. (1943). Sur l'espace des fonctions continues définies sur un bicompact de Hausdorff et ses sousespaces semiordonnés. *Rec. Math. (Mat. Sbornik), 13,* 1-38.
Kumar, V. (2007). On $\mathcal{I}$ and $\mathcal{I}^*$-convergence of double sequences. *Math. Commun., 12,* 171-181.
Kupka, J. (1974). A short proof and generalization of a measure theoretic disjointization lemma. *Proc. Amer. Math. Soc., 45,* 70-72.
Kupka, J. (1980). Uniform boundedness principles for regular Borel vector measures. *J. Austral. Math. Soc (Ser. A), 29,* 206-218.
Kuratowski, K. (1966 and 1968). *Topology. Voll. I and II.* New York, London: Acad. Press.
Kusraev, A. G., & Malyugin, S. A. (1988). *Some problems in the theory of vector measures (Russian).* Novosibirsk: Akad. Nauk SSSR Sibirsk. Otdel., Inst. Mat.
Labuda, I. (1972). Sur quelques généralisations des théorèmes de Nikodým et de Vitali-Hahn-Saks. *Bull. Acad. Polon. Sci. Sér. Sci. Math. Astronom. Phys., 20,* 447-456.
Labuda, I. (1973). Sur quelques théorèmes du type d'Orlicz-Pettis. *Bull. Acad. Polon. Sci. Sér. Sci. Math. Astronom. Phys., 21,* 127-132.
Labuda, I. (1975a). A note on exhaustive measures. *Annal. Soc. Math. Pol. Ser. I: Comment. Math., 18,* 217-221.
Labuda, I. (1975b). Sur les mesures exhaustives et certaines classes d'espaces vectoriels topologiques considérés par W. Orlicz and L. Schwartz. *C. R. Acad. Sci. Paris, 280,* 997-999.
Labuda, I. (1976). Exhaustive measures in arbitrary topological vector spaces. *Studia Math., 58,* 239-248.
Laczkovich, M. (2002). Paradoxes in Measure Theory. In: Pap, E. (Ed.), *Handbook of Measure Theory, vol. II* (pp. 1329-1379). Amsterdam: Elsevier.
Lahiri, B. K., & Das, P. (2003). Further results on $\mathcal{I}$-limit superior and limit inferior. *Math. Commun., 8,* 151-156.
Lahiri, B. K., & Das, P. (2005). $\mathcal{I}$ and $\mathcal{I}^*$-convergence in topological spaces. *Math. Bohemica, 130,* 153-160.
Lahiri, B. K., & Das, P. (2008). $\mathcal{I}$-and $\mathcal{I}^*$-convergence of nets. *Real Anal. Exchange, 33,* 431-442.
Landers. D., & Rogge, L. (1971a). The Hahn-Vitali-Saks and the uniform boundedness theorem in topological groups. *Manuscripta Math., 4,* 351-359.
Landers. D., & Rogge, L. (1971b). Equicontinuity and convergence of measures. *Manuscripta Math., 5,* 123-131.
Landers. D., & Rogge, L. (1972). Cauchy convergent sequences of regular measures with values in a topological group. *Z. Wahrscheinlichkeitstheorie verw. Gebiete, 21,* 188-196.
Lang, S. (1993). *Real and Functional Analysis.* Berlin, Heidelberg, New York: Springer-Verlag.
Leader, S. (1953). The theory of L^p-spaces for finitely additive set functions. *Ann. Math., 58,* 528-543.
Lebesgue, H. (1909). Sur les intégrales singulières. *Annales Fac. Sci. Toulouse, III Sér. 1,* 25-117.
Leblanc, L., & Fox, G. E. (1956). On the extension of measure by the method of Borel. *Canad. J. Math., 8,* 516-523.
Letavaj, P. (2011). $\mathcal{I}$-convergence to a set. *Acta Math. Univ. Comenian., 80,* 103-106.
Li, R., Gai, Y., & Lu, R. (1988). A generalization of Rosenthal's lemma. *J. Math. Anal. Appl. 135,* 346-353.
Li, R., & Swartz, C. (1993). A nonlinear Schur theorem. *Acta Sci. Math., 58,* 497-508.
Limaye, B. V., & Zeltser, M. (2009). On the Pringsheim convergence of double series. *Proc. Estonian Acad. Sci., 58,* 108-121.
Lindenstrauss, J. (1964). On the extension of operators with range in a *C*(*K*) space. *Proc. Amer. Math. Soc., 15,* 218-225.
Lipecki, Z. (1983). On unique extensions of positive additive set functions. *Arch. Math. (Basel), 41,* 71-79.
Lipecki, Z. (1988). On unique extensions of positive additive set functions. II. *Arch. Math. (Basel), 50,* 175-182.
Lipecki, Z. (1990). Maximal and tight extensions of positive additive set functions. *Math. Nachr., 146,* 167-173.
Lorentz, G. G. (1948). A contribution to the theory of divergent sequences. *Acta Math., 80,* 167-190.
Luu, D. Q. (1989). A short proof of biting lemma. *Semin. Anal. Convexe,* Montpellier 1989, Exposé n.1, 1-13.
Luxemburg, W. A. J., & Zaanen, A. C. (1971). *Riesz Spaces. I.* Amsterdam: North-Holland.

Mačaj, M., & Sleziak, M. (2011). $\mathcal{I}^{\mathcal{K}}$-convergence. *Real Anal. Exchange, 36,* 177-194.
Maddox, I. (1988). Statistical convergence in a locally convex space. *Math. Proc. Cambridge Phil. Soc., 104,* 141-145.
Maddox, I. (1989). A Tauberian teorem for statistical convergence. *Math. Proc. Cambridge Phil. Soc., 106,* 277-280.
Maeda, F., & Ogasawara, T. (1942). Representation of vector lattices (Japanese). *J. Sci. Hiroshima Univ. Ser. A, 12,* 17-35.
Maharam, D. (1976). Finitely additive measures on the integers. *Sankhy $\bar{a}$, 38,* 44-59.
Maharam, D. (1987). From finite to countable additivity. *Portug. Math., 44,* 265-282.
Malyugin, S. A. (1991). Quasi-Radon measures (Russian). *Sib. Math. J., 32*, 812-821.
Mantellini, I. (2013). On the asymptotic behaviour of linear combinations of Mellin-Picard type operators. *Math. Nachr., 286,* 1820-1832.
Martellotti, M. (2001). Finitely additive phenomena. *Rend. Ist. Mat. Univ. Trieste, 33,* 201-249.
Máté, L. (1991). Remarks on the Pták-Stein theorem. *Per. Math. Hungar., 23,* 59-63.
Matheron, E. (2001). A useful lemma concerning subseries convergence. *Bull. Austral. Math. Soc., 63,* 273-277.
Maurey, B., & Pisier, G. (1976). Séries de variables aléatoires vectorielles indépendantes et propriétés géométriques des espaces de Banach. *Studia Math., 58,* 45-90.
Mazur, S., & Orlicz, W. (1933). Über Folgen linearen Operationen. *Studia Math., 13,* 152-157.
Mcgill, P. (1975). Integration in vector lattices. *J. Lond. Math. Soc., 11,* 347-360.
Megginson, R. (1998). *An introduction to Banach space theory.* Berlin, Heidelberg, New York: Springer-Verlag.
Meyer-Nieberg, P. (1991). *Banach lattices.* Berlin, Heidelberg, New York: Springer-Verlag.
Meza Alcantara, D. (2009). *Ideal and filters on countable sets.* PhD thesis in Mathematics. Mexico City: Autonomous University of Mexico, Department of Mathematics.
Mikusiński, J. (1970). A theorem on vector matrices and its applications in measure theory and functional analysis. *Bull. Acad. Polon. Sci. Sér. Sci. Math. Astronom. Phys., 18,* 193-196.
Mikusiński, J. (1971). On a theorem of Nikodým on bounded measures. *Bull. Acad. Polon. Sci. Sér. Sci. Math. Astronom. Phys., 19,* 441-444.
Miller, H. (1995). A measure theoretical subsequence characterization of statistical convergence. *Trans. Amer. Math. Soc., 347,* 1811-1819.
Moltó, A. (1981a). On the Vitali-Hahn-Saks theorem. *Proc. Royal Soc. Edinburgh Sect. A, 90,* 163-173.
Moltó, A. (1981b). On uniform boundedness properties in exhausting additive set function spaces. *Proc. Royal Soc. Edinburgh Sect. A, 90,* 175-184.
Móricz, F. (2004). Ordinary convergence follows from statistical summability $(C,1)$ in the case of slowly decreasing or oscillating sequences. *Colloq. Math., 99,* 207-219.
Móricz, F. & Stadtmüller, U. (2004). Summability of double sequences by weighted mean methods and Tauberian conditions for convergence in Pringsheim's sense. *Int. J. Math. Math. Sci., 65,* 3499-3511.
Müller, C. H., & Kitsos, C. P. (2004). Optimal design criteria based on tolerance regions. In: Di Bucchianno, A., Lauter, H., Wynn, H. (Eds), *MODA7: Advances in Model-Oriented Design and Analysis* (pp. 107-115). Heidelberg: Physica-Verlag.
Mundici, D. (1986). Interpretation of *AF C**-algebras in Łukasiewicz sentential calculus. *J. Funct. Anal., 65,* 15-63.
Munkres, J. R. (1975). *Topology, a first course.* Englewood Cliffs, NJ: Prentice-Hall, Inc.
Mursaleen, M. (2000). λ-statistical convergence. *Math. Slovaca, 50*, 111-115.
Mursaleen, M., & Alotaibi, A. (2011). On $\mathcal{I}$-convergence in random 2-normed spaces. *Math. Slovaca, 61,* 933-940.
Mynard, F. (2013). A convergence-theoretic viewpoint of the Arzelà-Ascoli theorem. *Real Anal. Exchange, 38,* 431-444.
Nabiev, A., Pehlivan, S., & Gürdal, M. (2007). On *I*-Cauchy sequences. *Taiwanese J. Math., 11*, 569-576.
Nakamura, M. (1949). Notes on Banach Spaces. X: Vitali-Hahn-Saks' theorem and *K*-spaces. *Tôhoku Math. J., 1,* 100-108.
Nation, J. B. *Notes on lattice theory.* Unpublished course notes, avaliable at http://bigcheese.math.sc.edu/~mcnulty/alglatvar/lat.pdf . University of Hawaii.
Nikodým, O. M. (1930a). Sur les fonctions d'ensemble. *Comptes Rendus du I Congrès des Math. des Pays Slaves, Warszawa, 1929,* 304-313.

Nikodým, O. M. (1930b). Sur une généralisation des intégrales de M. J. Radon. *Fund. Math., 15,* 131-179.

Nikodým, O. M. (1933a). Sur les familles bornées des fonctions parfaitement additives d'ensemble abstrait. *Monatsh. Math. Phys., 40,* 418-426.

Nikodým, O. M. (1933b). Sur les suites convergentes de fonctions parfaitement additives d'ensemble abstrait. *Monatshefte für Math. und Physik, 40,* 427-432.

Nikodým, O. M. (1955). A theorem on infinite sequences of finitely additive real-valued measures. *Rend. Sem. Mat. Univ. Padova, 24,* 265-286.

Niven, I., Zuckerman, H. S., & Montgomery, H. L. (1991). *An Introduction to the Theory of Numbers.* London, New York: John Wiley & Sons.

Nuray, F., & Ruckle, W. H. (2000). Generalized statistical convergence and convergence free spaces. *J. Math. Anal. Appl., 245,* 513-527.

Oberle, R. A. (1974a). A note on equicontinuous families of volumes with an application to vector measures. *Canad. J. Math, 26,* 273-280.

Oberle, R. A. (1974b). Characterization of a class of equicontinuous sets of finitely additive measures with an application to vector valued Borel measures. *Canad. J. Math., 26,* 281-290.

Ohba, S. (1972). Some remarks on vector measures. *Bull. Math. Soc. Sci. Math. Roumanie, 16,* 217-223.

Okada, S., & Ricker, W. J. (1993). Non-weak compactness of the integration map for vector measures. *J. Austral. Math. Soc. (Series A), 54,* 287-303.

Okada, S., & Ricker, W. J. (1994). Criteria for weak compactness of vector-valued integration maps. *Comment. Math. Univ. Carolinae, 35,* 485-495.

Okada, S., & Ricker, W. J. (1995). The range of the integration map of a vector measure. *Arch. Math., 64,* 512-522.

Orlicz, W. (1929). Beiträge zur Theorie der Orthogonalentwicklungen. II. *Studia Math., 1,* 241-255.

Orlicz, W., & Urbański, R. (1980). A generalization of the Brooks-Jewett theorem. *Bull. Acad. Pol. Sci, Sér. Sci. Math., 28,* 55-59.

Pap, E. (1974). A generalization of the diagonal theorem on a block-matrix. *Mat. Vesnik, 11 (26),* 66-71.

Pap, E. (1976a). Uniform boundedness of a family of exhaustive set functions. *Mat. Vesnik, 13 (28),* 319-326.

Pap, E. (1976b). The additive exhaustive functions on *M*-lattice. *Publ. Inst. Math (N. S.), 20,* 203-207.

Pap, E. (1981). On semigroup valued additive exhaustive set functions. *Zb. Rad. Prirod.-Mat. Fak., Ser. Mat., 11,* 59-70.

Pap, E. (1986). A generalization of a theorem of Dieudonné for *k*-triangular set functions. *Acta Sci. Math., 50,* 159-167.

Pap, E. (1987). The Vitali-Hahn-Saks Theorems for *k*-triangular set functions. *Atti Sem. Mat. Fis. Univ. Modena, 35,* 21-32.

Pap, E. (1988). Nikodým type theorem for netric valued x_0-exhaustive set functions. *Zb. Rad. Prirod.-Mat. Fak., Ser. Mat., 18,* 101-109.

Pap, E. (1991). The Brooks-Jewett Theorem for non-additive set functions. *Zb. Rad. Prirod.-Mat. Fak., Ser. Mat., 21* 75-81.

Pap, E. (1994). Decompositions of supermodular functions and □-decomposable measures. *Fuzzy Sets Systems, 65,* 71-83.

Pap, E. (1995). *Null-Additive Set Functions.* Dordrecht: Kluwer Acad. Publ., and Bratislava: Ister Science.

Pap, E. (1997a). Nikodým type theorems. *Tatra Mt. Math. Publ., 10,* 147-157.

Pap, E. (1997b). Decompositions of null-additive set functions. *Atti Sem. Mat. Fis. Univ. Modena, 45,* 223-234.

Pap, E. (1998). The Lebesgue decomposition theorem for generalized measures. *Novi Sad J. Math., 28,* 99-106.

Pap, E. (Ed.) (2002a). *Handbook of Measure Theory, volls. I and II.* Amsterdam: Elsevier.

Pap, E. (2002b). Some Elements of the Classical Measure Theory. In: Pap, E. (Ed.), *Handbook of Measure Theory, vol. I* (pp. 27-82). Amsterdam: Elsevier.

Pap, E. (2002c). Pseudo-additive measures and their applications. In: Pap, E. (Ed.), *Handbook of Measure Theory, vol. II* (pp. 1403-1468). Amsterdam: Elsevier.

Paterson, A. L. (1988). *Amenability.* Mathematical Surveys and Monographs, *29,* American Mathematical Society. Rhode Island: Providence.

Paunić, D. (2002). History of Measure Theory. In: Pap, E. (Ed.), *Handbook of Measure Theory, vol. I* (pp. 1-26). Amsterdam: Elsevier.

Pavlakos, P. K. (1978). The Lebesgue decomposition theorem for partially ordered semigroup-valued measures. *Proc. Amer. Math. Soc., 71,* 207-211.

Peressini, P. L. (1967). *Ordered Topological Vector Spaces.* New York: Harper and Row.
Pervin, W. (1962). Quasi-uniformization of topological spaces. *Math. Ann., 147,* 316-317.
Petersen, G. M. (1966). *Regular matrix transformations.* McGraw-Hill, London, 1966.
Pettis, B. J. (1938). On integration in vector spaces. *Trans. Amer. Math. Soc., 44*, 277-304.
Pettis, B. J. (1951). Uniform Cauchy points and points of equicontinuity. *Amer. J. Math., 73*, 602-614.
Pfanzagl, J. (1971). Convergent sequences of regular maesures. *Manuscripta Math., 4,* 91-98.
Phillips, E. R. (1984). *An Introduction to Analysis and Integration Theory.* New York: Dover Publ., Inc.
Phillips, R. S. (1940a). Integration in a convex linear topological space. *Trans. Amer. Math. Soc., 47,* 114-145.
Phillips, R. S. (1940b). On linear transformations. *Trans. Amer. Math. Soc., 48,* 516-541.
Pier, J.-P. (1984). *Amenable locally compact groups.* New York: John Wiley & Sons.
Pietsch, A. (2007). *History of Banach spaces and linear operators*. Boston, Basel, Berlin: Birkhäuser.
Pontryagin, L. S. (1986). *Selected works, vol.2, Topological groups* (ed. R. V. Gamkrelidze). Transl. from the Russian and enlarged by A. Brown and P. S. V. Naidu, Classics of Soviet Mathematics. New York, London, Paris: Gordon and Breach.
Popa, D. (1992). On some classical theorems in measure theory. *Bull. Pol. Acad. Sci., Math., 40,* 255-263.
Popa, D. (1997). On Boolean algebras which have the Vitali-Hahn-Saks property. *Le Matematiche, 52,* 53-58.
Porcelli, P. (1960). Two embedding theorems with applications to weak convergence and compactness in spaces of additive type functions. *J. Math. Mech., 9,* 273-292.
Pringsheim, A. (1897). Elementare Theorie der unendliche Doppelreihen. *Sitsungsber. Akad. Wiss. München, 27,* 101-153.
Qu, W., & Wu, J. (2002). On Antosik's lemma and the Antosik-Mikusiński basic matrix theorem. *Proc. Amer. Math. Soc., 130,* 3283-3285.
Radon, J. (1919). Über lineare Funktionaltransformationen und Funktionalgleichungen. *Sitzsber. Akad. Wiss. Wien, 128,* 1083-1121.
Raimi, R. A. (1963). Convergence, density and τ-density of bounded sequences. *Proc. Amer. Math. Soc., 14,* 708-712.
Rainwater, J. (1959). A note on projective resolutions. *Proc. Am. Math. Soc., 10,* 734-735.
Randrianantoanina, N. & Saab, E. (1994). Weak compactness in the space of vector-valued measures of bounded variation. *Rocky Mountain J. Math., 24,* 681-688.
Rath, D., & Tripathy, B. C. (1994). On statistically convergent and statistically Cauchy sequences. *Indian J. Pure Appl. Math., 25,* 381-386.
Recław, I. (2012). Sets of filter convergence of sequences of continuous functions. *J. Math. Anal. Appl., 394,* 475-480.
Rickart, C. E. (1942). Integration in a convex linear topological space. *Trans. Amer. Math. Soc., 52,* 498-521.
Rickart, C. E. (1943). Decompositions of additive set functions. *Duke Math. J. 10,* 653-665.
Riečan, B. (2007). Probability theory on IF-events. In: *A volume in honor of Daniele Mundici's 60th birthday.* Lecture Notes in Computer Sciences. Berlin, Heidelberg, New York: Springer.
Riečan, B., & Mundici, D. (2002). Probability on *MV*-algebras. In: Pap., E. (Ed.), *Handbook of Measure Theory, vol. II* (pp. 869-909). Amsterdam: Elsevier.
Riečan, B., & Neubrunn, T. (1997). *Integral, Measure and Ordering.* Dordrecht: Kluwer Acd. Publ., and Bratislava: Ister Science.
Riečan, B., & Volauf, P. (1984). On a technical lemma in lattice ordered groups. *Acta Math. Univ. Comenianae, 44-45,* 31-35.
Riečan, B., & Vrábelová, M. (1988). The Kurzweil construction of an integral in ordered spaces. *Czech. Math. J. 48 (123)*, 565-574.
Rivera, M. J. (2000). The De La Vallée-Poussin theorem for vector-valued measure spaces. *Rocky Mt. J. Math., 30,* 715-722.
Robertson, A.P. (1972). Unconditional convergence and the Vitali-Hahn-Saks theorem. *Bull. Soc. Mat. France, 31-32,* 335-341.
Rogge, L. (1973). The convergence determining class of regular open sets. *Proc. Amer. Math. Soc., 37,* 581-585.
Rosenthal, H. P. (1968). On complemented and quasi-complemented subspaces of quotients of *C*(*S*) for Stonian *S*. *Proc. Nat. Acad. Sci. U. S. A., 60,* 1165-1169.
Rosenthal, H. P. (1970). On relatively disjoint families of measures, with some applications to Banach space theory. *Studia Math., 37,* 13-36.

Rosenthal, H. P. (1971). Correction to the paper:"On relatively disjoint families of measures, with some applications to Banach space theory". *Studia Math., 37,* 311-313.
Rudin, W. (1973). *Functional Analysis.* New York: McGraw-Hill.
Rüttiman, G. T. (1994). Weakly purely additive measures. *Canad. J. Math., 46,* 872-885.
Rybakov, V. I. (1970). Theorem of Bartle, Dunford and Schwartz concerning vector measures. *Math. Notes, 7,* 147-151.
Saeki, S. (1992). The Vitali-Hahn-Saks theorem and measuroids. *Proc. Amer. Math. Soc., 114,* 775-782.
Saks, S. (1933a). On some functionals. I. *Trans. Amer. Math. Soc., 35,* 549-556.
Saks, S. (1933b). Addition to the note: On some functionals. *Trans. Amer. Math. Soc. 35,* 965-970.
Saks, S. (1937). On some functionals. II. *Trans. Amer. Math. Soc., 41,* 160-170.
Šalát, T. (1980). On statistically convergent sequences of real numbers. *Math. Slovaca, 30,* 139-150.
Šalát, T., Tripathy, B. C., & Ziman, M. (2004). On some properties of $\mathcal{I}$-convergence. *Tatra Mt. Math. Publ., 28,* 274-286.
Šalát, T., Tripathy, B. C., & Ziman, M. (2005). On $\mathcal{I}$-convergence field. *Ital. J. Pure Appl. Math., 17,* 45-54.
Salvati, S. (2000). The Brooks-Jewett theorem for *k*-triangular functions. *Math. Slovaca 50,* 247-257.
Samaratunga, R. T., & Sember, J. J. (1988). Summability and substructures of $2^{\mathbb{N}}$. *Southeast Asian Bull. Math. 12,* 11-21.
Savaş, E. (1992). On strong almost *A*-summability with respect to a modulus and statistical convergence. *Indian J. Pure Appl. Math., 23,* 217-222.
Savaş, E. (2012). On generalized *A*-difference strongly summable sequence spaces defined by ideal convergence in a real *n*-normed space. *J. Inequal. Appl.,* Art. N. 87.
Savaş, E., Das, P., & Dutta, S. (2012). A note on strong matrix summability *via* ideals. *Appl. Math. Letters, 25,* 733-738.
Savaş, E., Das, P., & Dutta, S. (2013). A note on some generalized summability methods. *Acta Math. Univ. Comenianae, 82,* 297-304.
Sazhenkov, A. N. (1982). The principle of uniform boundedness for topological measures. *Math. Notes, 31,* 135-137.
Sazhenkov, A. N. (1986). Equivalence of the Nikodým theorems for group measures. *Sib. Math. J., 27,* 926-935.
Schachermayer, W. (1982). On some classical measure-theoretic theorems for non-sigma-complete Boolean algebras. *Dissertationes Math., 214,* 1-33.
Schaefer, H. H. (1971). *Topological vector spaces.* Berlin, Heidelberg, New York: Springer-Verlag.
Schaefer, H. H. (1974). *Banach lattices and positive operators.* Berlin, Heidelberg, New York: Springer-Verlag.
Schmidt, K. D. (1982). A general Jordan decomposition. *Arch. Math., 38,* 556-564.
Schmidt, K. D. (1985). A common abstraction of Boolean rings and lattice ordered groups. *Compositio Math., 54,* 51-62.
Schmidt, K. D. (1986). Decompositions of vector measures in Riesz spaces and Banach lattices. *Proc. Edinburgh Math. Soc., 29,* 23-39.
Schmidt, K. D. (1988). On the modulus of weakly compact operators and strongly additive vector measures. *Proc. Amer. Math. Soc., 102,* 862-866.
Schmidt, K. D. (1989). Jordan decompositions of generalized vector measures. *Pitman Research Motes Math. Ser., 214.*
Schmidt, K. D. (1998). Decomposition and extension of abstract measures in Riesz spaces. *Rend. Ist. Mat. Univ. Trieste, Suppl. 29,* 135-213.
Schoenberg, I. (1959). The integrability of certain functions and related summability methods. *Amer. Math. Monthly, 66,* 361-375. II. *Ibidem, 66,* 562-563.
Schur, J. (1920). Über lineare Transformationen in der Theorie der unendlichen Reihen. *J. Reine Angew. Math., 151,* 79-111.
Seever, G. L. (1968). Measures on *F*-spaces. *Trans. Amer. Math. Soc., 133,* 267-280.
Segal, I. (1947). Irreducible representations of operator algebras. *Bull. Amer. Math. Soc., 53,* 73-88. DOI:10.1090/S0002-9904-1947-08742-5
Shilkret, N. (1971). Maxitive measure and integration. *Indag. Math., 33,* 109-116.
Sikorski, R. (1964). *Boolean Algebras.* Berlin, Heidelberg, New York: Springer-Verlag.
Simons, S. (1978). Vector lattices and vector measures. *Rev. Roumaine Math. Pures Appl., 23,* 1101-1106.
Sion, M. (1969). Outer measures with values in a topological group. *Proc. London Math. Soc., 19,* 89-106.

Sion, M. (1973). *A theory of semigroup-valued measures.* Lecture Notes Math., *355.* Berlin, Heidelberg, New York: Springer-Verlag.
Sleziak, M. (2003). $\mathcal{I}$-continuity in topological spaces. *Acta Math. (Nitra), 6,* 115-122.
Solecki, S. (1999). Analytic ideals and their applications. *Ann. Pure Appl. Logic, 99,* 51-72.
Stein, J. D., Jr. (1972). A uniform-boundedness theorem for measures. *Michigan Math. J., 19,* 161-165.
Stein, J. D., Jr. (1974). Sequence of regular finitely additive set functions. *Trans. Amer. Math. Soc., 192*, 59-66.
Stein, J. D., Jr. (1975). Uniform absolute continuity in spaces of set functions. *Proc. Amer. Math. Soc., 51,* 137-140.
Stein, J. D., Jr. (1991). Some observations on local uniform boundedness principles. *Czech. Math. J., 41 (116),* 64-74.
Steinhaus, H. (1951). Sur la convergence ordinaire et la convergence asymptotique. *Colloq. Math., 2,* 73-74.
Stoilov, S. (1959). Continuous convergence. *Rev. Roumaine Math. Pures Appl., 4,* 341-344.
Stone, M. H. (1936). The theory of representations for Boolean algebras. *Trans. Amer. Math. Soc., 40,* 37-111.
Stone, M. H. (1938). The representation of Boolean algebras. *Bull. Amer. Math. Soc., 44,* 807-816.
Stone, M. H. (1940). A general theory of spectra. I. *Proc. Nat. Acad. Sci. U. S. A., 26,* 280-283.
Stone, M. H. (1941). A general theory of spectra. II. *Proc. Nat. Acad. Sci. U. S. A.* **27** (1941), 83-87.
Stone, M. H. (1949). Boundedness properties in function-lattices. *Canad. J. Math., 1,* 176-186.
Stuart, C. (1994). Interchanging the limit in a double series. *Southeast Asian Bull. Math., 18*, 81-84.
Stuart, C., & Abraham, P. (2004). Generalizations of the Nikodým boundedness and Vitali-Hahn-Saks theorem. *J. Math. Anal. Appl., 300,* 351-361.
Stuart, C., & Swartz, C. (2005). Generalizations of the Orlicz-Pettis theorem. *Proyecciones, 24,* 37-48.
Swartz, C. (1977). Nikodým theorems for set functions on algebras, mappings between lattices. *Rev. Roumaine Math. Pures Appl., 22*, 347-353.
Swartz, C. (1978). Applications of the Mikusiński diagonal theorem. *Bull. Acad. Polon. Sci. Ser. Sci. Math. Astronom. Phys., 26,* 421-424.
Swartz, C. (1982). A lemma of Labuda as a diagonal theorem. *Bull. Acad. Polon. Sci. Sér. Sci. Math. Astronom. Phys., 30,* 493-497.
Swartz, C. (1983). The Schur Lemma for bounded multiplier convergent series. *Math. Ann., 263,* 283-288.
Swartz, C. (1984). An abstract Orlicz-Pettis theorem. *Bull. Pol. Acad. Sci., Math., 32,* 433-437.
Swartz, C. (1985). The Schur and Hahn theorems for operator matrices. *Rocky Mt. J. Math., 15,* 61-73.
Swartz, C. (1987). The closed graph theorem without category. *Bull. Austral. Math. Soc., 36,* 283-287.
Swartz, C. (1989a). The Nikodým boundedness Theorem for lattice-valued measures. *Arch. Math., 53,* 390-393.
Swartz, C. (1989b). The uniform boundedness principle for order bounded operators. *Int. J. Math. and Math. Sci., 12,* 487-492.
Swartz, C. (1989c). Bilinear mappings between lattices. *Bull. Math. Soc. Sci. Math. Roumanie, 33,* 147-152.
Swartz, C. (1990). The evolution of the uniform boundedness principle. *Math. Chronicle, 19,* 1-18.
Swartz, C. (1991). A proof of Phillips' lemma. *Bull. Malaysian Math. Soc. Ser. II, 14,* 83-90.
Swartz, C. (1992a). Iterated series and the Hellinger-Toeplitz theorem. *Publ. Matem., 36,* 167-173.
Swartz, C. (1992b). *An introduction to Functional Analysis.* New York: Marcel Dekker.
Swartz, C. (1993). The gliding hump property in vector sequence spaces. *Monatsh. Math., 116,* 147-158.
Swartz, C. (1996a). A general Hahn-Schur theorem. *Southeast Asian Bull. Math., 20,* 57-58.
Swartz, C. (1996b). *Infinite matrices and the gliding hump.* Singapore: World Scientific Publ. Co.
Swartz, C. (2002). Interchanging orders of summation for multiplier convergent series. *Bol. Soc. Mat. Mexicana, 8*, 31-35.
Swartz, C. (2013). A weak Orlicz-Pettis theorem. *Bol. Soc. Mat. Mexicana, 19*, 85-89.
Swartz, C., & Stuart, C. (1998). Orlicz-Pettis theorems for multiplier convergent series. *Z. Anal. Anwend., 17,* 805-811.
Talagrand, M. (1984). Propriété de Nikodým and propriété de Grothendieck. *Studia Math., 78*, 165-171.
Thiago Bernardino, A. (2009). A simple natural approach to the uniform boundedness principle for multilinear mappings. *Proyecciones, 28,* 203-207.
Thomas, E. (1974). The Lebesgue-Nikodým theorem for vector-valued Radon measures. *Mem. Amer. Math. Soc., n. 139.*
Todorcevic. S. (1997). *Topics in topology.* Lecture Notes Math., *1652.* Berlin, Heidelberg, New York: Springer-Verlag.
Toeplitz, T. O. (1911). Über allgemeine lineare Mittelbildungen. *Prace Math.-Fiz., 22,* 113-119.

Topsøe, F. (1970). Compactness in spaces of measures. *Studia Math., 36,* 195-212.

Traynor, T. (1972a). A general Hewitt-Yosida decomposition. *Canad. J. Math., 24,* 1164-1169.

Traynor, T. (1972b). Decomposition of group-valued additive set functions. *Ann. Inst. Fourier (Grenoble), 22,* 131-140.

Traynor, T. (1973). (*s*)-bounded additive set functions. In: Tucker, D. H., & Maynard, H. B. (Eds.), *Vector and operator valued measures and applications* (pp. 355-365). New York, London: Acad. Press.

Traynor, T. (1976). The Lebesgue decomposition for group-valued set functions. *Trans. Amer. Math. Soc., 220,* 307-319.

Tripathy, B. C., Sen, M., & Nath, S. (2012). $\mathcal{I}$-convergence in probabilistic *n*-normed space. *Soft Comput., 16,* 1021-1027.

Tripathy, B., & Tripathy, B. C. (2005). On $\mathcal{I}$-convergent double sequences. *Soochow J. Math., 31*, 549-560.

Turpin, P. (1978). *Boundedness problems for finitely additive measures*. Lecture Notes Math., *645*. Berlin, Heidelberg, New York: Springer-Verlag.

Uhl, J. J. (1967). Orlicz spaces of finitely additive set functions. *Studia Math., 29,* 19-58.

Velasco, M. V., & Rodríguez-Palacios, A. (Eds.) (2007). *Advanced courses of Mathematical Analysis II. Proceedings of the Second International School.* Singapore: World Scientific Publ. Co.

Ventriglia, F. (2007). Cafiero and Brooks-Jewett theorems for Vitali spaces. *Ricerche Mat., 56*, 209-216. DOI 10.1007/s11587-007-0014-5

Ventriglia, F. (2008a). Asymptotically exhaustive functions in Vitali spaces. *Rend. Circ. Mat. Palermo, 57,* 203-212.

Ventriglia, F. (2008b). Cafiero theorem for *k*-triangular functions on an orthomodular lattice. *Rend. Accad. Sci. Fis. Mat. Napoli, 75,* 45-52.

Ventriglia, F. (2010). Dieudonné's theorem for non-additive set functions. *J. Math. Anal. Appl., 367,* 296-303.

Vinti, C. (1998). A survey on recent results of the Mathematical Seminar in Perugia, inspired by the work of Professor P. L. Butzer. *Results Math., 34,* 32-55.

Vinti, G. (2001). A general approximation result for nonlinear integral operators and applications to signal processing. *Appl. Anal., 79,* 217-238.

Vitali, G. (1905). *On the problem of the measure of the groups of points of a line (Italian).* Bologna: Tip. Gamberini e Parmeggiani.

Vitali, G. (1907). On the integration by series (Italian). *Rend. Circ. Mat. Palermo, 23,* 137-155.

Volauf, P. (1993). Alexandrov and Kolmogorov consistency theorem for values in partially ordered groups. *Tatra Mt. Math. Publ., 3,* 237-244.

von Neumann, J. (1930). Zur Algebra der Funktionaloperationen und Theorie der normalen Operatoren. *Math. Ann., 102,* 370-427. DOI:10.1007/BF01782352.

Vrábel, P., & Vrábelová, M. (2013). $\mathcal{I}$-convergence and $\mathcal{I}$-continuity of the fuzzy number-valued functions. *Acta Math. (Nitra), 16,* 237-243.

Vrábelová, M., & Riečan, B. (1996). On the Kurzweil integral for functions with values in ordered spaces. III. *Tatra Mt. Math. Publ., 8,* 93-100.

Vulikh, B. Z. (1967). *Introduction to the theory of partially ordered spaces.* Groningen: Wolters – Noordhoff Sci. Publ.

Wagon, S. (1981). Invariance properties of finitely additive measures in $\mathbb{R}^n$. *Ill. J. Math., 25*, 74-86.

Wagon, S. (1985). *The Banach-Tarski Paradox.* Cambridge: Cambridge Univ. Press.

Wang, Z., & Klir, G. J. (2009). *Generalized Measure Theory.* New York: Spinger.

Weber, H. (1976). Fortsetzung von Massen mit Werten in uniformen Halbgruppen. *Arch. Math., 27,* 412-423.

Weber, H. (1985). Pointwise sequential compactness and weak compactness in spaces of contents. *Rend. Circ. Mat. Palermo Ser. II, 34,* 56-78.

Weber, H. (1986). Compactness in spaces of group-valued contents, the Vitali-Hahn-Saks theorem and Nikodým's boundedness theorem. *Rocky Mt. J. Math., 16,* 253-275.

Weber, H. (2002). *FN*-topologies and group-valued measures. In: Pap. E. (Ed.), *Handbook on Measure Theory, vol. I* (pp. 703-743). Amsterdam: Elsevier.

Weber, H. (2009). Boolean algebras of lattice uniformities and decompositions of modular functions. *Ricerche Mat., 58,* 15-32.

Wells, B. B. (1969). Weak compactness of measures. *Proc. Amer. Math. Soc., 20,* 124-130.

Willard, S. (1970). *General Topology.* Massachusets, London, Don Mills: Addison-Wesley Publ. Company, Reading.

Wnuk, W. (2009). The strong Schur property in Banach lattices. *Positivity, 13,* 435-441.
Wolff, K. E. (1994). A first course in Formal Concept Analysis, How to understand line diagrams. In: Faulbaum, F. (Ed.), *SoftStat '93, Advances in Statistical Software 4* (pp. 429-438). Stuttgart: Gustav Fisher Verlag.
Wolk, E. (1975). Continuous convergence in partially ordered sets. *Gen. Topology Appl., 5,* 221-234.
Wright, J. D. M. (1969). Stone-algebra-valued measures and integrals. *Proc. London Math. Soc., 19,* 107-122.
Wright, J. D. M. (1971). The measure extension problem for vector lattices. *Ann. Inst. Fourier (Grenoble), 21,* 65-85.
Wu, J. D., Luo, J. W., & Lu, S. J. (2005). A unified convergence theorem. *Acta Math. Sinica (Engl. Ser.), 21*, 315-322. DOI: 10.1007/s10114-004-0481-5
Wyler, O. (1966). Clans. *Compositio Math., 17,* 172-189.
Yosida, K. (1941). On vector lattice with a unit. *Proc. Imp. Acad. Tokyo, 17,* 121-124.
Yosida, K., & Hewitt, E. (1952). Finitely additive measures. *Trans. Amer. Math. Soc., 72,* 46-66.
Zaanen, A. C. (1983). *Riesz Spaces. II.* Amsterdam: North-Holland.
Zabrejko, P. P. (1997). *K*-metric and *K*-normed linear spaces: a survey. *Collect. Math., 48*, 825-859.
Zakrzewski, P. (2002). Measures on Algebraic-Topological Structures. In: Pap, E. (Ed.), *Handbook of Measure Theory, vol. II* (pp. 1091-1130). Amsterdam: Elsevier.
Zhang, X.-D. (1997). On weak compactness in spaces of measures. *J. Funct. Anal., 143,* 1-9.
Zygmund, A. (1968). *Trigonometric series, Vol. I.* Cambridge: Cambridge Univ. Press.

Index

Made in the USA
Las Vegas, NV
04 November 2026

57425982R00326